14.95

SAS USER'S GUIDE: Statistics 1982 Edition

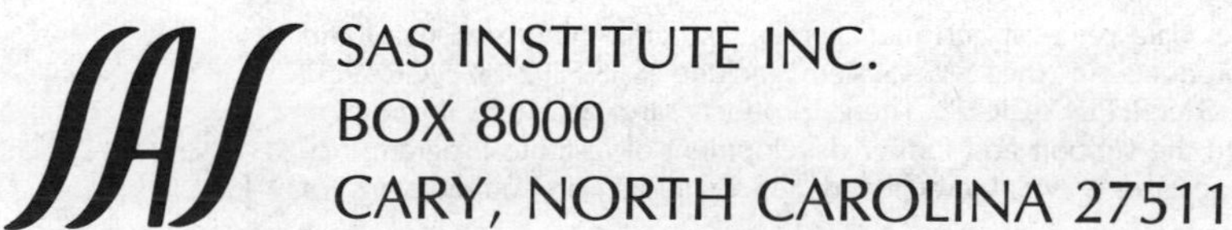

SAS INSTITUTE INC.
BOX 8000
CARY, NORTH CAROLINA 27511

Alice Allen Ray edited the *SAS User's Guide: Statistics, 1982 Edition*. **John P. Sall** was contributing editor. **Marian Saffer** was copy editor with assistance from **Stephenie P. Joyner** and **Judith K. Whatley**.

Data entry and text-editing were accomplished by **Andrea U. Littleton**.

Composition and production were provided by **Stephen K. Douglas**, **Elisabeth C. Smith**, and **June L. Woodward** under the direction of **W. Wayne Lindsey**.

Program authorship includes design, programming, debugging, support, and preliminary documentation. The SAS Institute staff member listed first currently has primary responsibility for the procedure; others give specific assistance.

ANOVA J.H. Goodnight
CANCORR W.S. Sarle
CANDISC W.S. Sarle
CLUSTER W.S. Sarle
DISCRIM J.H. Goodnight
FACTOR W.S. Sarle, J.P. Sall
FASTCLUS W.S. Sarle
FUNCAT J.P. Sall
GLM J.H. Goodnight, J.P. Sall, W.S. Sarle
MATRIX J.P. Sall
NEIGHBOR J.H. Goodnight, W.S. Sarle
NESTED J.P. Sall
NLIN J.H. Goodnight, J.P. Sall
NPAR1WAY J.P. Sall
PLAN J.P. Sall
PRINCOMP W.S. Sarle
PROBIT D.M. DeLong, J.H. Goodnight
RANK K.D. Kumar, J.P. Sall
REG J.P. Sall
RSQUARE W.S. Sarle, J.H. Goodnight
RSREG J.P. Sall
SCORE J.P. Sall
STANDARD J.H. Goodnight
STEPDISC W.S. Sarle
STEPWISE J.H. Goodnight
TREE G.K. Howell, W.S. Sarle
TTEST J.H. Goodnight
VARCLUS W.S. Sarle
VARCOMP J.H. Goodnight

Probability Routines D.M. DeLong, J.H. Goodnight
Multiple Comparisons Routines W.S. Sarle, D.M. DeLong
Multivariate Routines W.S. Sarle
Other Numerical Routines W.S. Sarle, D.M. DeLong, J.P. Sall, J.H. Goodnight
Other Library Routines W.S. Sarle, R.D. Langston, J.P. Sall
Consulting Statisticians D.M. DeLong, J.H. Goodnight, W.S. Sarle, J.P. Sall, H.J. Kirk, F.W. Young

(If you have questions or encounter problems, call SAS Institute and ask for technical support rather than an individual staff member.)

The correct bibliographic information for this manual is:

SAS Institute Inc. *SAS User's Guide: Statistics, 1982 Edition*. Cary, NC: SAS Institute Inc., 1982. 584 pp.

SAS User's Guide: Statistics, 1982 Edition

ISBN 0-917382-32-3

84 83 7 6 5 4 3

SAS,® the basis of the SAS System, provides data retrieval and management, programming, statistical, and reporting capabilities. Other software products in the SAS System include SAS/FSP,® SAS/GRAPH,® SAS/IMS-DL/I,® SAS/ETS,™ SAS/OR,™ and SAS/REPLAY-CICS.™ These products are available from SAS Institute Inc., a private company devoted to the support and further development of Institute Program Products and related services. *SAS Communications*®, *SAS Views*®, and *SAS Training*™ are publications of SAS Institute Inc.

Acknowledgments

Hundreds of people have helped SAS® in many ways since its inception. The individuals that we remember here have been especially helpful in the development of statistical procedures. An acknowledgment for SAS generally is in *SAS User's Guide: Basics, 1982 Edition*.

Anthony James Barr, Barr Systems
Wilbert P. Byrd, Clemson University
George Chao, Arnar-Stone Laboratories
Daniel Chilko, University of West Virginia
Richard Cooper, USDA
Sandra Donaghy, North Carolina State University
David B. Duncan, Johns Hopkins University
R.J. Freund, Texas A & M University
Wayne Fuller, Iowa State University
A. Ronald Gallant, North Carolina State University
Charles Gates, Texas A & M University
Thomas M. Gerig, North Carolina State University
Francis Giesbrecht, North Carolina State University
Harvey J. Gold, North Carolina State University
Harold Gugel, General Motors Corporation
Donald Guthrie, University of California at Los Angeles
Gerald Hajian, Burroughs Wellcome Company
Frank Harrell, Duke University
Walter Harvey, Ohio State University
Ronald Helms, University of North Carolina at Chapel Hill
Jane T. Helwig, Seasoned Systems Inc., Chapel Hill
Don Henderson, ORI
Harold Huddleston, Data Collection & Analysis, Inc., Falls Church, Virginia
David Hurst, University of Alabama at Birmingham
Emilio A. Icaza, Louisiana State University
William Kennedy, Iowa State University
Gary Koch, University of North Carolina at Chapel Hill
Kenneth Koonce, Louisiana State University
Clyde Y. Kramer (deceased), Virginia Polytechnic Institute, University of Kentucky, and Upjohn
Ardell C. Linnerud, North Carolina State University
Ramon C. Littell, University of Florida
H.L. Lucas (deceased), North Carolina State University
David D. Mason, North Carolina State University
J. Philip Miller, Washington University
Robert J. Monroe, North Carolina State University
Robert D. Morrison, Oklahoma State University
Kenneth Offord, Mayo Clinic
Robert Parks, Washington University
Richard M. Patterson, Auburn University

Virginia Patterson, University of Tennessee
C.H. Proctor, North Carolina State University
Dana Quade, University of North Carolina at Chapel Hill
William L. Sanders, University of Tennessee
Robert Schechter, Scott Paper Company
Shayle Searle, Cornell University
Jolayne Service, University of California at Irvine
Roger Smith, USDA
Michael Speed, formerly of Louisiana State University
Robert Teichman, ICI Americas Inc.
Glenn Ware, University of Georgia
Love Casanova Webb, Santa Fe International
Edward W. Whitehorne, U.S. Forest Service
William Wigton, USDA
Forrest W. Young, University of North Carolina at Chapel Hill

The final responsibility for SAS lies with SAS Institute alone. We hope that you will always let us know your feelings about the system and its documentation. It is through such communications that the progress of SAS has been accomplished.

THE STAFF OF SAS INSTITUTE

Preface

Two volumes—*SAS User's Guide: Basics, 1982 Edition* and this volume, *SAS User's Guide: Statistics, 1982 Edition*—contain primary documentation for the 1982 release of the SAS System (SAS82). Many SAS users need only the data processing, summarizing, and reporting features of the system and do not need the advanced statistical procedures. Your choice of one or both manuals will depend on whether you are using the SAS System primarily for data processing or for statistics.

If you are new to the SAS System, the first book to read is the *SAS Introductory Guide*. *SAS User's Guide: Basics* contains the fundamentals of the SAS System: the DATA step, macros, system options, and procedures for descriptive statistics, report writing, and utilities. Other background and applications literature is available in draft form as technical reports in the A series (for example, Technical Report A-102, "SAS Regression Applications") and in the SAS Series in Statistical Applications, inaugurated with *SAS for Linear Models: A Guide to the ANOVA and GLM Procedures* (1981).

Organization

How to use this manual The chapters in *SAS User's Guide: Statistics* are organized to serve both as a reference for experienced SAS users and as a learning tool for new users.

Each major section contains an introductory chapter summarizing the available procedures, followed by a chapter devoted to each procedure. The procedures are classified by subject to aid the new SAS user or statistician new to a method in choosing the appropriate SAS procedure.

The introductory chapter in each major section briefly describes the procedures available for that statistical category and what they do. Each procedure description is self-contained; you need to be familiar with only the most basic features of the SAS System and SAS terminology to use most procedures. The statements and syntax necessary to run each procedure are presented in a uniform format throughout the manual. You can duplicate the examples by using the same statements and data to run a SAS job. The examples are also useful as models for writing your own programs if you are a new SAS user.

Sections The procedures in this volume are divided into the following major sections:

- Regression
- Analysis of Variance
- Categorical Data Analysis
- Multivariate Methods
- Discriminant Analysis
- Clustering
- Scoring
- MATRIX

Chapters Each procedure description is divided into the following major parts:

ABSTRACT: a short paragraph describing what the procedure does.

INTRODUCTION: introductory and background material, including definitions and occasional introductory examples.

SPECIFICATIONS: reference section for the syntax of the control language for the procedure.

DETAILS: expanded descriptions of features, internal operations, output, treatment of missing values, computational methods, required computational resources, and usage notes.

EXAMPLES: examples using the procedure, including data, SAS statements, and printed output. You can reproduce these examples by copying the statements and data and running the job.

REFERENCES: a selected bibliography.

New Features in SAS Release 82 Statistical Procedures

Procedures Changes in SAS Release 82 that were not in SAS Release 79 include these new procedures and other features:

REG is a new procedure for general purpose regression, including diagnostics.

NLIN now has the DUD method, which does not need derivatives, and the NOHALVE and SIGSQ options.

RSREG is a new procedure for response-surface regression.

STEPWISE and RSQUARE now print Mallow's C_p.

GLM and ANOVA have many new means comparisons methods.

NPAR1WAY and RANK now have Savage scores.

FUNCAT has NOGLS options and a new CONTRAST statement.

CANCORR now reads CORR data sets, outputs statistics to a data set, has redundancy statistics, more multivariate statistics, and better labeling.

FACTOR has residual correlations, weighting, printing options, revised ML factoring, unweighted least-squares factoring, revised image analysis, Harris-Kaiser orthoblique rotation, oblique Procrustean rotation, and scree plots.

PRINCOMP is a new procedure for principal components.

STEPDISC is a new procedure for stepwise discriminant analysis.

FASTCLUS is a new procedure for fast disjoint clustering for a large number of observations.

CANDISC is a new procedure for canonical discriminant analysis.

VARCLUS is a new procedure for clustering variables by oblique component analysis.

TREE is a new procedure that prints tree diagrams of hierarchical clusterings.

CLUSTER has been rewritten to use Ward's method, the centroid method, or the average linkage on squared Euclidean distances method. Two of these methods can take far less space and cputime to run than the method implemented in SAS79. An output data set that can be fed into PROC TREE is provided.

PROC MATRIX has several new functions.

Documentation Some procedures, familiar from SAS Release 79, are now documented elsewhere.

These procedures are documented in *SAS User's Guide: Basics, 1982 Edition*: APPEND, BMDP, CALENDAR, CHART, CONTENTS, CONVERT, COPY, CORR,

DATASETS, EDITOR, FORMAT, FREQ, MEANS, OPTIONS, PDS, PDSCOPY, PLOT, PRINT, PRINTTO, RELEASE, SORT, SOURCE, SUMMARY, TABULATE, TAPECOPY, TAPELABEL, TRANSPOSE, and UNIVARIATE.

The AUTOREG, SPECTRA, and SYSREG procedures have been moved to the *SAS/ETS User's Guide, 1982 Edition*, since they concern econometric and time-series features. SYSREG features for ordinary least squares are found in the REG procedure.

The GUTTMAN and DUNCAN procedures have been moved to the *SAS Supplemental Library User's Guide*, where they will be supported as contributed works.

Documents now obsolete This volume and the *SAS User's Guide: Basics, 1982 Edition* include or supersede these earlier documents:

SAS User's Guide, 1979 Edition

SAS Technical Report P-111 (for the SAS Release 79.3 and SAS Release 79.4 releases), April 1980

SAS Technical Report P-115 (for the SAS Release 79.5 release),February 1981.

SAS Release 82? To find out which release of SAS you are using, run any SAS job and look at the first line on the log printout. The release number should start with the digits 82. If you have a later release, you should get new documentation. If you have an earlier release (for example, SAS Release 79.5 or SAS Release 79.6), your data center has not installed the latest version and you should use older documentation until they do.

MORE ABOUT THE SAS SYSTEM

What is the SAS System?

The SAS System is a computer system of software products for data analysis. The goal of SAS Institute is to provide data analysts one system to meet all their computing needs. When your computing needs are met, you are free to concentrate on results rather than on the mechanics of getting them. Instead of learning programming languages, several statistical packages, and utility programs, you only need to learn the SAS System.

Originally developed for statistical needs, the SAS System became an all-purpose data analysis system in response to the changing needs of its user community. To the basic SAS System, users can add tools for graphics, forecasting, data entry, operations research, and interfaces to other data bases to provide one total system. SAS software runs on IBM 360/370/30xx/43xx and compatible machines in batch and interactively under OS, VM/CMS, DOS/VSE, and TSO; on Digital Equipment Corporation VAXS™ 11/7xx Series under VMSS™, Data General ECLIPSE® MV Series under AOS/VS; and Prime Series 50 under PRIMOS® (Note: not all products are available for all systems.)

The basic SAS System provides tools for:

- information storage and retrieval
- data modification and programming
- report writing
- statistical analysis
- file handling.

Information storage and retrieval The SAS System reads data values in virtually any form from cards, disk, or tape and then organizes the values into a SAS data set.

The data can be combined with other SAS data sets using the file-handling operations described below. The data can be analyzed statistically, and its contents can produce reports. SAS data sets are automatically self- documenting since they contain both the data values and their descriptions. The special structure of a SAS data library minimizes maintenance.

Data modification and programming A complete set of SAS statements and functions is available for modifying data. Some program statements perform standard operations such as creating new variables, accumulating totals, and checking for errors; others are powerful programming tools such as DO/END and IF-THEN/ELSE statements. The data-handling features of the SAS System are so valuable that SAS software is used by many as a data base management system.

Report writing Just as SAS software reads data in almost any form, it can write data in almost any form. In addition to the preformatted reports that SAS procedures produce, SAS users can design and produce printed reports in any form, as well as punched cards and output files.

Statistical analysis The statistical analysis procedures in the SAS System are among the finest available. They range from simple descriptive statistics to complex multivariate techniques. Their designs are based on our belief that you should never need to tell SAS anything it can figure out by itself. Statistical integrity is thus accompanied by ease of use. Two especially noteworthy statistical features are the linear model procedures, of which GLM (**G**eneral **L**inear **M**odels) is the flagship, and the MATRIX procedure, which gives users the ability to handle any problem that can be expressed in traditional matrix notation.

File handling Combining values and observations from several data sets is often necessary for data analysis. SAS software has tools for editing, subsetting, concatenating, merging, and updating data sets. Multiple input files can be processed simultaneously, and several reports can be produced in one pass of the data.

Computer work usually involves related chores: data sets must be copied, tape contents investigated, program libraries moved. To help users cope with these needs, the SAS System includes a group of utility procedures.

Other SAS Institute Products

With the basic SAS System, you can integrate Institute products for graphics, forecasting, data entry, and interfaces to other data bases to provide one total system:

- SAS/GRAPH® —device-intelligent color graphics for business and research applications
- SAS/ETS™—expanded tools for business analysis, forecasting, and financial planning
- SAS/FSP® —interactive, menu-driven facilities for data entry, editing, retrieval of SAS files, and letter writing on IBM 327x-series terminals
- SAS/OR™—decision support tools for operations research and project management
- SAS/IMS-DL/I® —interface for reading, updating, and writing IMS/VS DB, IMS/VS DD/DC, CICS/OS/VS, CICS/DOS/VS, and DL/I DOS/VS data bases.
- SAS/REPLAY-CICS™—provides CICS/OS/VS and CICS/DOS/VS a facility to store, manage, and replay SAS/GRAPH displays

SAS Documentation

Using SAS Software Because the SAS System was designed as an all-purpose data analysis tool, many SAS features are needed only for complex problems. You don't need to know everything in the SAS manuals in order to use SAS software. In fact, 90% of SAS jobs use only 10% of the information.

The *SAS Introductory Guide* contains this first 10% and gets you started quickly using the SAS System. Another helpful manual is the *SAS Applications Guide*, which deals with common data-handling applications. *SAS User's Guide: Basics* provides thorough descriptions of the system's data management features.

If you have any problem with this SAS manual, please take time to complete the review page at the end of this book and send it to SAS Institute. We will consider your suggestions for future editions. In the meantime, ask your installation's SAS consultant for help.

Other SAS manuals Below is a list of other SAS manuals that can be obtained by contacting the SAS Institute Publication Sales Department:

SAS User's Guide: Basics, 1982 Edition
SAS Introductory Guide
Guiá Introductoria al SAS
Eine Einfuehrung in das SAS
Guide d'Introduction à SAS
SAS Applications Guide
SUGI Supplemental Library User's Guide, 1983 Edition
SAS Programmer's Guide, 1981 Edition
SAS Views: SAS Basics, 1983 Edition
SAS Views: SAS Color Graphics, 1983 Edition
SAS Companion for the VSE Operating System, 1983 Edition
SAS/GRAPH User's Guide, 1981 Edition
SAS/ETS User's Guide, 1982 Edition
SAS/IMS-DL/I User's Guide, 1981 Edition
SAS/FSP User's Guide, 1982 Edition
SAS/OR User's Guide, 1983 Edition
SAS/REPLAY-CICS User's Guide, 1984 Edition
SAS Video Training Basics 100-Series Workbook, 1981 Edition
SAS Color Graphics 100-Series Video Training Course Workbook, 1983 Edition
SAS Color Graphics 100-Series Video Training Course Instructional Guide, 1983 Edition
1981 SUGI Proceedings
1982 SUGI Proceedings
1983 SUGI Proceedings
1984 SUGI Proceedings
SAS for Linear Models: A Guide to the ANOVA and GLM Procedures
Merrill's Expanded Guide to CPE

Technical reports and Data Library Series The SAS Technical Report Series documents work-in-progress, describes new procedures, and covers a variety of application areas. Some of the features described in these reports are still in experimental form and are not yet available as SAS procedures. The SAS Data Library series consists of data on tape with accompanying documentation.

Write to SAS Institute for the current list of reports in the Technical Report Series and their costs.

SAS Services to Users

Technical support SAS Institute supports users through the Technical Support Group. If you have a problem running a SAS job, you should contact the individual at your site who is responsible for maintaining and supporting SAS. If the problem cannot be resolved locally, you or your local support personnel should call the Institute's Technical Support Group at (919) 467-8000 on weekdays between 9:00 AM and 5:00 PM Eastern Standard Time. A brochure describing the services provided by the Technical Support Group is available from SAS Institute.

SAS training SAS Institute sponsors a comprehensive training program, including programs of study for novice data processors, statisticians, applications programmers, systems programmers, and local support personnel. *SAS Training,* a semi-annual training publication, describes the total training program and each course currently being offered by SAS Institute. Courses are taught in major metropolitan areas, at the Institute's training center in Cary, NC, and at customer locations. Video training tapes are also produced at the studio located on the Institute's campus.

News magazine *SAS Communications* is the quarterly news magazine of SAS Institute. Each issue contains ideas for more effective use of the SAS System, preliminary documentation of new features, information about research and development underway at SAS Institute, the current training schedule, new publications, and news of the SAS Users Group International (SUGI).

To receive a copy of *SAS Communications,* send your name and complete address to:

SAS Communications Mailing List
SAS Institute Inc.
SAS Circle
Box 8000
Cary, NC 27511-8000

Supplemental library You can write your own SAS procedure in PL/I or FORTRAN taking advantage of the SAS supervisor for data input, manipulation, and output formatting. The *SAS Programmer's Guide*, available from SAS Institute, gives directions for writing SAS procedures.

Sample Library One of the SAS data sets included on the SAS installation tape is called SAS.SAMPLE. This data set contains sample SAS applications to illustrate features of SAS procedures and creative SAS programming techniques that can help you gain an in-depth knowledge of SAS capabilites.

Here are just a few examples of programs included:

ANOVA	analyzing a Latin-square split-plot design
CENSUS	reading hierarchical files of the U.S. Census Bureau Public Use Sample tapes
HARRIS	reading Harris Poll tapes coded in column-binary format
TEACH	teaching arithmetic to your child.

The library is on the installation tape in unloaded-PDS form. Check with the person who installed SAS software at your site to find out how to access the library since it may have been put on disk.

SUGI The SAS Users Group International (SUGI) is a non-profit association of professionals who are interested in how others are using the SAS System. Although SAS

Institute provides administrative support, SUGI is independent from the Institute. Membership is open to all users at SAS sites, and there is no membership fee.

Annual conferences are structured to allow many avenues of discussion. Users present invited and contributed papers on various topics, for example:

- computer performance evaluation and systems software
- econometrics, time series, and operations research
- graphics
- information systems
- interactive techniques
- statistics
- tutorials written as SAS programs.

Proceedings of the annual conferences are distributed free to SUGI registrants. Extra copies are available from SAS Institute.

SUGI also sponsors a code-critiquing service to assist SAS users who are unsure about how to approach a problem or who want help in writing programs.

SASware ballot SAS users provide valuable input toward the direction of future SAS development by ranking their priorities on the annual SASware ballot. The top vote-getters are announced in the spring *SAS Communications*. In this listing, completed projects, projects planned for the coming year, and long-term plans are noted. Complete results of the SASware ballot are also printed in the *SUGI Proceedings*.

Licensing the SAS System

The SAS System is licensed to customers in the Western Hemisphere from the Institute's headquarters in Cary, NC. To better serve the needs of our international customers, the Institute maintains subsidiaries in the United Kingdom, New Zealand, Australia, and Germany. In addition, agents in other countries are licensed distributors for the SAS System. For a complete list of offices, write or call:

SAS Institute Inc.
SAS Circle
Box 8000
Cary, NC 27511-8000
(919) 467-8000

Contents

SCORING

MATRIX

REGRESSION

Introduction to SAS Regression Procedures

This chapter reviews the SAS procedures that are used for regression analysis: REG, RSQUARE, STEPWISE, NLIN, and RSREG.

Many procedures in SAS perform regression analysis, each with special features. The following SAS procedures have similar specifications and computations:

REG	performs general-purpose regression with many diagnostic and input/output capabilities
RSQUARE	builds models and shows fitness measures for all possible models
STEPWISE	implements several stepping methods for selecting models
NLIN	builds nonlinear regression models
RSREG	builds quadratic response-surface regression models.

Several other procedures also perform regression:

GLM	performs an analysis of general linear models including models containing categorical terms and polynomials (documented with other analysis-of-variance procedures)
AUTOREG	implements regression models using time-series data where the errors are autocorrelated (documented in the *SAS/ETS User's Guide*)
SYSREG	handles linear simultaneous systems of equations, such as econometric models (documented in the *SAS/ETS User's Guide*)
SYSNLIN	handles nonlinear simultaneous systems of equations, such as econometric models (documented in the *SAS/ETS User's Guide*)

Other regression procedures contributed by SAS users are documented in the *SAS Supplemental Library User's Guide*.

These procedures perform regression analysis, which is the fitting of an equation to a set of values. The equation predicts a *response variable* from a function of *regressor variables* and *parameters*, adjusting the parameters such that a measure of fit is optimized. For example, the equation for the ith observation might be:

$$y_i = \beta_0 + \beta_1 x_i + \varepsilon_i$$

where y_i is the response variable, x_i is a regressor variable, β_0 and β_1 are unknown parameters to be estimated, and ε_i is an error term.

For example, you might use regression analysis to find out how well you can predict a person's weight if you know his height. Suppose you collect your data by measuring heights and weights of 20 school children. You need to estimate the intercept β_0 and the slope β_1 of a line of fit described by the equation:

$$\text{WEIGHT} = \beta_0 + \beta_1 \text{ HEIGHT} + \varepsilon$$

where

WEIGHT	is the response variable (also called the *dependent variable*)
β_0, β_1	are the unknown parameters
HEIGHT	is the regressor variable (also called the *independent variable, predictor, explanatory variable, factor, carrier*)
ε	is the unknown error.

A plot of your data is:

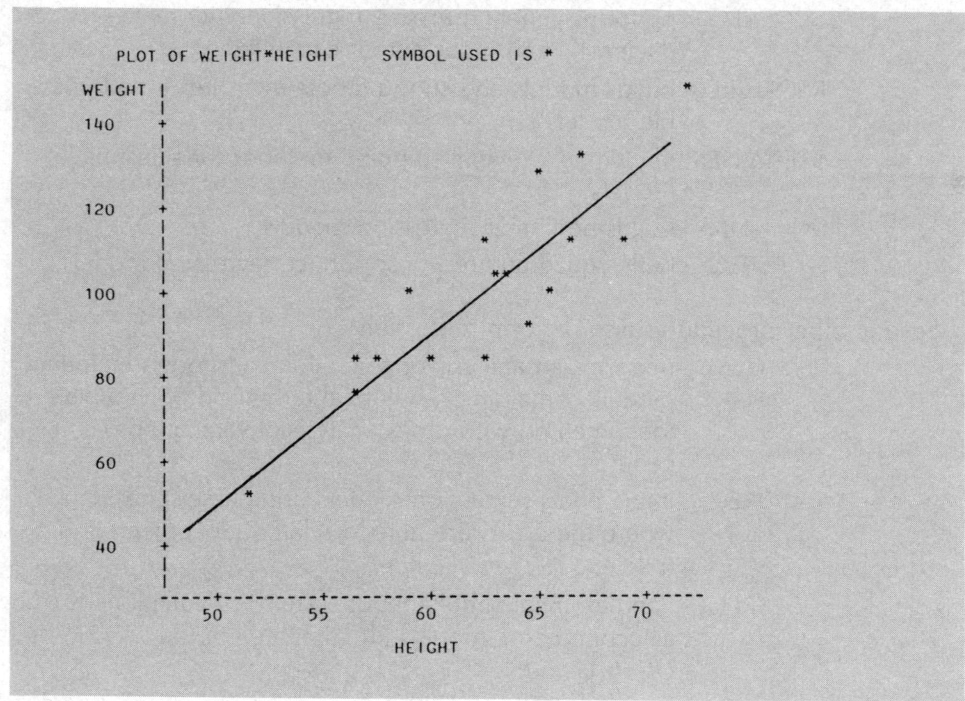

Regression estimates are $b_0 = -143$ and $b_1 = 3.9$, so the line of fit is:

$$\text{WEIGHT} = -143 + 3.9*\text{HEIGHT}$$

Regression is often used in an exploratory fashion to look for empirical relationships like the relationship between HEIGHT and WEIGHT. In this example HEIGHT is not the cause of WEIGHT. We do not even have evidence that the two variables change together over time, since these data are across subjects (cross-sectional) rather than across time (longitudinal). (We would need a controlled experiment to confirm the relationship scientifically.)

The method used to estimate the parameters is to minimize the sum of squares of the differences between the actual response value and the value predicted by the equation. The estimates are called *least-squares estimates* and the criterion value is called the *sum-of-squares error*:

$$SSE = \Sigma (y_i - b_0 - b_1 x_i)^2$$

where b_0 and b_1 are the values for β_0 and β_1 that minimize SSE.

For a general discussion of the theory of least-squares estimation of linear models and its application to regression and analysis of variance, see one of the applied regression texts including Draper and Smith (1981), Daniel and Wood (1980), and Johnston (1972).

SAS regression procedures produce the following information for a typical regression analysis:

- parameter estimates using the least-squares criterion
- estimates of the variance of the error term
- estimates of the variance or standard deviation of the parameter estimates
- hypotheses tests about the parameters
- predicted values and residuals using the estimates
- evaluation of the fit or lack of fit.

Besides the usual statistics of fit produced for a regression, SAS regression procedures can produce many other specialized diagnostic statistics, including:

- collinearity diagnostics to measure how much regressors are related to other regressors and how this affects the stability and variance of the estimates. (REG)
- influence diagnostics to measure how each individual observation contributes to determining the parameter estimates, the SSE, and the fitted values. (REG, RSREG)
- lack-of-fit diagnostics that measure the lack of fit of the regression model by comparing the error variance estimate to another pure error variance that is not dependent on the form of the model. (RSREG)
- time-series diagnostics for equally spaced time-series data that measure how much errors may be related across neighboring observations. These diagnostics can also measure functional goodness of fit for data sorted by regressor or response. (REG)

Other diagnostic statistics can be produced by programming a sequence of runs. For example, tests to measure structural change in a model over time can be performed by calculating items from several regressions or by writing a program with PROC MATRIX.

Comparison of Procedures

The REG Procedure PROC REG is a general-purpose procedure for regression with these capabilities:

- handles multiple MODEL statements
- can use correlations or crossproducts for input
- prints predicted values, residuals, studentized residuals, and confidence limits and can output these items to an output SAS data set
- prints special influence statistics
- produces partial regression leverage plots
- estimates parameters subject to linear restrictions
- tests linear hypotheses

- tests multivariate hypotheses
- writes estimates to an output SAS data set
- writes the crossproducts matrix to an output SAS data set
- computes special collinearity diagnostics.

The RSQUARE Procedure PROC RSQUARE fits all possible combinations of a list of variables specified in a MODEL statement. The procedure prints R^2 and C_p statistics and reports no estimates. PROC RSQUARE is useful when you want to look at alternative models. Since the number of possible models (2^n) gets large quickly, you should use the RSQUARE procedure only when you have fewer than 12 regressors to consider.

The STEPWISE Procedure PROC STEPWISE selects regressors for a model by various stepping strategies; you can request five different methods to search for good models. The FORWARD method starts with an empty model and at each step selects the variable that would maximize the fit. The BACKWARD method starts with a full model and at each step removes the variable that contributes least to the fit. There are three other variations: STEPWISE, MAXR, and MINR. PROC STEPWISE gives an analysis of variance and parameter estimates, but cannot produce predicted and residual values.

The NLIN Procedure PROC NLIN implements iterative methods that attempt to find least-squares estimates for nonlinear models. The default method is Gauss-Newton, although several other methods are available. You must specify parameter names and starting values, expressions for the model, and expressions for derivatives of the model with respect to the parameters (except for METHOD=DUD). A grid search is also available to select starting values of the parameters. Since nonlinear models are often tricky to estimate, NLIN may not always find the least-squares estimates.

The RSREG Procedure PROC RSREG fits a quadratic response-surface model, which is useful in searching for factor values that optimize a response. The three features in RSREG that make it preferable to other regression procedures for analyzing response surfaces are:

- automatic generation of quadratic effects
- a lack-of-fit test
- solutions for critical values of the surface.

The GLM Procedure PROC GLM for linear models can handle regression, analysis-of-variance, and analysis of covariance. (GLM is documented with the analysis-of-variance procedures.) The features for regression that distinguish GLM from other regression procedures are:

- ease of specifying categorical effects (GLM automatically generates dummy variables for class variables)
- direct specification of polynomial effects.

Statistical Background

The rest of this chapter outlines the way many SAS regression procedures calculate various regression quantities. Exceptions and further details are documented with individual procedures.

In matrix algebra notation, a linear model is

$$\mathbf{y} = \mathbf{X}\boldsymbol{\beta} + \boldsymbol{\varepsilon} \quad ,$$

where **X** is the $n \times k$ design matrix (rows are observations and columns are the regressors), β is the $k \times 1$ vector of unknown parameters, and ε is the $n \times 1$ vector of unknown errors. The first column of **X** is usually a vector of 1s used in estimating the intercept term.

The statistical theory of linear models is based on some strict classical assumptions. Ideally, the response is measured with all the factors controlled in an experimentally determined environment. Or, if you cannot control the factors experimentally, you must assume that the factors are fixed with respect to the response variable.

Other assumptions are that:

- the form of the model is correct
- regressor variables are measured without error
- the expected value of the errors is zero
- the variance of the errors (and thus the dependent variable) is a constant across observations called σ^2
- the errors are uncorrelated across observations.

When hypotheses are tested, the additional assumption is made that:

- the errors are normally distributed.

Statistical Model If the model satisfies all the necessary assumptions, the least-squares estimates are the best linear unbiased estimates (BLUE); in other words, the estimates have minimum variance among the class of estimators that are a linear function of the responses. If the additional assumption that the error term is normally distributed is also satisfied, then:

- the statistics that are computed have the proper sampling distributions for hypothesis testing
- parameter estimates will be normally distributed
- various sums of squares are distributed proportional to chi-square, at least under proper hypotheses
- ratios of estimates to standard errors are distributed as Student's *t* under certain hypotheses
- appropriate ratios of sums of squares are distributed as *F* under certain hypotheses.

When regression analysis is used to model data that do not meet the assumptions, the results should be interpreted in a cautious, exploratory fashion, with discounted credence in the significance probabilities.

Box (1966) and Tukey and Mosteller (1977, Chapters 12-13) discuss the problems that are encountered with regression data, especially when the data are not under experimental control.

Parameter Estimates and Associated Statistics

Parameter estimates are formed using least-squares criteria by solving the normal equations:

$$(\mathbf{X'X})\,\mathbf{b} = \mathbf{X'y} \quad ,$$

yielding

$$\mathbf{b} = (\mathbf{X'X})^{-1}\mathbf{X'y} \quad .$$

Assume for the present that (**X′X**) is full rank (we relax this later). The variance of the error σ^2 is estimated by the mean square error:

$$s^2 = \text{MSE} = \text{SSE}/(n-k) = \Sigma(y_i - X_i b)^2/(n-k) \quad .$$

The parameter estimates are unbiased:

$$E(\mathbf{b}) = \beta$$

$$E(s^2) = \sigma^2 \ .$$

The estimates have the variance-covariance matrix:

$$\text{Var}(\mathbf{b}) = (\mathbf{X'X})^{-1}\sigma^2 \ .$$

The estimate of the variance matrix replaces σ^2 with s^2 in the formula above:

$$\text{COVB} = (\mathbf{X'X})^{-1}s^2 \ .$$

The correlations of the estimates are derived by scaling to 1s on the diagonal. Let:

$$\mathbf{S} = \text{diag}((\mathbf{X'X})^{-1})^{-.5}$$

$$\text{CORRB} = \mathbf{S}(\mathbf{X'X})^{-1}\mathbf{S} \ .$$

Standard errors of the estimates are computed using the equation:

$$\text{STDERR}(b_i) = \sqrt{((\mathbf{X'X})^{ii}s^2)}$$

where $(\mathbf{X'X})^{ii}$ is the ith diagonal element of $(\mathbf{X'X})^{-1}$. The ratio

$$t = \mathbf{b}_i/\text{stderr}(\mathbf{b}_i)$$

is distributed as Student's t under the hypothesis that β_i is zero. Regression procedures print the t ratio and the significance probability, the probability under the hypothesis $\beta_i = 0$ of a larger absolute t value than was actually obtained. When the probability is less than some small level, the event is considered so unlikely that the hypothesis is rejected.

Type I SS and Type II SS measure the contribution of a variable to the reduction in SSE. Type I SS measure the reduction in SSE as that variable is entered into the model in sequence. Type II SS are the increment in SSE that results from removing the variable from the full model. Type II SS are equivalent to the Type III and Type IV SS reported in the GLM procedure. If Type II SS are used in the numerator of an F test, the test is equivalent to the t test for the hypothesis that the parameter is zero. In polynomial models, Type I SS measure the contribution of each polynomial term as if it is orthogonal to the previous terms in the model. The four types of SS are described in a more general context for the GLM procedure and in Chapter 15, "The Four Types of Estimable Functions."

Standardized estimates are defined as the estimates that result when all variables are standardized to a mean of 0 and a variance of 1. Standardized estimates are computed by multiplying the original estimates by the standard deviation of the regressor variable and dividing by the sample standard deviation of the dependent variable.

Tolerances and variance inflation factors measure the strength of interrelationships among the regressor variables in the model. If all variables are orthogonal to each other, both tolerance and variance inflation are 1. If a variable is very closely related to other variables, the tolerance goes to 0 and the variance inflation gets very large. Tolerance (TOL) is 1 minus the R^2 that results from the regression of the other variables in the model on that regressor. Variance inflation (VIF) is the diagonal of $(\mathbf{X'X})^{-1}$ if $(\mathbf{X'X})$ is scaled to correlation form. The statistics are related as shown below:

$$\text{VIF} = 1/\text{TOL} \quad .$$

Models not of full rank If the model is not full rank, then a generalized inverse can be used to solve the normal equations to minimize the SSE:

$$\mathbf{b} = (\mathbf{X'X})^{-}\mathbf{X'y} \, .$$

However, these estimates are not unique, since there are an infinite number of solutions using different generalized inverses. REG and other regression procedures choose a nonzero solution for all variables that are linearly independent of previous variables and a zero solution for other variables. This corresponds to using a generalized inverse in the normal equations, and the expected values of the "estimates" are the Hermite normal form of **X'X** times the true parameters:

$$\text{E}(\mathbf{b}) = (\mathbf{X'X})^{-}(\mathbf{X'X})\beta \quad .$$

Degrees of freedom for the zeroed estimates are reported as zero. The hypotheses that are not testable have *t* tests printed as missing. The message that the model is not full rank includes a printout of the relations that exist in the matrix.

Predicted Values and Residuals

After the model has been fit, predicted values and residuals are usually calculated and output. The predicted values are calculated from the estimated regression equation; the residuals are calculated as actual minus predicted. Some procedures can calculate standard errors.

Consider the ith observation where $\mathbf{x}_i$ is the row of regressors, **b** is the vector of parameter estimates, and s^2 is the mean squared error.

Let:

$$\mathbf{h}_i = \mathbf{x}_i(\mathbf{X'X})^{-1}\mathbf{x}_i' \text{ (the leverage)}$$

Then

$$\hat{y}_i = \mathbf{x}_i\mathbf{b} \text{ (the predicted value)}$$

$$\text{STDERR}(\hat{y}_i) = \sqrt{(\mathbf{h}_i s^2)} \text{ (the standard error of the predicted value)}$$

$$\text{resid}_i = \mathbf{y}_i - \mathbf{x}_i\mathbf{b} \text{ (the residual)}$$

$$\text{STDERR}(\text{resid}_i) = \sqrt{((1-\mathbf{h}_i)s^2)} \text{ (the standard error of the residual).}$$

The ratio of the residual to its standard error, called the *studentized residual*, is sometimes shown as:

$$\text{Student} = \text{resid} \,/\, \text{STDERR(resid)}.$$

There are two kinds of confidence intervals for predicted values. One type of confidence interval is an interval for the expected value of the response. The other type of confidence interval is an interval for the actual value of a response, which is the expected value plus error.

For example, construct for the ith observation a confidence interval that contains the true expected value of the response with probability $1 - \alpha$. The upper and lower limits of the confidence interval for the expected value are:

$$\text{LowerM} = \mathbf{x}_i\mathbf{b} - t_{\alpha/2}\sqrt{(\mathbf{h}_i s^2)}$$

$$\text{UpperM} = \mathbf{x}_i\mathbf{b} + t_{\alpha/2}\sqrt{(\mathbf{h}_i s^2)} \quad .$$

The limits for the confidence interval for an actual individual response (forecasting interval) are:

$$\text{LowerI} = \mathbf{x}_i\mathbf{b} - t_{\alpha/2}\sqrt{(\mathbf{h}_i s^2 + s^2)}$$

$$\text{UpperI} = \mathbf{x}_i\mathbf{b} + t_{\alpha/2}\sqrt{(\mathbf{h}_i s^2 + s^2)} \quad .$$

One measure of influence, Cook's D, measures the change to the estimates that results from deleting each observation.

$$\text{COOKD} = \text{Student}^2\ (\text{STDERR}(\hat{y})/\text{STDERR(resid)})^2/k.$$

For more information, see Cook (1977, 1979).

The *predicted residual* for observation i is defined as the residual for the ith observation that results from dropping the ith observation from the parameter estimates. The sum of squares of predicted residual errors is called the *press* statistic:

$$\text{presid}_i = \text{resid}/(1 + \text{h}_i)$$

$$\text{press} = \Sigma \text{presid}_i^2$$

Testing Linear Hypotheses

The general form of a linear hypothesis for the parameters is:

$$\mathbf{H}_0{:}\mathbf{L}\beta = \mathbf{c} \ ,$$

where $\mathbf{L}$ is $q \times k$, $\mathbf{b}$ is $k \times 1$, and $\mathbf{c}$ is $q \times 1$. To test this hypothesis, the linear function is taken with respect to the parameters:

$$(\mathbf{Lb} - \mathbf{c}) \quad .$$

This has variance:

$$\text{Var}(\mathbf{Lb} - \mathbf{c}) = \mathbf{L}\ \text{Var}(\mathbf{b})\mathbf{L}' = \mathbf{L}(\mathbf{X}'\mathbf{X})^{-}\mathbf{L}'\sigma^2 \quad .$$

A quadratic form called the *sum of squares due to the hypothesis* is calculated:

$$\text{SS}(\mathbf{Lb} - \mathbf{c}) = (\mathbf{Lb} - \mathbf{c})'(\mathbf{L}(\mathbf{X}'\mathbf{X})^{-}\mathbf{L}')^{-1}(\mathbf{Lb} - \mathbf{c}) \quad .$$

Assuming that this is testable, the SS can be used as a numerator of the F test:

$$F = \mathrm{SS}(\mathbf{Lb} - \mathbf{c})/q/s^2 \ .$$

This is referred to an F distribution with q and *dfe* degrees of freedom, where *dfe* is the degrees of freedom for residual error.

Multivariate Tests

Multivariate hypotheses involve several dependent variables in the form:

$$\mathbf{H_0{:}L\beta M} = \mathbf{d} \ ,$$

where $\mathbf{L}$ is a linear function on the regressor side, β is a matrix of parameters, $\mathbf{M}$ is a linear function on the dependent side, and $\mathbf{d}$ is a matrix of constants.

The special case (handled by REG) where the constants are the same for each dependent variable is written:

$$(\mathbf{L\beta - cj})\mathbf{M} = 0 \ ,$$

where $\mathbf{c}$ is a column vector of constants, and $\mathbf{j}$ is a row vector of 1s. The special case where the constants are 0 is

$$\mathbf{L\beta M} = 0 \ .$$

These multivariate tests are covered in detail in Morrison (1976); Timm (1975); Mardia, Kent, and Bibby (1979); Bock (1975); and other works cited in Chapter 20, "Introduction to SAS Multivariate Procedures."

To test this hypothesis, construct two matrices, $\mathbf{H}$ and $\mathbf{E}$, that correspond to the numerator and denominator of a univariate F test:

$$\mathbf{H} = \mathbf{M'(LB-cj)'(L(X'X)^{-1}L')^{-1}(LB-cj)M}$$

$$\mathbf{E} = \mathbf{M'(Y'Y-B'(X'X)B)M} \ .$$

Four test statistics, based on the eigenvalues of $\mathbf{E^{-1}H}$ or $\mathbf{(E+H)^{-1}H}$, are formed. Let λ_i be the ordered eigenvalues of $\mathbf{E^{-1}H}$ (if the inverse exists), and let ξ_i be the ordered eigenvalues of $\mathbf{(E+H)^{-1}H}$. It happens that $\xi_i = \lambda_i/(1+\lambda_i)$ and $\lambda_i = \xi_i/(1-\xi_i)$, and it turns out that $\varrho_i = \sqrt{(\xi_i)}$ is the ith canonical correlation.

Let p be the rank of $\mathbf{(H+E)}$, which is less than or equal to the number of columns of $\mathbf{M}$. Let q be the rank of $\mathbf{L(X'X)^{-1}L'}$. Let v be the degrees of freedom for error. Let $s = \min(p,q)$. Let $m = .5(|p-q|-1)$, and let $n = .5(v-p-1)$. Then the statistics below have the approximate F statistics as shown:

- **Wilks' Lambda**

  $$\Lambda = \det(\mathbf{E})/\det(\mathbf{H+E}) = \Pi\ 1/(1+\lambda_i) = \Pi(1-\xi_i) \ .$$

  $F = (1-\Lambda^{1/t})/(\Lambda^{1/t})\ (rt-2u)/pq$ is approximately F, where $r = v-(p+q+1)/2$ and $u = (pq-2)/4$, and $t = \sqrt{((p^2q^2-4)/(p^2+q^2-5))}$ if $(p^2+q^2-5)>0$ or 1 otherwise. The degrees of freedom are pq and $rt-2u$. This approximation is exact if $\min(p,q) <= 2$. (See Rao, 1973, p. 556.)

- **Pillai's Trace**

$$\mathbf{V} = \text{trace}(\mathbf{H}(\mathbf{H}+\mathbf{E})^{-1}) = \Sigma\lambda_i/(1+\lambda_i) = \Sigma\,\xi_i \ .$$

$F=(2n+s+1)/(2m+s+1)\mathbf{V}/(s-\mathbf{V})$ is approximately F with $s(2m+s+1)$ and $s(2n+s+1)$ degrees of freedom.

- **Hotelling-Lawley Trace**

$$\mathbf{U} = \text{trace}(\mathbf{E}^{-1}\mathbf{H}) = \Sigma\,\lambda_i = \Sigma\,\xi_i/(1-\xi_i) \ .$$

$F=2(sn+1)\mathbf{U}/(s^2(2m+s+1))$ is approximately F with $s(2m+s+1)$ and $2(sn+1)$ degrees of freedom.

- **Roy's Maximum Root**

$$\Theta = \xi_1 \ .$$

$F=\Theta(v-r-1)/r$ where $r=\max(p,q)$ is an upper bound on F that yields a lower bound on the significance level.

Tables of critical values for these statistics are found in Pillai (1960).

REFERENCES

Allen, D.M. (1971), "Mean Square Error of Prediction as a Criterion for Selecting Variables," *Technometrics*, 13, 469-475.

Allen, D.M. and Cady, F.B. (1982), *Analyzing Experimental Data by Regression*, Belmont, CA: Lifetime Learning Publications.

Belsley, D.A., Kuh, E., and Welsch, R.E. (1980), *Regression Diagnostics*, New York: John Wiley & Sons.

Bock, R.D. (1975), *Multivariate Statistical Methods in Behavioral Research*, New York: McGraw-Hill.

Box, G.E.P. (1966), "The Use and Abuse of Regression," *Technometrics*, 8, 625-629.

Cook, R.D. (1977), "Detection of Influential Observations in Linear Regression," *Technometrics*, 19, 15-18.

Cook, R.D. (1979), "Influential Observations in Linear Regression," *Journal of the American Statistical Association*, 74, 169-174.

Daniel, C. and Wood, F. (1980), *Fitting Equations to Data*, Revised Edition, New York: John Wiley & Sons.

Draper, N. and Smith, H. (1981), *Applied Regression Analysis*, Second Edition, New York: John Wiley & Sons.

Durbin, J. and Watson, G.S. (1951), "Testing for Serial Correlation in Least Squares Regression," *Biometrika*, 37, 409-428.

Freund, R.J. and Littell, R.C. (1981), *SAS for Linear Models*, Cary, NC: SAS Institute.

Goodnight, J.H. (1979), "A Tutorial on the SWEEP Operator," *The American Statistician*, 33, 149-158.

Johnston, J. (1972), *Econometric Methods*, New York: McGraw-Hill.

Kennedy, W.J. and Gentle, J.E. (1980), *Statistical Computing*, New York: Marcel Dekker.

Mallows, C.L. (1973), "Some Comments on Cp," *Technometrics*, 15, 661-675.

Mardia, K.V., Kent, J.T., and Bibby, J.M. (1979), *Multivariate Analysis*, London: Academic Press.

Morrison, D.F. (1976), *Multivariate Statistical Methods*, Second Edition, New York: McGraw-Hill.

Mosteller, F. and Tukey, J.W. (1977), *Data Analysis and Regression*, Reading, MA: Addison-Wesley.

Neter, J. and Wasserman, W. (1974), *Applied Linear Statistical Models*, Homewood, IL: Irwin.

Pillai, K.C.S. (1960), *Statistical Table for Tests of Multivariate Hypotheses*, Manila: The Statistical Center, University of Philippines.

Pindyck, R.S. and Rubinfeld, D.L. (1981), *Econometric Models and Econometric Forecasts*, Second Edition, New York: McGraw-Hill.

Rao, C.R. (1973), *Linear Statistical Inference and Its Applications*, Second Edition, New York: John Wiley & Sons.

Sall, J.P. (1981), "SAS Regression Application," Revised Edition, SAS Technical Report A-102, Cary, NC: SAS Institute.

Timm, N.H. (1975), *Multivariate Analysis with Applications in Education and Psychology*, Monterey, CA: Brooks/Cole.

Weisberg, S. (1980), *Applied Linear Regression*, New York: John Wiley & Sons.

The NLIN Procedure

ABSTRACT

The NLIN (NonLINear regression) procedure produces least-squares or weighted least-squares estimates of the parameters of a nonlinear model.

INTRODUCTION

PROC NLIN fits nonlinear regression models by least squares. Nonlinear models are more difficult to specify and estimate than linear models. Instead of simply listing regressor variables, you must write the regression expression, declare parameter names, guess starting values for them, and specify derivatives of the model with respect to the parameters. Some models are difficult to fit, and there is no guarantee that the procedure will be able to fit the model successfully.

NLIN first examines the starting value specifications of the parameters. If a grid of values is specified, NLIN evaluates the residual sum of squares at each combination of values to determine the best set of values to start the iterative algorithm. Then NLIN uses one of these four iterative methods:

- modified Gauss-Newton method
- Marquardt method
- gradient or steepest-descent method
- multivariate secant or false position (DUD).

The Gauss-Newton and Marquardt iterative methods involve regressing the residuals on the partial derivatives of the model with respect to the parameters until the iterations converge.

For each nonlinear model to be analyzed, you must specify:

- the names and starting values of the parameters to be estimated
- the model (using a single dependent variable)
- partial derivatives of the model with respect to each parameter (except for METHOD= DUD).

You may also:

- confine the estimation procedure to a certain range of values of the parameters by imposing bounds on the estimates
- adjust the convergence criterion
- produce a new SAS data set containing predicted values, residuals, estimates of parameters, and the residual sum of squares.

NLIN can be used for segmented models (see **Example 4**) and for computing maximum-likelihood estimates for certain models (see Jennrich and Moore, 1975).

SPECIFICATIONS

The following statements are used with PROC NLIN:

PROC NLIN *options;*
PARMS *parameter=values...;*
BOUNDS *expressions...;*
other programming statements
MODEL *dependent=expression;*
DER.*parameter=expression;*
OUTPUT *options;*

The PARMS and MODEL statements are required; the other statements are optional.

PROC NLIN Statement

PROC NLIN *options;*

The option below may appear in the PROC NLIN statement:

DATA=*SASdataset* names the SAS data set containing the data to be analyzed by PROC NLIN. If DATA= is omitted, the most recently created SAS data set is used.

Grid search options

BEST=*n* requests that NLIN print the residual sums of squares only for the best *n* combinations of possible starting values from the grid. When BEST= is not specified, NLIN prints the residual sum of squares for every combination of possible starting values of parameters.

PLOT requests that NLIN print a contour plot of the residual sums of squares. The two axes of the plot represent a grid of starting values for the two parameters. PLOT is appropriate only when the PARMS statement includes exactly two parameters, each with a grid of values.

Method options

METHOD=GAUSS
METHOD=MARQUARDT
METHOD=GRADIENT
METHOD=DUD
specifies the iterative method NLIN uses. METHOD=GAUSS is the default if DER statements are present; METHOD=DUD is the default if not. See the section on **Computational Method** for details.

NOHALVE turns off the step-halving during iteration. This is used with some types of weighted regression problems.

SIGSQ=*value* specifies a value to replace the mean square error for computing the standard errors of the estimates. SIGSQ= is used with maximum-likelihood estimation.

Tuning options

EFORMAT requests that NLIN print all numeric values in scientific E-notation. This is useful if your parameters have very different scales.

MAXITER=*i* places a limit on the number of iterations NLIN performs until it gives up trying to converge. The *i* value must be a positive integer. The default is 50.

CONVERGE=*c* specifies the relative convergence criterion. The iterations are said to have converged if:

$$(SSE_i - SSE_{i-1})/(SSE_i + 10^{-6}) < c .$$

The default is 10^{-8}. The constant *c* should be a small positive number.

Programming Statements with PROC NLIN

Any number of SAS programming statements after the PROC NLIN statement may be used. These statements should follow the PARMS statement and precede the MODEL statement.

Assignment statements, ARRAY statements, IF statements, and program control statements all can be executed by NLIN. Program statements can be used to create new SAS variables for the duration of the procedure (these variables are not permanently included in the data set to which NLIN is applied). Program statements may include variables in the data set specified by DATA=, parameter names, and variables created by preceding program statements within PROC NLIN.

LAG and DIF are the only SAS functions that do not operate in PROC NLIN the same way that they operate in a DATA step. LAG and DIF operate correctly across observations but are not reset at the beginning of a new iteration.

Consult the section **Special Variables** for information on special variables available to the NLIN procedure.

PARMS Statement

PARMS *parameter=values ...;*
PARAMETERS *parameter=values ...;*

A PARAMETERS (or PARMS) statement must follow the PROC NLIN statement. Several parameter names and values may appear. The parameter names must each be valid SAS names and must not duplicate the names of any variables in the data set to which NLIN is applied.

parameter=values

The parameter names identify the parameters to be estimated, both in subsequent procedure statements and in NLIN's printed output. The values specify the possible starting values of the parameters.

Usually only one value is specified for each parameter. If you specify several values for each parameter, NLIN evaluates the model at each point on the grid. The value specifications may take any of several forms:

m	a single value
*m*1,*m*2,...,*mn*	several values
m TO *n*	a sequence: starting, ending, increment=1
m TO *n* BY *i*	a sequence: starting, ending, increment
*m*1, *m*2 TO *m*3	mixed values and sequences.

This PARMS statement names five parameters and sets their possible starting values as shown:

```
PARMS  B0=0
       B1=4 TO 8
       B2=0 TO .6 BY .2
       B3=1, 10, 100
       B4=0, .5, 1 TO 4;
```

possible starting values	B0	B1	B2	B3	B4
	0	4	0	1	0
		5	.2	10	.5
		6	.4	100	1
		7	.6		2
		8			3
					4

Residual sums of squares are calculated for each of the $1\times5\times4\times3\times6=360$ combinations of possible starting values. (This can be expensive.)

See the section **Special Variables** for information on programming parameter starting values.

BOUNDS Statement

BOUNDS *expression...*;

The BOUNDS statement restrains the parameter estimates within specified bounds. In each BOUNDS statement, you can specify a series of bounds separated by commas. Each bound contains an expression consisting of a parameter name, an inequality comparison operator, and a value. Double-bounded expressions are also permitted. For example:

```
BOUNDS A<=20, 0<=B<=10, 20>C;
```

If you need to restrict an expression involving several parameters, for example, $A+B<1$, you can reparameterize the model so that the expression becomes a parameter.

The computational method used does not guarantee that if the iteration procedure sticks at the boundary of a constrained parameter, the other parameter estimates finally obtained will produce the restricted minimum residual sum of squares (Jennrich and Sampson, 1968).

MODEL Statement

MODEL *dependent=expression*;

The MODEL statement defines the prediction equation by declaring the dependent variable and defining an expression that evaluates predicted values. The expression can be any valid SAS expression yielding a numeric result. NLIN uses the same compiler as the DATA step, so any operators or functions defined there are available here. The expression may include parameter names, variables in the data set, and variables created by program statements. A MODEL statement must appear.

DER Statements for Derivatives

DER.*parameter*=*expression*;

For most of the computational methods, a DER statement must be included for each parameter to be estimated. The expression must be an algebraic representation of the partial derivative of the expression in the MODEL statement with respect to the parameter whose name immediately follows DER. The expression in the DER statement must conform to the rules for a valid SAS expression and may include any quantities that the MODEL statement expression contains.

The set of statements below specifies that a model:

$$Y = \beta_0(1 - e^{-\beta_1 x})$$

is to be fitted by the modified Gauss-Newton method where observed values of the dependent and independent variables are contained in the SAS variables Y and X, respectively.

```
PROC NLIN;
  PARMS B0=0 TO 10
    B1=.01 TO .09 BY .005;
  MODEL Y=B0*(1-EXP(-B1*X));
  DER.B0=1-EXP(-B1*X);
  DER.B1=B0*X*EXP(-B1*X);
```

Replacing the last three statements above with the statements

```
TEMP=EXP(-B1*X);
MODEL Y=B0*(1-TEMP);
DER.B0=1-TEMP;
DER.B1=B0*X*TEMP;
```

saves computer time, since the expression EXP(-B1*X) is evaluated only once per program execution rather than three times, as in the earlier example. If necessary, numerical rather than analytical derivatives can be used.

Weighted Regression

WEIGHT=*expression*;

To get weighted least-squares estimates of parameters, the _WEIGHT_ variable can be given a value in an assignment statement. When it is included, the expression is evaluated for each observation in the data set to be analyzed, and the values obtained are taken as inverse elements of the diagonal variance-covariance matrix of the dependent variable. When a variable name is given after the equal sign, the values of the variable are taken as the inverse elements of the variance-covariance matrix. The larger the _WEIGHT_ value, the more importance the observation is given.

OUTPUT Statement

OUTPUT OUT=*SASdataset keyword*=*names*...;

OUT=*SASdataset* names the SAS data set to be created by PROC NLIN when an OUTPUT statement is included. The new data set includes all the variables in the data set to which NLIN is applied, plus new variables whose names are given in the OUTPUT statement. If you want to create a permanent SAS data set, you must specify a two-level name (see Chapter 12, "SAS Data Sets," in *SAS User's Guide: Basics, 1982 Edition*, for more information on permanent SAS Data sets).

All of the names appearing in the OUTPUT statement must be valid SAS names, and **none of the new variable names may match a variable already existing in the data set to which NLIN is applied**.

PREDICTED=*name*
P=*name*
: names a variable on the output data set to contain the predicted values of the dependent variable.

RESIDUAL=*name*
R=*name*
: names a variable on the output data set to contain the residuals (actual-predicted).

PARMS=*names*
: names variables on the output data set to contain parameter estimates. These are normally the same names as listed on the PARMS statement; however, you can pick new names for the parameters identified in the sequence from the PARMS statement. Note that for each of these new variables, the values are the same for every observation in the new data set.

SSE=*name*
ESS=*name*
: specifies the name of a variable to be included in the new data set whose values are the residual sums of squares finally determined by the procedure. The values of the variable are the same for every observation in the new data set.

DETAILS

Missing Values

If the value of any one of the SAS variables involved in the model is missing from an observation, that observation is omitted from the analysis. If only the value of the dependent variable is missing, that observation has a predicted value calculated for it when an OUTPUT statement with PREDICTED= is present.

If an observation includes a missing value for one of the independent variables, both the predicted value and the residual value are missing for that observation. If the iterations fail to converge, all the values of all the variables named in the OUTPUT statement are missing values.

Troubleshooting

This section describes a number of problems that can occur in your analysis with PROC NLIN.

Time exceeded If you specify a grid of starting values that contains many points, then the job may run out of time, since the procedure must go through the entire data set for each point on the grid. The job may also run out of time if your problem takes many iterations to converge, since each iteration requires as much time as a linear regression with predicted values and residuals calculated.

Dependencies The matrix of partial derivatives may be singular, possibly indicating an over-parameterized model. For example, if B0 starts at zero in the following model, the derivatives for B1 are all zero for the first iteration.

```
PARMS B0=0 B1=.022;
MODEL POP=B0*EXP(B1*(YEAR-1790));
DER.B0=EXP(B1*(YEAR-1790));
DER.B1=(YEAR-1790)*B0*EXP(B1*(YEAR-1790));
```

The first iteration changes a subset of the parameters; then the procedure can make progress in succeeding iterations. This singularity problem is local. The next example shows a global problem.

You may have an add-factor B2 in the exponential that is non-identifiable, since it trades roles with B0.

```
PARMS B0=3.9 B1=.022 B2=0;
MODEL POP=B0*EXP(B1*(YEAR-1790)+B2);
DER.B0=EXP(B1*(YEAR-1790)+B2);
DER.B1=(YEAR-1790)*B0*EXP(B1*(YEAR-1790)+B2);
DER.B2=B0*EXP(B1*(YEAR-1790)+B2);
```

Unable to improve The method may lead to steps that do not improve the estimates, even after a series of step-halvings. If this happens, the procedure states that it was unable to make further progress, but it then prints the message "CONVERGENCE ASSUMED," and prints out the results. This often means that you have not converged at all. You should check the derivatives very closely and check the sum-of-squares error surface before proceeding. If you have not converged, try a different set of starting values or a different METHOD=.

Divergence The program may diverge and blow up with overflows and into areas that make arguments to functions illegal. For example, consider the model:

```
PARMS B=0;
MODEL Y=X / B;
```

Suppose that Y happens to be all zero and X is nonzero. There is no least-squares estimate for B, since the SSE declines as B approaches infinity or minus infinity. The same model could be parameterized with no problem into Y=A*X.

If you actually run the model, the procedure claims to converge after awhile, since it measures convergence with respect to the sum-of-squares error rather than to the parameter estimates. If you have divergence problems, try reparameterizing, different starting values, or a BOUNDS statement.

Local minimum The program may converge very nicely to a local rather than a global minimum. For example, consider the model:

```
PARMS A=1 B=-1;
MODEL Y=(1-A*X)*(1-B*X);
DER.A=-X*(1-B*X);
DER.B=-X*(1-A*X);
```

Once a solution is found, an equivalent solution with the same SSE is to switch the values between A and B.

Discontinuities The computational methods assume that the model is a continuous and smooth function of the parameters. If this is not true, the method does not work. For example, the following models will not work:

```
MODEL Y=A+INT(B*X);
MODEL Y=A+B*X+4*(Z>C);
```

Responding to trouble NLIN does not necessarily produce a good solution the first time. Much depends on specifying good initial values for the parameters. You can specify a grid of values on the PARMS statement to search for good starting values. While most practical models should give you no trouble, other models may require switching to a different iteration method. METHOD=MARQUARDT sometimes works when the default method (Gauss-Newton) does not work.

Computational Methods

For the nonlinear model,

$$Y=F(\beta_0,\beta_1,\ldots,\beta_k,X_1,X_2,\ldots,X_n)+\varepsilon = F(\beta) + \varepsilon \quad ,$$

the nonlinear "normal" equations are

$$\mathbf{X}'F(\beta) = \mathbf{X}'\mathbf{y}$$

where

$$\mathbf{X} = \partial F/\partial\beta .$$

In the nonlinear situation, both $\mathbf{X}$ and $F(\beta)$ are functions of β, and a closed-form solution generally does not exist. Thus NLIN uses an iterative process: a starting value for β is chosen and continually improved until the error sum-of-squares $\varepsilon'\varepsilon$ (SSE) is minimized.

The iterative techniques NLIN uses are similar to a series of linear regressions involving the matrix $\mathbf{X}$ evaluated for the current values of β and $\mathbf{y}=Y-\hat{Y}$, where

$$\hat{Y} = F(\beta)$$

are the predicted values evaluated for the current values of β.

The iterative process begins at some point β_0. Then $\mathbf{X}$ and $\mathbf{Y}$ are employed to compute a $\mathbf{\Delta}$ such that

$$SSE(\beta_0+\Delta)<SSE(\beta_0).$$

The three methods differ in how $\mathbf{\Delta}$ is computed to change the vector of parameters.

Steepest descent	$\mathbf{\Delta} = \mathbf{X}'\mathbf{y}$ (direction)
Gauss-Newton	$\mathbf{\Delta} = (\mathbf{X}'\mathbf{X})^{-1}\mathbf{X}'\mathbf{y}$ (direction)
Marquardt	$\mathbf{\Delta} = (\mathbf{X}'\mathbf{X}+\lambda\mathbf{I})^{-1}(\mathbf{X}'\mathbf{y})$ (direction and distance).

Steepest descent (gradient) The steepest descent method is defined by:

$$.5\ \partial\varepsilon'\varepsilon/\partial\beta = -\mathbf{X}'\mathbf{Y} + \mathbf{X}'F(\beta) = -\mathbf{X}'\mathbf{y}.$$

$-\mathbf{X'y}$ is the gradient along which $\varepsilon'\varepsilon$ increases. Thus $\Delta = \mathbf{X'y}$ is the direction of steepest descent.

Using the method of steepest descent, let

$$\beta_{i+1} = \beta_i + k\Delta$$

where the scalar **k** is chosen such that

$$SSE(\beta_i + k\Delta) < SSE(\beta_i).$$

Note: the steepest descent method converges very slowly and is therefore not recommended.

Gauss-Newton The Gauss-Newton method uses the Taylor series

$$F(\beta) = F(\beta_0) + \mathbf{X}(\beta - \beta_0) + \ldots$$

where $\mathbf{X} = \partial F / \partial \beta$ is evaluated at $\beta = \beta_0$.

Substituting the first two terms of this series into the "normal" equations:

$$\begin{aligned}
\mathbf{X'}F(\beta) &= \mathbf{X'Y} \\
\mathbf{X'}(F(\beta_0) + \mathbf{X}(\beta - \beta_0)) &= \mathbf{X'Y} \\
\mathbf{X'}F(\beta_0) + \mathbf{X'X}(\beta - \beta_0) &= \mathbf{X'Y} \\
(\mathbf{X'X})(\beta - \beta_0) &= \mathbf{X'Y} - \mathbf{X'}F(\beta_0) \\
(\mathbf{X'X})\Delta &= \mathbf{X'y} \text{ where } \Delta = (\mathbf{X'X})^{-1}\mathbf{X'y}.
\end{aligned}$$

Note: if $SSE(\beta_0 + \Delta) > SSE(\beta_0)$, then compute $SSE(\beta_0 + .5\Delta)$, $SSE(\beta_0 + .25\Delta)$,..., until a smaller SSE is found. When step-halving is employed, the method is known as modified Gauss-Newton (Hartley, 1961).

Marquardt The Marquardt updating formula is:

$$\Delta = (\mathbf{X'X} + \lambda \text{diag}(\mathbf{X'X}))^{-1}\mathbf{X'y}$$

The Marquardt method is a compromise between Gauss-Newton and steepest descent (Marquardt, 1963). As $\lambda \to 0$, the direction approaches Gauss-Newton. As $\lambda \to \infty$, the direction approaches steepest descent.

Marquardt's studies indicate that the average angle between Gauss-Newton and steepest descent directions is about 90°. A choice of λ between 0 and ∞ produces a compromise direction.

NLIN chooses $\lambda = 10^{-8}$ to start and computes a Δ. If $SSE(\beta_0 + \Delta) < SSE(\beta_0)$, then $\lambda = \lambda/10$ for the next iteration. Each time $SSE(\beta + \Delta) > SSE(\beta_0)$, then $\lambda = \lambda \times 10$.

Note: if the SSE improves on each iteration, then $\lambda \to 0$, and you are essentially using Gauss-Newton. If SSE does not improve, then λ is increased until you are moving in the steepest descent direction.

Marquardt's method is equivalent to performing a series of ridge regressions and is most useful when the parameter estimates are highly correlated.

Secant Method (DUD) The multivariate secant method is like Gauss-Newton, except that the derivatives are estimated from the history of iterations rather than being supplied analytically. The method is also called the *method of false position*, or the DUD method (Ralston and Jennrich, 1979). If only one parameter is being estimated, the derivative for iteration $i+1$ can be estimated from the previous two iterations:

$$\text{der}_{i+1} = (\widehat{Y}_i - \widehat{Y}_{i-1})/(b_i - b_{i-1})$$

When k parameters are to be estimated, the method uses the last $k+1$ iterations to estimate the derivatives.

Special Variables

Several special variables are created automatically and may be used in NLIN program statements. The values of these special variables are set by NLIN and should not be reset to a different value by programming statements.

_N_ indicates the number of times the program has been entered. It is never reset for successive passes through the data set.

ERROR is set to 1 if a numerical error or invalid argument to a function occurs during the current execution of the program. It is reset to 0 before each new execution.

OBS indicates the observation number in the data set for the current program execution. It is reset to 1 to start each pass through the data set (unlike _N_).

ITER represents the current iteration number. _ITER_ is set to -1 during the grid search phase.

MODEL is set to 1 for passes through the data when only the predicted values are needed, not the derivatives. It is 0 when both predicted values and derivatives are needed. If your derivative calculations consume a lot of time, you can save resources by coding:

```
IF _MODEL_ THEN RETURN;
```

after your MODEL statement, but before your derivative calculations.

SSE has the sum of squares error of the last iteration. During the grid search phase _SSE_ is set to 0; it is set to 1E70 for iteration 0.

For the derivative methods (GAUSS, MARQUARDT, and GRADIENT), the parameter values in the procedure are updated from the program data vector after the first observation of iteration 0. If you want to supply starting parameter values in your program (rather than using the values on the PARMS statement), then follow this example:

```
PROC NLIN;
  PARMS B0=1 B1=1;
  IF _ITER_=0 THEN IF _OBS_=1 THEN DO;
    B0=B0START;
    B1=B1START;
    END;
  MODEL Y=expression;
  DER.B0=expression;
  DER.B1=expression;
```

where B0START and B1START are in the input data set or calculated with program statements.

Printed Output

In addition to the output data set, NLIN also produces:

1. the estimates of the parameters and the residual sums of squares determined in each iteration
2. a list of the residual sums of squares associated with all or some of the combinations of possible starting values of parameters
3. for two-parameter models, a contour plot of residual sums of squares associated with possible starting values of parameters.

If the convergence criterion is met, NLIN prints:

4. an analysis-of-variance table including as sources of variation REGRESSION, RESIDUAL, UNCORRECTED TOTAL, and CORRECTED TOTAL
5. parameter estimates
6. ASYMPTOTIC STD. ERROR, an asymptotically valid standard error of the estimate
7. ASYMPTOTIC 95% CONFIDENCE INTERVAL for the estimate of the parameter
8. ASYMPTOTIC CORRELATION MATRIX OF THE PARAMETERS.

EXAMPLES

Negative Exponential Growth Curve: Example 1

This example demonstrates typical NLIN specifications for Marquardt's method and a grid of starting values. The predicted values and residuals are output for plotting.

```
TITLE NEGATIVE EXPONENTIAL: Y=B0*(1-EXP(-B1*X));

DATA A;
  INPUT X Y @@;
  CARDS;
020 0.57 030 0.72 040 0.81 050 0.87 060 0.91 070 0.94
080 0.95 090 0.97 100 0.98 110 0.99 120 1.00 130 0.99
140 0.99 150 1.00 160 1.00 170 0.99 180 1.00 190 1.00
200 0.99 210 1.00
;
PROC NLIN BEST=10 PLOT METHOD=MARQUARDT;
  PARMS B0=0 TO 2 BY .5 B1=.01 TO .09 BY .01;
  MODEL Y=B0*(1-EXP(-B1*X));
  DER.B0=1-EXP(-B1*X);
  DER.B1=B0*X*EXP(-B1*X);
  OUTPUT OUT=B P=YHAT R=YRESID;
PROC PLOT DATA=B;
  PLOT Y*X='A' YHAT*X='P' / OVERLAY VPOS=25;
  PLOT YRESID*X / VREF=0 VPOS=25;
```

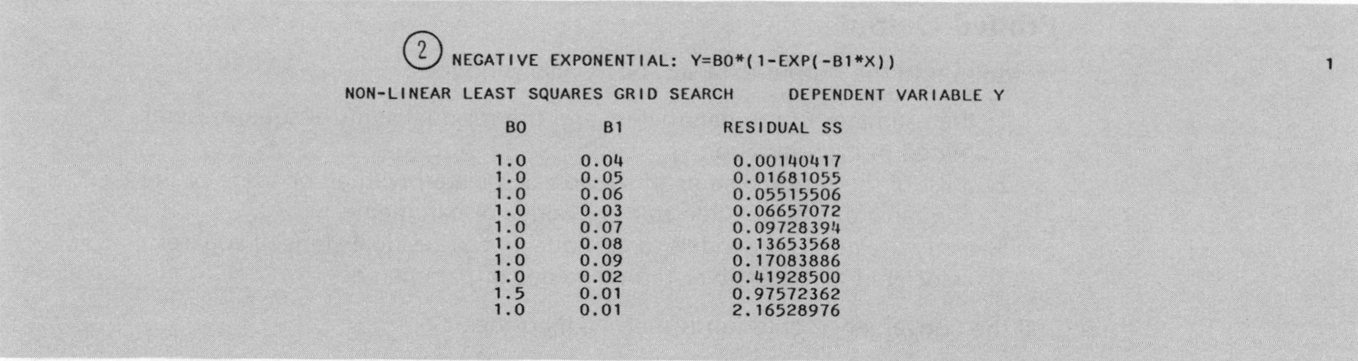

② NEGATIVE EXPONENTIAL: Y=B0*(1-EXP(-B1*X)) 1

NON-LINEAR LEAST SQUARES GRID SEARCH DEPENDENT VARIABLE Y

B0	B1	RESIDUAL SS
1.0	0.04	0.00140417
1.0	0.05	0.01681055
1.0	0.06	0.05515506
1.0	0.03	0.06657072
1.0	0.07	0.09728394
1.0	0.08	0.13653568
1.0	0.09	0.17083886
1.0	0.02	0.41928500
1.5	0.01	0.97572362
1.0	0.01	2.16528976

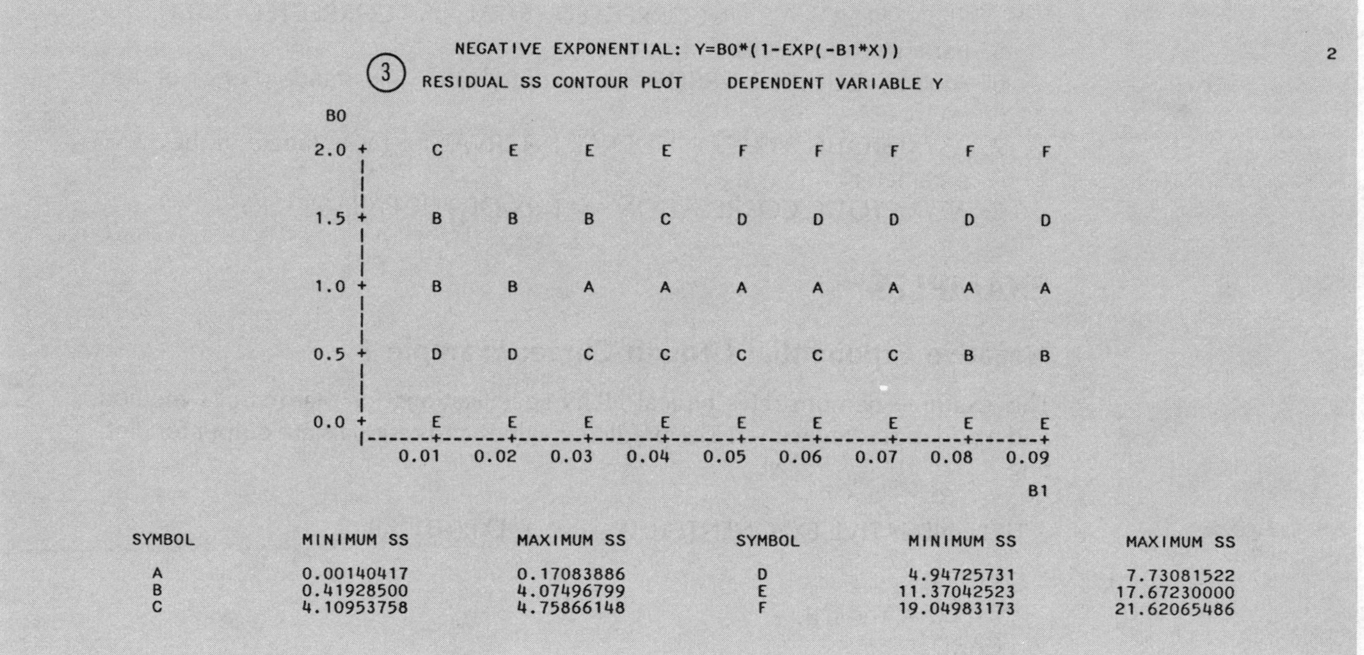

NEGATIVE EXPONENTIAL: Y=B0*(1-EXP(-B1*X)) 2

③ RESIDUAL SS CONTOUR PLOT DEPENDENT VARIABLE Y

```
 B0
2.0 +    C      E      E      E      F      F      F      F      F
    |
    |
1.5 +    B      B      B      C      D      D      D      D      D
    |
    |
1.0 +    B      B      A      A      A      A      A      A      A
    |
    |
0.5 +    D      D      C      C      C      C      C      B      B
    |
    |
0.0 +    E      E      E      E      E      E      E      E      E
    |------+------+------+------+------+------+------+------+------+
        0.01   0.02   0.03   0.04   0.05   0.06   0.07   0.08   0.09
                                                                  B1
```

SYMBOL	MINIMUM SS	MAXIMUM SS	SYMBOL	MINIMUM SS	MAXIMUM SS
A	0.00140417	0.17083886	D	4.94725731	7.73081522
B	0.41928500	4.07496799	E	11.37042523	17.67230000
C	4.10953758	4.75866148	F	19.04983173	21.62065486

NEGATIVE EXPONENTIAL: Y=B0*(1-EXP(-B1*X)) 3

NON-LINEAR LEAST SQUARES ITERATIVE PHASE

DEPENDENT VARIABLE: Y METHOD: MARQUARDT

①

ITERATION	B0	B1	RESIDUAL SS
0	1.00000000	0.04000000	0.00140417
1	0.99615180	0.04185364	0.00058005
2	0.99619193	0.04195211	0.00057681
3	0.99618863	0.04195386	0.00057681
4	0.99618857	0.04195389	0.00057681

NOTE: CONVERGENCE CRITERION MET.

NEGATIVE EXPONENTIAL: Y=B0*(1-EXP(-B1*X)) 4

NON-LINEAR LEAST SQUARES SUMMARY STATISTICS DEPENDENT VARIABLE Y

④

SOURCE	DF	SUM OF SQUARES	MEAN SQUARE
REGRESSION	2	17.67172319	8.83586159
RESIDUAL	18	0.00057681	0.00003205
UNCORRECTED TOTAL	20	17.67230000	
(CORRECTED TOTAL)	19	0.24385500	

(continued on next page)

(continued from previous page)

(5) PARAMETER	ESTIMATE	(6) ASYMPTOTIC STD. ERROR	(7) ASYMPTOTIC 95 % CONFIDENCE INTERVAL LOWER	UPPER
B0	0.99618857	0.00161380	0.99279812	0.99957901
B1	0.04195389	0.00039823	0.04111724	0.04279053

ASYMPTOTIC CORRELATION MATRIX OF THE PARAMETERS

(8)	B0	B1
B0	1.000000	-0.555896
B1	-0.555896	1.000000

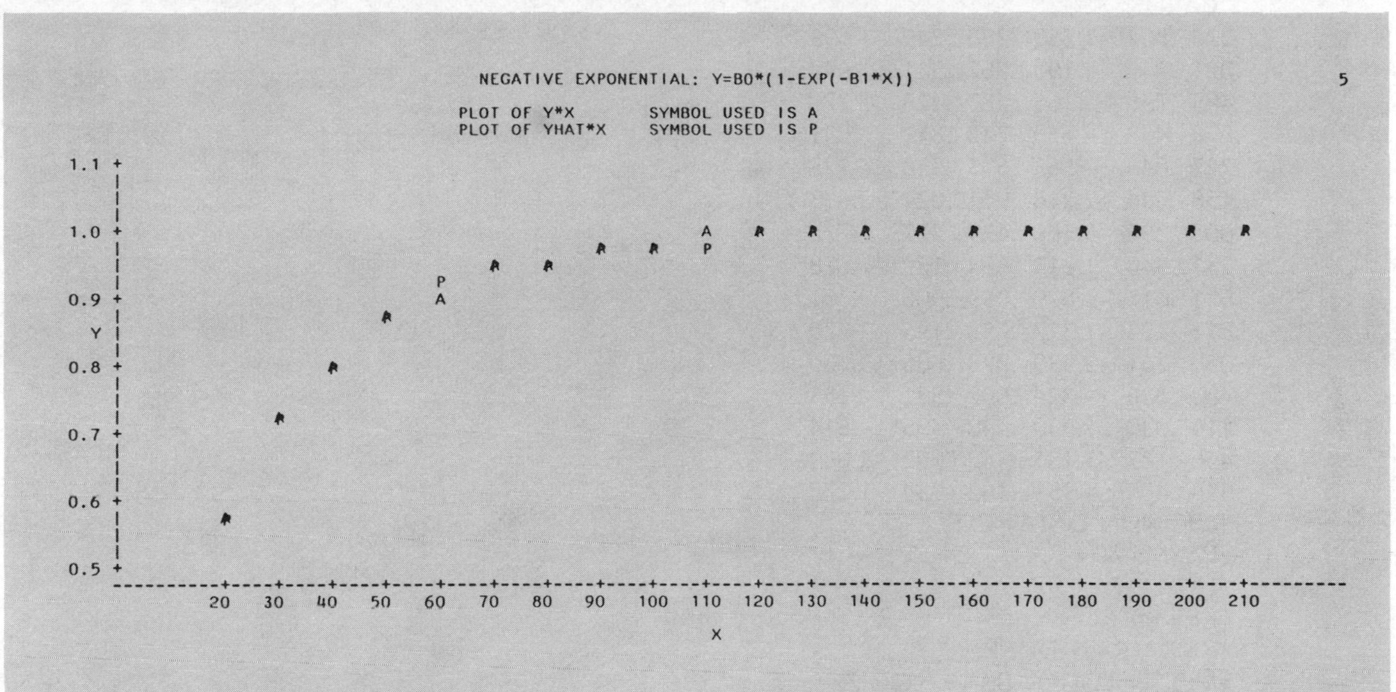

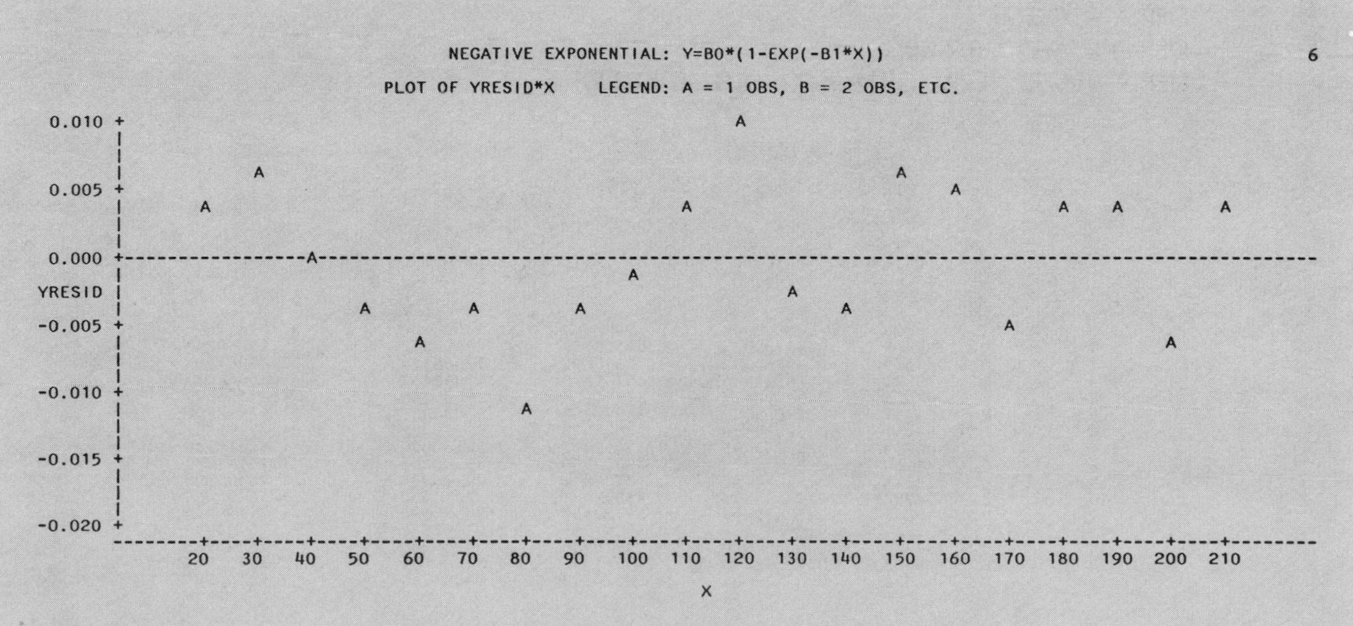

CES Production Function: Example 2

The CES production function in economics models the quantity produced as a function of inputs such as capital and labor. Arrow, Chenery, Minhas, and Solow developed the CES production function and named it for its property of constant elasticity of substitution. A is the efficiency parameter, D is the distribution or factor share parameter, and R is the substitution parameter. This example was described by Lutkepohl in the encyclopedic work by Judge et al. (1980).

```
TITLE CES MODEL: LOGQ = B0 + A*LOG(D*L**R+(1-D)*K**R);
DATA CES;
  INPUT L K LOGQ @@;
  CARDS;
.228 .802 -1.359 .258 .249 -1.695
.821 .771   .193 .767 .511  -.649
.495 .758  -.165 .487 .425  -.270
.678 .452  -.473 .748 .817   .031
.727 .845  -.563 .695 .958  -.125
.458 .084 -2.218 .981 .021 -3.633
.002 .295 -5.586 .429 .277  -.773
.231 .546 -1.315 .664 .129 -1.678
.631 .017 -3.879 .059 .906 -2.301
.811 .223 -1.377 .758 .145 -2.270
.050 .161 -2.539 .823 .006 -5.150
.483 .836  -.324 .682 .521  -.253
.116 .930 -1.530 .440 .495  -.614
.456 .185 -1.151 .342 .092 -2.089
.358 .485  -.951 .162 .934 -1.275  ;
PROC NLIN DATA=CES;
  PARMS B0=1 A=-1 D=.5 R=-1;
  LR=L**R;
  KR=K**R;
  Z=D*LR+(1-D)*KR;
  MODEL LOGQ=B0+A*LOG(Z);
  DER.B0=1;
  DER.A =LOG(Z);
  DER.D =(A/Z) * (LR-KR);
  DER.R =(A/Z)* (D*LOG(L)*LR+(1-D)*LOG(K)*KR);
```

```
                    CES MODEL: LOGQ = B0 + A*LOG(D*L**R+(1-D)*K**R)                               7

                         NON-LINEAR LEAST SQUARES ITERATIVE PHASE

                    DEPENDENT VARIABLE: LOGQ      METHOD: GAUSS-NEWTON

     ITERATION            B0            A             D             R          RESIDUAL SS

         0        1.00000000   -1.00000000    0.50000000   -1.00000000     37.09647696
         1        0.53348846   -0.48109035    0.45060133   -1.49993613     35.48664241
         2        0.32051558   -0.30765599    0.38316020   -2.30968355     22.69061393
         3        0.12479022   -0.28742856    0.30140828   -3.41818117      1.84546513
         4        0.12404406   -0.30792127    0.31714961   -3.20435160      1.83336003
         5        0.12293317   -0.35563222    0.34972967   -2.80035238      1.82033702
         6        0.12508517   -0.32429484    0.33021412   -3.08911305      1.77400355
         7        0.12401054   -0.34250499    0.34052971   -2.95160409      1.76210788
         8        0.12471260   -0.33275379    0.33459590   -3.03898265      1.76117653
         9        0.12434641   -0.33824362    0.33784883   -2.99373468      1.76105698
        10        0.12456252   -0.33519741    0.33602421   -3.02017060      1.76104269
        11        0.12444618   -0.33689004    0.33703533   -3.00586988      1.76104014
        12        0.12451237   -0.33594673    0.33647149   -3.01396582      1.76103953
        13        0.12447601   -0.33647084    0.33678488   -3.00950550      1.76103936
        14        0.12449640   -0.33617890    0.33661040   -3.01200194      1.76103931
        15        0.12448510   -0.33634124    0.33670746   -3.01061741      1.76103929

NOTE: CONVERGENCE CRITERION MET.
```

```
               CES MODEL: LOGQ = B0 + A*LOG(D*L**R+(1-D)*K**R)                                        8

         NON-LINEAR LEAST SQUARES SUMMARY STATISTICS      DEPENDENT VARIABLE LOGQ

              SOURCE                  DF        SUM OF SQUARES          MEAN SQUARE

              REGRESSION               4          130.00369371          32.50092343
              RESIDUAL                26            1.76103929           0.06773228
              UNCORRECTED TOTAL       30          131.76473300

              (CORRECTED TOTAL)       29           61.28965430

PARAMETER          ESTIMATE            ASYMPTOTIC                    ASYMPTOTIC 95 %
                                       STD. ERROR                 CONFIDENCE INTERVAL
                                                                LOWER              UPPER
B0               0.12448510            0.07834296          -0.03654989         0.28552010
A               -0.33634124            0.27218006          -0.89581094         0.22312847
D                0.33670746            0.13608506           0.05698283         0.61643208
R               -3.01061741            2.32290326          -7.78537569         1.76414086

               ASYMPTOTIC CORRELATION MATRIX OF THE PARAMETERS

                                 B0          A          D          R

               B0          1.000000   0.296490  -0.176550  -0.326696
               A           0.296490   1.000000  -0.783557  -0.999130
               D          -0.176550  -0.783557   1.000000   0.783363
               R          -0.326696  -0.999130   0.783363   1.000000
```

Probit Model With Numerical Derivatives: Example 3

This example fits the population of the U.S. across time to the inverse of the cumulative distribution function of the normal. Numerical derivatives are coded since the analytic derivatives are messy.

The C parameter is the upper population limit. The A and B parameters scale time.

```
TITLE U.S. POPULATION GROWTH;
TITLE2 PROBIT MODEL WITH NUMERICAL DERIVATIVES;

DATA USPOP;
  INPUT POP :6.3 @@;
  RETAIN YEAR 1780;
  YEAR=YEAR+10;
  YEARSQ=YEAR*YEAR;
  CARDS;
3929 5308 7239 9638 12866 17069 23191 31443 39818 50155
62947 75994 91972 105710 122775 131669 151325 179323 203211
;

PROC NLIN DATA=USPOP;
  PARMS A=-2.4 B=.012 C=400;
  DELTA=.0001;
  X=YEAR-1790;
  POPHAT=C*PROBNORM(A+B*X);
  MODEL POP=POPHAT;
  DER.A=(POPHAT-C*PROBNORM((A-DELTA)+B*X ))/DELTA;
  DER.B=(POPHAT-C*PROBNORM(A+(B-DELTA)*X ))/DELTA;
  DER.C=POPHAT/C;
  OUTPUT OUT=P P=PREDICT;
PROC PLOT DATA=P;
  PLOT POP*YEAR PREDICT*YEAR='P' /OVERLAY VPOS=30;
RUN;
```

```
                                   U.S. POPULATION GROWTH                                         9
                          PROBIT MODEL WITH NUMERICAL DERIVATIVES

                           NON-LINEAR LEAST SQUARES ITERATIVE PHASE

                      DEPENDENT VARIABLE: POP        METHOD: GAUSS-NEWTON

            ITERATION             A               B               C                RESIDUAL SS

                 0       -2.40000000      0.01200000    400.00000000         7174.59080509
                 1       -2.27190836      0.01262292    399.06649867          209.32792709
                 2       -2.30242518      0.01266078    404.80474237          177.39206407
                 3       -2.30278759      0.01262822    407.07275056          177.37004391
                 4       -2.30281850      0.01262856    407.07980055          177.36980352
                 5       -2.30281826      0.01262851    407.08266771          177.36980304

NOTE: CONVERGENCE CRITERION MET.
```

```
                                   U.S. POPULATION GROWTH                                        10
                          PROBIT MODEL WITH NUMERICAL DERIVATIVES

              NON-LINEAR LEAST SQUARES SUMMARY STATISTICS      DEPENDENT VARIABLE POP

                 SOURCE                  DF       SUM OF SQUARES          MEAN SQUARE

                 REGRESSION               3     164227.89925296       54742.63308432
                 RESIDUAL                16        177.36980304          11.08561269
                 UNCORRECTED TOTAL       19     164405.26905600

                 (CORRECTED TOTAL)       18      71922.76175474

       PARAMETER          ESTIMATE          ASYMPTOTIC                      ASYMPTOTIC 95 %
                                            STD. ERROR                    CONFIDENCE INTERVAL
                                                                      LOWER              UPPER
       A               -2.30281826          0.03283271           -2.37242015        -2.23321637
       B                0.01262851          0.00095699            0.01059980         0.01465722
       C              407.08266771         61.78489847          276.10518493       538.06015048

                         ASYMPTOTIC CORRELATION MATRIX OF THE PARAMETERS

                                              A           B           C

                                   A    1.000000   -0.007910   -0.219798
                                   B   -0.007910    1.000000   -0.972273
                                   C   -0.219798   -0.972273    1.000000
```

```
                                   U.S. POPULATION GROWTH                                        11
                          PROBIT MODEL WITH NUMERICAL DERIVATIVES

                    PLOT OF POP*YEAR          LEGEND: A = 1 OBS, B = 2 OBS, ETC.
                    PLOT OF PREDICT*YEAR      SYMBOL USED IS P

 POP |
     I
     I
     I                                                                                                                  P
     I
 200 +                                                                                                            R
     I
     I                                                                                                      A
     I                                                                                                      P
     I
     I                                                                                                P
 150 +                                                                                                A
     I                                                                                          P
     I                                                                                          A
     I                                                                                    A
     I                                                                                    P
     I                                                                              A
 100 +                                                                              P
     I                                                                        R
     I
     I                                                                  R
     I                                                            A
     I                                                            P
  50 +                                                      R
     I                                                R
     I                                          R
     I                                    R
     I                        R     R
     IP     R     R     R
   0 +A
     -+-----+-----+-----+-----+-----+-----+-----+-----+-----+-----+-----+-----+-----+-----+-----+-----+-----+-----+-----+
    1790  1800  1810  1820  1830  1840  1850  1860  1870  1880  1890  1900  1910  1920  1930  1940  1950  1960  1970  1980

                                                         YEAR

NOTE:     1 OBS HAD MISSING VALUES
```

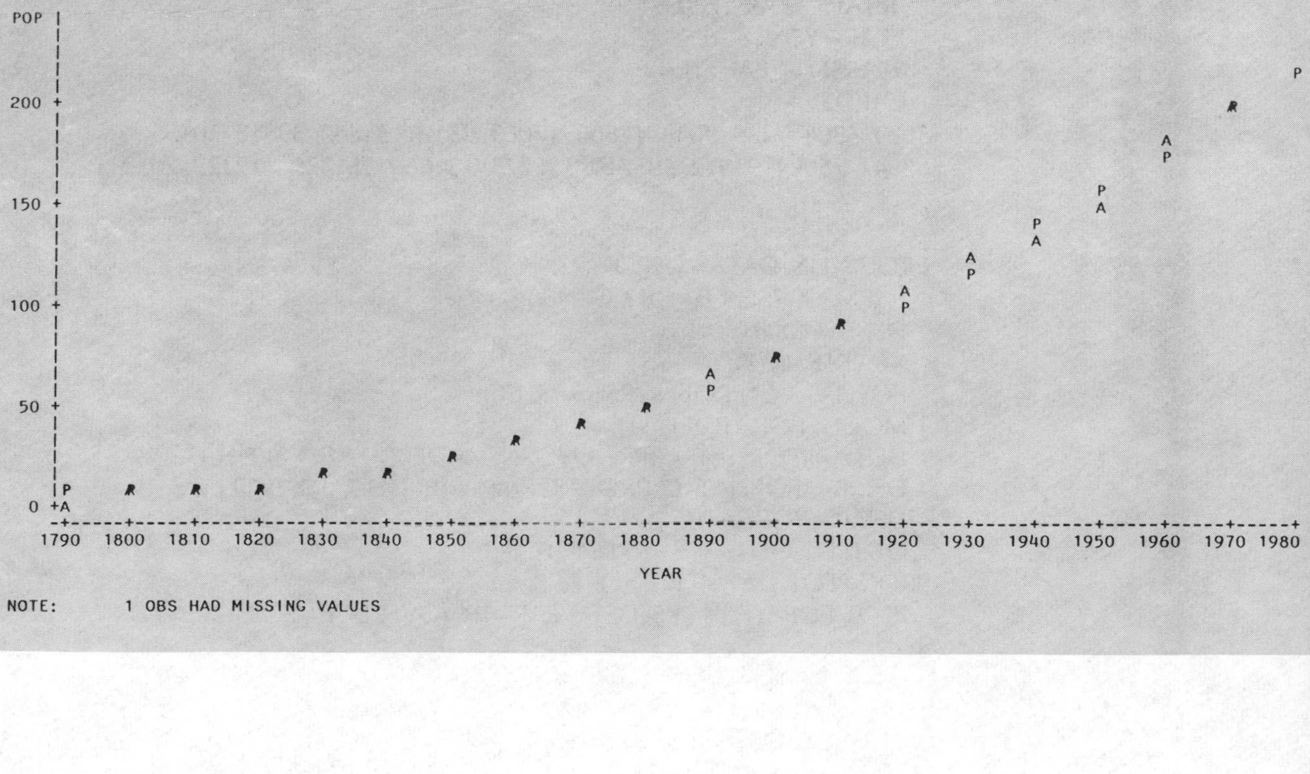

Segmented Model: Example 4

From theoretical considerations we can hypothesize that

$$y = a + bx + cx^2 \quad \text{if } x < x_0$$
$$y = p \quad \text{if } x > x_0 \; .$$

That is, for values of x less than x_0, the equation relating y and x is quadratic (a parabola) and, for values of x greater than x_0, the equation is constant (a horizontal line). NLIN can fit such a segmented model even when the joint point, x_0, is unknown.

The curve must be continuous (the two sections must meet at x_0), and the curve must be smooth (the first derivatives with respect to x are the same at x_0).

These conditions imply that

$$x_0 = -b/2c$$
$$p = a - b^2/4c \quad .$$

The segmented equation includes only three parameters; however, the equation is nonlinear with respect to these parameters.

You can write program statements with NLIN to conditionally execute different sections of code for the two parts of the model, depending on whether x is less than x_0.

We also use a PUT statement to print the constrained parameters every time the program is executed for the first observation (where $x=1$).

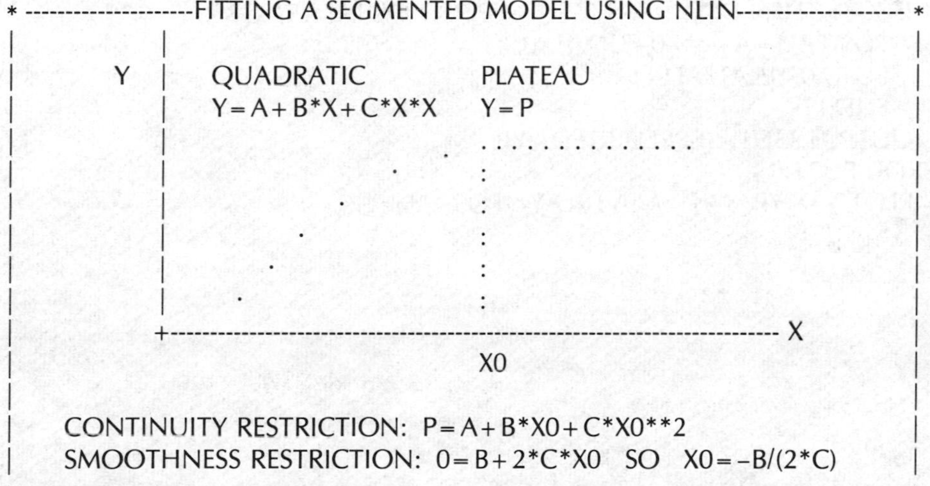

```
TITLE QUADRATIC MODEL WITH PLATEAU;
DATA A;
  INPUT Y X @@;
  CARDS;
.46 1 .47 2 .57 3 .61 4 .62 5 .68 6 .69 7
.78 8 .70 9 .74 10 .77 11 .78 12 .74 13 .80 13
.80 15 .78 16
;
PROC NLIN;
  PARMS A=.45 B=.05 C=-.0025;
  X0=-.5*B/C;                 * ESTIMATE JOIN POINT;
  DB=-.5/C;                   * DERIV OF X0 WRT B;
  DC=.5*B/C**2;               * DERIV OF X0 WRT C;
  IF X<X0 THEN DO;            * QUADRATIC PART OF MODEL;
    MODEL Y=A+B*X+C*X*X;
    DER.A=1;
    DER.B=X;
    DER.C=X*X;
    END;
  ELSE DO;                    * PLATEAU PART OF MODEL;
    MODEL Y=A+B*X0 +C*X0*X0;
    DER.A=1;
    DER.B=X0 +B*DB+2*C*X0*DB;
    DER.C=B*DC +X0*X0+2*C*X0*DC;
    END;
  IF X=1 THEN DO;             * PRINT OUT IF 1ST OBS;
    PLATEAU=A+B*X0+C*X0*X0;
    PUT X0=PLATEAU=;
    END;
  OUTPUT OUT=B PREDICTED=YP;
PROC PLOT;
  PLOT Y*X YP*X='*' / OVERLAY VPOS=35;
```

```
                              QUADRATIC MODEL WITH PLATEAU                                     12

                          NON-LINEAR LEAST SQUARES ITERATIVE PHASE

                     DEPENDENT VARIABLE: Y          METHOD: GAUSS-NEWTON

                 ITERATION            A            B              C             RESIDUAL SS
X0=10 PLATEAU=0.7
                         0       0.45000000   0.05000000   -0.00250000        0.05623125
X0=13.16594 PLATEAU=0.7936622
X0=13.16594 PLATEAU=0.7936622
                         1       0.38811776   0.06160510   -0.00233956        0.01176434
X0=12.8223 PLATEAU=0.7780506
X0=12.8223 PLATEAU=0.7780506
                         2       0.39304039   0.06005321   -0.00234175        0.01006840
X0=12.75562 PLATEAU=0.7775531
X0=12.75562 PLATEAU=0.7775531
                         3       0.39221643   0.06041831   -0.00236830        0.01006602
X0=12.74846 PLATEAU=0.7775026
X0=12.74846 PLATEAU=0.7775026
                         4       0.39212579   0.06045854   -0.00237121        0.01006599
X0=12.74774 PLATEAU=0.7774979
X0=12.74774 PLATEAU=0.7774979
                         5       0.39211632   0.06046272   -0.00237151        0.01006599
X0=12.74767 PLATEAU=0.7774974
X0=12.74767 PLATEAU=0.7774974
                         6       0.39211537   0.06046314   -0.00237154        0.01006599

NOTE: CONVERGENCE CRITERION MET.
```

```
                              QUADRATIC MODEL WITH PLATEAU                                     13

              NON-LINEAR LEAST SQUARES SUMMARY STATISTICS      DEPENDENT VARIABLE Y

                SOURCE                  DF       SUM OF SQUARES            MEAN SQUARE

                REGRESSION               3           7.72563401             2.57521134
                RESIDUAL                13           0.01006599             0.00077431
                UNCORRECTED TOTAL       16           7.73570000

                (CORRECTED TOTAL)       15           0.18694375

     PARAMETER             ESTIMATE          ASYMPTOTIC                    ASYMPTOTIC 95 %
                                             STD. ERROR                 CONFIDENCE INTERVAL
                                                                    LOWER              UPPER
     A                   0.39211537          0.02667415        0.33448941         0.44974132
     B                   0.06046314          0.00842304        0.04226628         0.07866001
     C                  -0.00237154          0.00055132       -0.00356259        -0.00118049

                          ASYMPTOTIC CORRELATION MATRIX OF THE PARAMETERS

                                          A            B            C

                            A         1.000000    -0.902025     0.812433
                            B        -0.902025     1.000000    -0.978795
                            C         0.812433    -0.978795     1.000000

X0=12.74767  PLATEAU=0.7774974
```

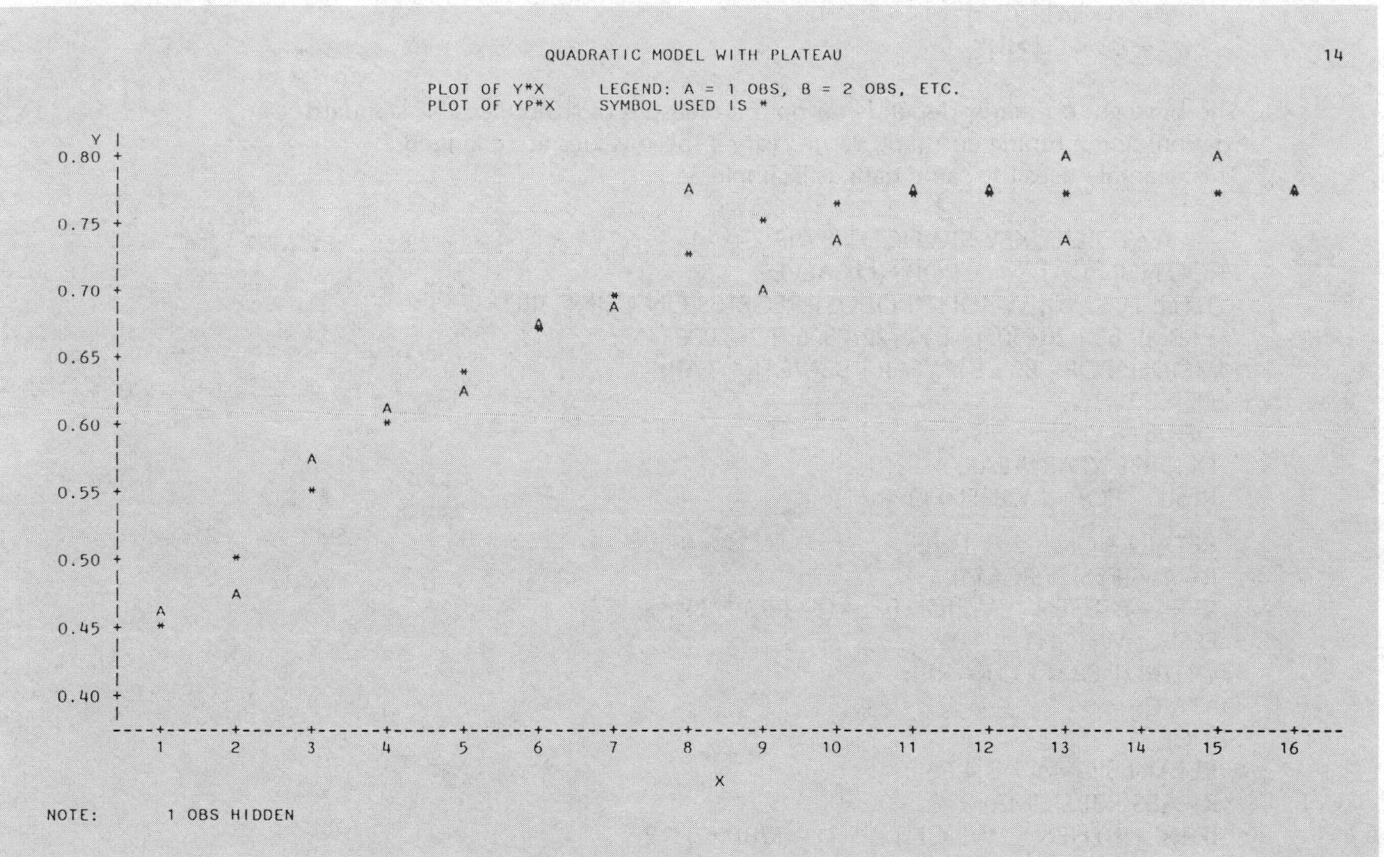

Iteratively Reweighted Least Squares: Example 5

The NLIN procedure is suited to methods that make the weight a function of the parameters in each iteration, since the _WEIGHT_ variable can be computed with program statements. The NOHALVE option is used because we are modifying the SSE definition at each iteration and are thus circumventing the step-shortening criteria.

Iteratively reweighted least squares (IRLS) can produce estimates for many of the robust regression criteria suggested in the literature. These methods act like automatic outlier rejectors, since large residual values lead to very small weights. Holland and Welsch (1977) outline several of these robust methods. For example, the biweight criterion suggested by Beaton and Tukey (1974) tries to minimize:

$$S_{biweight} = \Sigma\, \varrho(r),$$

where

$$\varrho(r) = (B^2/2)(1-(1-(r/B)^2)^3) \quad \text{if } |r|<=B\,, \quad \text{or otherwise}$$
$$\varrho(r) = (B^2/2)$$

where

r is abs(residual)/σ
σ is a measure of the scale of the error
B is a tuning constant (we use B=4.685).

The weighting function for the biweight is:

$$w_i = (1-(r/B)^2)^2 \quad \text{if } |r|<=B\,, \quad \text{or}$$
$$w_i = 0 \quad \text{if } |r|>B\ .$$

The biweight estimator depends on both a measure of scale (like the standard deviation) and a tuning constant; results vary if these values are changed.

This example uses the same data as Example 3.

```
*——BEATON/TUKEY BIWEIGHT BY IRLS——;
PROC NLIN DATA=USPOP NOHALVE;
  TITLE TUKEY BIWEIGHT ROBUST REGRESSION USING IRLS;
  PARMS B0=20450.43 B1=-22.7806  B2=.0063456;
  MODEL POP=B0+B1*YEAR+B2*YEAR*YEAR;
  DER.B0=1;
  DER.B1=YEAR;
  DER.B2=YEAR*YEAR;
  RESID=POP-MODEL.POP;

  RETAIN SIGMA 2 B 4.685;
  R=ABS(RESID/SIGMA);
  IF R<=B THEN _WEIGHT_=(1-(R/B)**2)**2;
  ELSE _WEIGHT_=0;
  OUTPUT OUT=C R=RBI;
DATA C;
SET C;
  RETAIN SIGMA 2 B 4.685;
  R=ABS(RBI/SIGMA);
  IF R<=B THEN _WEIGHT_=(1-(R/B)**2)**2;
  ELSE _WEIGHT_=0;
PROC PRINT;
```

The printout of the computed weights shows that the observations for 1940 and 1950 are highly discounted because of their large residuals.

SAS prints a note that missing values were propagated in 32 places. This happens when the last observation with a missing value for POP is handled. Since there are 15 iterations plus an initial iteration and the program is executed twice for each iteration (for each observation), these propagations occurred 32 times.

TUKEY BIWEIGHT ROBUST REGRESSION USING IRLS 15

NON-LINEAR LEAST SQUARES ITERATIVE PHASE

DEPENDENT VARIABLE: POP METHOD: GAUSS-NEWTON

ITERATION	B0	B1	B2	WEIGHTED RESIDUAL SS
0	20450.430000	-22.78060000	0.00634560	57.26481685
1	20711.580896	-23.06894014	0.00642514	31.31634829
2	20889.771439	-23.26396764	0.00647849	19.79450867
3	20950.186052	-23.33029155	0.00649669	16.75487545
4	20966.814401	-23.34856824	0.00650171	16.05727921
5	20970.962721	-23.35312851	0.00650296	15.89534769
6	20971.960688	-23.35422556	0.00650326	15.85719793
7	20972.198207	-23.35448666	0.00650333	15.84816740
8	20972.254576	-23.35454863	0.00650335	15.84602718
9	20972.267943	-23.35456332	0.00650335	15.84551981
10	20972.271113	-23.35456681	0.00650335	15.84539951
11	20972.271864	-23.35456763	0.00650336	15.84537100
12	20972.272042	-23.35456783	0.00650336	15.84536423
13	20972.272084	-23.35456787	0.00650336	15.84536263
14	20972.272094	-23.35456788	0.00650336	15.84536225
15	20972.272097	-23.35456789	0.00650336	15.84536216

NOTE: CONVERGENCE CRITERION MET.

TUKEY BIWEIGHT ROBUST REGRESSION USING IRLS 16

NON-LINEAR LEAST SQUARES SUMMARY STATISTICS DEPENDENT VARIABLE POP

SOURCE	DF	WEIGHTED SS	WEIGHTED MS
REGRESSION	3	122571.96279258	40857.32093086
RESIDUAL	16	15.84536216	0.99033514
UNCORRECTED TOTAL	19	122587.80815475	
(CORRECTED TOTAL)	18	59465.92678107	

PARAMETER	ESTIMATE	ASYMPTOTIC STD. ERROR	ASYMPTOTIC 95 % CONFIDENCE INTERVAL LOWER	UPPER
B0	20972.27209683	309.61766720	20315.91522523	21628.62896844
B1	-23.35456789	0.32987833	-24.05387523	-22.65526055
B2	0.00650336	0.00008781	0.00631720	0.00668951

ASYMPTOTIC CORRELATION MATRIX OF THE PARAMETERS

	B0	B1	B2
B0	1.000000	-0.999904	0.999613
B1	-0.999904	1.000000	-0.999903
B2	0.999613	-0.999903	1.000000

NOTE: MISSING VALUES WERE GENERATED AS A RESULT OF PERFORMING
AN OPERATION ON MISSING VALUES.
EACH PLACE IS GIVEN BY: (NUMBER OF TIMES) AT (LINE):(COLUMN).

32 AT 170:10 32 AT 173:6 32 AT 174:26

TUKEY BIWEIGHT ROBUST REGRESSION USING IRLS 17

OBS	POP	YEAR	YEARSQ	RBI	SIGMA	B	R	_WEIGHT_
1	3.929	1790	3204100	-1.0673	2	4.685	0.53364	0.974220
2	5.308	1800	3240000	0.3869	2	4.685	0.19347	0.996592
3	7.239	1810	3276100	1.0925	2	4.685	0.54625	0.972996
4	9.638	1820	3312400	0.9654	2	4.685	0.48269	0.978883
5	12.866	1830	3348900	0.3666	2	4.685	0.18330	0.996941
6	17.069	1840	3385600	-0.5579	2	4.685	0.27894	0.992923
7	23.191	1850	3422500	-0.8640	2	4.685	0.43200	0.983067
8	31.443	1860	3459600	-0.3408	2	4.685	0.17040	0.997356
9	39.818	1870	3496900	-0.9953	2	4.685	0.49764	0.977562
10	50.155	1880	3534400	-0.9884	2	4.685	0.49421	0.977868
11	62.947	1890	3572100	0.1728	2	4.685	0.08638	0.999320
12	75.994	1900	3610000	0.2883	2	4.685	0.14414	0.998108
13	91.972	1910	3648100	2.0341	2	4.685	1.01706	0.907966
14	105.710	1920	3686400	0.2393	2	4.685	0.11964	0.998696
15	122.775	1930	3724900	0.4708	2	4.685	0.23539	0.994957
16	131.669	1940	3763600	-8.7694	2	4.685	4.38469	0.015399
17	151.325	1950	3802500	-8.5482	2	4.685	4.27411	0.028128
18	179.323	1960	3841600	-1.2857	2	4.685	0.64287	0.962697
19	203.211	1970	3880900	0.5661	2	4.685	0.28304	0.992714
20	.	1980	3920400	.	2	4.685	.	.

Maximum Likelihood for Logistic Model: Example 6

As described by Jennrich and Moore (1975), maximum-likelihood estimates can be computed for exponential distributions by iteratively reweighted least squares, where the weights are the reciprocals of the variances. In this case you maximize the binomial likelihood in a logistic regression model. The same data and model are used in an example in the PROC FUNCAT chapter.

Use the SIGSQ option to scale the standard errors to the proper values. Since each observation is weighted by the reciprocal of the variance (estimated from the parameter estimates), you must scale the NLIN MSE variance to 1.

```
* -------------------------------INGOT DATA------------------------------- *
|                                                                          |
|   INGOTS ARE TESTED FOR READINESS TO ROLL AFTER DIFFERENT                |
|   TREATMENTS OF HEATING TIME AND SOAKING TIME.                           |
|   FROM COX (1970, PP. 67-68).                                            |
* ------------------------------------------------------------------------ * ;

DATA INGOTS;
   INPUT HEAT SOAK NREADY NTOTAL @@;
   COUNT=NREADY;      Y=1;    OUTPUT;
   COUNT=NTOTAL-NREADY;      Y=0;   OUTPUT;
   CARDS;
   7 1.0 0 10 14 1.0 0 31 27 1.0 1 56 51 1.0 3 13
   7 1.7 0 17 14 1.7 0 43 27 1.7 4 44 51 1.7 0  1
   7 2.2 0  7 14 2.2 2 33 27 2.2 0 21 51 2.2 0  1
   7 2.8 0 12 14 2.8 0 31 27 2.8 1 22
   7 4.0 0  9 14 4.0 0 19 27 4.0 1 16 51 4.0 0  1
;
PROC NLIN NOHALVE SIGSQ=1;
   TITLE LOGISTIC REGRESSION OF INGOT DATA USING PROC NLIN;
   PARMS B0=0 B1=0 B2=0;
   E=EXP(B0+B1*HEAT+B2*SOAK);
   P=E/(1+E);                  * DEFIN OF P WRT. LOGISTIC;
   MODEL Y=P;                  * EXPECTED VALUE OF Y IS P;
   W=1/(P*(1-P));              * WEIGHT IS 1/VARIANCE ;
   _WEIGHT_=W*COUNT;           * TIMES THE COUNT;
   DER=E/(1+E)**2;             * DERIV OF LINEAR PARAMETERS PARTIAL;
   DER.B0=DER;
   DER.B1=DER*HEAT;
   DER.B2=DER*SOAK;
```

```
                 LOGISTIC REGRESSION OF INGOT DATA USING PROC NLIN                  18

                      NON-LINEAR LEAST SQUARES ITERATIVE PHASE

                 DEPENDENT VARIABLE: Y             METHOD: GAUSS-NEWTON

         ITERATION            B0             B1             B2   WEIGHTED RESIDUAL SS

                 0  0.000000E+00   0.000000E+00   0.000000E+00           387.00000000
                 1   -2.15940610     0.01387843     0.00373270           126.00348700
                 2   -3.53344044     0.03631539     0.01197337           168.16292287
                 3   -4.74889861     0.06400132     0.02992008           251.25790395
                 4   -5.41381728     0.07902723     0.04981996           324.91659548
                 5   -5.55393071     0.08192764     0.05643949           344.37970151
                 6   -5.55915959     0.08203067     0.05677077           345.11457310
                 7   -5.55916646     0.08203080     0.05677131           345.11549849
                 8   -5.55916646     0.08203080     0.05677131           345.11549849
NOTE: CONVERGENCE CRITERION MET.
```

```
              LOGISTIC REGRESSION OF INGOT DATA USING PROC NLIN                                  19

          NON-LINEAR LEAST SQUARES SUMMARY STATISTICS      DEPENDENT VARIABLE Y

            SOURCE                  DF           WEIGHTED SS          WEIGHTED MS

            REGRESSION               3           13.01188816           4.33729605
            RESIDUAL                22          345.11549849          15.68706811
            UNCORRECTED TOTAL       25          358.12738665

            (CORRECTED TOTAL)       24          352.65340692

  PARAMETER            ESTIMATE           ASYMPTOTIC                  ASYMPTOTIC 95 %
                                          STD. ERROR               CONFIDENCE INTERVAL
                                                                LOWER              UPPER
  B0                -5.55916646           1.11969470       -7.88125264        -3.23708028
  B1                 0.08203080           0.02373448        0.03280889         0.13125272
  B2                 0.05677131           0.33121314       -0.63011723         0.74365985

NOTE: STANDARD ERRORS COMPUTED USING SIGSQ=1.

                  ASYMPTOTIC CORRELATION MATRIX OF THE PARAMETERS

                                  B0          B1          B2

                  B0        1.000000   -0.811522   -0.759767
                  B1       -0.811522    1.000000    0.333826
                  B2       -0.759767    0.333826    1.000000
```

REFERENCES

Bard, J. (1970), "Comparison of Gradient Methods for the Solution of the Nonlinear Parameter Estimation Problem," *SIAM Journal of Numerical Analysis*, 7, 157–186.

Bard, J. (1974), *Nonlinear Parameter Estimation*, New York: Academic Press.

Cox, D.R. (1970), *Analysis of Binary Data*, London: Chapman and Hall.

Gallant, A.R. (1975), "Nonlinear Regression," *American Statistician*, 29, 73–81.

Hartley, H.O. (1961), "The Modified Gauss-Newton Method for the Fitting of Non-Linear Regression Functions by Least Squares," *Technometrics*, 3, 269–280.

Hartley, H.O. (1961), "Least Squares Estimators," *Annals of Mathematical Statistics*, 40, 633–643.

Holland, P.H. and Welsch, R.E. (1977), "Robust Regression Using Iteratively Reweighted Least-Squares," *Communications Statistics: Theory and Methods*, 6, 813–827.

Jennrich, R.I. and Moore, R.H. (1975), "Maximum Likelihood Estimation by Means of Nonlinear Least Squares," American Statistical Association, *1975 Proceedings of the Statistical Computing Section*, 57–65.

Jennrich, R.I. and Sampson, P.F. (1968), "Application of Stepwise Regression to Non-Linear Estimation," *Technometrics*, 10, 63–72.

Judge, G.G., Griffiths, W.E., Hill, R.C., and Lee, T. (1980), *The Theory and Practice of Econometrics*, New York: John Wiley & Sons.

Marquardt, Donald W. (1963), "An Algorithm for Least-Squares Estimation of Nonlinear Parameters," *Journal for the Society of Industrial and Applied Mathematics*, 11, 431–441.

Ralston, M.L. and Jennrich, R.I. (1978), "DUD, A Derivative-Free Algorithm for Nonlinear Least Squares," *Technometrics*, 1, 7–14.

The REG Procedure

ABSTRACT

The REG procedure fits least-squares estimates to linear regression models.

INTRODUCTION

Suppose that a response variable Y can be predicted by a linear combination of some regressor variables X1 and X2. Then you can fit the β parameters in the equation:

$$Y_i = \beta_0 + \beta_1 X1_i + \beta_2 X2_i + \varepsilon_i$$

for the observations $i=1...n$. To fit this model with the REG procedure, specify:

```
PROC REG;
  MODEL Y=X1 X2;
```

REG uses the principle of least squares to produce estimates that are the best linear unbiased estimates (BLUE) under classical statistical assumptions (Gauss, 1809; Markov, 1900).

For example, we can apply regression techniques to the behavior of gross investment in General Electric as a linear function of lagged capital stock and the value of outstanding shares (Grunfeld, 1958). We use PROC REG to fit the model:

$$I = \beta_0 + \beta_1 C + \beta_2 F + \varepsilon \quad .$$

The SAS statements that read the data and perform the regression are:

```
TITLE GRUNFELD''S INVESTMENT MODEL;
DATA GRUNFELD;
  INPUT YEAR I F C @@;
  LABEL I=GROSS INVESTMENT GE
        C=CAPITAL STOCK LAGGED GE
        F=VALUE OF SHARES GE LAGGED;
  CARDS;
1935  33.1 1170.6  97.8  1936  45.0 2015.8 104.4
1937  77.2 2803.3 118.0  1938  44.6 2039.7 156.2
1939  48.1 2256.2 172.6  1940  74.4 2132.2 186.6
1941 113.0 1834.1 220.9  1942  91.9 1588.0 287.8
1943  61.3 1749.4 319.9  1944  56.8 1687.2 321.3
1945  93.6 2007.7 319.6  1946 159.9 2208.3 346.0
1947 147.2 1656.7 456.4  1948 146.3 1604.4 543.4
1949  98.3 1431.8 618.3  1950  93.5 1610.5 647.4
1951 135.2 1819.4 671.3  1952 157.3 2079.7 726.1
1953 179.5 2371.6 800.3  1954 189.6 2759.9 888.9
;
PROC REG;
  MODEL I=F C;
```

The results:

```
                                        GRUNFELD'S INVESTMENT MODEL                                        1

DEP VARIABLE: I          GROSS INVESTMENT GE

                        SUM OF         MEAN
SOURCE      DF         SQUARES       SQUARE        F VALUE        PROB>F

MODEL        2       31632.030    15816.015         20.344        0.0001
ERROR       17       13216.588      777.446
C TOTAL     19       44848.618

      ROOT MSE       27.882725     R-SQUARE         0.7053
      DEP MEAN         102.290     ADJ R-SQ         0.6706
      C.V.             27.2585

                     PARAMETER     STANDARD    T FOR H0:                         VARIABLE
VARIABLE    DF        ESTIMATE        ERROR  PARAMETER=0    PROB > |T|           LABEL

INTERCEP     1       -9.956306    31.374249       -0.317        0.7548   INTERCEPT
F            1        0.026551     0.015566        1.706        0.1063   VALUE OF SHARES GE LAGGED
C            1        0.151694     0.025704        5.902        0.0001   CAPITAL STOCK LAGGED GE
```

PROC REG is one of many regression procedures in SAS. REG is a general-purpose procedure for regression, while other SAS regression procedures have more specialized applications. GLM is designed to handle classification effects that occur in analysis-of-variance problems. STEPWISE and RSQUARE choose variables for building regression models. NLIN handles nonlinear models. SAS/ETS procedures are specialized for applications in time-series or simultaneous systems. Other SAS regression procedures are discussed in Chapter 1, "Introduction to SAS Regression Procedures."

PROC REG:

- handles multiple MODEL statements
- can use either correlations or crossproducts for input
- prints predicted values, residuals, studentized residuals, and confidence limits, and can output these items to an output SAS data set
- prints special influence statistics
- produces partial regression leverage plots
- estimates parameters subject to linear restrictions
- tests linear hypotheses
- tests multivariate hypotheses
- writes estimates to an output data set
- writes the crossproducts matrix to an output SAS data set
- computes special collinearity diagnostics.

SPECIFICATIONS

The following statements are used with the REG procedure:

PROC REG *options*;
  MODEL *dependents=regressors / options*;
  VAR *variables*;
  FREQ *variable*;
  WEIGHT *variable*;
  ID *variable*;
  OUTPUT OUT=*SASdataset keyword=names...*;
  RESTRICT *linear_equation,...*;
  TEST *linear_equation,...*;
  MTEST *linear_equation,...*;
  BY *variables*;

The PROC REG statement is always accompanied by one or more MODEL statements to specify regression models. One OUTPUT statement may follow each MODEL statement. Several RESTRICT, TEST, and MTEST statements may follow each MODEL. WEIGHT, FREQ, and ID statements are optionally specified once for the entire PROC step. The purposes of the statements are:

- The MODEL statement specifies the dependent and independent variables in the regression model.
- The OUTPUT statement requests an output data set and names the variables to contain predicted values, residuals, and other output values.
- The RESTRICT statement places linear restrictions on the parameter estimates.
- The TEST statement composes an *F* test on linear functions of the parameters.

- The MTEST statement composes multivariate tests across multiple dependent variables.
- The ID statement names a variable to identify observations in the printout.
- The WEIGHT and FREQ statements declare variables to weight observations.
- The VAR statement, rarely used with REG, lists variables for which crossproducts are to be computed.
- The BY statement specifies variables to define subgroups for the analysis.

PROC REG Statement

PROC REG *options*;

These options may be specified on the PROC REG statement:

DATA=*SASdataset* names the SAS data set to be used by PROC REG. If DATA= is not specified, REG uses the most recently created SAS data set.

OUTEST=*SASdataset* requests that parameter estimates be output to this data set. See **Output Data Sets** later in this chapter for details. If you want to create a permanent SAS data set, you must specify a two-level name (see Chapter 12, "SAS Data Sets," in *SAS User's Guide: Basics, 1982 Edition*, for more information on permanent SAS data sets).

OUTSSCP=*SASdataset* requests that the crossproducts matrix be output to this TYPE=SSCP data set. See **Output Data Sets** for details. If you want to create a permanent SAS data set, you must specify a two-level name (see Chapter 12, "SAS Data Sets," in *SAS User's Guide: Basics, 1982 Edition*, for more information on permanent SAS data sets).

NOPRINT suppresses the normal printed output. Using this option on the PROC REG statement is equivalent to specifying NOPRINT on each MODEL statement.

SIMPLE prints the "simple" descriptive statistics for each variable used in REG.

USSCP prints the (uncorrected) sums-of-squares and crossproducts matrix for all variables used in the procedure.

ALL requests many different printouts. Using ALL on the PROC REG statement is equivalent to specifying ALL on every MODEL statement. ALL also implies SIMPLE and SSCP.

COVOUT outputs the covariance matrices for the parameter estimates to the OUTEST data set. This option is valid only if OUTEST= is also specified. See **Output Data Sets** later in this chapter.

EPSILON=*n* tunes the mechanism used to check for singularities. The default value is 1E-8. This option is rarely needed. Singularity checking is described in **Computational Method**.

MODEL Statement

label: MODEL *dependents*=*regressors* / *options*;

After the keyword MODEL, the dependent (response) variables are specified, followed by an equal sign and the regressor variables. Variables specified in the MODEL statement must be variables in the data set being analyzed. The label is optional.

General options:

NOPRINT suppresses the normal printout of regression results.

NOINT suppresses the intercept term that is normally included in the model automatically.

ALL requests all the features of these options: XPX, SS1, SS2, STB, TOL, COVB, CORRB, SEQB, P, R, CLI, CLM.

Options to request regression calculations:

XPX prints the **X'X** crossproducts matrix for the model. The crossproducts matrix is bordered with the **X'Y** and **Y'Y** matrices.

I prints the $(\mathbf{X'X})^{-1}$ matrix. The inverse of the crossproducts matrix is bordered with the parameter estimates and SSE matrices.

Options for details on the estimates:

SS1 prints the sequential sums of squares (Type I SS) along with the parameter estimates for each term in the model.

SS2 prints the partial sums of squares (Type II SS) along with the parameter estimates for each term in the model.

STB prints standardized regression coefficients. A standardized regression coefficient is computed by dividing a parameter estimate by the ratio of the sample standard deviation of the dependent variable to the sample standard deviation of the regressor.

TOL prints tolerance values for the estimates. Tolerance is defined as $1-R^2$ for a variable with respect to all other regressor variables in the model.

VIF prints variance inflation factors with the parameter estimates. Variance inflation is the reciprocal of tolerance.

COVB prints the estimated covariance matrix of the estimates. This matrix is $(\mathbf{X'X})^{-1}S^2$ where S^2 is the mean squared error.

CORRB prints the correlation matrix of the estimates. This is the $(\mathbf{X'X})^{-1}$ matrix scaled to unit diagonals.

SEQB prints a sequence of parameter estimates as each variable is entered into the model. This is printed as a lower triangular matrix where each row is a set of parameter estimates.

COLLIN requests a detailed analysis of collinearity among the regressors. This includes eigenvalues, condition indices, and decomposition of the variances of the estimates with respect to each eigenvalue. Currently the number of variables printed in the analysis is limited to 35. (See **Collinearity Diagnostics** below.)

COLLINOINT requests the same analysis as the COLLIN option with the intercept variable adjusted out rather than included in the diagnostics. (Also see **Parameter Estimates and Associated Statistics** later in this chapter.)

Options for predicted values and residuals:

P calculates predicted values from the input data and the estimated model. The printout includes the observation number, the first ID variable if specified, the actual and predicted values, and the residual.

R requests that the residual be analyzed. The printed output includes everything requested by the P option plus the standard errors of the predicted and residual values, the studentized residual, and Cook's *D* statistic to measure the influence of each observation on the parameter estimates.

CLM prints the 95% upper and lower confidence limits for the expected value of the dependent variable (mean) for each observation. This is not a prediction interval (see the CLI option), because it takes into account only the variation in the parameter estimates, not the variation in the error term.

CLI requests the 95% upper and lower confidence limits for an individual predicted value. The confidence limits reflect variation in the error as well as variation in the parameter estimates.

DW calculates a Durbin-Watson statistic to test whether or not the errors have first-order autocorrelation. (This test is only appropriate for time-series data.) The sample autocorrelation of the residuals is also printed. (See **Autocorrelation in Time-Series Data**.)

INFLUENCE requests a detailed analysis of the influence of each observation on the estimates and the predicted values. (See **Influence Diagnostics**.)

PARTIAL requests partial regression leverage plots for each regressor. (See **Influence Diagnostics** and also **Predicted Values and Residuals**.)

VAR Statement

VAR *variables*;

The VAR statement is used to include variables in the crossproducts matrix that are not specified on any MODEL statement. This statement is rarely used with REG, and only then with the OUTSSCP= feature.

FREQ Statement

FREQ *variable*;

If a variable in your data set represents the frequency of occurrence for the other values in the observation, include the variable's name in a FREQ statement. The procedure then treats the data set as if each observation appears *n* times, where *n* is the value of the FREQ variable for the observation. The total number of observations will be considered equal to the sum of the FREQ variable when the procedure determines degrees of freedom for significance probabilities.

The WEIGHT and FREQ statements have similar effects, except in the calculation of degrees of freedom.

WEIGHT Statement

WEIGHT *variable*;

A WEIGHT statement names a variable on the input data set whose values are relative weights for a weighted least-squares fit. If the weight value is proportional to the reciprocal of the variance for each observation, then the weighted estimates are the best linear unbiased estimates (BLUE).

ID Statement

ID *variable*;

The ID statement specifies one variable to identify observations as output from the MODEL options P, R, CLM, CLI, and INFLUENCE.

OUTPUT Statement

The OUTPUT statement specifies an output data set to contain statistics calculated for each observation. For each statistic, specify the keyword, an equal sign, and a variable name for the statistic on the output data set. If the MODEL has several dependent variables, then a list of output variable names can be specified after each keyword to correspond to the list of dependent variables.

OUTPUT OUT=*SASdataset*
PREDICTED=*names* or P=*names*
RESIDUAL=*names* or R=*names*
L95M=*names*
U95M=*names*
L95=*names*
U95=*names*
STDP=*names*
STDR=*names*
STUDENT=*names*
COOKD=*names*
H=*names*
PRESS=*names*
RSTUDENT=*names*
DFFITS=*names*
COVRATIO=*names*;

The output data set named with OUT= contains all the variables in the input data set, including any BY variables, any ID variables, and variables named in the OUTPUT statement that contain statistics.

For example, the SAS statements

```
PROC REG DATA=A;
  MODEL Y Z=X1 X2;
  OUTPUT OUT=B
         P=YHAT ZHAT
         R=YRESID ZRESID;
```

create an output data set named B. In addition to the variables on the input data set, B contains the variable YHAT, whose values are predicted values of the dependent variable Y; ZHAT, whose values are predicted values of the dependent variable Z; YRESID, whose values are the residual values of Y; and ZRESID, whose values are the residual values of Z.

These statistics may be output to the new data set:

PREDICTED= P=	predicted values.
RESIDUAL= R=	residuals, calculated as ACTUAL minus PREDICTED.
L95M=	lower bound of a 95% confidence interval for the expected value (mean) of the dependent variable.
U95M=	upper bound of a 95% confidence interval for the expected value (mean) of the dependent variable.
L95=	lower bound of a 95% confidence interval for an individual prediction. This includes the variance of the error as well as the variance of the parameter estimates.
U95=	upper bound of a 95% confidence interval for an individual prediction.
STDP=	standard error of the mean predicted value.
STDR=	standard error of the residual.
STUDENT=	studentized residuals, the residual divided by its standard error.
COOKD=	Cook's *D* influence statistic.
H=	leverage, $x_i(\mathbf{X'X})^{-1}x_i'$.
PRESS=	residual for estimates dropping this observation, which is the residual divided by $(1-h)$ where h is leverage above.
RSTUDENT=	studentized residual defined slightly differently than above.
DFFITS=	standard influence of observation on predicted value.
COVRATIO=	standard influence of observation on covariance of betas, as discussed with INFLUENCE option.

(See **Predicted Values and Residuals** and **Influence Diagnostics** for details.)

RESTRICT Statement

RESTRICT *equation1,*
equation2,
⋮
equationk;

A RESTRICT statement is used to place restrictions on the parameter estimates in the MODEL preceding it. If more than one restriction appears, restrictions should be separated by commas.

Each restriction is written as a linear equation.

The form of an equation is:

±*term*[±*term*...][= ±*term*[±*term*...]]

where *term* is a *variable*|*number*|*number***variable* and brackets [ ] mean zero or more.

When no equal sign appears, the linear combination is set equal to zero. Each variable name mentioned must be a variable in the MODEL statement to which the RESTRICT statement refers. The keyword INTERCEPT can also be used as a variable name and refers to the intercept parameter in the regression model.

Note that the parameters associated with the variables are restricted, not the variables themselves. Restrictions should be consistent and not redundant.

Examples of valid RESTRICT statements:

```
RESTRICT A+B=1;
RESTRICT 2*F=G+H, INTERCEPT+F=0;
```

You cannot write

```
RESTRICT F-G=0,
         F-INTERCEPT=0,
         G-INTERCEPT=1;
```

since the three restrictions are not consistent. If these restrictions are included in a RESTRICT statement, one of the restrict parameters is zero and has zero degrees of freedom, indicating that REG is unable to apply a restriction.

The restrictions usually operate even if the model is not of full rank. Check that DF=1 for each restriction.

The parameter estimates are those that minimize the quadratic criterion (SSE) subject to the restrictions. If a restriction cannot be applied, its parameter value and degrees of freedom are listed as zero.

The method used for restricting the parameter estimates is to introduce a Lagrangian parameter for each restriction (Pringle and Raynor, 1971). The estimates of these parameters are printed with test statistics. The Lagrangian parameter λ measures the sensitivity of the SSE to the restriction constant. If the restriction constant is changed by a small amount, ε, the SSE is changed by $2\lambda\varepsilon$. The t ratio tests the significance of the restrictions. If λ is zero, the restricted estimates are the same as the unrestricted, and a change in the restriction constant in either direction increases the SSE.

TEST Statement

label: TEST *equation1*,
equation2,
⋮
equationk;

label: TEST *equation1*,..., *equationk* / *options*;

The TEST statement, which has the same syntax as the RESTRICT statement except for options, tests hypotheses about the parameters estimated in the preceding MODEL statement. Each equation specifies a linear hypothesis to be tested. The rows of the hypothesis are separated by commas. The syntax of a hypothesis is identical to the form of a restriction in a RESTRICT statement. Variable names must correspond to regressors, and each variable name represents the coefficient of the corresponding variable in the model. An optional label is useful to identify each test with a name. Again, INTERCEPT may be used instead of a variable name to refer to the model's intercept.

REG performs an *F* test for the joint hypotheses specified in a single TEST statement. More than one TEST statement may accompany a MODEL statement. The numerator is the usual quadratic form of the estimates; the denominator is the mean squared error. If hypotheses can be represented by

$$\mathbf{L}\beta = \mathbf{c} \quad ,$$

then the numerator of the *F* test is

$$\mathbf{Q} = (\mathbf{Lb}-\mathbf{c})'(\mathbf{L}(\mathbf{X}'\mathbf{X})^{-}\mathbf{L}')^{-1}(\mathbf{Lb}-\mathbf{c})$$

divided by degrees of freedom, where **b** is the estimate of β.

For example:

```
        MODEL Y=A1 A2 B1 B2;
APLUS:  TEST A1+A2=1;
B1:     TEST B1=0, B2=0;
B2:     TEST B1, B2;
```

The last two tests are equivalent; since no constant is specified, zero is assumed.

One option may be specified in the TEST statement after a slash (/):

PRINT prints intermediate calculations. This includes $\mathbf{L}(\mathbf{X}'\mathbf{X})^{-}\mathbf{L}'$ bordered by $\mathbf{Lb}-\mathbf{c}$, and $(\mathbf{L}(\mathbf{X}'\mathbf{X})^{-}\mathbf{L}')^{-1}$ bordered by $(\mathbf{L}(\mathbf{X}'\mathbf{X})^{-}\mathbf{L}')^{-1}(\mathbf{Lb}-\mathbf{c})$.

MTEST Statement

The MTEST statement is used to test hypotheses in multivariate regression models where there are several dependent variables fit to the same regressors. The hypotheses that can be estimated are of the form:

$$(\mathbf{L}\beta - \mathbf{cj})\mathbf{M}=0$$

where **L** is a linear function on the regressor side, β is a matrix of parameters, **c** is a column vector of constants, **j** is a row vector of ones, and **M** is a linear function on the dependent side. The special case where the constants are zero is

$$\mathbf{L}\beta\mathbf{M}=0 \quad .$$

(See **Multivariate Tests** later in this chapter.)

The MTEST statement has the same syntax as the TEST and RESTRICT statements:

label: MTEST *equation1*,
equation2,
⋮
equationk;

label: MTEST *equation1*,..., *equationk* / *options*;

where the equations are linear functions composed of coefficients and variable names.

Each linear function extends across either the regressor variables or the dependent variables. If the equation is across the dependent variables, then the constant term, if specified, must be zero. The equations for the regressor variables form the **L** matrix and **c** vector in the above formula; the equations for dependent variables form the **M** matrix. If no equations for the dependent variables are given, REG uses an identity matrix for **M**, testing the same hypothesis across all dependent variables. If no equations for the regressor variables are given, REG forms a linear function that tests all non-intercept parameters.

For example:

```
MODEL Y1 Y2=X1 X2 X3;
```

The statement

```
MTEST X1,X2;
```

tests the hypothesis that the X1 and X2 parameters are zero in both Y1 and Y2.

The statement

```
MTEST Y1-Y2, X1;
```

tests the hypothesis that the X1 parameter is the same for both dependent variables.

The statement

```
MTEST Y1-Y2;
```

tests the hypothesis that all parameters except the intercept are the same for both dependent variables.

The statement

```
MTEST;
```

tests the hypothesis that all non-intercept parameters for all dependent variables are zero.

These options are available on the MTEST statement:

PRINT prints the **H** and **E** matrices.

CANPRINT prints the canonical correlations for the hypothesis combinations and the dependent variable combinations. If you specify

```
MTEST / CANPRINT;
```

the canonical correlations between the regressors and the dependent variables are printed. See CANCORR for a description of the statistics printed.

DETAILS prints the **M** matrix and various intermediate calculations.

BY Statement

BY *variables;*

A BY statement may be used with PROC REG to obtain separate analyses on observations in groups defined by the BY variables. When a BY statement appears, the procedure expects the input data set to be sorted in order of the BY variables. If your input data set is not sorted in ascending order, use the SORT procedure with a similar BY statement to sort the data, or, if appropriate, use the BY statement options NOTSORTED or DESCENDING. For more information, see the discussion of the BY statement in Chapter 8, "Statements Used in the PROC Step," in *SAS User's Guide: Basics, 1982 Edition*.

DETAILS

Missing Values

REG constructs only one crossproducts matrix for the variables in all regressions. If any variable needed for any regression is missing, the observation is excluded from all estimates.

Input Data Set

The input data set for most applications of PROC REG contains standard rectangular data, but special TYPE=CORR or TYPE=SSCP data sets can also be used. TYPE=CORR data sets created by PROC CORR contain means, standard deviations, and correlations. TYPE=SSCP data sets created in previous runs of PROC REG contain the sums of squares and crossproducts of the variables. See Chapter 12, "SAS Data Sets," in *SAS User's Guide: Basics, 1982 Edition* for more information on special SAS data sets.

An example using PROC CORR:

```
PROC CORR DATA=FITNESS OUTP=R;
  VAR OXY RUNTIME AGE WEIGHT RUNPULSE MAXPULSE RSTPULSE;
PROC PRINT DATA=R;
PROC REG DATA=R;
  MODEL OXY=RUNTIME AGE WEIGHT;
```

The data set containing the correlation matrix is printed by the PRINT procedure:

1

VARIABLE	N	MEAN	STD DEV	SUM	MINIMUM	MAXIMUM
OXY	31	47.37580645	5.32723050	1468.65000000	37.38800000	60.05500000
RUNTIME	31	10.58612903	1.38741409	328.17000000	8.17000000	14.03000000
AGE	31	47.67741935	5.21144316	1478.00000000	38.00000000	57.00000000
WEIGHT	31	77.44451613	8.32856764	2400.78000000	59.08000000	91.63000000
RUNPULSE	31	169.64516129	10.25198643	5259.00000000	146.00000000	186.00000000
MAXPULSE	31	173.77419355	9.16409544	5387.00000000	155.00000000	192.00000000
RSTPULSE	31	53.45161290	7.61944315	1657.00000000	40.00000000	70.00000000

CORRELATION COEFFICIENTS / PROB > |R| UNDER HO:RHO=0 / N = 31

	OXY	RUNTIME	AGE	WEIGHT	RUNPULSE	MAXPULSE	RSTPULSE
OXY	1.00000	-0.86219	-0.30459	-0.16275	-0.39797	-0.23674	-0.39936
	0.0000	0.0001	0.0957	0.3817	0.0266	0.1997	0.0260

(continued on next page)

(continued from previous page)

```
RUNTIME  -0.86219  1.00000  0.18875  0.14351  0.31365  0.22610  0.45038
           0.0001   0.0000   0.3092   0.4412   0.0858   0.2213   0.0110

AGE      -0.30459  0.18875  1.00000 -0.23354 -0.33787 -0.43292 -0.16410
           0.0957   0.3092   0.0000   0.2061   0.0630   0.0150   0.3777

WEIGHT   -0.16275  0.14351 -0.23354  1.00000  0.18152  0.24938  0.04397
           0.3817   0.4412   0.2061   0.0000   0.3284   0.1761   0.8143

RUNPULSE -0.39797  0.31365 -0.33787  0.18152  1.00000  0.92975  0.35246
           0.0266   0.0858   0.0630   0.3284   0.0000   0.0001   0.0518

MAXPULSE -0.23674  0.22610 -0.43292  0.24938  0.92975  1.00000  0.30512
           0.1997   0.2213   0.0150   0.1761   0.0001   0.0000   0.0951

RSTPULSE -0.39936  0.45038 -0.16410  0.04397  0.35246  0.30512  1.00000
           0.0260   0.0110   0.3777   0.8143   0.0518   0.0951   0.0000
```

```
                                                                                          2

 OBS   _TYPE_   _NAME_       OXY      RUNTIME      AGE      WEIGHT   RUNPULSE  MAXPULSE  RSTPULSE

   1   MEAN                47.3758    10.5861   47.6774    77.4445   169.645   173.774   53.4516
   2   STD                  5.3272     1.3874    5.2114     8.3286    10.252     9.164    7.6194
   3   N                   31.0000    31.0000   31.0000    31.0000    31.000    31.000   31.0000
   4   CORR     OXY         1.0000    -0.8622   -0.3046    -0.1628    -0.398    -0.237   -0.3994
   5   CORR     RUNTIME    -0.8622     1.0000    0.1887     0.1435     0.314     0.226    0.4504
   6   CORR     AGE        -0.3046     0.1887    1.0000    -0.2335    -0.338    -0.433   -0.1641
   7   CORR     WEIGHT     -0.1628     0.1435   -0.2335     1.0000     0.182     0.249    0.0440
   8   CORR     RUNPULSE   -0.3980     0.3136   -0.3379     0.1815     1.000     0.930    0.3525
   9   CORR     MAXPULSE   -0.2367     0.2261   -0.4329     0.2494     0.930     1.000    0.3051
  10   CORR     RSTPULSE   -0.3994     0.4504   -0.1641     0.0440     0.352     0.305    1.0000
```

```
                                                                                          3
DEP VARIABLE: OXY

                        SUM OF          MEAN
SOURCE       DF        SQUARES        SQUARE       F VALUE       PROB>F

MODEL         3        656.271       218.757        30.272       0.0001
ERROR        27        195.111      7.226318
C TOTAL      30        851.382

      ROOT MSE        2.688181      R-SQUARE        0.7708
      DEP MEAN       47.375806      ADJ R-SQ        0.7454
      C.V.            5.674165

                     PARAMETER      STANDARD     T FOR H0:
VARIABLE     DF       ESTIMATE         ERROR   PARAMETER=0     PROB > |T|

INTERCEP      1      93.126150      7.559156        12.320       0.0001
RUNTIME       1      -3.140387      0.367380        -8.548       0.0001
AGE           1      -0.173877      0.099546        -1.747       0.0921
WEIGHT        1      -0.054437      0.061809        -0.881       0.3862
```

An example using the saved crossproducts matrix:

```
PROC REG DATA=FITNESS OUTSSCP=SSCP;
  MODEL OXY=RUNTIME AGE WEIGHT RUNPULSE
    MAXPULSE RSTPULSE;
PROC PRINT DATA=SSCP;
PROC REG DATA=SSCP;
  MODEL OXY=RUNTIME AGE WEIGHT;
```

The SSCP printout from PROC PRINT:

1

DEP VARIABLE: OXY

SOURCE	DF	SUM OF SQUARES	MEAN SQUARE	F VALUE	PROB>F
MODEL	6	722.544	120.424	22.433	0.0001
ERROR	24	128.838	5.368247		
C TOTAL	30	851.382			

ROOT MSE	2.316948	R-SQUARE	0.8487
DEP MEAN	47.375806	ADJ R-SQ	0.8108
C.V.	4.890572		

VARIABLE	DF	PARAMETER ESTIMATE	STANDARD ERROR	T FOR H0: PARAMETER=0	PROB > |T|
INTERCEP	1	102.934	12.403258	8.299	0.0001
RUNTIME	1	-2.628653	0.384562	-6.835	0.0001
AGE	1	-0.226974	0.099837	-2.273	0.0322
WEIGHT	1	-0.074177	0.054593	-1.359	0.1869
RUNPULSE	1	-0.369628	0.119853	-3.084	0.0051
MAXPULSE	1	0.303217	0.136495	2.221	0.0360
RSTPULSE	1	-0.021534	0.066054	-0.326	0.7473

2

OBS	_NAME_	INTERCEP	OXY	RUNTIME	AGE	WEIGHT	RUNPULSE	MAXPULSE	RSTPULSE
1	OXY	1468.65	70430	15356.1	69768	113522	248497	254867	78015
2	RUNTIME	328.17	15356	3531.8	15687	25465	55806	57114	17684
3	AGE	1478.00	69768	15687.2	71282	114159	250194	256218	78806
4	WEIGHT	2400.78	113522	25464.7	114159	188008	407746	417765	128409
5	RUNPULSE	5259.00	248497	55806.3	250194	407746	895317	916499	281928
6	MAXPULSE	5387.00	254867	57113.7	256218	417765	916499	938641	288583
7	RSTPULSE	1657.00	78015	17684.0	78806	128409	281928	288583	90311
8	INTERCEP	31.00	1469	328.2	1478	2401	5259	5387	1657

3

DEP VARIABLE: OXY

SOURCE	DF	SUM OF SQUARES	MEAN SQUARE	F VALUE	PROB>F
MODEL	3	656.271	218.757	30.272	0.0001
ERROR	27	195.111	7.226318		
C TOTAL	30	851.382			

ROOT MSE	2.688181	R-SQUARE	0.7708
DEP MEAN	47.375806	ADJ R-SQ	0.7454
C.V.	5.674165		

VARIABLE	DF	PARAMETER ESTIMATE	STANDARD ERROR	T FOR H0: PARAMETER=0	PROB > |T|
INTERCEP	1	93.126150	7.559156	12.320	0.0001
RUNTIME	1	-3.140387	0.367380	-8.548	0.0001
AGE	1	-0.173877	0.099546	-1.747	0.0921
WEIGHT	1	-0.054437	0.061809	-0.881	0.3862

These summary files save cputime. It takes nk^2 operations (n = number of observations, k = number of variables) to calculate crossproducts; the regressions are of the order k^3. When n is in the thousands and k in units, you may save 99 percent of the cputime by reusing the SSCP matrix rather than recomputing it.

When special SAS data sets are used, REG must be informed by the data set TYPE parameter. PROC CORR and PROC REG automatically set the type for output data sets; however, if you create the data set by some other means, you will need to specify its type with the data set option TYPE= .

```
PROC REG DATA=A(TYPE=CORR);
```

When data sets of TYPE=CORR or TYPE=SSCP are used with REG, options that require predicted values or residuals have no effect. The OUTPUT statement and the MODEL options P, R, CLM, CLI, DW, INFLUENCE, and PARTIAL are disabled.

REG does not compute new values for regressors. For example, if you need a lagged variable, you should create it when you prepare the input data.

Parameter Estimates and Associated Statistics

The following example shows the parameter estimates using all optional features.

```
PROC REG DATA=FITNESS;
  MODEL OXY=RUNTIME AGE WEIGHT RUNPULSE MAXPULSE RSTPULSE
      / SS1 SS2 STB TOL VIF COVB CORRB;
```

For further discussion of the parameters and statistics, see Chapter 1, "Introduction to SAS Regression Procedures," and the **Printed Output** section later in this chapter.

DEP VARIABLE: OXY   1

SOURCE	DF	SUM OF SQUARES	MEAN SQUARE	F VALUE	PROB>F
MODEL	6	722.544	120.424	22.433	0.0001
ERROR	24	128.838	5.368247		
C TOTAL	30	851.382			

ROOT MSE	2.316948	R-SQUARE	0.8487
DEP MEAN	47.375806	ADJ R-SQ	0.8108
C.V.	4.890572		

VARIABLE	DF	PARAMETER ESTIMATE	STANDARD ERROR	T FOR H0: PARAMETER=0	PROB > |T|	TYPE I SS	TYPE II SS	STANDARDIZED ESTIMATE	TOLERANCE
INTERCEP	1	102.934	12.403258	8.299	0.0001	69578.478	369.728	0.000000	.
RUNTIME	1	-2.628653	0.384562	-6.835	0.0001	632.900	250.822	-0.684601	0.628588
AGE	1	-0.226974	0.099837	-2.273	0.0322	17.765633	27.745771	-0.222041	0.661010
WEIGHT	1	-0.074177	0.054593	-1.359	0.1869	5.605217	9.910588	-0.115969	0.865554
RUNPULSE	1	-0.369628	0.119853	-3.084	0.0051	38.875742	51.058058	-0.711330	0.118522
MAXPULSE	1	0.303217	0.136495	2.221	0.0360	26.826403	26.491424	0.521605	0.114366
RSTPULSE	1	-0.021534	0.066054	-0.326	0.7473	0.570513	0.570513	-0.030799	0.706420

VARIABLE	DF	VARIANCE INFLATION
INTERCEP	1	0.000000
RUNTIME	1	1.590868
AGE	1	1.512836
WEIGHT	1	1.155329
RUNPULSE	1	8.437274
MAXPULSE	1	8.743848
RSTPULSE	1	1.415589

COVARIANCE OF ESTIMATES

COVB	INTERCEP	RUNTIME	AGE	WEIGHT	RUNPULSE	MAXPULSE	RSTPULSE
INTERCEP	153.8408	0.7678374	-0.902049	-0.178238	0.2807965	-0.832762	-0.147955
RUNTIME	0.7678374	0.1478881	-0.0141917	-0.00441767	-0.00904778	0.00462495	-0.0109152
AGE	-0.902049	-0.0141917	0.009967521	0.00102191	-0.00120391	0.003582384	0.001489753
WEIGHT	-0.178238	-0.00441767	0.00102191	0.002980413	0.0009644683	-0.00137224	0.0003799295
RUNPULSE	0.2807965	-0.00904778	-0.00120391	0.0009644683	0.01436473	-0.0149525	-0.000764507
MAXPULSE	-0.832762	0.00462495	0.003582384	-0.00137224	-0.0149525	0.01863094	0.0003425724
RSTPULSE	-0.147955	-0.0109152	0.001489753	0.0003799295	-0.000764507	0.0003425724	0.004363167

CORRELATION OF ESTIMATES   2

CORRB	INTERCEP	RUNTIME	AGE	WEIGHT	RUNPULSE	MAXPULSE	RSTPULSE
INTERCEP	1.0000	0.1610	-0.7285	-0.2632	0.1889	-0.4919	-0.1806
RUNTIME	0.1610	1.0000	-0.3696	-0.2104	-0.1963	0.0881	-0.4297
AGE	-0.7285	-0.3696	1.0000	0.1875	-0.1006	0.2629	0.2259
WEIGHT	-0.2632	-0.2104	0.1875	1.0000	0.1474	-0.1842	0.1054
RUNPULSE	0.1889	-0.1963	-0.1006	0.1474	1.0000	-0.9140	-0.0966
MAXPULSE	-0.4919	0.0881	0.2629	-0.1842	-0.9140	1.0000	0.0380
RSTPULSE	-0.1806	-0.4297	0.2259	0.1054	-0.0966	0.0380	1.0000

If the model is not full rank, there are an infinite number of least-squares solutions for the estimates. REG chooses a nonzero solution for all variables that are linearly independent of previous variables and a zero solution for other variables. This solution corresponds to using a generalized inverse in the normal equations, and the expected values of the "estimates" are the HERMITE NORMAL FORM of **X** times the true parameters:

$$E(\mathbf{b}) = (\mathbf{X'X})^{-}(\mathbf{X'X})\beta \quad .$$

Degrees of freedom for the zeroed estimates are reported as zero. The hypotheses that are not testable have t tests printed as missing. The message that the model is not full rank includes a printout of the relations that exist in the matrix.

In this example, we introduce another term DIF=RUNPULSE−RSTPULSE into the model to show how this problem is diagnosed.

```
DATA FIT2;
  SET FITNESS;
  DIF=RUNPULSE-RSTPULSE;
PROC REG DATA=FIT2;
  MODEL OXY=RUNTIME AGE WEIGHT RUNPULSE MAXPULSE
    RSTPULSE DIF;
```

```
                                                                                  3

DEP VARIABLE: OXY

                      SUM OF           MEAN
SOURCE     DF        SQUARES         SQUARE       F VALUE          PROB>F

MODEL       6        722.544        120.424        22.433          0.0001
ERROR      24        128.838       5.368247
C TOTAL    30        851.382

    ROOT MSE        2.316948       R-SQUARE        0.8487
    DEP MEAN       47.375806       ADJ R-SQ        0.8108
    C.V.            4.890572

NOTE: MODEL IS NOT FULL RANK. LEAST SQUARES SOLUTIONS FOR THE
      PARAMETERS ARE NOT UNIQUE. SOME STATISTICS WILL BE
      MISLEADING. A REPORTED DF OF 0 OR B MEANS THAT THE
      ESTIMATE IS BIASED. THE FOLLOWING PARAMETERS HAVE BEEN
      SET TO 0, SINCE THE VARIABLES ARE A LINEAR COMBINATION
      OF OTHER VARIABLES AS SHOWN.

DIF        =+RUNPULSE-1*RSTPULSE

                     PARAMETER       STANDARD    T FOR H0:
VARIABLE   DF         ESTIMATE          ERROR  PARAMETER=0     PROB > |T|

INTERCEP    1          102.934      12.403258        8.299         0.0001
RUNTIME     1        -2.628653       0.384562       -6.835         0.0001
AGE         1        -0.226974       0.099837       -2.273         0.0322
WEIGHT      1        -0.074177       0.054593       -1.359         0.1869
RUNPULSE    B        -0.369628       0.119853       -3.084         0.0051
MAXPULSE    1         0.303217       0.136495        2.221         0.0360
RSTPULSE    B        -0.021534       0.066054       -0.326         0.7473
DIF         0                0              .            .              .
```

Collinearity Diagnostics

When a regressor is nearly a linear combination of other regressors in the model, the affected estimates are unstable and have high standard errors. This problem is called *collinearity* or *multicollinearity*. It is a good idea to find out which variables are nearly collinear with which other variables. The approach in PROC REG follows that of Belsley, Kuh, and Welsch (1980).

The COLLIN option on the MODEL statement requests that a collinearity analysis be done. First, **X′X** is scaled to have 1s on the diagonal. If COLLINOINT is specified, the intercept variable is adjusted out first. Then the eigenvalues and eigenvectors are extracted. The analysis in REG is reported with eigenvalues of **X′X** rather than singular values of **X**. The eigenvalues of **X′X** are the squares of the singular values of **X**.

The condition indices are the square roots of the ratio of the largest eigenvalue to each individual eigenvalue. The largest condition index is the condition number of the scaled **X** matrix. When this number is large, the problem is said to be ill-conditioned. When this number is extremely large, the estimates may have a fair amount of numerical error (although the statistical standard error almost always is much greater than the numerical error).

For each variable, REG prints the proportion of the variance of the estimate accounted for by each principal component. A collinearity problem occurs when a component associated with a high condition index contributes strongly to the variance of two or more variables.

Here is an example using the COLLIN option on the exercise data:

```
PROC REG DATA=FITNESS;
  MODEL OXY=RUNTIME AGE WEIGHT RUNPULSE
    MAXPULSE RSTPULSE / TOL VIF COLLIN;
```

```
DEP VARIABLE: OXY                                                                                          1

                     SUM OF           MEAN
SOURCE      DF      SQUARES         SQUARE         F VALUE        PROB>F

MODEL        6      722.544        120.424          22.433        0.0001
ERROR       24      128.838       5.368247
C TOTAL     30      851.382

     ROOT MSE      2.316948       R-SQUARE          0.8487
     DEP MEAN     47.375806       ADJ R-SQ          0.8108
     C.V.          4.890572

                  PARAMETER       STANDARD     T FOR H0:                                      VARIANCE
VARIABLE   DF      ESTIMATE          ERROR   PARAMETER=0     PROB > |T|    TOLERANCE         INFLATION

INTERCEP    1       102.934      12.403258         8.299         0.0001         .              0.000000
RUNTIME     1     -2.628653       0.384562        -6.835         0.0001    0.628588            1.590868
AGE         1     -0.226974       0.099837        -2.273         0.0322    0.661010            1.512836
WEIGHT      1     -0.074177       0.054593        -1.359         0.1869    0.865554            1.155329
RUNPULSE    1     -0.369628       0.119853        -3.084         0.0051    0.118522            8.437274
MAXPULSE    1      0.303217       0.136495         2.221         0.0360    0.114366            8.743848
RSTPULSE    1     -0.021534       0.066054        -0.326         0.7473    0.706420            1.415589

COLLINEARITY DIAGNOSTICS          VARIANCE PROPORTIONS

                       CONDITION     PORTION   PORTION   PORTION   PORTION   PORTION   PORTION   PORTION
NUMBER EIGENVALUE          INDEX    INTERCEP   RUNTIME       AGE    WEIGHT  RUNPULSE  MAXPULSE  RSTPULSE

     1      6.950          1.000      0.0000    0.0002    0.0002    0.0002    0.0000    0.0000    0.0003
     2   0.018676         19.291      0.0022    0.0252    0.1463    0.0104    0.0000    0.0000    0.3906
     3   0.015034         21.501      0.0006    0.1286    0.1501    0.2357    0.0012    0.0012    0.0281
     4   0.009110         27.621      0.0064    0.6090    0.0319    0.1831    0.0015    0.0012    0.1903
     5   0.006073         33.829      0.0013    0.1250    0.1128    0.4444    0.0151    0.0083    0.3648
     6   0.001018         82.638      0.7997    0.0975    0.4966    0.1033    0.0695    0.0056    0.0203
     7 0.00017947        196.786      0.1898    0.0146    0.0621    0.0228    0.9128    0.9836    0.0057
```

Predicted Values and Residuals

The printout of the predicted values and residuals is controlled by the P, R, CLM, and CLI options. The P option causes REG to print out the observation number, the ID value (if an ID statement is used), the actual value, the predicted value, and the residual. The R, CLI, and CLM options also produce the items under the P option. Thus P is unnecessary if you use one of the other options.

The R option requests more detail, especially about the residuals. The standard errors of the predicted value and the residual are printed. The studentized residual, which is the residual divided by its standard error, is both printed and plotted. A measure of influence, Cook's *D*, is printed. Cook's *D* measures the change to the estimates that results from deleting each observation. See Cook (1977,1979). (This statistic is very similar to DFFITS.)

The CLM option requests REG to print the 95% lower and upper confidence limits for the predicted values. This accounts for the variation due to estimating the parameters only. If you want a 95% confidence interval for observed values, then you would use the CLI option, which adds in the variability of the error term. Here is an example using U.S. population data:

```
PROC REG DATA=USPOP;
  ID YEAR;
  MODEL POP=YEAR YEARSQ / P R CLI CLM;
  OUTPUT OUT=C P=PRED L95=L95 U95=U95 R=RESID
    COOKD=COOKD;
PROC PLOT DATA=C;
  PLOT POP*YEAR='A' PRED*YEAR='P' U95*YEAR='U' L95*YEAR='L'
     / OVERLAY VPOS=32 HPOS=80;
  PLOT RESID*YEAR / VREF=0 VPOS=18 HPOS=60;
  PLOT COOKD*YEAR / VREF=0 VPOS=18 HPOS=60;
```

DEP VARIABLE: POP 1

SOURCE	DF	SUM OF SQUARES	MEAN SQUARE	F VALUE	PROB>F
MODEL	2	71799.016	35899.508	4641.719	0.0001
ERROR	16	123.746	7.734098		
C TOTAL	18	71922.762			

ROOT MSE	2.781025	R-SQUARE	0.9983
DEP MEAN	69.767474	ADJ R-SQ	0.9981
C.V.	3.986133		

VARIABLE	DF	PARAMETER ESTIMATE	STANDARD ERROR	T FOR H0: PARAMETER=0	PROB > |T|
INTERCEP	1	20450.434	843.475	24.245	0.0001
YEAR	1	-22.780606	0.897849	-25.372	0.0001
YEARSQ	1	0.006345585	0.0002387695	26.576	0.0001

OBS	ID	ACTUAL	PREDICT VALUE	STD ERR PREDICT	LOWER95% MEAN	UPPER95% MEAN	LOWER95% PREDICT	UPPER95% PREDICT	RESIDUAL	STD ERR RESIDUAL	STUDENT RESIDUAL
1	1790	3.929	5.038	1.729	1.373	8.703	-1.903	11.980	-1.109	2.178	-0.509
2	1800	5.308	5.039	1.391	2.090	7.987	-1.553	11.631	0.269101	2.408	0.112
3	1810	7.239	6.308	1.130	3.912	8.705	-.055392	12.672	0.930534	2.541	0.366
4	1820	9.638	8.847	0.957106	6.818	10.876	2.612	15.082	0.790849	2.611	0.303
5	1830	12.866	12.655	0.872081	10.806	14.504	6.476	18.833	0.211047	2.641	0.080
6	1840	17.069	17.732	0.857844	15.913	19.550	11.562	23.901	-.662872	2.645	-0.251
7	1850	23.191	24.078	0.883512	22.205	25.951	17.892	30.264	-.886908	2.637	-0.336
8	1860	31.443	31.693	0.920177	29.742	33.644	25.483	37.903	-.250061	2.624	-0.095
9	1870	39.818	40.577	0.948730	38.566	42.589	34.348	46.806	-.759331	2.614	-0.290
10	1880	50.155	50.731	0.959248	48.697	52.764	44.494	56.967	-.575718	2.610	-0.221
11	1890	62.947	62.153	0.948730	60.142	64.164	55.924	68.382	0.793778	2.614	0.304
12	1900	75.994	74.845	0.920177	72.894	76.796	68.635	81.055	1.149	2.624	0.438
13	1910	91.972	88.806	0.883512	86.933	90.679	82.620	94.991	3.166	2.637	1.201
14	1920	105.710	104.035	0.857844	102.217	105.854	97.866	110.205	1.675	2.645	0.633
15	1930	122.775	120.534	0.872081	118.686	122.383	114.356	126.713	2.241	2.641	0.848
16	1940	131.669	138.302	0.957106	136.274	140.331	132.068	144.537	-6.633	2.611	-2.540
17	1950	151.325	157.340	1.130	154.943	159.736	150.976	163.704	-6.015	2.541	-2.367
18	1960	179.323	177.646	1.391	174.698	180.595	171.054	184.238	1.677	2.408	0.696
19	1970	203.211	199.221	1.729	195.556	202.886	192.280	206.163	3.990	2.178	1.831
20	1980	.	222.066	2.135	217.540	226.592	214.634	229.498	.	.	.
21	1990	.	246.180	2.602	240.664	251.695	238.106	254.253	.	.	.
22	2000	.	271.562	3.126	264.936	278.189	262.693	280.432	.	.	.

```
                                  COOK'S
 OBS      ID   -2-1-0 1 2              D

   1    1790 |      *|     |       0.054
   2    1800 |       |     |       0.001
   3    1810 |       |     |       0.009
   4    1820 |       |     |       0.004
```

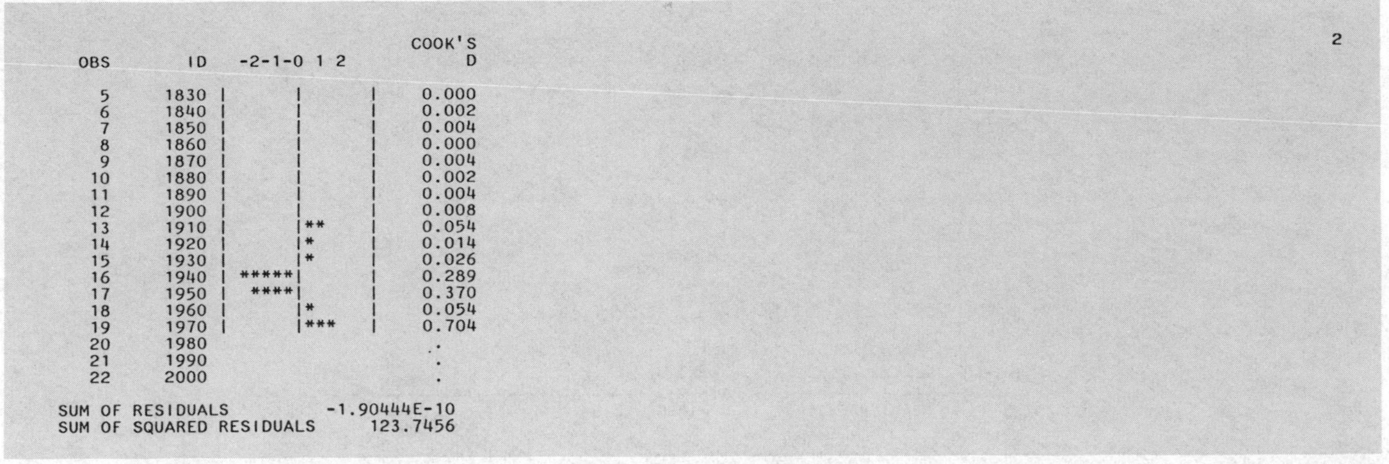

```
                                                           2
                                  COOK'S
OBS        ID   -2-1-0 1 2            D

  5      1830 |     |     |       0.000
  6      1840 |     |     |       0.002
  7      1850 |     |     |       0.004
  8      1860 |     |     |       0.000
  9      1870 |     |     |       0.004
 10      1880 |     |     |       0.002
 11      1890 |     |     |       0.004
 12      1900 |     |     |       0.008
 13      1910 |     |**   |       0.054
 14      1920 |     |*    |       0.014
 15      1930 |     |*    |       0.026
 16      1940 |*****|     |       0.289
 17      1950 | ****|     |       0.370
 18      1960 |     |*    |       0.054
 19      1970 |     |***  |       0.704
 20      1980                         .
 21      1990                         .
 22      2000                         .

SUM OF RESIDUALS              -1.90444E-10
SUM OF SQUARED RESIDUALS          123.7456
```

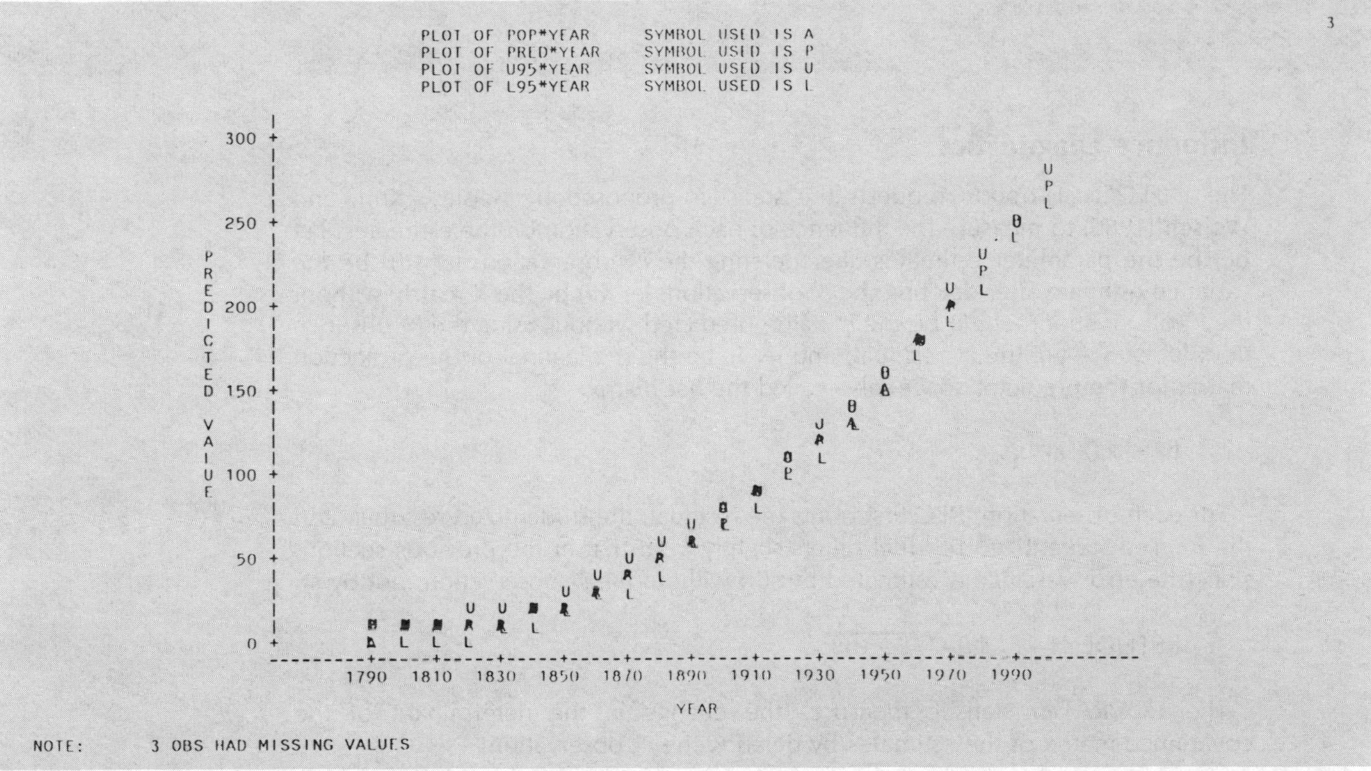

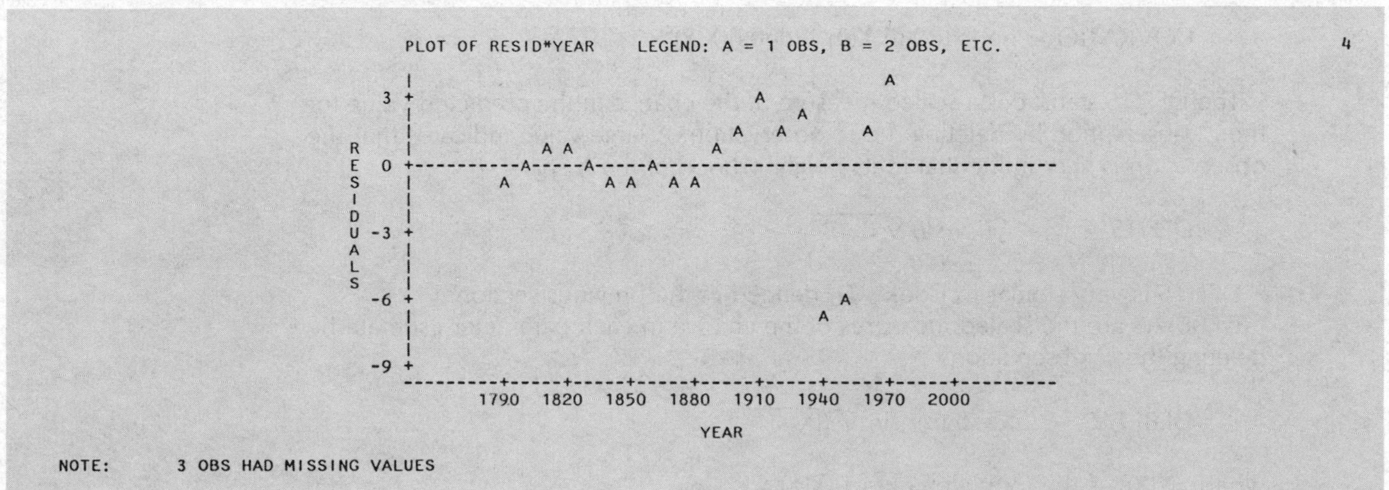

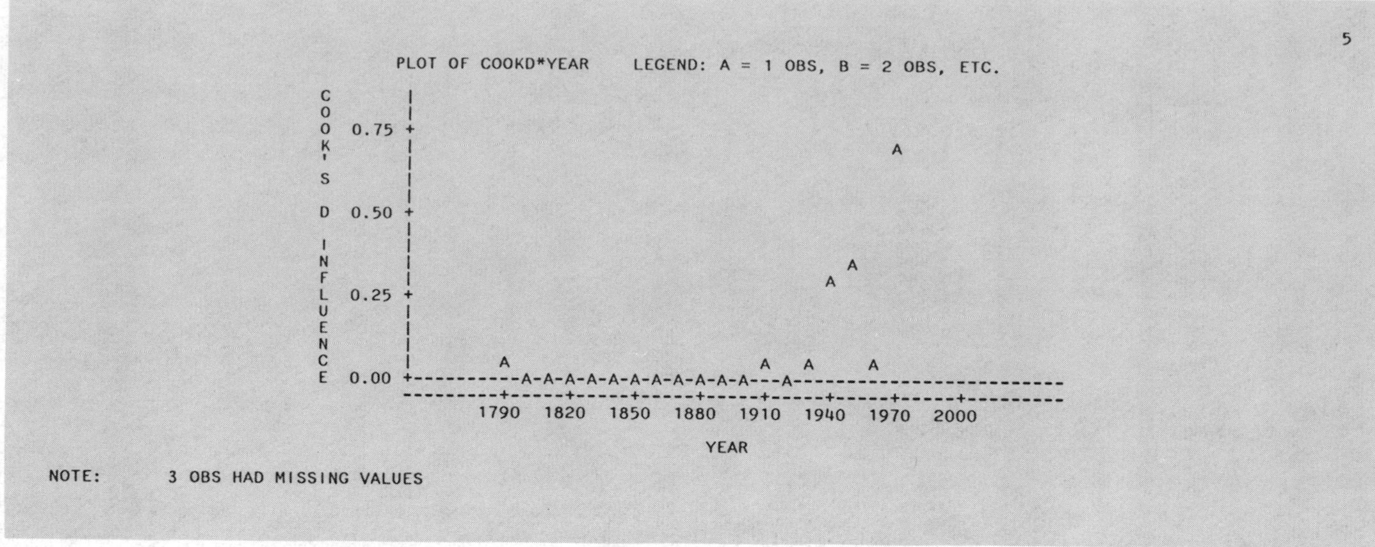

Influence Diagnostics

The INFLUENCE option requests the statistics proposed by Belsley, Kuh, and Welsch (1980) to measure the influence of each observation on the estimates. Let $\mathbf{b}(i)$ be the parameter estimates after deleting the ith observation; let $s(i)^2$ be the variance estimate after deleting the ith observation; let $\mathbf{X}(i)$ be the $\mathbf{X}$ matrix without the ith observation; let $\hat{y}(i)$ be the ith value predicted without using the ith observation; let $r_i = y_i - \hat{y}_i$, the ith residual; and let $\mathbf{h}_i$ be the ith diagonal of the projection matrix for the predictor space, also called the *hat matrix*:

$$\mathbf{h}_i = \mathbf{x}_i(\mathbf{X}'\mathbf{X})^{-1}\mathbf{x}'_i \quad .$$

For each observation, REG first prints the residual, the studentized residual, and the $\mathbf{h}_i$. The studentized residual differs slightly from that in the previous section, since the error variance is estimated by $\mathbf{s}(i)^2$ without the ith observation, not by $\mathbf{s}^2$:

$$\text{RSTUDENT} = \mathbf{r}_i / \, \mathbf{s}(i) \sqrt{(1 - \mathbf{h}_i)} \, .$$

The COVRATIO statistic measures the change in the determinant of the covariance matrix of the estimates by deleting the ith observation:

$$\text{COVRATIO} = \det(\mathbf{s}^2(i)(\mathbf{X}(i)'\mathbf{X}(i))^{-1})/\det(\mathbf{s}^2(\mathbf{X}'\mathbf{X})^{-1}).$$

The DFFITS statistic is a scaled measure of the change in the predicted value for the ith observation by deleting the ith observation. A large value indicates that the observation is very influential in its neighborhood of the $\mathbf{X}$ space.

$$\text{DFFITS} = (\hat{y}_i - \hat{y}(i)) / \, \mathbf{s}(i) \sqrt{(\mathbf{h}(i))} \, .$$

DFFITS is very similar to Cook's *D*, defined in the previous section.

DFBETAS are the scaled measures of the change in each parameter estimate by deleting the ith observation:

$$\text{DFBETAS}_j = (b_j - b_j(i)) / \, s(i) \sqrt{((X'X)^{jj})} \, ,$$

where $(\mathbf{X}'\mathbf{X})^{jj}$ is the $(j,j)^{th}$ element of $(\mathbf{X}'\mathbf{X})^{-1}$.

This printout results from the INFLUENCE option used with the population example:

```
PROC REG DATA=USPOP;
  MODEL POP=YEAR YEARSQ / INFLUENCE;
```

```
                                                                                                   1

DEP VARIABLE: POP

                        SUM OF            MEAN
SOURCE      DF         SQUARES          SQUARE        F VALUE         PROB>F

MODEL        2       71799.016       35899.508       4641.719         0.0001
ERROR       16         123.746        7.734098
C TOTAL     18       71922.762

     ROOT MSE         2.781025      R-SQUARE          0.9983
     DEP MEAN        69.767474      ADJ R-SQ          0.9981
     C.V.             3.986133

                     PARAMETER        STANDARD    T FOR H0:
VARIABLE    DF        ESTIMATE           ERROR  PARAMETER=0     PROB > |T|

INTERCEP     1       20450.434         843.475       24.245         0.0001
YEAR         1      -22.780606        0.897849      -25.372         0.0001
YEARSQ       1     0.006345585    0.0002387695       26.576         0.0001

                              HAT DIAG        COV                 DFBETAS   DFBETAS   DFBETAS
   OBS RESIDUAL RSTUDENT             H      RATIO     DFFITS     INTERCEP      YEAR    YEARSQ

     1 -1.10945   -0.4972       0.3865     1.8834    -0.3946      -0.2842    0.2810   -0.2779
     2 0.269101    0.1082       0.2501     1.6147     0.0625       0.0376   -0.0370    0.0365
     3 0.930534    0.3561       0.1652     1.4176     0.1584       0.0666   -0.0651    0.0636
     4 0.790849    0.2941       0.1184     1.3531     0.1078       0.0182   -0.0172    0.0161
     5 0.211047    0.0774       0.0983     1.3444     0.0256      -0.0030    0.0033   -0.0035
     6 -.662872   -0.2431       0.0951     1.3255    -0.0788       0.0296   -0.0302    0.0307
     7 -.886908   -0.3268       0.1009     1.3214    -0.1095       0.0609   -0.0616    0.0621
     8 -.250061   -0.0923       0.1095     1.3605    -0.0324       0.0216   -0.0217    0.0218
     9 -.759331   -0.2820       0.1164     1.3519    -0.1023       0.0743   -0.0745    0.0747
    10 -.575718   -0.2139       0.1190     1.3650    -0.0786       0.0586   -0.0587    0.0587
    11 0.793778    0.2949       0.1164     1.3499     0.1070      -0.0784    0.0783   -0.0781
    12  1.14916    0.4265       0.1095     1.3144     0.1496      -0.1018    0.1014   -0.1009
    13  3.16642    1.2189       0.1009     1.0168     0.4084      -0.2357    0.2338   -0.2318
    14  1.67456    0.6207       0.0951     1.2430     0.2013      -0.0811    0.0798   -0.0784
    15  2.24059    0.8407       0.0983     1.1724     0.2776      -0.0427    0.0404   -0.0380
    16  -6.6335   -3.1845       0.1184     0.2924    -1.1673      -0.1531    0.1636   -0.1747
    17 -6.01471   -2.8433       0.1652     0.3989    -1.2649      -0.4843    0.4958   -0.5076
    18  1.67697    0.6847       0.2501     1.4757     0.3954       0.2240   -0.2274    0.2308
    19  3.98953    1.9947       0.3865     0.9766     1.5831       1.0902   -1.1025    1.1151
    20        .         .            .          .          .            .         .         .
    21        .         .            .          .          .            .         .         .
    22        .         .            .          .          .            .         .         .
```

The PARTIAL option produces partial regression leverage plots. One plot is printed for each regressor. For a given regressor, the partial regression leverage plot is the plot of the dependent variable and the regressor after they have been made orthogonal to the other regressors in the model. These can be obtained by plotting the residuals of the dependent variable omitting the selected regressor against the residuals of the selected regressor on all the other regressors. A line fit to the points has a slope equal to the parameter estimate in the full model. In the plot, each observation is represented by a plus sign (+) or up to 8 characters of the ID value. For overlapping points, an asterisk (*) is printed. If the points go outside the plot, a greater than (>) or less than (<) sign is printed on the side.

```
PROC REG DATA=FITNESS;
  MODEL OXY=RUNTIME WEIGHT AGE / PARTIAL;
```

```
                                                                                 1

DEP VARIABLE: OXY

                       SUM OF           MEAN
SOURCE       DF       SQUARES         SQUARE        F VALUE        PROB>F

MODEL         3       656.271        218.757         30.272        0.0001
ERROR        27       195.111       7.226318
C TOTAL      30       851.382

     ROOT MSE        2.688181       R-SQUARE         0.7708
     DEP MEAN       47.375806       ADJ R-SQ         0.7454
     C.V.            5.674165

                    PARAMETER       STANDARD    T FOR H0:
VARIABLE     DF      ESTIMATE          ERROR  PARAMETER=0    PROB > |T|

INTERCEP      1     93.126150       7.559156       12.320        0.0001
RUNTIME       1     -3.140387       0.367380       -8.548        0.0001
WEIGHT        1     -0.054437       0.061809       -0.881        0.3862
AGE           1     -0.173877       0.099546       -1.747        0.0921
```

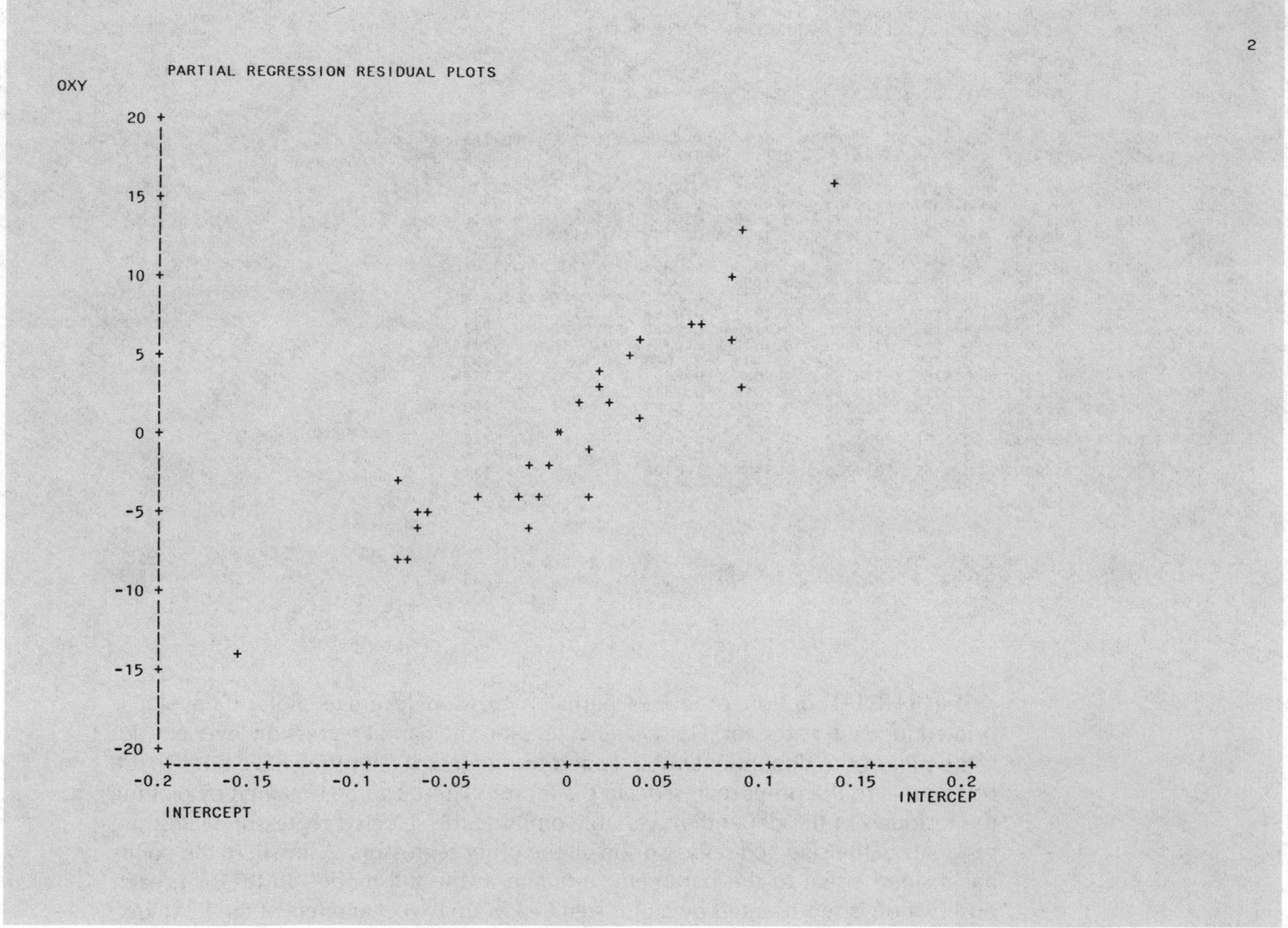

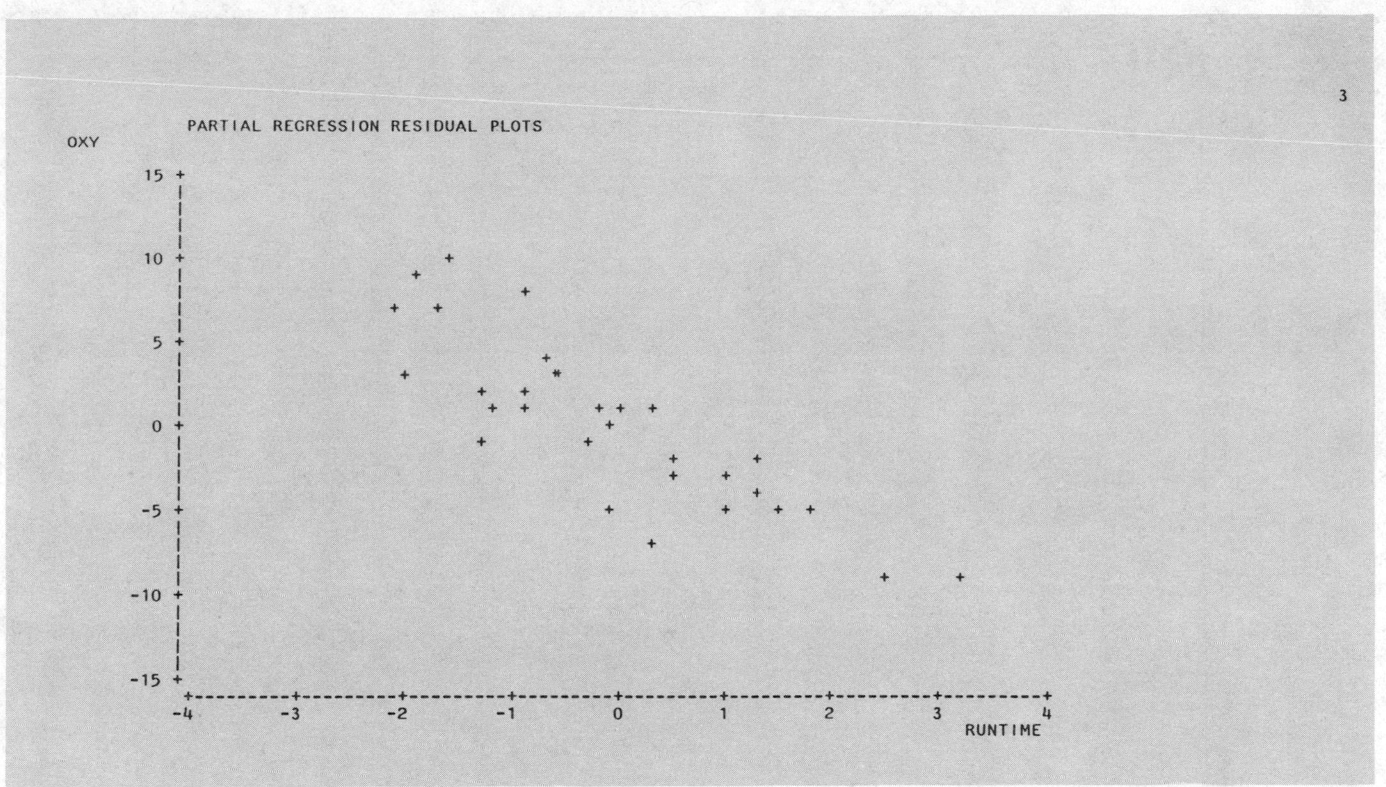

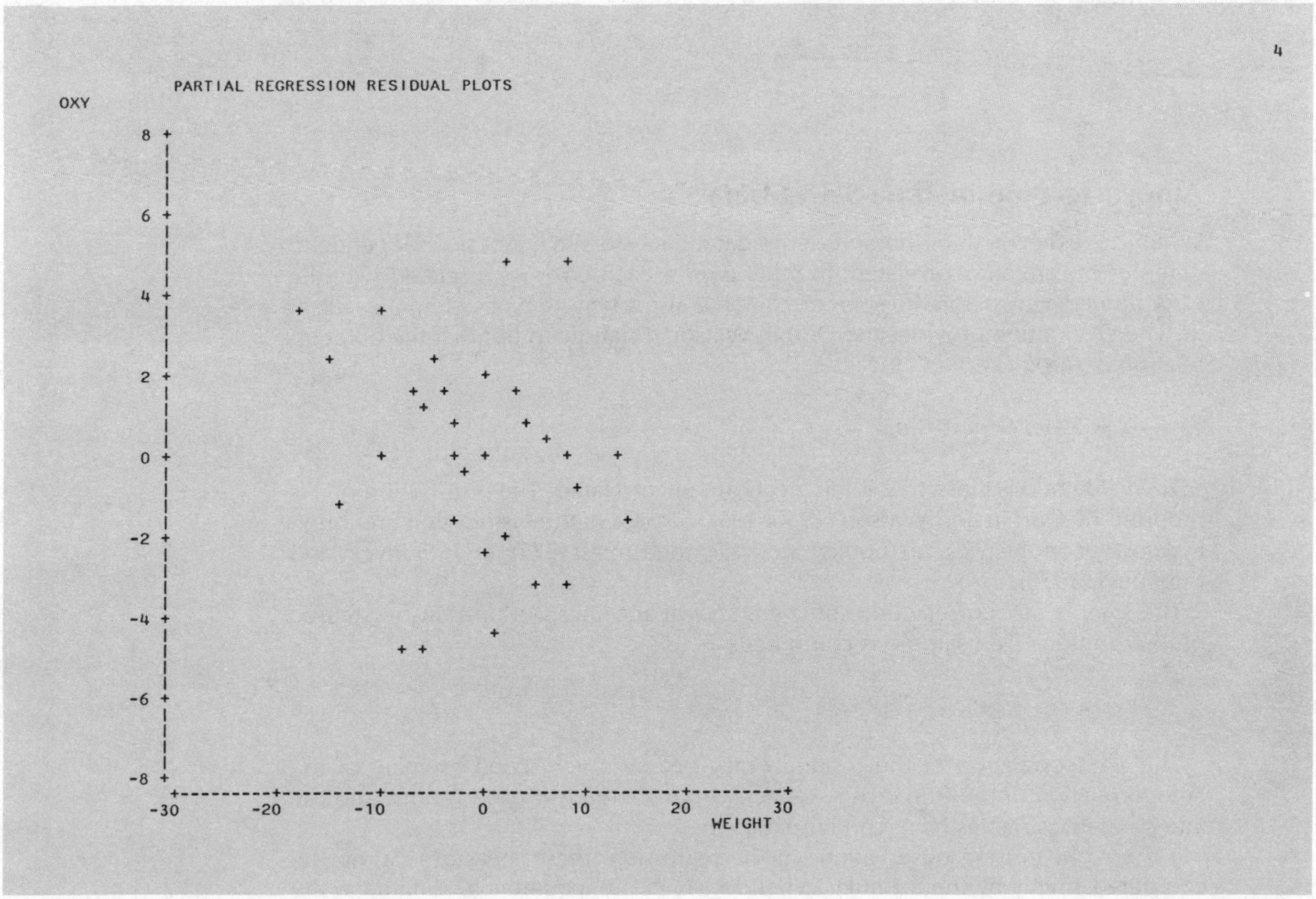

4

PARTIAL REGRESSION RESIDUAL PLOTS

OXY

8
6
4
2
0
-2
-4
-6
-8

-30 -20 -10 0 10 20 30

WEIGHT

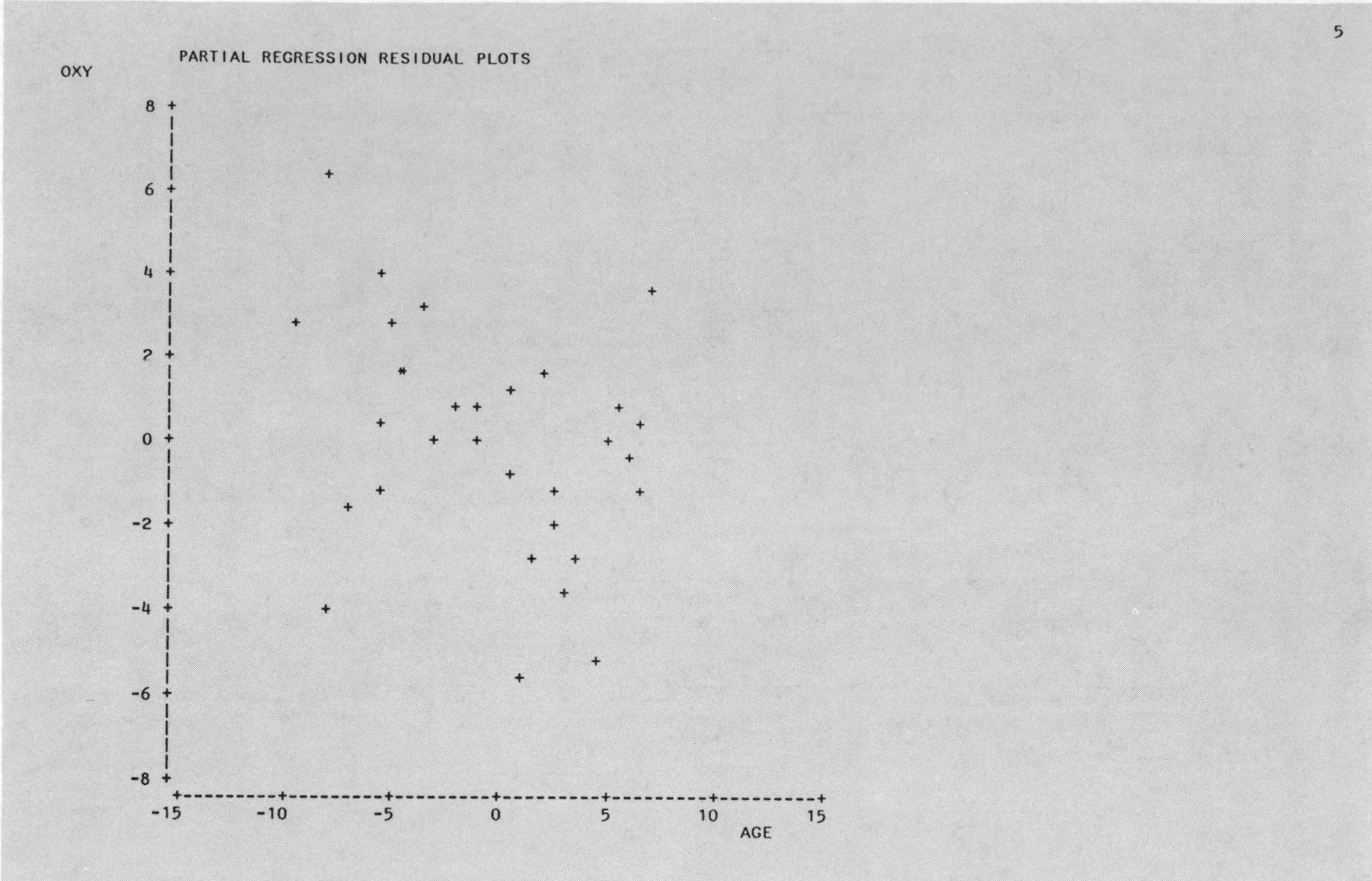

Autocorrelation in Time-Series Data

When regression is done on time-series data, the errors may not be independent. Often errors are autocorrelated; in other words, each error is correlated with the error immediately before it. Autocorrelation is also a symptom of systematic lack of fit. The DW option provides the Durbin-Watson d statistic to test that the autocorrelation is zero:

$$d = \Sigma_2{}^n(e_i - e_{i-1})^2/\Sigma e_i^2 \quad .$$

The value of d is close to 2 if the errors are uncorrelated. The distribution of d is reported by Durbin and Watson (1950, 1951). Tables of the distribution are found in most econometrics textbooks, such as Johnston (1972) and Pindyck and Rubinfeld (1976).

The sample autocorrelation estimate is shown after the Durbin-Watson statistic on the printout. The sample is computed:

$$r = \Sigma e_i e_{i-1}/\Sigma e_i^2 \quad .$$

This autocorrelation of the residuals may not be a very good estimate of the autocorrelation of the true errors, especially if there are few observations and the independent variables have certain patterns.

If there are missing observations in the regression, these measures cannot be computed strictly by the formula. When PROC REG encounters missing data, the statistics are calculated based on all adjacent non-missing pairs of observations.

If autocorrelation is present, the residual error estimate is inflated and tests on the parameters are less able to reject the null hypothesis. There are several better estimation methods available in this case. (Consult the *SAS/ETS User's Guide*.)

The following SAS statements request the DW option for the population data:

```
PROC REG DATA=USPOP;
  MODEL POP=YEAR YEARSQ / DW;
```

```
                                                                                          2

DEP VARIABLE: POP

                        SUM OF          MEAN
SOURCE      DF         SQUARES        SQUARE        F VALUE       PROB>F

MODEL        2       71799.016     35899.508       4641.719       0.0001
ERROR       16         123.746      7.734098
C TOTAL     18       71922.762

    ROOT MSE          2.781025      R-SQUARE         0.9983
    DEP MEAN         69.767474      ADJ R-SQ         0.9981
    C.V.              3.986133

                     PARAMETER      STANDARD     T FOR H0:
VARIABLE    DF        ESTIMATE         ERROR   PARAMETER=0     PROB > |T|

INTERCEP     1       20450.434       843.475        24.245         0.0001
YEAR         1      -22.780606      0.897849       -25.372         0.0001
YEARSQ       1  0.006345585 0.0002387695            26.576         0.0001

DURBIN-WATSON D                   1.264
1ST ORDER AUTOCORRELATION         0.299
```

Multivariate Tests

The MTEST statement described above can test hypotheses involving several dependent variables in the form:

$$(\mathbf{L}\beta - \mathbf{cj})\mathbf{M} = 0$$

where **L** is a linear function on the regressor side, β is a matrix of parameters, **c** is a column vector of constants, **j** is a row vector of ones, and **M** is a linear function on the dependent side. The special case where the constants are zero is

$$\mathbf{L}\beta\mathbf{M} = 0 \quad .$$

To test this hypothesis, REG constructs two matrices called **H** and **E** that correspond to the numerator and denominator of a univariate *F* test:

$$\mathbf{H} = \mathbf{M}'(\mathbf{LB} - \mathbf{cj})'(\mathbf{L}(\mathbf{X}'\mathbf{X})^{-}\mathbf{L}')^{-1}(\mathbf{LB} - \mathbf{cj})\mathbf{M}$$

$$\mathbf{E} = \mathbf{M}'(\mathbf{Y}'\mathbf{Y} - \mathbf{B}'(\mathbf{X}'\mathbf{X})\mathbf{B})\mathbf{M}.$$

These matrices are printed for each MTEST statement if the PRINT option is specified.

Four test statistics, based on the eigenvalues of $\mathbf{E}^{-1}\mathbf{H}$ or $(\mathbf{E}+\mathbf{H})^{-1}\mathbf{H}$, are formed. These are Wilks' Lambda, Pillai's Trace, the Hotelling-Lawley Trace, and Roy's maximum root. These are discussed in Chapter 1, "Introduction to SAS Regression Procedures."

Here is an example of a multivariate analysis of variance:

```
* MANOVA DATA FROM MORRISON'S MULTIVARIATE TEXT, 2ND EDITION,
PAGE 190;
DATA A;
  INPUT SEX $ DRUG $ @;
  DO REP=1 TO 4;
    INPUT Y1 Y2 @;
    OUTPUT;
    END;
  CARDS;
M A   5  6  5  4  9  9  7  6
M B   7  6  7  7  9 12  6  8
M C  21 15 14 11 17 12 12 10
F A   7 10  6  6  9  7  8 10
F B  10 13  8  7  7  6  6  9
F C  16 12 14  9 14  8 10  5
DATA B;
  SET A;
    SEXCODE=(SEX='M')-(SEX='F');
    DRUG1=(DRUG='A')-(DRUG='C');
    DRUG2=(DRUG='B')-(DRUG='C');
    SEXDRUG1=SEXCODE*DRUG1;
    SEXDRUG2=SEXCODE*DRUG2;
PROC REG;
  MODEL Y1 Y2=SEXCODE DRUG1 DRUG2 SEXDRUG1 SEXDRUG2;
SEX:        MTEST SEXCODE;
DRUG:       MTEST DRUG1,DRUG2;
SEXDRUG:    MTEST SEXDRUG1,SEXDRUG2;
Y1MY2:      MTEST Y1-Y2;
Y1Y2DRUG:   MTEST Y1=Y2, DRUG1,DRUG2;
DRUGSHOW:   MTEST DRUG1,DRUG2 / PRINT CANPRINT;
```

```
                                                                                        1
DEP VARIABLE: Y1

                       SUM OF          MEAN
SOURCE       DF       SQUARES        SQUARE        F VALUE        PROB>F

MODEL         5       316.000     63.200000         12.038        0.0001
ERROR        18     94.500000      5.250000
C TOTAL      23       410.500

    ROOT MSE         2.291288      R-SQUARE         0.7698
    DEP MEAN         9.750000      ADJ R-SQ         0.7058
    C.V.             23.50039

                    PARAMETER      STANDARD     T FOR HO:
VARIABLE     DF      ESTIMATE         ERROR   PARAMETER=0    PROB > |T|

INTERCEP      1      9.750000      0.467707        20.846        0.0001
SEXCODE       1      0.166667      0.467707         0.356        0.7257
DRUG1         1     -2.750000      0.661438        -4.158        0.0006
DRUG2         1     -2.250000      0.661438        -3.402        0.0032
SEXDRUG1      1     -0.666667      0.661438        -1.008        0.3269
SEXDRUG2      1     -0.416667      0.661438        -0.630        0.5366
```

2

DEP VARIABLE: Y2

SOURCE	DF	SUM OF SQUARES	MEAN SQUARE	F VALUE	PROB>F
MODEL	5	69.333333	13.866667	2.189	0.1008
ERROR	18	114.000	6.333333		
C TOTAL	23	183.333			

ROOT MSE	2.516611	R-SQUARE	0.3782
DEP MEAN	8.666667	ADJ R-SQ	0.2055
C.V.	29.03782		

VARIABLE	DF	PARAMETER ESTIMATE	STANDARD ERROR	T FOR H0: PARAMETER=0	PROB > |T|
INTERCEP	1	8.666667	0.513701	16.871	0.0001
SEXCODE	1	0.166667	0.513701	0.324	0.7493
DRUG1	1	-1.416667	0.726483	-1.950	0.0669
DRUG2	1	-0.166667	0.726483	-0.229	0.8211
SEXDRUG1	1	-1.166667	0.726483	-1.606	0.1257
SEXDRUG2	1	-0.416667	0.726483	-0.574	0.5734

3

MULTIVARIATE TEST: SEX

MULTIVARIATE TEST STATISTICS AND EXACT F STATISTICS

STATISTIC	VALUE	F	NUM DF	DEN DF	PROB>F
WILKS' LAMBDA	0.9925369	0.06391302	2	17	0.9383109
PILLAI'S TRACE	0.007463063	0.06391302	2	17	0.9383109
HOTELLING-LAWLEY TRACE	0.007519179	0.06391302	2	17	0.9383109
ROY'S GREATEST ROOT	0.007519179	0.06391302	2	17	0.9383109

MULTIVARIATE TEST: DRUG

MULTIVARIATE TEST STATISTICS AND F APPROXIMATIONS

STATISTIC	VALUE	F	NUM DF	DEN DF	PROB>F
WILKS' LAMBDA	0.1686295	12.19913	4	34	.00000295855
PILLAI'S TRACE	0.8803781	7.076856	4	36	0.0002601809
HOTELLING-LAWLEY TRACE	4.639537	18.55815	4	32	5.59676E-08
ROY'S GREATEST ROOT	4.576027	41.18424	2	18	1.91901E-07

NOTE: F STATISTIC FOR ROY'S GREATEST ROOT IS AN UPPER BOUND
F STATISTIC FOR WILKS' LAMBDA IS EXACT

MULTIVARIATE TEST: SEXDRUG

MULTIVARIATE TEST STATISTICS AND F APPROXIMATIONS

STATISTIC	VALUE	F	NUM DF	DEN DF	PROB>F
WILKS' LAMBDA	0.7743623	1.159326	4	34	0.3459007
PILLAI'S TRACE	0.2269491	1.151993	4	36	0.3480795
HOTELLING-LAWLEY TRACE	0.2896916	1.158766	4	32	0.3472767
ROY'S GREATEST ROOT	0.2837227	2.553505	2	18	0.1056231

NOTE: F STATISTIC FOR ROY'S GREATEST ROOT IS AN UPPER BOUND
F STATISTIC FOR WILKS' LAMBDA IS EXACT

MULTIVARIATE TEST: Y1MY2

MULTIVARIATE TEST STATISTICS AND EXACT F STATISTICS

STATISTIC	VALUE	F	NUM DF	DEN DF	PROB>F
WILKS' LAMBDA	0.2749794	9.491892	5	18	0.0001448038
PILLAI'S TRACE	0.7250206	9.491892	5	18	0.0001448038
HOTELLING-LAWLEY TRACE	2.636637	9.491892	5	18	0.0001448038
ROY'S GREATEST ROOT	2.636637	9.491892	5	18	0.0001448038

MULTIVARIATE TEST: Y1Y2DRUG

```
                                                                                                          4

                        MULTIVARIATE TEST STATISTICS AND EXACT F STATISTICS

STATISTIC                          VALUE              F             NUM DF          DEN DF           PROB>F

WILKS' LAMBDA                  0.2805392       23.08108                 2              18      .00001076321
PILLAI'S TRACE                 0.7194608       23.08108                 2              18      .00001076321
HOTELLING-LAWLEY TRACE          2.564565       23.08108                 2              18      .00001076321
ROY'S GREATEST ROOT             2.564565       23.08108                 2              18      .00001076321

MULTIVARIATE TEST: DRUGSHOW

                                     E, THE ERROR MATRIX


                                          94.5             76.5
                                          76.5              114

                                   H, THE HYPOTHESIS MATRIX


                                           301             97.5
                                          97.5         36.33333

CANONICAL CORRELATIONS AND TESTS OF H0: THE CANONICAL CORRELATION IN THE CURRENT ROW AND ALL THAT FOLLOW ARE ZERO

   CANONICAL     ADJUSTED       APPROX        VARIANCE   CANONICAL     LIKELIHOOD
  CORRELATION    CAN CORR     STD ERROR         RATIO    R-SQUARED       RATIO      F STATISTIC  NUM DF  DEN DF  PROB>F

1  0.905903324  0.880130527  0.040101457       4.5760  0.820660832  0.168629523       12.1991       4      34  0.0000
2  0.244371174 -0.131953124  0.210253610       0.0635  0.059717271  0.940282729        1.1432       1      18  0.2991

                                       EIGENVECTORS


                                       0.06261         -0.03258
                                      -0.03510          0.11177

                        MULTIVARIATE TEST STATISTICS AND F APPROXIMATIONS

STATISTIC                          VALUE              F             NUM DF          DEN DF           PROB>F

WILKS' LAMBDA                  0.1686295       12.19913                 4              34      .00000295855
PILLAI'S TRACE                 0.8803781       7.076856                 4              36      0.0002601809
HOTELLING-LAWLEY TRACE          4.639537       18.55815                 4              32       5.59676E-08
ROY'S GREATEST ROOT             4.576027       41.18424                 2              18       1.91901E-07

     NOTE: F STATISTIC FOR ROY'S GREATEST ROOT IS AN UPPER BOUND
           F STATISTIC FOR WILKS' LAMBDA IS EXACT
```

Output Data Sets

Estimates The OUTEST= specification produces a TYPE=EST output SAS data set containing estimates from the regression models. For each BY group on each dependent variable occurring in each MODEL statement, REG outputs an observation to the OUTEST data set. The parameter estimates are stored under the names of the variables associated with the estimates. A special variable, INTERCEP, is created for the intercept parameter. The dependent variable is coded as -1. Variables not in the model are coded missing. A special variable, __SIGMA__, stores the root mean squared error, the estimate of the standard deviation of the error term. The name of the dependent variable is stored in a special variable, __DEPVAR__. If the MODEL is identified with a label, the label is stored in a special variable, __MODEL__. If the COVOUT option is used, the covariance matrix of the estimates is output after the estimates; the names of the rows are identified by a special variable, __NAME__.

Here is an example with a printout of the OUTEST data set:

```
PROC REG DATA=USPOP OUTEST=EST;
M1: MODEL POP=YEAR;
M2: MODEL POP=YEAR YEARSQ;
PROC PRINT DATA=EST;
```

```
MODEL: M1                                                                                    1
DEP VARIABLE: POP

                     SUM OF            MEAN
SOURCE      DF      SQUARES          SQUARE       F VALUE        PROB>F

MODEL        1    66336.469       66336.469       201.873        0.0001
ERROR       17     5586.293         328.605
C TOTAL     18    71922.762

    ROOT MSE      18.127478       R-SQUARE         0.9223
    DEP MEAN      69.767474       ADJ R-SQ         0.9178
    C.V.           25.98271

                  PARAMETER        STANDARD     T FOR HO:
VARIABLE    DF     ESTIMATE           ERROR   PARAMETER=0     PROB > |T|

INTERCEP     1    -1958.366         142.805       -13.714         0.0001
YEAR         1     1.078795        0.075928        14.208         0.0001
```

```
MODEL: M2                                                                                    2
DEP VARIABLE: POP

                     SUM OF            MEAN
SOURCE      DF      SQUARES          SQUARE       F VALUE        PROB>F

MODEL        2    71799.016       35899.508      4641.719        0.0001
ERROR       16      123.746        7.734098
C TOTAL     18    71922.762

    ROOT MSE       2.781025       R-SQUARE         0.9983
    DEP MEAN      69.767474       ADJ R-SQ         0.9981
    C.V.           3.986133

                  PARAMETER        STANDARD     T FOR HO:
VARIABLE    DF     ESTIMATE           ERROR   PARAMETER=0     PROB > |T|

INTERCEP     1    20450.434         843.475        24.245         0.0001
YEAR         1   -22.780606        0.897849       -25.372         0.0001
YEARSQ       1  0.006345585    0.0002387695        26.576         0.0001
```

```
  OBS    _TYPE_    _MODEL_    _DEPVAR_    _SIGMA_    POP      YEAR       YEARSQ       INTERCEP     3

   1     OLS       M1         POP         18.1275    -1       1.079      .            -1958.4
   2     OLS       M2         POP          2.7810    -1     -22.781      0.00634559   20450.4
```

Sums of squares and crossproducts OUTSSCP= produces a TYPE=SSCP output SAS data set containing sums of squares and crossproducts. A special row (observation) and column (variable) of the matrix called INTERCEP contains the number of observations and sums. Observations are identified by the eight-byte character variable, _NAME_. The data set contains all variables used in MODEL statements. You specify additional variables that you want included in the crossproducts matrix with a VAR statement.

The SSCP data set is used when a large number of observations are to be explored in many different runs. The SSCP can be saved and used for subsequent runs, which are much less expensive, since REG never reads the original data again. On the step that creates the SSCP data set, you should include in the regression all the variables that you will ever need.

Here is an example of the OUTSSCP feature used with the exercise data:

```
PROC REG DATA=FITNESS OUTSSCP=SSCP;
  VAR OXY RUNTIME AGE WEIGHT RSTPULSE RUNPULSE MAXPULSE;
PROC PRINT DATA=SSCP;
```

1

OBS	_NAME_	INTERCEP	OXY	RUNTIME	AGE	WEIGHT	RSTPULSE	RUNPULSE	MAXPULSE
1	OXY	1468.65	70430	15356.1	69768	113522	78015	248497	254867
2	RUNTIME	328.17	15356	3531.8	15687	25465	17684	55806	57114
3	AGE	1478.00	69768	15687.2	71282	114159	78806	250194	256218
4	WEIGHT	2400.78	113522	25464.7	114159	188008	128409	407746	417765
5	RSTPULSE	1657.00	78015	17684.0	78806	128409	90311	281928	288583
6	RUNPULSE	5259.00	248497	55806.3	250194	407746	281928	895317	916499
7	MAXPULSE	5387.00	254867	57113.7	256218	417765	288583	916499	938641
8	INTERCEP	31.00	1469	328.2	1478	2401	1657	5259	5387

Computational Methods

The REG procedure first composes a crossproducts matrix. The matrix can be calculated from input data, reformed from an input correlation matrix, or read in from an SSCP data set. For each model, the procedure selects the appropriate crossproducts from the main matrix. The normal equations formed from the crossproducts are solved using a sweep algorithm (Goodnight, 1979). The method is accurate for data that are reasonably scaled and not too collinear.

The mechanism PROC REG uses to check for singularity involves the diagonal (pivot) elements of **X'X** as it is being swept. If a pivot is less than EPSILON*CSS, then a singularity is declared, and the pivot is not swept (where CSS is the corrected sum of squares for the regressor, and EPSILON is 1E-8 or reset in the PROC statement).

Computer Resources

The REG procedure is efficient for ordinary regression; however, requests for optional features can multiply the costs several times.

The major computational expense is the collection of the crossproducts matrix. For p variables and n observations, the time required is proportional to np^2. For each model run, REG needs time roughly proportional to k^3, where k is the number of regressors in the model. Add an additional nk^2 for one of the R, CLM, or CLI options and another nk^2 for the INFLUENCE option.

Most of the memory REG needs to solve large problems is used for crossproducts matrices. REG requires $4p^2$ bytes for the main crossproducts matrix plus $4k^2$ bytes for the largest model. If several output data sets are requested, memory is needed for buffers also.

Printed Output

Many of the more specialized printouts are described in detail in the sections above. Most of the formulas for the statistics are in Chapter 1, "Introduction to SAS Regression Procedures."

The analysis-of-variance table is always printed and includes:

1. The SOURCE of the variation, MODEL for the fitted regression, ERROR for the residual error, and C TOTAL for the total variation after correcting for the mean
2. the degrees of freedom (DF) associated for the source
3. the SUM OF SQUARES for the term
4. the MEAN SQUARE, the sum of squares divided by the degrees of freedom
5. the F VALUE for testing the hypothesis that all parameters are zero except for the intercept. This is formed by dividing the mean square for MODEL by the mean square for ERROR.

6. the PROB>F, the probability of getting a greater F statistic than that observed if the hypothesis is true. This is the significance probability.

Other statistics are printed:

7. ROOT MSE is an estimate of the standard deviation of the error term. It is calculated as the square root of the mean square error.
8. DEP MEAN is the sample mean of the dependent variable
9. C.V. is the coefficient of variation, computed as 100 times ROOT MSE divided by DEP MEAN. This expresses the variation in unitless values.
10. R-SQUARE is a measure between 0 and 1 that indicates the portion of the (corrected) total variation that is attributed to the fit rather than left to residual error. It is calculated as SS(MODEL) divided by SS(TOTAL). It is also called the *coefficient of determination*. It is the square of the multiple correlation, in other words, the square of the correlation between the dependent variable and the predicted values.
11. ADJ R-SQ, the adjusted R^2, is a version of R^2 that has been adjusted for degrees of freedom. It is calculated:

    $$\bar{R}^2 = 1 - (1 - R^2)(n - 1)/dfe$$

    where *dfe* is the degrees of freedom for error.

The parameter estimates and associated statistics are then printed:

12. the VARIABLE used as the regressor, including the name INTERCEP to estimate the intercept parameter
13. the degree of freedom (DF) for the variable. There is one degree of freedom unless the model is not full rank.
14. the PARAMETER ESTIMATE
15. the STANDARD ERROR, the estimate of the standard deviation of the parameter estimate
16. T FOR H0: PARAMETER=0, the t test that the parameter is zero. This is computed as PARAMETER ESTIMATE divided by the STANDARD ERROR.
17. the PROB>|T|, the probability that a t statistic would obtain a greater absolute value than that observed given that the true parameter is zero. This is the two-tailed significance probability.

EXAMPLES

Population Growth Trends: Example 1

In the following example, the population of the U.S. from 1790 to 1970 is fit to linear and quadratic functions of time. The statements request influence diagnostic options and plots of the output data set.

```
DATA USPOP;
  INPUT POP 6.3 @@;
  RETAIN YEAR 1780;
  YEAR=YEAR+10;
  YEARSQ=YEAR*YEAR;
  CARDS;
3929 5308 7239 9638 12866 17069 23191 31443 39818 50155
62947 75994 91972 105710 122775 131669 151325 179323 203211  . . .
;
```

```
PROC REG DATA=USPOP;
  MODEL POP=YEAR / R CLI CLM INFLUENCE DW;
  MODEL POP=YEAR YEARSQ / R CLI CLM INFLUENCE DW;
  OUTPUT OUT=C P=PRED L95=L95 U95=U95 R=RESID
    COOKD=COOKD;
PROC PLOT DATA=C;
  PLOT POP*YEAR='A' PRED*YEAR='P' U95*YEAR='U' L95*YEAR='L'
    / OVERLAY VPOS=40 HPOS=80;
  PLOT RESID*YEAR / VREF=0 VPOS=30 HPOS=80;
  PLOT COOKD*YEAR / VREF=0 VPOS=30 HPOS=80;
```

1

DEP VARIABLE: POP

(1) SOURCE	(2) DF	(3) SUM OF SQUARES	(4) MEAN SQUARE	(5) F VALUE	(6) PROB>F
(7) MODEL	1	66336.469	66336.469	201.873	0.0001
ERROR	17	5586.293	328.605		
C TOTAL	18	71922.762			

(8) ROOT MSE	18.127478	(10) R-SQUARE	0.9223
DEP MEAN	69.767474	(11) ADJ R-SQ	0.9178
(9) C.V.	25.98271		

(12) VARIABLE	(13) DF	(14) PARAMETER ESTIMATE	(15) STANDARD ERROR	(16) T FOR H0: PARAMETER=0	(17) PROB > |T|
INTERCEP	1	-1958.366	142.805	-13.714	0.0001
YEAR	1	1.078795	0.075928	14.208	0.0001

OBS	ACTUAL	PREDICT VALUE	STD ERR PREDICT	LOWER95% MEAN	UPPER95% MEAN	LOWER95% PREDICT	UPPER95% PREDICT	RESIDUAL	STD ERR RESIDUAL	STUDENT RESIDUAL	-2-1-0 1 2	COOK'S D
1	3.929	-27.324	7.999	-44.201	-10.447	-69.128	14.480	31.253	16.267	1.921	***	0.446
2	5.308	-16.536	7.361	-32.067	-1.005	-57.815	24.743	21.844	16.565	1.319	**	0.172
3	7.239	-5.748	6.749	-19.986	8.490	-46.558	35.062	12.987	16.824	0.772	*	0.048
4	9.638	5.040	6.168	-7.974	18.054	-35.359	45.439	4.598	17.046	0.270		0.005
5	12.866	15.828	5.631	3.948	27.708	-24.220	55.876	-2.962	17.231	-0.172		0.002
6	17.069	26.616	5.150	15.751	37.480	-13.143	66.374	-9.547	17.381	-0.549	*	0.013
7	23.191	37.404	4.742	27.400	47.408	-2.129	76.936	-14.213	17.496	-0.812	*	0.024
8	31.443	48.192	4.427	38.851	57.532	8.822	87.561	-16.749	17.579	-0.953	*	0.029
9	39.818	58.980	4.227	50.060	67.899	19.708	98.251	-19.162	17.628	-1.087	**	0.034
10	50.155	69.767	4.159	60.993	78.542	30.529	109.006	-19.612	17.644	-1.112	**	0.034
11	62.947	80.555	4.227	71.636	89.475	41.284	119.827	-17.608	17.628	-0.999	*	0.029
12	75.994	91.343	4.427	82.003	100.684	51.974	130.713	-15.349	17.579	-0.873	*	0.024
13	91.972	102.131	4.742	92.127	112.135	62.599	141.663	-10.159	17.496	-0.581	*	0.012
14	105.710	112.919	5.150	102.054	123.784	73.161	152.678	-7.209	17.381	-0.415		0.008
15	122.775	123.707	5.631	111.827	135.587	83.659	163.755	-.932202	17.231	-0.054		0.000
16	131.669	134.495	6.168	121.481	147.509	94.096	174.894	-2.826	17.046	-0.166		0.002
17	151.325	145.283	6.749	131.045	159.521	104.473	186.093	6.042	16.824	0.359		0.010
18	179.323	156.071	7.361	140.540	171.602	114.792	197.350	23.252	16.565	1.404	**	0.195
19	203.211	166.859	7.999	149.982	183.736	125.055	208.663	36.352	16.267	2.235	****	0.604
20	.	177.647	8.657	159.382	195.912	135.264	220.030	.	.	.		.
21	.	188.435	9.330	168.750	208.120	145.421	231.449	.	.	.		.
22	.	199.223	10.016	178.092	220.354	155.528	242.917	.	.	.		.

SUM OF RESIDUALS -9.91207E-13
SUM OF SQUARED RESIDUALS 5586.293

2

OBS	RESIDUAL	RSTUDENT	HAT DIAG H	COV RATIO	DFFITS	DFBETAS INTERCEP	DFBETAS YEAR
1	31.253	2.1066	0.1947	0.8592	1.0359	0.9002	-0.8849
2	21.8441	1.3502	0.1649	1.0894	0.6000	0.5048	-0.4951
3	12.9871	0.7624	0.1386	1.2203	0.3058	0.2462	-0.2408
4	4.5982	0.2623	0.1158	1.2658	0.0949	0.0719	-0.0701
5	-2.96175	-0.1669	0.0965	1.2451	-0.0545	-0.0379	0.0368
6	-9.54669	-0.5377	0.0807	1.1848	-0.1593	-0.0977	0.0940
7	-14.2126	-0.8038	0.0684	1.1196	-0.2178	-0.1102	0.1046
8	-16.7486	-0.9501	0.0596	1.0757	-0.2393	-0.0886	0.0821
9	-19.1615	-1.0932	0.0544	1.0336	-0.2622	-0.0546	0.0471
10	-19.6125	-1.1198	0.0526	1.0247	-0.2639	-0.0077	-0.0000
11	-17.6084	-0.9988	0.0544	1.0578	-0.2395	0.0361	-0.0430

(continued on next page)

(continued from previous page)

```
         12 -15.3494   -0.8668    0.0596    1.0952  -0.2183    0.0689  -0.0749
         13 -10.1593   -0.5690    0.0684    1.1642  -0.1542    0.0701  -0.0741
         14 -7.20926   -0.4045    0.0807    1.2033  -0.1198    0.0678  -0.0707
         15 -.932202   -0.0525    0.0965    1.2490  -0.0172    0.0112  -0.0116
         16 -2.82615   -0.1610    0.1158    1.2726  -0.0583    0.0419  -0.0430
         17  6.04191    0.3497    0.1386    1.2907   0.1403   -0.1079   0.1105
         18   23.252    1.4482    0.1649    1.0567   0.6436   -0.5202   0.5310
         19   36.352    2.5798    0.1947    0.6992   1.2686   -1.0641   1.0837
         20        .         .         .         .        .         .        .
         21        .         .         .         .        .         .        .
         22        .         .         .         .        .         .        .

       DURBIN-WATSON D              0.180
       1ST ORDER AUTOCORRELATION    0.704
```

```
                                                                                        3

DEP VARIABLE: POP

                       SUM OF          MEAN
SOURCE       DF       SQUARES        SQUARE      F VALUE       PROB>F

MODEL         2     71799.016     35899.508     4641.719       0.0001
ERROR        16       123.746      7.734098
C TOTAL      18     71922.762

    ROOT MSE         2.781025     R-SQUARE      0.9983
    DEP MEAN        69.767474     ADJ R-SQ      0.9981
    C.V.             3.986133

                    PARAMETER      STANDARD    T FOR H0:
VARIABLE   DF        ESTIMATE         ERROR  PARAMETER=0    PROB > |T|

INTERCEP    1       20450.434       843.475       24.245        0.0001
YEAR        1      -22.780606      0.897849      -25.372        0.0001
YEARSQ      1  0.006345585 0.0002387695          26.576        0.0001

                PREDICT  STD ERR LOWER95% UPPER95% LOWER95% UPPER95%           STD ERR  STUDENT                      COOK'S
 OBS   ACTUAL     VALUE  PREDICT     MEAN     MEAN  PREDICT  PREDICT RESIDUAL RESIDUAL RESIDUAL    -2-1-0 1 2             D

   1    3.929     5.038    1.729    1.373    8.703   -1.903   11.980   -1.109    2.178   -0.509 |      *|      |      0.054
   2    5.308     5.039    1.391    2.090    7.987   -1.553   11.631 0.269101    2.408    0.112 |       |      |      0.001
   3    7.239     6.308    1.130    3.912    8.705 -.055392   12.672 0.930534    2.541    0.366 |       |      |      0.009
   4    9.638     8.847 0.957106    6.818   10.876    2.612   15.082 0.790849    2.611    0.303 |       |      |      0.004
   5   12.866    12.655 0.872081   10.806   14.504    6.476   18.833 0.211047    2.641    0.080 |       |      |      0.000
   6   17.069    17.732 0.857844   15.913   19.550   11.562   23.901 -.662872    2.645   -0.251 |       |      |      0.002
   7   23.191    24.078 0.883512   22.205   25.951   17.892   30.264 -.886908    2.637   -0.336 |       |      |      0.004
   8   31.443    31.693 0.920177   29.742   33.644   25.483   37.903 -.250061    2.624   -0.095 |       |      |      0.000
   9   39.818    40.577 0.948730   38.566   42.589   34.348   46.806 -.759331    2.614   -0.290 |       |      |      0.004
  10   50.155    50.731 0.959248   48.697   52.764   44.494   56.967 -.575718    2.610   -0.221 |       |      |      0.002
  11   62.947    62.153 0.948730   60.142   64.164   55.924   68.382 0.793778    2.614    0.304 |       |      |      0.004
  12   75.994    74.845 0.920177   72.894   76.796   68.635   81.055    1.149    2.624    0.438 |       |      |      0.008
  13   91.972    88.806 0.883512   86.933   90.679   82.620   94.991    3.166    2.637    1.201 |       |**    |      0.054
  14  105.710   104.035 0.857844  102.217  105.854   97.866  110.205    1.675    2.645    0.633 |       |*     |      0.014
  15  122.775   120.534 0.872081  118.686  122.383  114.356  126.713    2.241    2.641    0.848 |       |*     |      0.026
  16  131.669   138.302 0.957106  136.274  140.331  132.068  144.537   -6.633    2.611   -2.540 |  *****|      |      0.289
  17  151.325   157.340    1.130  154.943  159.736  150.976  163.704   -6.015    2.541   -2.367 |   ****|      |      0.370
  18  179.323   177.646    1.391  174.698  180.595  171.054  184.238    1.677    2.408    0.696 |       |*     |      0.054
  19  203.211   199.221    1.729  195.556  202.886  192.280  206.163    3.990    2.178    1.831 |       |***   |      0.704
  20        .   222.066    2.135  217.540  226.592  214.634  229.498        .        .        .                           .
  21        .   246.180    2.602  240.664  251.695  238.106  254.253        .        .        .                           .
  22        .   271.562    3.126  264.936  278.189  262.693  280.432        .        .        .                           .

SUM OF RESIDUALS             -1.90444E-10
SUM OF SQUARED RESIDUALS         123.7456
```

```
                                                                                        4

                           HAT DIAG      COV             DFBETAS   DFBETAS   DFBETAS
 OBS RESIDUAL  RSTUDENT          H     RATIO    DFFITS  INTERCEP      YEAR    YEARSQ

   1 -1.10945   -0.4972     0.3865    1.8834   -0.3946   -0.2842    0.2810   -0.2779
   2 0.269101    0.1082     0.2501    1.6147    0.0625    0.0376   -0.0370    0.0365
   3 0.930534    0.3561     0.1652    1.4176    0.1584    0.0666   -0.0651    0.0636
   4 0.790849    0.2941     0.1184    1.3531    0.1078    0.0182   -0.0172    0.0161
   5 0.211047    0.0774     0.0983    1.3444    0.0256   -0.0030    0.0033   -0.0035
   6 -.662872   -0.2431     0.0951    1.3255   -0.0788    0.0296   -0.0302    0.0307
   7 -.886908   -0.3268     0.1009    1.3214   -0.1095    0.0609   -0.0616    0.0621
   8 -.250061   -0.0923     0.1095    1.3605   -0.0324    0.0216   -0.0217    0.0218
   9 -.759331   -0.2820     0.1164    1.3519   -0.1023    0.0743   -0.0745    0.0747
  10 -.575718   -0.2139     0.1190    1.3650   -0.0786    0.0586   -0.0587    0.0587
  11 0.793778    0.2949     0.1164    1.3499    0.1070   -0.0784    0.0783   -0.0781
  12  1.14916    0.4265     0.1095    1.3144    0.1496   -0.1018    0.1014   -0.1009
  13  3.16642    1.2189     0.1009    1.0168    0.4084   -0.2357    0.2338   -0.2318
```

(continued on next page)

(continued from previous page)

14	1.67456	0.6207	0.0951	1.2430	0.2013	-0.0811	0.0798	-0.0784
15	2.24059	0.8407	0.0983	1.1724	0.2776	-0.0427	0.0404	-0.0380
16	-6.6335	-3.1845	0.1184	0.2924	-1.1673	-0.1531	0.1636	-0.1747
17	-6.01471	-2.8433	0.1652	0.3989	-1.2649	-0.4843	0.4958	-0.5076
18	1.67697	0.6847	0.2501	1.4757	0.3954	0.2240	-0.2274	0.2308
19	3.98953	1.9947	0.3865	0.9766	1.5831	1.0902	-1.1025	1.1151
20	.	.	.	.	.	.	.	.
21	.	.	.	.	.	.	.	.
22	.	.	.	.	.	.	.	.

DURBIN-WATSON D 1.264
1ST ORDER AUTOCORRELATION 0.299

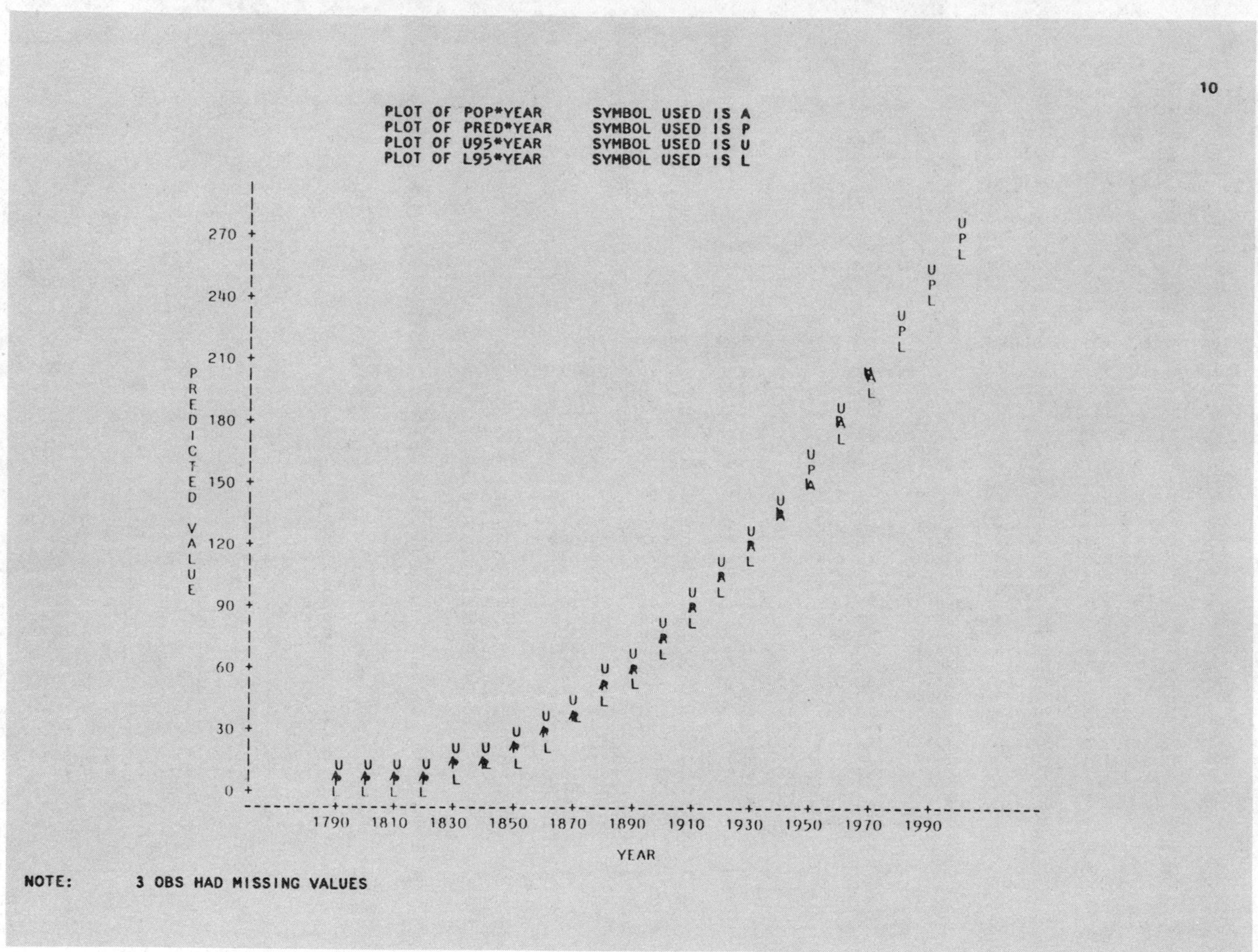

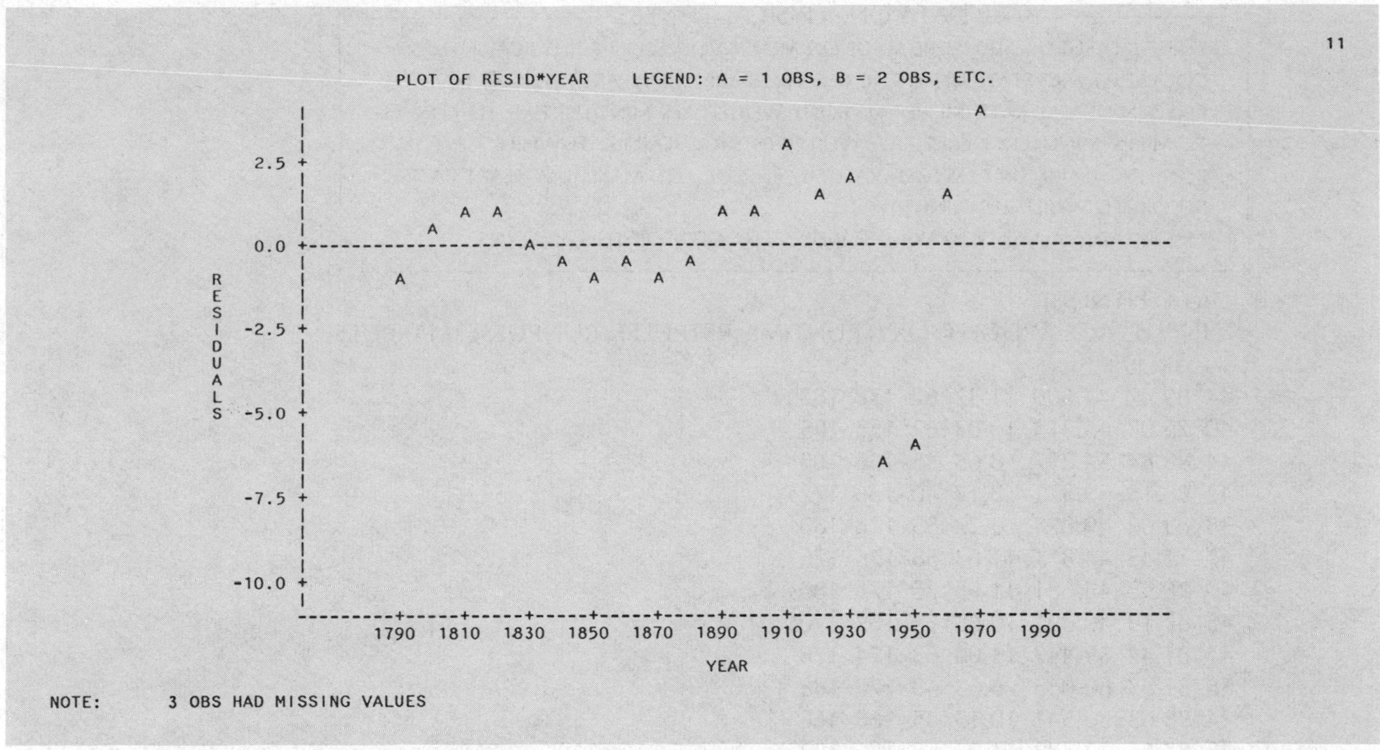

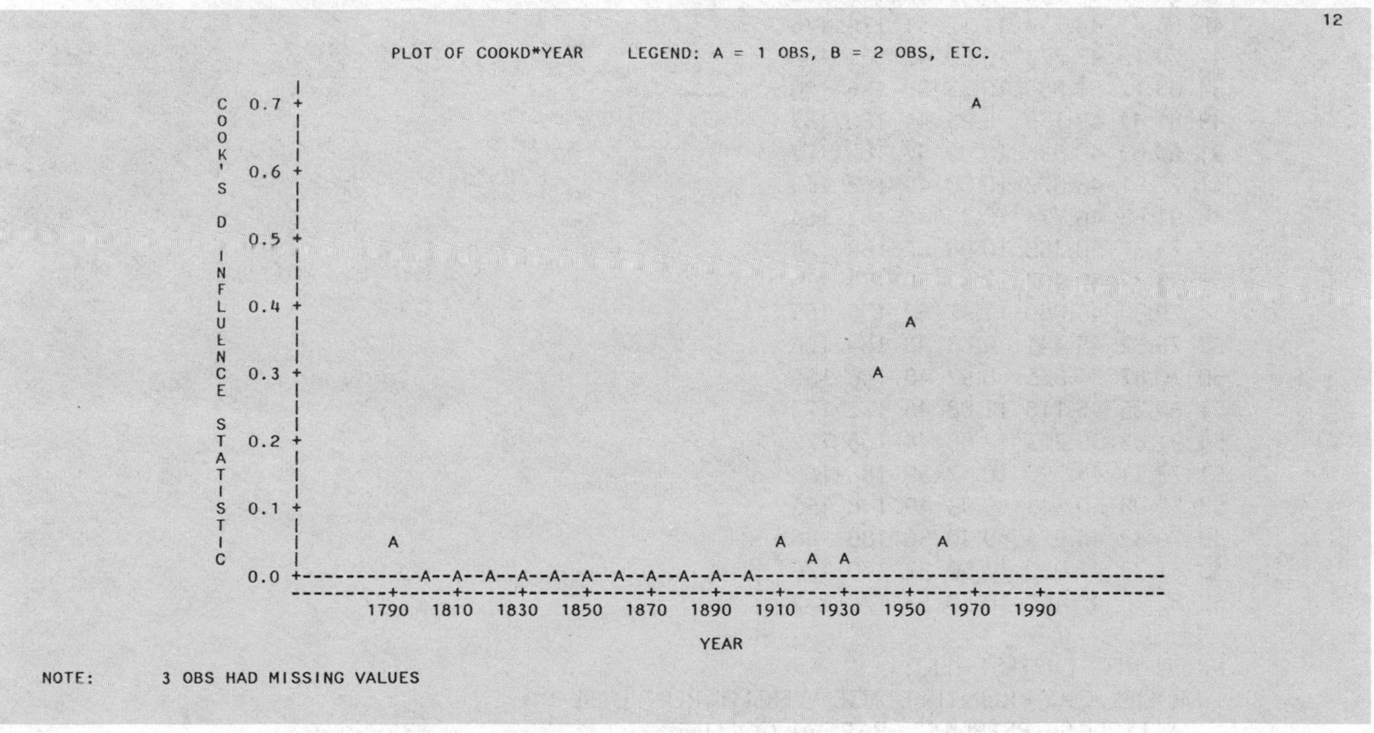

Aerobic Fitness Prediction: Example 2

Aerobic fitness (measured by the ability to consume oxygen) is fit to some simple exercise tests. The goal is to develop an equation to predict fitness based on the exercise tests rather than on expensive and cumbersome oxygen consumption measurements. Since regressors are correlated, collinearity diagnostics are requested.

```
* --------------------------------DATA ON PHYSICAL FITNESS-------------------------------- *
|  THESE MEASUREMENTS WERE MADE ON MEN INVOLVED IN A PHYSICAL FITNESS                       |
|  COURSE AT N.C. STATE UNIV. THE VARIABLES ARE AGE (YEARS), WEIGHT (KG),                  |
|  OXYGEN UPTAKE RATE (ML PER KG BODY WEIGHT PER MINUTE), TIME TO RUN                       |
|  1.5 MILES (MINUTES), HEART RATE WHILE RESTING, HEART RATE WHILE                          |
|  RUNNING (SAME TIME OXYGEN RATE MEASURED), AND MAXIMUM HEART RATE                         |
|  RECORDED WHILE RUNNING.                                                                  |
|  ***CERTAIN VALUES OF MAXPULSE WERE CHANGED FOR THIS ANALYSIS                             |
* ----------------------------------------------------------------------------------------- * ;
DATA FITNESS;
  INPUT AGE WEIGHT OXY RUNTIME RSTPULSE RUNPULSE MAXPULSE;
  CARDS;
44 89.47 44.609 11.37 62 178 182
40 75.07 45.313 10.07 62 185 185
44 85.84 54.297  8.65 45 156 168
42 68.15 59.571  8.17 40 166 172
38 89.02 49.874  9.22 55 178 180
47 77.45 44.811 11.63 58 176 176
40 75.98 45.681 11.95 70 176 180
43 81.19 49.091 10.85 64 162 170
44 81.42 39.442 13.08 63 174 176
38 81.87 60.055  8.63 48 170 186
44 73.03 50.541 10.13 45 168 168
45 87.66 37.388 14.03 56 186 192
45 66.45 44.754 11.12 51 176 176
47 79.15 47.273 10.60 47 162 164
54 83.12 51.855 10.33 50 166 170
49 81.42 49.156  8.95 44 180 185
51 69.63 40.836 10.95 57 168 172
51 77.91 46.672 10.00 48 162 168
48 91.63 46.774 10.25 48 162 164
49 73.37 50.388 10.08 67 168 168
57 73.37 39.407 12.63 58 174 176
54 79.38 46.080 11.17 62 156 165
52 76.32 45.441  9.63 48 164 166
50 70.87 54.625  8.92 48 146 155
51 67.25 45.118 11.08 48 172 172
54 91.63 39.203 12.88 44 168 172
51 73.71 45.790 10.47 59 186 188
57 59.08 50.545  9.93 49 148 155
49 76.32 48.673  9.40 56 186 188
48 61.24 47.920 11.50 52 170 176
52 82.78 47.467 10.50 53 170 172
;
PROC REG OUTEST=EST;
  MODEL OXY=RUNTIME AGE WEIGHT RUNPULSE
    MAXPULSE RSTPULSE / PARTIAL COLLIN;
```

1

DEP VARIABLE: OXY

SOURCE	DF	SUM OF SQUARES	MEAN SQUARE	F VALUE	PROB>F
MODEL	6	722.544	120.424	22.433	0.0001
ERROR	24	128.838	5.368247		
C TOTAL	30	851.382			

ROOT MSE	2.316948	R-SQUARE	0.8487
DEP MEAN	47.375806	ADJ R-SQ	0.8108
C.V.	4.890572		

VARIABLE	DF	PARAMETER ESTIMATE	STANDARD ERROR	T FOR H0: PARAMETER=0	PROB > |T|
INTERCEP	1	102.934	12.403258	8.299	0.0001
RUNTIME	1	-2.628653	0.384562	-6.835	0.0001
AGE	1	-0.226974	0.099837	-2.273	0.0322
WEIGHT	1	-0.074177	0.054593	-1.359	0.1869
RUNPULSE	1	-0.369628	0.119853	-3.084	0.0051
MAXPULSE	1	0.303217	0.136495	2.221	0.0360
RSTPULSE	1	-0.021534	0.066054	-0.326	0.7473

COLLINEARITY DIAGNOSTICS VARIANCE PROPORTIONS

NUMBER	EIGENVALUE	CONDITION INDEX	PORTION INTERCEP	PORTION RUNTIME	PORTION AGE	PORTION WEIGHT	PORTION RUNPULSE	PORTION MAXPULSE	PORTION RSTPULSE
1	6.950	1.000	0.0000	0.0002	0.0002	0.0002	0.0000	0.0000	0.0003
2	0.018676	19.291	0.0022	0.0252	0.1463	0.0104	0.0000	0.0000	0.3906
3	0.015034	21.501	0.0006	0.1286	0.1501	0.2357	0.0012	0.0012	0.0281
4	0.009110	27.621	0.0064	0.6090	0.0319	0.1831	0.0015	0.0012	0.1903
5	0.006073	33.829	0.0013	0.1250	0.1128	0.4444	0.0151	0.0083	0.3648
6	0.001018	82.638	0.7997	0.0975	0.4966	0.1033	0.0695	0.0056	0.0203
7	0.00017947	196.786	0.1898	0.0146	0.0621	0.0228	0.9128	0.9836	0.0057

2

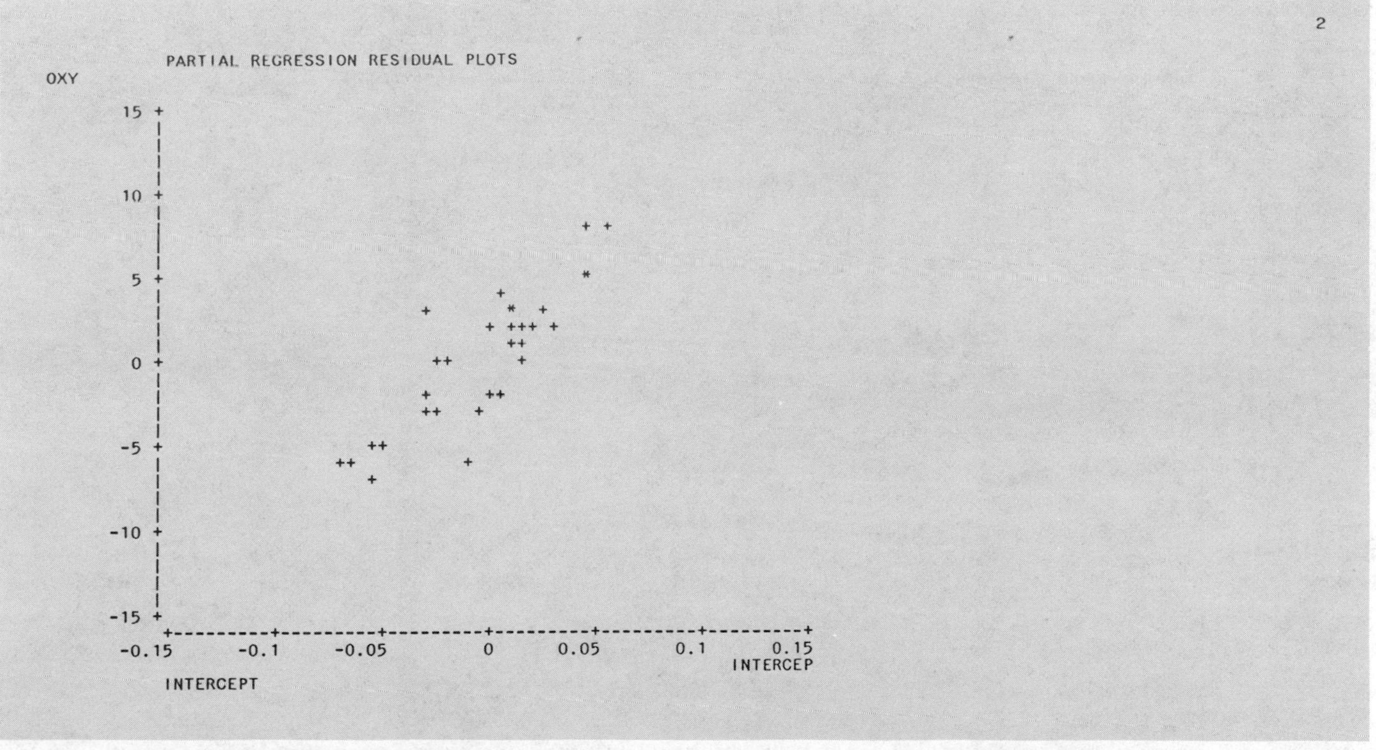

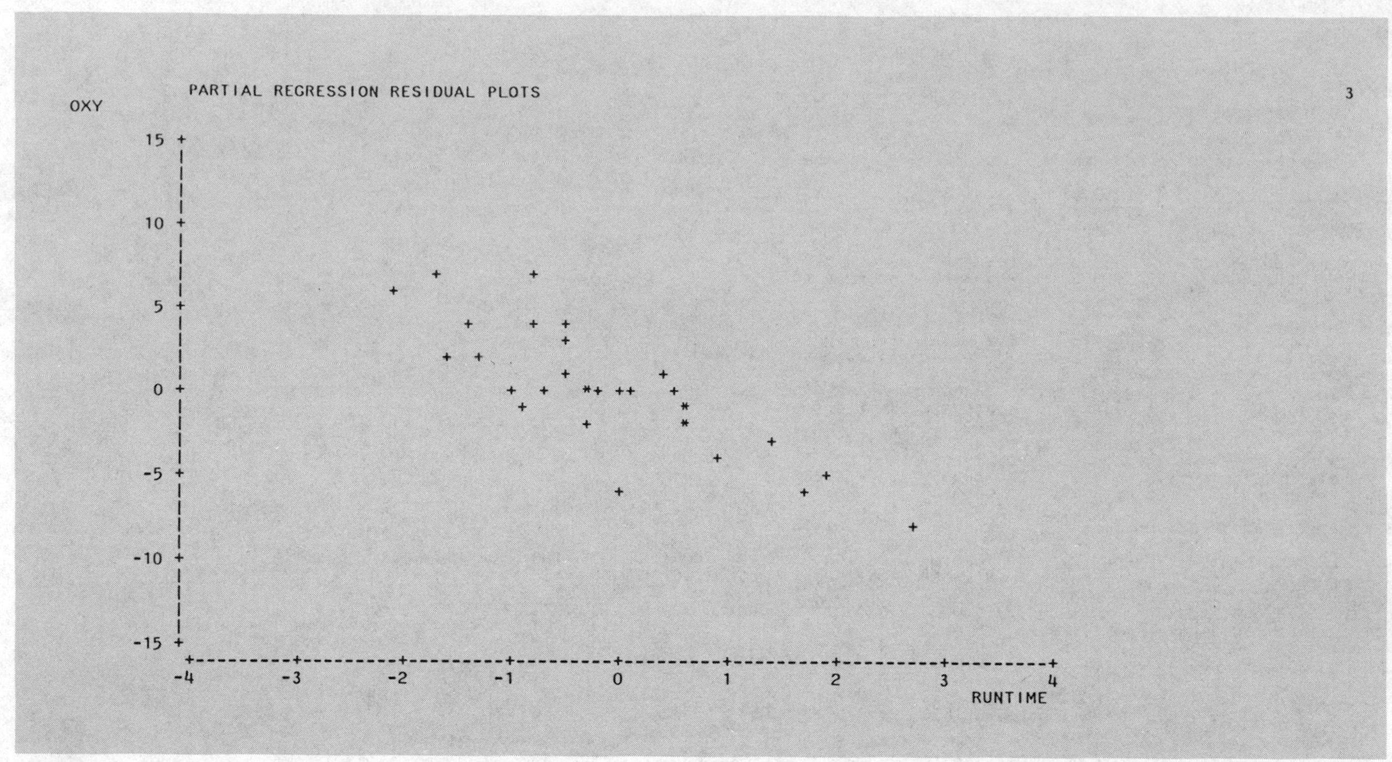

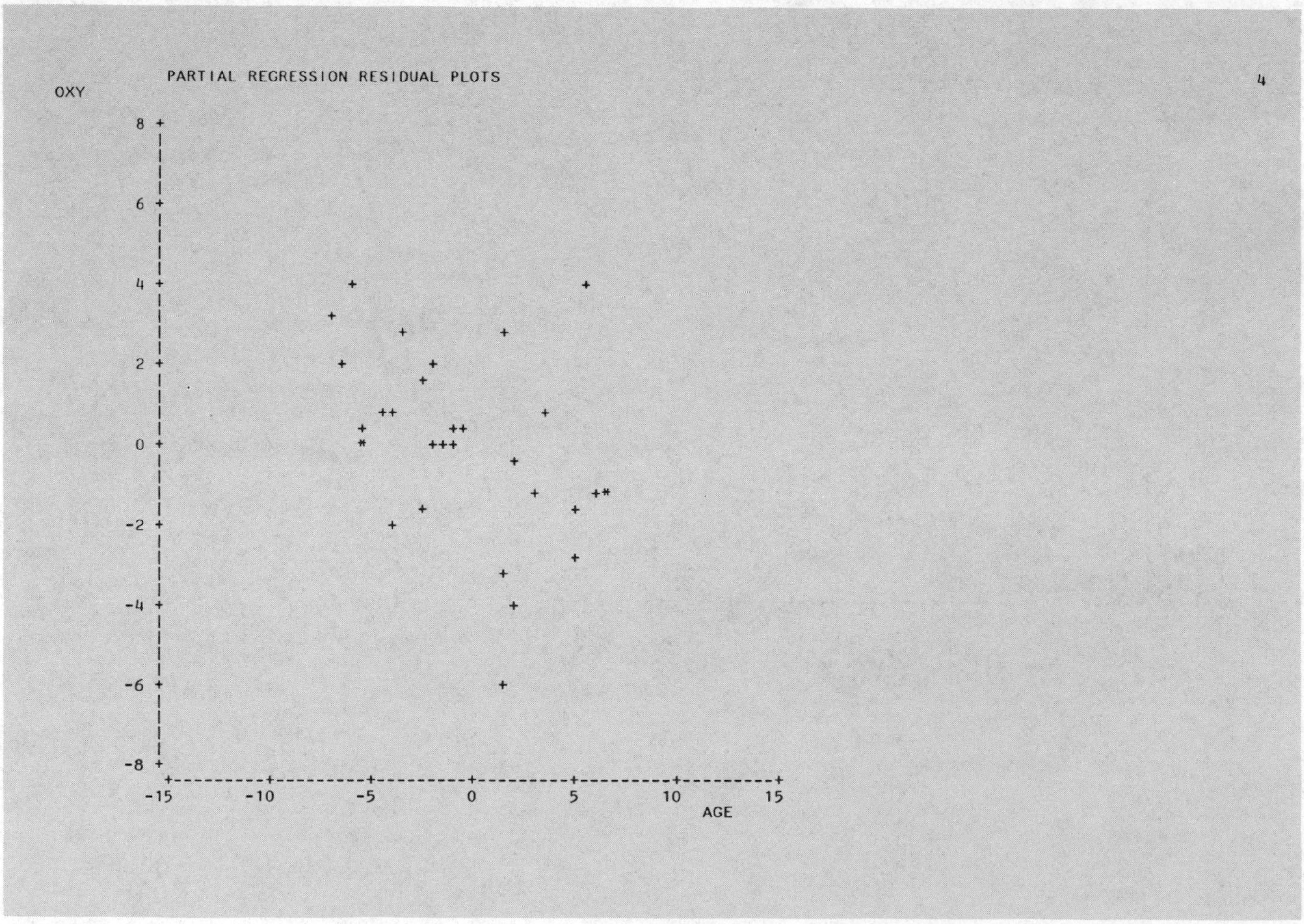

PARTIAL REGRESSION RESIDUAL PLOTS
3
OXY
15
10
5
0
-5
-10
-15
-4
-3
-2
-1
0
1
2
3
4
RUNTIME

PARTIAL REGRESSION RESIDUAL PLOTS
4
OXY
8
6
4
2
0
-2
-4
-6
-8
-15
-10
-5
0
5
10
15
AGE

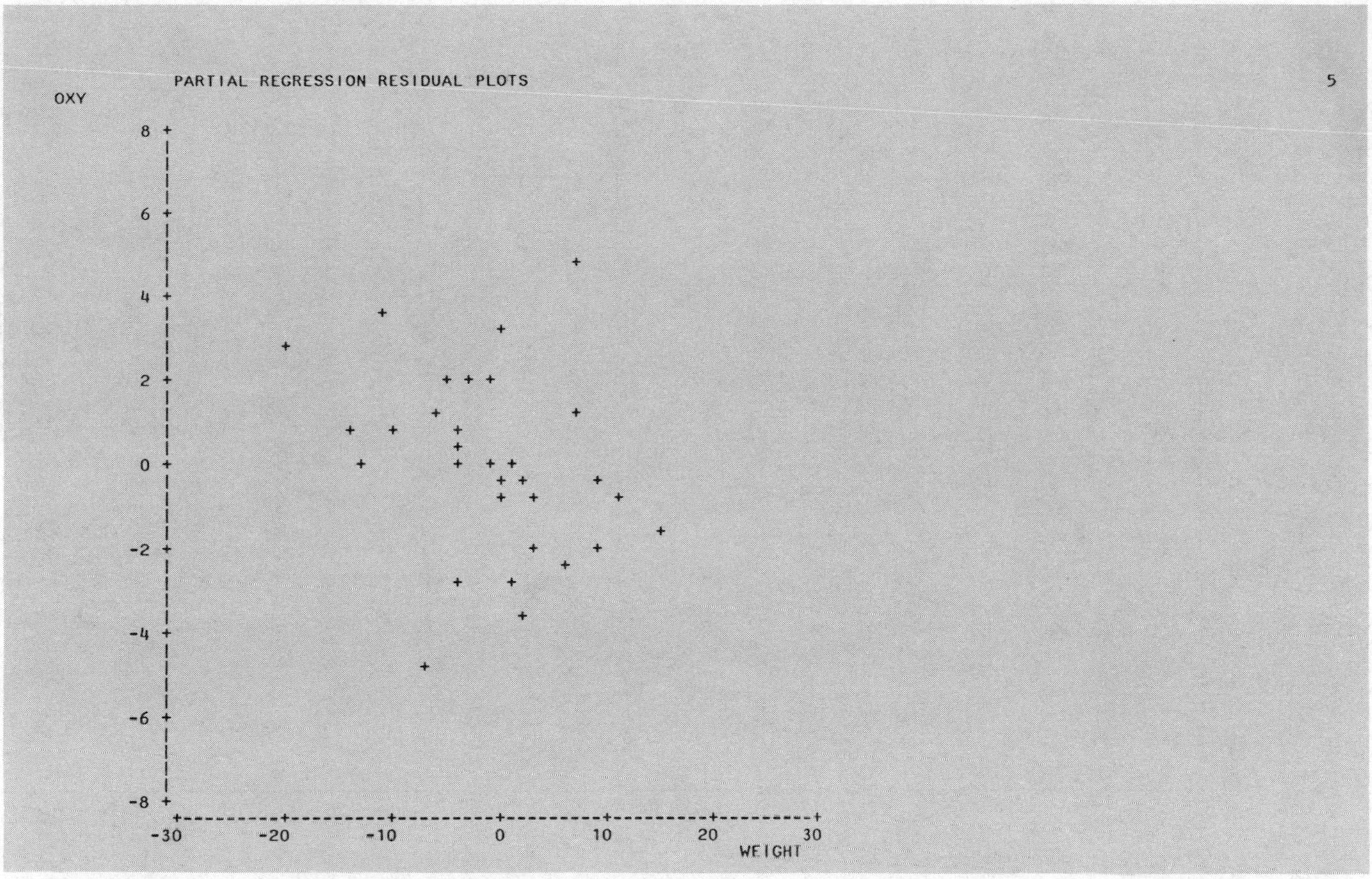

PARTIAL REGRESSION RESIDUAL PLOTS
5
OXY
8
6
4
2
0
-2
-4
-6
-8
-30 -20 -10 0 10 20 30
WEIGHT

PARTIAL REGRESSION RESIDUAL PLOTS
6
OXY
8
6
4
2
0
-2
-4
-6
-8
-15 -10 -5 0 5 10 15
RUNPULSE

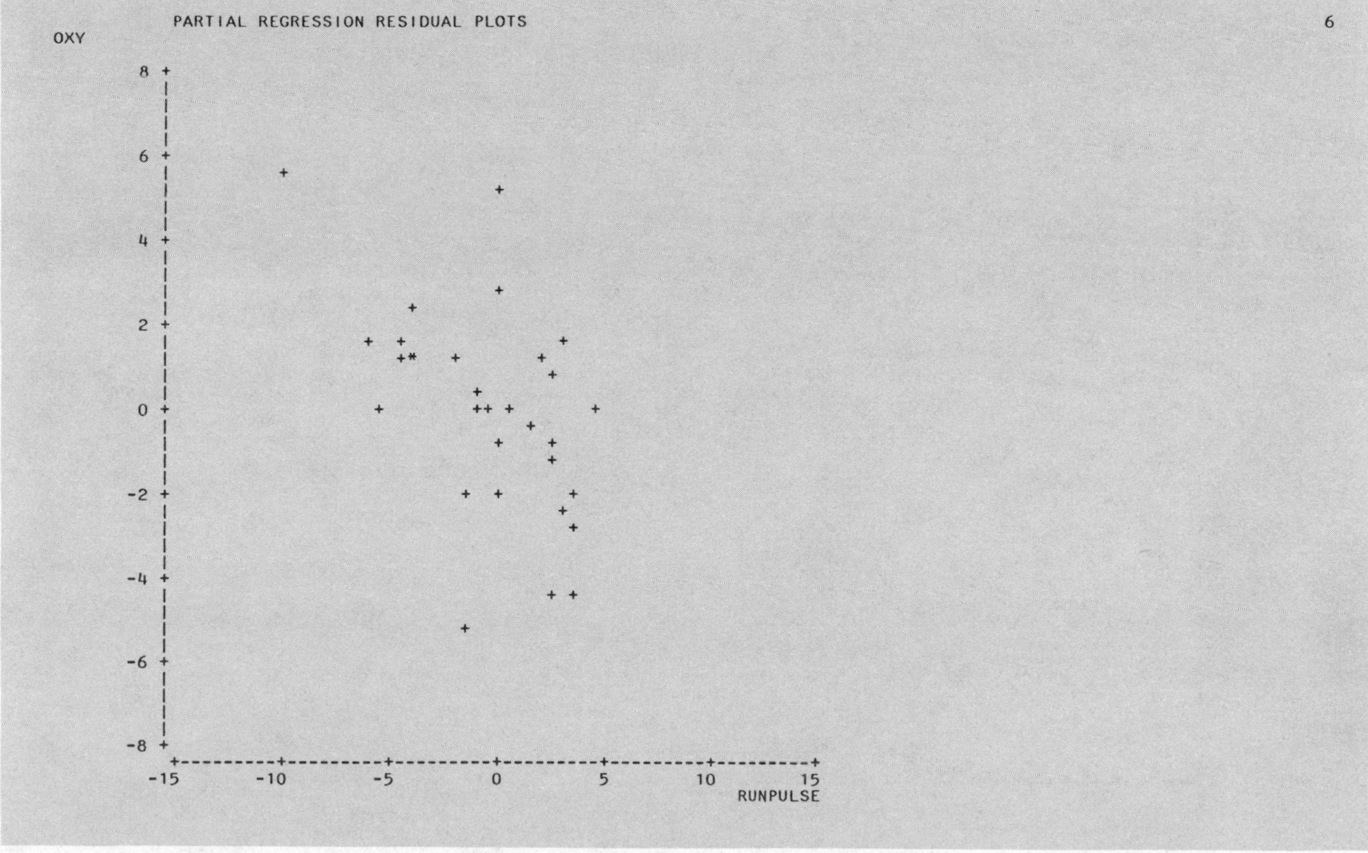

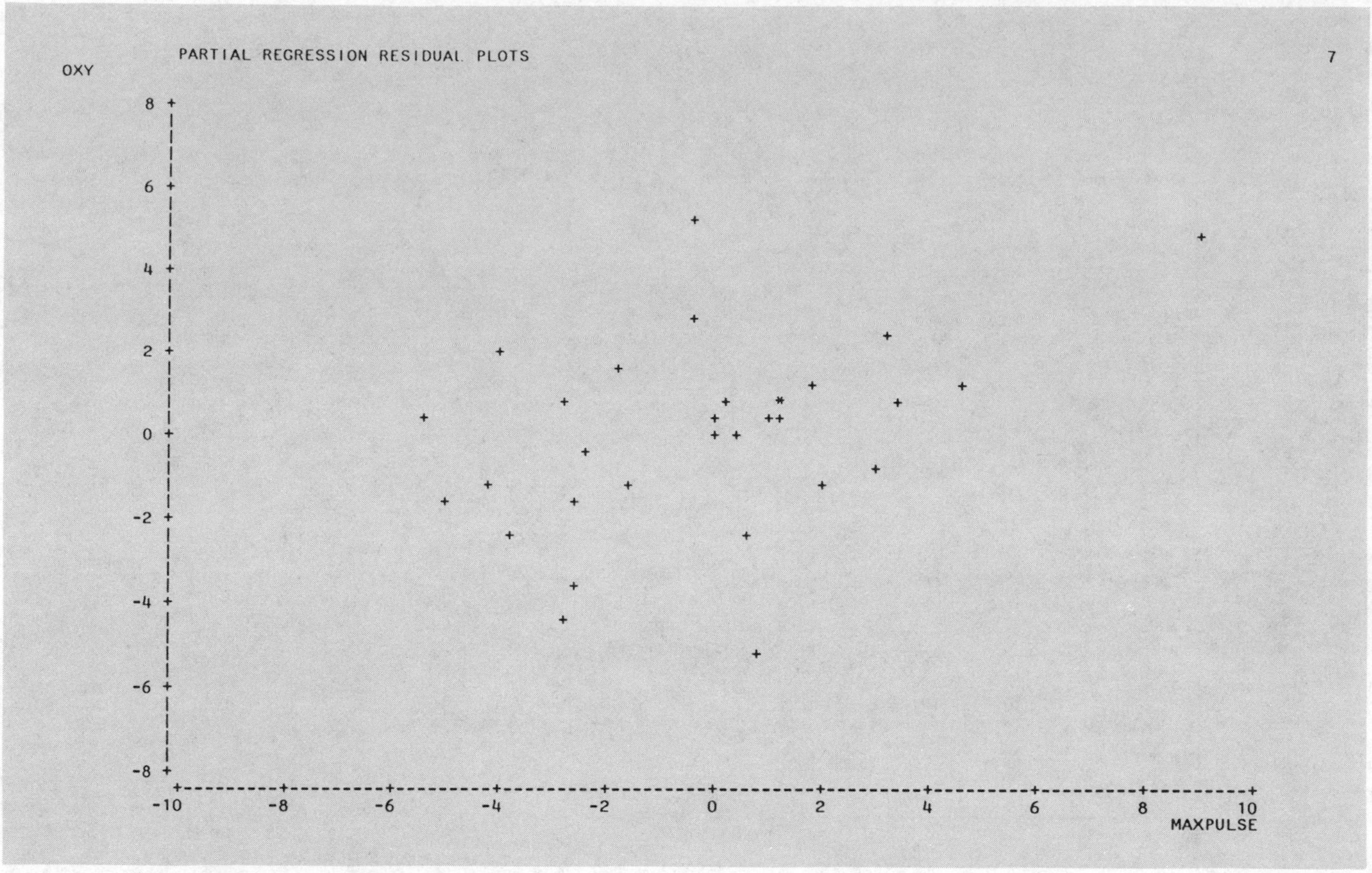

```
        PARTIAL REGRESSION RESIDUAL PLOTS                                          7
OXY
```

```
        PARTIAL REGRESSION RESIDUAL PLOTS                                          8
OXY
```

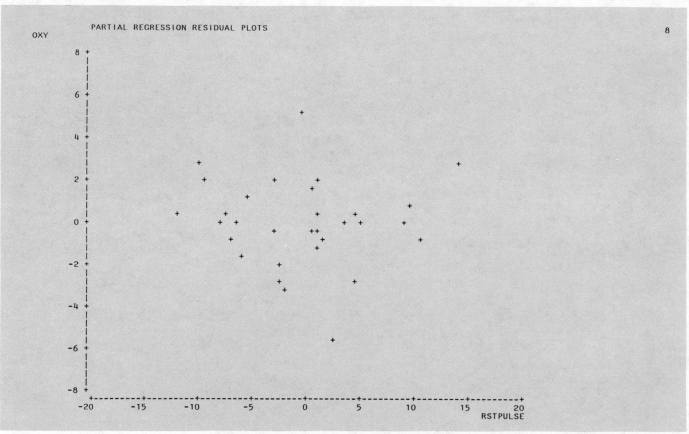

Height and Weight Examples: Example 3

To illustrate multiple MODEL statements, BY groups, and the OUTEST and OUTSSCP features, we predict WEIGHT by HEIGHT and AGE in this group of students.

```
* -------------DATA ON AGE, WEIGHT, AND HEIGHT OF CHILDREN------------- *
|   AGE (MONTHS), HEIGHT (INCHES), AND WEIGHT (POUNDS) WERE RECORDED FOR   |
|   A GROUP OF SCHOOLCHILDREN.                                             |
|   FROM T. LEWIS & L.R. TAYLOR, "INTRODUCTION TO EXPERIMENTAL ECOLOGY."   |
* ----------------------------------------------------------------------- * ;
DATA HTWT;
  INPUT SEX $ AGE :3.1 HEIGHT WEIGHT @@;
  CARDS;
F 143 56.3  85.0 F 155 62.3 105.0 F 153 63.3 108.0 F 161 59.0  92.0
F 191 62.5 112.5 F 171 62.5 112.0 F 185 59.0 104.0 F 142 56.5  69.0
F 160 62.0  94.5 F 140 53.8  68.5 F 139 61.5 104.0 F 178 61.5 103.5
F 157 64.5 123.5 F 149 58.3  93.0 F 143 51.3  50.5 F 145 58.8  89.0
F 191 65.3 107.0 F 150 59.5  78.5 F 147 61.3 115.0 F 180 63.3 114.0
F 141 61.8  85.0 F 140 53.5  81.0 F 164 58.0  83.5 F 176 61.3 112.0
F 185 63.3 101.0 F 166 61.5 103.5 F 175 60.8  93.5 F 180 59.0 112.0
F 210 65.5 140.0 F 146 56.3  83.5 F 170 64.3  90.0 F 162 58.0  84.0
F 149 64.3 110.5 F 139 57.5  96.0 F 186 57.8  95.0 F 197 61.5 121.0
F 169 62.3  99.5 F 177 61.8 142.5 F 185 65.3 118.0 F 182 58.3 104.5
F 173 62.8 102.5 F 166 59.3  89.5 F 168 61.5  95.0 F 169 62.0  98.5
F 150 61.3  94.0 F 184 62.3 108.0 F 139 52.8  63.5 F 147 59.8  84.5
F 144 59.5  93.5 F 177 61.3 112.0 F 178 63.5 148.5 F 197 64.8 112.0
F 146 60.0 109.0 F 145 59.0  91.5 F 147 55.8  75.0 F 145 57.8  84.0
F 155 61.3 107.0 F 167 62.3  92.5 F 183 64.3 109.5 F 143 55.5  84.0
F 183 64.5 102.5 F 185 60.0 106.0 F 148 56.3  77.0 F 147 58.3 111.5
F 154 60.0 114.0 F 156 54.5  75.0 F 144 55.8  73.5 F 154 62.8  93.5
F 152 60.5 105.0 F 191 63.3 113.5 F 190 66.8 140.0 F 140 60.0  77.0
F 148 60.5  84.5 F 189 64.3 113.5 F 143 58.3  77.5 F 178 66.5 117.5
F 164 65.3  98.0 F 157 60.5 112.0 F 147 59.5 101.0 F 148 59.0  95.0
F 177 61.3  81.0 F 171 61.5  91.0 F 172 64.8 142.0 F 190 56.8  98.5
F 183 66.5 112.0 F 143 61.5 116.5 F 179 63.0  98.5 F 186 57.0  83.5
F 182 65.5 133.0 F 182 62.0  91.5 F 142 56.0  72.5 F 165 61.3 106.5
F 165 55.5  67.0 F 154 61.0 122.5 F 150 54.5  74.0 F 155 66.0 144.5
F 163 56.5  84.0 F 141 56.0  72.5 F 147 51.5  64.0 F 210 62.0 116.0
F 171 63.0  84.0 F 167 61.0  93.5 F 182 64.0 111.5 F 144 61.0  92.0
F 193 59.8 115.0 F 141 61.3  85.0 F 164 63.3 108.0 F 186 63.5 108.0
F 169 61.5  85.0 F 175 60.3  86.0 F 180 61.3 110.5 M 165 64.8  98.0
M 157 60.5 105.0 M 144 57.3  76.5 M 150 59.5  84.0 M 150 60.8 128.0
M 139 60.5  87.0 M 189 67.0 128.0 M 183 64.8 111.0 M 147 50.5  79.0
M 146 57.5  90.0 M 160 60.5  84.0 M 156 61.8 112.0 M 173 61.3  93.0
M 151 66.3 117.0 M 141 53.3  84.0 M 150 59.0  99.5 M 164 57.8  95.0
M 153 60.0  84.0 M 206 68.3 134.0 M 250 67.5 171.5 M 176 63.8  98.5
M 176 65.0 118.5 M 140 59.5  94.5 M 185 66.0 105.0 M 180 61.8 104.0
M 146 57.3  83.0 M 183 66.0 105.5 M 140 56.5  84.0 M 151 58.3  86.0
M 151 61.0  81.0 M 144 62.8  94.0 M 160 59.3  78.5 M 178 67.3 119.5
M 193 66.3 133.0 M 162 64.5 119.0 M 164 60.5  95.0 M 186 66.0 112.0
M 143 57.5  75.0 M 175 64.0  92.0 M 175 68.0 112.0 M 175 63.5  98.5
M 173 69.0 112.5 M 170 63.8 112.5 M 174 66.0 108.0 M 164 63.5 108.0
M 144 59.5  88.0 M 156 66.3 106.0 M 149 57.0  92.0 M 144 60.0 117.5
M 147 57.0  84.0 M 188 67.3 112.0 M 169 62.0 100.0 M 172 65.0 112.0
```

```
M 150 59.5  84.0 M 193 67.8 127.5 M 157 58.0  80.5 M 168 60.0  93.5
M 140 58.5  86.5 M 156 58.3  92.5 M 156 61.5 108.5 M 158 65.0 121.0
M 184 66.5 112.0 M 156 68.5 114.0 M 144 57.0  84.0 M 176 61.5  81.0
M 168 66.5 111.5 M 149 52.5  81.0 M 142 55.0  70.0 M 188 71.0 140.0
M 203 66.5 117.0 M 142 58.8  84.0 M 189 66.3 112.0 M 188 65.8 150.5
M 200 71.0 147.0 M 152 59.5 105.0 M 174 69.8 119.5 M 166 62.5  84.0
M 145 56.5  91.0 M 143 57.5 101.0 M 163 65.3 117.5 M 166 67.3 121.0
M 182 67.0 133.0 M 173 66.0 112.0 M 155 61.8  91.5 M 162 60.0 105.0
M 177 63.0 111.0 M 177 60.5 112.0 M 175 65.5 114.0 M 166 62.0  91.0
M 150 59.0  98.0 M 150 61.8 118.0 M 188 63.3 115.5 M 163 66.0 112.0
M 171 61.8 112.0 M 162 63.0  91.0 M 141 57.5  85.0 M 174 63.0 112.0
M 142 56.0  87.5 M 148 60.5 118.0 M 140 56.8  83.5 M 160 64.0 116.0
M 144 60.0  89.0 M 206 69.5 171.5 M 159 63.3 112.0 M 149 56.3  72.0
M 193 72.0 150.0 M 194 65.3 134.5 M 152 60.8  97.0 M 146 55.0  71.5
M 139 55.0  73.5 M 186 66.5 112.0 M 161 56.8  75.0 M 153 64.8 128.0
M 196 64.5  98.0 M 164 58.0  84.0 M 159 62.8  99.0 M 178 63.8 112.0
M 153 57.8  79.5 M 155 57.3  80.5 M 178 63.5 102.5 M 142 55.0  76.0
M 164 66.5 112.0 M 189 65.0 114.0 M 164 61.5 140.0 M 167 62.0 107.5
M 151 59.3  87.0
;

TITLE-------------- DATA ON AGE, WEIGHT, AND HEIGHT OF CHILDREN -------------;
PROC REG OUTEST=EST1 OUTSSCP=SSCP1;
  BY SEX;
  EQ1: MODEL WEIGHT=HEIGHT;
  EQ2: MODEL WEIGHT=HEIGHT AGE;
PROC PRINT DATA=SSCP1;
  TITLE2 SSCP TYPE DATA SET;
PROC PRINT DATA=EST1;
  TITLE2 EST TYPE DATA SET;
```

```
                      ---------- DATA ON AGE, WEIGHT, AND HEIGHT OF CHILDREN ----------                    1
                                                SEX=F

MODEL: EQ1
DEP VARIABLE: WEIGHT

                       SUM OF          MEAN
SOURCE     DF         SQUARES        SQUARE       F VALUE       PROB>F

MODEL       1       21506.523     21506.523       141.094       0.0001
ERROR     109       16614.585       152.427
C TOTAL   110       38121.108

    ROOT MSE        12.346149      R-SQUARE        0.5642
    DEP MEAN        98.878378      ADJ R-SQ        0.5602
    C.V.              12.4862

                    PARAMETER      STANDARD     T FOR H0:
VARIABLE   DF        ESTIMATE         ERROR   PARAMETER=0   PROB > |T|

INTERCEP    1        -153.129     21.248143        -7.207       0.0001
HEIGHT      1        4.163612      0.350523        11.878       0.0001
```

```
---------- DATA ON AGE, WEIGHT, AND HEIGHT OF CHILDREN ----------                    2
                                 SEX=F

MODEL: EQ2
DEP VARIABLE: WEIGHT

                      SUM OF          MEAN
SOURCE       DF      SQUARES        SQUARE       F VALUE       PROB>F

MODEL         2    22432.272     11216.136        77.210       0.0001
ERROR       108    15688.836       145.267
C TOTAL     110    38121.108

     ROOT MSE      12.052676      R-SQUARE        0.5884
     DEP MEAN      98.878378      ADJ R-SQ        0.5808
     C.V.           12.18939

                   PARAMETER      STANDARD    T FOR H0:
VARIABLE     DF     ESTIMATE         ERROR  PARAMETER=0    PROB > |T|

INTERCEP      1     -150.597     20.767300        -7.252       0.0001
HEIGHT        1     3.603780      0.407768         8.838       0.0001
AGE           1     1.907026      0.755428         2.524       0.0130
```

```
---------- DATA ON AGE, WEIGHT, AND HEIGHT OF CHILDREN ----------                    3
                                 SEX=M

MODEL: EQ1
DEP VARIABLE: WEIGHT

                      SUM OF          MEAN
SOURCE       DF      SQUARES        SQUARE       F VALUE       PROB>F

MODEL         1    31126.060     31126.060       206.239       0.0001
ERROR       124    18714.355       150.922
C TOTAL     125    49840.415

     ROOT MSE      12.285040      R-SQUARE        0.6245
     DEP MEAN        103.448      ADJ R-SQ        0.6215
     C.V.           11.87552

                   PARAMETER      STANDARD    T FOR H0:
VARIABLE     DF     ESTIMATE         ERROR  PARAMETER=0    PROB > |T|

INTERCEP      1     -125.698     15.993625        -7.859       0.0001
HEIGHT        1     3.689771      0.256929        14.361       0.0001
```

```
---------- DATA ON AGE, WEIGHT, AND HEIGHT OF CHILDREN ----------                    4
                                 SEX=M

MODEL: EQ2
DEP VARIABLE: WEIGHT

                      SUM OF          MEAN
SOURCE       DF      SQUARES        SQUARE       F VALUE       PROB>F

MODEL         2    32974.750     16487.375       120.241       0.0001
ERROR       123    16865.664       137.119
C TOTAL     125    49840.415

     ROOT MSE      11.709792      R-SQUARE        0.6616
     DEP MEAN        103.448      ADJ R-SQ        0.6561
     C.V.           11.31945

                   PARAMETER      STANDARD    T FOR H0:
VARIABLE     DF     ESTIMATE         ERROR  PARAMETER=0    PROB > |T|

INTERCEP      1     -113.713     15.590214        -7.294       0.0001
HEIGHT        1     2.680749      0.368091         7.283       0.0001
AGE           1     3.081672      0.839274         3.672       0.0004
```

---------- DATA ON AGE, WEIGHT, AND HEIGHT OF CHILDREN ---------- 5

SSCP TYPE DATA SET

OBS	SEX	_NAME_	INTERCEP	WEIGHT	HEIGHT	AGE
1	F	WEIGHT	10975.5	1123361	669470	182445
2	F	HEIGHT	6718.4	669470	407879	110818
3	F	AGE	1824.9	182445	110818	30364
4	F	INTERCEP	111.0	10976	6718	1825
5	M	WEIGHT	13034.5	1398239	817920	217717
6	M	HEIGHT	7825.0	817920	488244	129433
7	M	AGE	2072.1	217717	129433	34516
8	M	INTERCEP	126.0	13035	7825	2072

---------- DATA ON AGE, WEIGHT, AND HEIGHT OF CHILDREN ---------- 6

EST TYPE DATA SET

OBS	SEX	_TYPE_	_MODEL_	_DEPVAR_	_SIGMA_	WEIGHT	HEIGHT	AGE	INTERCEP
1	F	OLS	EQ1	WEIGHT	12.3461	-1	4.16361		-153.13
2	F	OLS	EQ2	WEIGHT	12.0527	-1	3.60378	1.90703	-150.60
3	M	OLS	EQ1	WEIGHT	12.2850	-1	3.68977	.	-125.70
4	M	OLS	EQ2	WEIGHT	11.7098	-1	2.68075	3.08167	-113.71

REFERENCES

Allen, D.M. (1971), "Mean Square Error of Prediction as a Criterion for Selecting Variables," *Technometrics*, 13, 469–475.

Allen, D.M. and Cady, F.B. (1982), *Analyzing Experimental Data by Regression*, Belmont, CA: Lifetime Learning Publications.

Belsley, D.A., Kuh, E., and Welsch, R.E. (1980), *Regression Diagnostics*, New York: John Wiley & Sons.

Bock, R.D. (1975), *Multivariate Statistical Methods in Behavioral Research*, New York: McGraw-Hill.

Box, G.E.P. (1966), "The Use and Abuse of Regression," *Technometrics*, 8, 625–629.

Cook, R.D. (1977), "Detection of Influential Observations in Linear Regression," *Technometrics*, 19, 15–18.

Cook, R.D. (1979), "Influential Observations in Linear Regression," *Journal of the American Statistical Association*, 74, 169–174.

Daniel, C. and Wood, F. (1980), *Fitting Equations to Data*, Revised Edition, New York: John Wiley & Sons.

Draper, N. and Smith, H. (1981), *Applied Regression Analysis*, Second Edition, New York: John Wiley & Sons.

Durbin, J. and Watson, G.S. (1951), "Testing for Serial Correlation in Least Squares Regression," *Biometrika*, 37, 409–428.

Gauss, K.F. (1809), *Werke*, 4, 1–93.

Goodnight, J.H. (1979), "A Tutorial on the SWEEP Operator," *The American Statistician*, 33, 149–158.

Grunfeld, Y. (1958), "The Determinants of Corporate Investment," Unpublished Thesis, Chicago, discussed in Boot, J.C.G. (1960), "Investment Demand: An Empirical Contribution to the Aggregation Problem," *International Economic Review*, 1, 3–30.

Johnston, J. (1972), *Econometric Methods*, New York: McGraw-Hill.

Kennedy, W.J. and Gentle, J.E. (1980), *Statistical Computing*, New York: Marcel Dekker.

Lewis, T. and Taylor, L.R. (1967), *Introduction to Experimental Ecology*, New York: Academic Press.

Mallows, C.L. (1973), "Some Comments on Cp," *Technometrics*, 15, 661–675.

Mardia, K.V., Kent, J.T., and Bibby, J.M. (1979), *Multivariate Analysis*, London: Academic Press.

Markov, A.A. (1900), *Wahrscheinlichkeitscrechnung*, Tebrer, Leipzig.

Morrison, D.F. (1976), *Multivariate Statistical Methods*, Second Edition, New York: McGraw-Hill.

Mosteller, F. and Tukey, J.W. (1977), *Data Analysis and Regression*, Reading, MA: Addison-Wesley.

Neter, J. and Wasserman, W. (1974), *Applied Linear Statistical Models*, Homewood, IL: Irwin.

Pillai, K.C.S. (1960), *Statistical Table for Tests of Multivariate Hypotheses*, Manila: The Statistical Center, University of Philippines.

Pindyck, R.S. and Rubinfeld, D.L. (1981), *Econometric Models and Econometric Forecasts*, Second Edition, New York: McGraw-Hill.

Pringle, R.M. and Raynor, A.A. (1971), *Generalized Inverse Matrices with Applications to Statistics*, New York: Hafner Publishing Company.

Rao, C.R. (1973), *Linear Statistical Inference and Its Applications*, Second Edition, New York: John Wiley & Sons.

Sall, J.P. (1981), "SAS Regression Applications," Revised Edition, SAS Technical Report A-102, Cary, NC: SAS Institute.

Timm, N.H. (1975), *Multivariate Analysis with Applications in Education and Psychology*, Monterey, CA: Brooks/Cole.

Weisberg, S. (1980), *Applied Linear Regression*, New York: John Wiley & Sons.

The RSQUARE Procedure

ABSTRACT

The RSQUARE procedure performs all possible regressions for one or more dependent variables and a collection of independent variables, printing the R^2 value and, optionally, Mallow's C_p statistic for each model.

INTRODUCTION

RSQUARE is useful when you want to investigate the behavior of many regression models. RSQUARE evaluates each combination of a dependent variable with the independent variables. If k independent variables are specified, RSQUARE evaluates each of the 2^{k-1} linear models: k of the models include one independent variable, $k(k-1)/2$ of the models include two independent variables, and so on. For each model evaluated, RSQUARE prints the unadjusted R^2 value and Mallow's C_p statistics if requested.

SPECIFICATIONS

The statements for RSQUARE are:

PROC RSQUARE *options*;
  MODEL *dependents=independents/options*;
  BY *variables*;

Many MODEL statements are allowed. The BY statement can be placed anywhere.

PROC RSQUARE Statement

PROC RSQUARE *options*;

The following options may be specified on the PROC statement:

DATA=*SASdataset* names the SAS data set to be used. If the DATA= option is omitted RSQUARE uses the most recently created SAS data set.

CP prints Mallow's C_p statistic (Hocking, 1976) for each model. See PROC STEPWISE for further discussion of C_p.

MODEL Statement

MODEL *dependents=independents/options;*

The MODEL statement specifies the variables to use in the analysis. On the left side of the equal sign specify the dependent variable or variables; on the right side of the equal sign specify the independent variables.

When more than one dependent variable is used, RSQUARE performs separate analyses for each dependent variable specified. No multivariate analyses are performed.

Any number of MODEL statements may follow the PROC RSQUARE statement.
The options below may appear in the MODEL statement after a slash (/).

NOINT suppresses the intercept parameter that is normally added automatically to the model.

START=*n* specifies the smallest number of independent variables you want reported in the model. Otherwise all one-variable models, all two-variable models, and so on are considered.

STOP=*n* specifies the largest number of independent variables you want reported in the model. Otherwise, RSQUARE continues to evaluate models until all the independent variables are used in a model.

PRINT=*p* requests printed results for only the *p* models of each size that produce the highest R^2. When PRINT= is omitted, RSQUARE prints results for all the $n!/(k!(n\text{-}k)!)$ models of a given size *k*, where *n* is the number of independent variables, unless more than ten independent variables are included in the MODEL statement. For more than ten variables PRINT=*n* is implied automatically.

INCLUDE=*i* requests that the first *i* variables after the equal sign in the MODEL statement always be included in the regression model; the other variables are then switched in and out.

BY Statement

BY *variables;*

A BY statement may be used with PROC RSQUARE to obtain separate analyses on observations in groups defined by the BY variables. When a BY statement appears, the procedure expects the input data set to be sorted in order of the BY variables. If your input data set is not sorted in ascending order, use the SORT procedure with a similar BY statement to sort the data, or, if appropriate, use the BY statement options NOTSORTED or DESCENDING. For more information, see the discussion of the BY statement in Chapter 8, "Statements Used in the PROC Step," in *SAS User's Guide: Basics, 1982 Edition*.

DETAILS

Missing Values

When RSQUARE is evaluating a model, it does not include observations that have a missing value for any of the variables in the model. However, these observations are used for models that do not include the variables with the missing values.

Limitations

There is no built-in limit on the number of independent variables, but the calculations for a large number of independent variables can be lengthy. Therefore, we recommend that no more than 14 independent variables be used for a single analysis.

If the INCLUDE option is used to include certain variables in all the models, the recommended limit is $14+i$, where i is the INCLUDE value.

There is no limit on the number of dependent variables that may be included in a MODEL statement.

Printed Output

RSQUARE prints its results beginning with the model containing the fewest independent variables and producing the smallest R^2. Results for other models with the same number of variables are then printed in order of increasing R^2, and so on for models with larger numbers of variables.

For each model considered, RSQUARE prints:

1. NUMBER in MODEL, the number of independent variables used in each model
2. RSQUARE. R^2, the square of the multiple correlation coefficient, can be expressed as the ratio of the regression sum of squares to the corrected total sum of squares is not adjusted for degrees of freedom
3. C(P), Mallow's C_p if the CP option is used
4. VARIABLES IN MODEL, the names of the independent variables included in the model.

EXAMPLES

All Possible Regressions: Example 1

The example below requests R^2 and C_p statistics for all possible combinations of the six independent variables.

```
* --------------------------------DATA ON PHYSICAL FITNESS-------------------------------- *
|  THESE MEASUREMENTS WERE MADE ON MEN INVOLVED IN A PHYSICAL FITNESS                      |
|  COURSE AT N.C. STATE UNIV. THE VARIABLES ARE AGE(YEARS), WEIGHT(KG),                    |
|  OXYGEN UPTAKE RATE(ML PER KG BODY WEIGHT PER MINUTE), TIME TO RUN                       |
|  1.5 MILES(MINUTES), HEART RATE WHILE RESTING, HEART RATE WHILE                          |
|  RUNNING (SAME TIME OXYGEN RATE MEASURED), AND MAXIMUM HEART RATE                        |
|  RECORDED WHILE RUNNING. CERTAIN VALUES OF MAXPULSE WERE MODIFIED                        |
|  FOR CONSISTENCY. DATA COURTESY DR. A.C. LINNERUD                                        |
* ---------------------------------------------------------------------------------------- *  ;
```

```
DATA FITNESS;
  INPUT AGE WEIGHT OXY RUNTIME RSTPULSE RUNPULSE MAXPULSE @@;
  CARDS;
44 89.47 44.609 11.37 62 178 182     40 75.07 45.313 10.07 62 185 185
44 85.84 54.297  8.65 45 156 168     42 68.15 59.571  8.17 40 166 172
38 89.02 49.874  9.22 55 178 180     47 77.45 44.811 11.63 58 176 176
40 75.98 45.681 11.95 70 176 180     43 81.19 49.091 10.85 64 162 170
44 81.42 39.442 13.08 63 174 176     38 81.87 60.055  8.63 48 170 186
44 73.03 50.541 10.13 45 168 168     45 87.66 37.388 14.03 56 186 192
45 66.45 44.754 11.12 51 176 176     47 79.15 47.273 10.60 47 162 164
54 83.12 51.855 10.33 50 166 170     49 81.42 49.156  8.95 44 180 185
51 69.63 40.836 10.95 57 168 172     51 77.91 46.672 10.00 48 162 168
48 91.63 46.774 10.25 48 162 164     49 73.37 50.388 10.08 67 168 168
57 73.37 39.407 12.63 58 174 176     54 79.38 46.080 11.17 62 156 165
52 76.32 45.441  9.63 48 164 166     50 70.87 54.625  8.92 48 146 155
51 67.25 45.118 11.08 48 172 172     54 91.63 39.203 12.88 44 168 172
51 73.71 45.790 10.47 59 186 188     57 59.08 50.545  9.93 49 148 155
49 76.32 48.673  9.40 56 186 188     48 61.24 47.920 11.50 52 170 176
52 82.78 47.467 10.50 53 170 172
;
PROC RSQUARE CP;
  MODEL OXY=AGE WEIGHT RUNTIME RUNPULSE RSTPULSE MAXPULSE;
```

```
                         S T A T I S T I C A L   A N A L Y S I S   S Y S T E M                       1

N=      31      REGRESSION MODELS FOR DEPENDENT VARIABLE OXY

NUMBER IN  (2) R-SQUARE        (3) C(P)     VARIABLES IN MODEL
(1) MODEL                              (4)

     1          0.02648849    127.39484639    WEIGHT
     1          0.05604592    122.70716165    MAXPULSE
     1          0.09277653    116.88184063    AGE
     1          0.15838344    106.47686032    RUNPULSE
     1          0.15948531    106.30210845    RSTPULSE
     1          0.74338010     13.69884048    RUNTIME
-----------------------------------------------------
     2          0.06751590    122.88807048    WEIGHT MAXPULSE
     2          0.15063534    109.70567678    AGE WEIGHT
     2          0.16685536    107.13324845    WEIGHT RUNPULSE
     2          0.17403933    105.99390096    RSTPULSE MAXPULSE
     2          0.18060672    104.95234098    WEIGHT RSTPULSE
     2          0.23503072     96.32092307    RUNPULSE RSTPULSE
     2          0.25998174     92.36379575    AGE MAXPULSE
     2          0.28941948     87.69509301    RUNPULSE MAXPULSE
     2          0.30027026     85.97420424    AGE RSTPULSE
     2          0.37599543     73.96451003    AGE RUNPULSE
     2          0.74353296     15.67459836    RUNTIME RSTPULSE
     2          0.74493479     15.45227429    WEIGHT RUNTIME
     2          0.74522106     15.40687170    RUNTIME MAXPULSE
     2          0.76142381     12.83718362    RUNTIME RUNPULSE
     2          0.76424693     12.38944895    AGE RUNTIME
-----------------------------------------------------
     3          0.18823207    105.74299140    WEIGHT RSTPULSE MAXPULSE
     3          0.24465116     96.79516014    WEIGHT RUNPULSE RSTPULSE
     3          0.29021246     89.56932988    AGE WEIGHT MAXPULSE
     3          0.32077932     84.72155302    WEIGHT RUNPULSE MAXPULSE
     3          0.35377183     79.48908047    RUNPULSE RSTPULSE MAXPULSE
     3          0.35684729     79.00132416    AGE WEIGHT RSTPULSE
     3          0.39000680     73.74236530    AGE RSTPULSE MAXPULSE
     3          0.40912553     70.71021483    AGE WEIGHT RUNPULSE
     3          0.42227346     68.62500839    AGE RUNPULSE MAXPULSE
     3          0.46664844     61.58732295    AGE RUNPULSE RSTPULSE
     3          0.74511138     17.42426679    WEIGHT RUNTIME RSTPULSE
     3          0.74522683     17.40595758    RUNTIME RSTPULSE MAXPULSE
     3          0.74615485     17.25877645    WEIGHT RUNTIME MAXPULSE
     3          0.76182904     14.77291649    WEIGHT RUNTIME RUNPULSE
     3          0.76189848     14.76190332    RUNTIME RUNPULSE RSTPULSE
     3          0.76734943     13.89740598    AGE RUNTIME RSTPULSE
     3          0.77083060     13.34530613    AGE WEIGHT RUNTIME
     3          0.78173017     11.61668036    AGE RUNTIME MAXPULSE
     3          0.80998844      7.13503673    RUNTIME RUNPULSE MAXPULSE
     3          0.81109446      6.95962673    AGE RUNTIME RUNPULSE
-----------------------------------------------------
     4          0.38579687     76.41004295    WEIGHT RUNPULSE RSTPULSE MAXPULSE
     4          0.42560710     70.09630639    AGE WEIGHT RSTPULSE MAXPULSE
     4          0.47171966     62.78304777    AGE WEIGHT RUNPULSE MAXPULSE
     4          0.50245083     57.90921343    AGE RUNPULSE RSTPULSE MAXPULSE
     4          0.50339774     57.75903763    AGE WEIGHT RUNPULSE RSTPULSE
     4          0.74617854     19.25501923    WEIGHT RUNTIME RSTPULSE MAXPULSE
```

```
                     S T A T I S T I C A L   A N A L Y S I S   S Y S T E M                          2

N=      31      REGRESSION MODELS FOR DEPENDENT VARIABLE OXY

NUMBER IN       R-SQUARE            C(P)    VARIABLES IN MODEL
  MODEL

    4         0.76225238     16.70577655    WEIGHT RUNTIME RUNPULSE RSTPULSE
    4         0.77503285     14.67884787    AGE WEIGHT RUNTIME RSTPULSE
    4         0.78343214     13.34675469    AGE RUNTIME RSTPULSE MAXPULSE
    4         0.78622430     12.90393067    AGE WEIGHT RUNTIME MAXPULSE
    4         0.81040041      9.06969999    RUNTIME RUNPULSE RSTPULSE MAXPULSE
    4         0.81167015      8.86832440    AGE RUNTIME RUNPULSE RSTPULSE
    4         0.81584902      8.20557304    WEIGHT RUNTIME RUNPULSE MAXPULSE
    4         0.81649255      8.10351211    AGE WEIGHT RUNTIME RUNPULSE
    4         0.83681815      4.87995808    AGE RUNTIME RUNPULSE MAXPULSE
--------------------------------------------------------------------------------
    5         0.55406593     51.72327517    AGE WEIGHT RUNPULSE RSTPULSE MAXPULSE
    5         0.78870109     14.51112242    AGE WEIGHT RUNTIME RSTPULSE MAXPULSE
    5         0.81608280     10.16849716    WEIGHT RUNTIME RUNPULSE RSTPULSE MAXPULSE
    5         0.81755611      9.93483665    AGE WEIGHT RUNTIME RUNPULSE RSTPULSE
    5         0.83703132      6.84614970    AGE RUNTIME RUNPULSE RSTPULSE MAXPULSE
    5         0.84800181      5.10627546    AGE WEIGHT RUNTIME RUNPULSE MAXPULSE
--------------------------------------------------------------------------------------
    6         0.84867192      7.00000000    AGE WEIGHT RUNTIME RUNPULSE RSTPULSE MAXPULSE
----------------------------------------------------------------------------------------------
```

Including a Variable in All Regressions: Example 2

Now the independent variable AGE is included in all the models by listing it first among the independent variables and specifying INCLUDE=1 in the MODEL statement.

```
PROC RSQUARE CP;
  MODEL OXY=AGE WEIGHT RUNTIME RUNPULSE RSTPULSE MAXPULSE
    / INCLUDE=1;
```

```
                     S T A T I S T I C A L   A N A L Y S I S   S Y S T E M                          3

N=      31      REGRESSION MODELS FOR DEPENDENT VARIABLE OXY

NUMBER IN       R-SQUARE            C(P)    VARIABLES IN MODEL
  MODEL

    1         0.09277653    116.88184063    AGE

NOTE: THE ABOVE VARIABLES ARE INCLUDED IN ALL MODELS TO FOLLOW.

    2         0.15063534    109.70567678    WEIGHT
    2         0.25998174     92.36379575    MAXPULSE
    2         0.30027026     85.97420424    RSTPULSE
    2         0.37599543     73.96451003    RUNPULSE
    2         0.76424693     12.38944895    RUNTIME
---------------------------------------------------------
    3         0.29021246     89.56932988    WEIGHT MAXPULSE
    3         0.35684729     79.00132416    WEIGHT RSTPULSE
    3         0.39000680     73.74236530    RSTPULSE MAXPULSE
    3         0.40912553     70.71021483    WEIGHT RUNPULSE
    3         0.42227346     68.62500839    RUNPULSE MAXPULSE
    3         0.46664844     61.58732295    RUNPULSE RSTPULSE
    3         0.76734943     13.89740598    RUNTIME RSTPULSE
    3         0.77083060     13.34530613    WEIGHT RUNTIME
    3         0.78173017     11.61668036    RUNTIME MAXPULSE
    3         0.81109446      6.95962673    RUNTIME RUNPULSE
------------------------------------------------------------------
    4         0.42560710     70.09630639    WEIGHT RSTPULSE MAXPULSE
    4         0.47171966     62.78304777    WEIGHT RUNPULSE MAXPULSE
    4         0.50245083     57.90921343    RUNPULSE RSTPULSE MAXPULSE
    4         0.50339774     57.75903763    WEIGHT RUNPULSE RSTPULSE
    4         0.77503285     14.67884787    WEIGHT RUNTIME RSTPULSE
    4         0.78343214     13.34675469    RUNTIME RSTPULSE MAXPULSE
    4         0.78622430     12.90393067    WEIGHT RUNTIME MAXPULSE
    4         0.81167015      8.86832440    RUNTIME RUNPULSE RSTPULSE
    4         0.81649255      8.10351211    WEIGHT RUNTIME RUNPULSE
    4         0.83681815      4.87995808    RUNTIME RUNPULSE MAXPULSE
------------------------------------------------------------------------
    5         0.55406593     51.72327517    WEIGHT RUNPULSE RSTPULSE MAXPULSE
    5         0.78870109     14.51112242    WEIGHT RUNTIME RSTPULSE MAXPULSE
    5         0.81755611      9.93483665    WEIGHT RUNTIME RUNPULSE RSTPULSE
    5         0.83703132      6.84614970    RUNTIME RUNPULSE RSTPULSE MAXPULSE
    5         0.84800181      5.10627546    WEIGHT RUNTIME RUNPULSE MAXPULSE
--------------------------------------------------------------------------------
    6         0.84867192      7.00000000    WEIGHT RUNTIME RUNPULSE RSTPULSE MAXPULSE
-----------------------------------------------------------------------------------------
```

Three- and Four-Variable Models Only: Example 3

In this example, only R^2 statistics for the three- and four-variable models are requested. The START and STOP options are used to limit the calculations.

```
PROC RSQUARE CP;
   MODEL OXY=AGE WEIGHT RUNTIME RUNPULSE RSTPULSE MAXPULSE
      /START=3 STOP=4;
```

```
                       S T A T I S T I C A L   A N A L Y S I S   S Y S T E M                    4

N=      31      REGRESSION MODELS FOR DEPENDENT VARIABLE OXY

NUMBER IN       R-SQUARE            C(P)    VARIABLES IN MODEL
  MODEL

     3         0.18823207    105.74299140   WEIGHT RSTPULSE MAXPULSE
     3         0.24465116     96.79516014   WEIGHT RUNPULSE RSTPULSE
     3         0.29021246     89.56932988   AGE WEIGHT MAXPULSE
     3         0.32077932     84.72155302   WEIGHT RUNPULSE MAXPULSE
     3         0.35377183     79.48908047   RUNPULSE RSTPULSE MAXPULSE
     3         0.35684729     79.00132416   AGE WEIGHT RSTPULSE
     3         0.39000680     73.74236530   AGE RSTPULSE MAXPULSE
     3         0.40912553     70.71021483   AGE WEIGHT RUNPULSE
     3         0.42227346     68.62500839   AGE RUNPULSE MAXPULSE
     3         0.46664844     61.58732295   AGE RUNPULSE RSTPULSE
     3         0.74511138     17.42426679   WEIGHT RUNTIME RSTPULSE
     3         0.74522683     17.40595758   RUNTIME RSTPULSE MAXPULSE
     3         0.74615485     17.25877645   WEIGHT RUNTIME MAXPULSE
     3         0.76182904     14.77291649   WEIGHT RUNTIME RUNPULSE
     3         0.76189848     14.76190332   RUNTIME RUNPULSE RSTPULSE
     3         0.76734943     13.89740598   AGE RUNTIME RSTPULSE
     3         0.77083060     13.34530613   AGE WEIGHT RUNTIME
     3         0.78173017     11.61668036   AGE RUNTIME MAXPULSE
     3         0.80998844      7.13503673   RUNTIME RUNPULSE MAXPULSE
     3         0.81109446      6.95962673   AGE RUNTIME RUNPULSE
-----------------------------------------------------------------------
     4         0.38579687     76.41004295   WEIGHT RUNPULSE RSTPULSE MAXPULSE
     4         0.42560710     70.09630639   AGE WEIGHT RSTPULSE MAXPULSE
     4         0.47171966     62.78304777   AGE WEIGHT RUNPULSE MAXPULSE
     4         0.50245083     57.90921343   AGE RUNPULSE RSTPULSE MAXPULSE
     4         0.50339774     57.75903763   AGE WEIGHT RUNPULSE RSTPULSE
     4         0.74617854     19.25501923   WEIGHT RUNTIME RSTPULSE MAXPULSE
     4         0.76225238     16.70577655   WEIGHT RUNTIME RUNPULSE RSTPULSE
     4         0.77503285     14.67884787   AGE WEIGHT RUNTIME RSTPULSE
     4         0.78343214     13.34675469   AGE RUNTIME RSTPULSE MAXPULSE
     4         0.78622430     12.90393067   AGE WEIGHT RUNTIME MAXPULSE
     4         0.81040041      9.06969999   RUNTIME RUNPULSE RSTPULSE MAXPULSE
     4         0.81167015      8.86832440   AGE RUNTIME RUNPULSE RSTPULSE
     4         0.81584902      8.20557304   WEIGHT RUNTIME RUNPULSE MAXPULSE
     4         0.81649255      8.10351211   AGE WEIGHT RUNTIME RUNPULSE
     4         0.83681815      4.87995808   AGE RUNTIME RUNPULSE MAXPULSE
-----------------------------------------------------------------------
```

REFERENCES

Cuthbert, D. and Wood, F.S. (1980), *Fitting Equations to Data*, Second Edition. New York: John Wiley and Sons, Inc.

Draper, N.R. and Smith, H. (1981), *Applied Regression Analysis,* Second Edition, New York: John Wiley and Sons, Inc.

Hocking, R.R. (1976), "The Analysis and Selection of Variables in a Linear Regression," *Biometrics*, 32, 1-50.

Mallows, C.L. (1964), "Some Comments on C(p)," *Technometrics*, 15, 661-675.

Shatzoff, M., Tsao, R., and Fienberg, S., (1968), "Efficient Calculation of All Possible Regression," *Technometrics*, 10, 769-779.

The RSREG Procedure

ABSTRACT

The RSREG procedure fits the parameters of a complete quadratic response surface and then determines critical values to optimize the response with respect to the factors in the model.

INTRODUCTION

Many industrial experiments are conducted to discover which factor values optimize a response. If each factor variable is measured at three or more values, a quadratic response surface can be estimated by least-squares regression. The predicted optimal value can be found from the estimated surface if the surface is shaped appropriately.

Suppose that a response variable y is measured at combinations of values of two factor variables, x_1 and x_2. The quadratic response-surface model for this is written:

$$y = \beta_0 + \beta_1 x_1 + \beta_2 x_2 + \beta_3 x_1^2 + \beta_4 x_2^2 + \beta_5 x_1 x_2 + \varepsilon.$$

The parameters in the model are estimated by least-squares regression using the statements

```
PROC RSREG;
  MODEL Y=X1 X2;
```

The results from RSREG can answer these questions:

1. How much does each type of effect contribute to the statistical fit? (The types are linear, quadratic, and crossproduct.)
2. Is part of the residual error due to lack-of-fit? Does the quadratic response model adequately represent the true response surface?
3. How much does each factor variable contribute to the statistical fit? Can the response be predicted as well if the variable is removed?
4. What combination of factor values yields the maximum or minimum response? Where is the optimum?
5. Is the surface shaped like a hill, a valley, a saddle-surface, or a flat surface?
6. For a grid of factor values, what are the predicted responses? Use them for plotting or selecting purposes.

Other procedures in SAS can be used for these response-surface problems, but RSREG is more specialized. The RSREG MODEL statement

```
MODEL Y=X1 X2 X3;
```

is more compact than the MODEL statement for other regression procedures in SAS. For example, GLM's MODEL statement appears below:

```
MODEL Y=X1 X1*X1
        X2 X1*X2 X2*X2
        X3 X1*X3 X2*X3 X3*X3;
```

Variables are used according to the following conventions:

factor variables	the factor variables for which the quadratic response surface is constructed. Variables must be numeric. For the necessary parameters to be estimated, each variable should have at least three distinct values in the data.
response variables	the response (or dependent) variables must be numeric.
covariates	additional independent variables to be included in the regression but not considered in the response surface.
WEIGHT variable	a numeric variable for weighting the observations in the regression.
ID variables	variables not in the above list that you want transferred to an output data set.

SPECIFICATIONS

The RSREG procedure allows one of each of the following statements:

PROC RSREG *options*;
  MODEL *response=independents / options*;
  WEIGHT *variable*;
  ID *variables*;
  BY *variables*;

PROC RSREG Statement

PROC RSREG *options*;

The PROC RSREG statement can have the following options:

DATA=*SASdataset*	specifies the input SAS data set that contains the data to analyze. If not specified, RSREG uses the most recently created SAS data set.
OUT=*SASdataset*	names an output SAS data set to contain the BY variables, ID variables, WEIGHT variable, and variables in the MODEL statement. If you do not specify OUT=, the default name is OUT=_DATA_, which produces a name like DATA*n*. If you want to create a permanent SAS data set, you must specify a two-level name (see Chapter 12, "SAS Data Sets," in *SAS User's Guide: Basics, 1982 Edition* for more information on permanent SAS data sets). For details on the data set created by RSREG, see **Output Data Set** below.
NOPRINT	suppresses the printed results when you want an output data set only.

MODEL Statement

The MODEL statement has the following form:

MODEL *response*=*independents* / *options*;

where any of the following options can be specified:

LACKFIT	specifies a lack-of-fit test is to be performed. If you specify LACKFIT, you must first sort your data on the independent variables so that observations repeating the same values are grouped together.
NOOPT NOOPTIMAL	suppresses the feature that finds the critical values for the quadratic response surface.
COVAR=*n* COVARIATES=*n*	declares that the first *n* variables on the independent side of the model are simple regressors (covariates) rather than factors in the quadratic response surface. If you do not specify COVAR, then RSREG will form quadratic and crossproduct effects for all regressor variables in the MODEL statement.

Output options The following options control which types of statistics are output to the OUT= data set. The option keywords become values of the special variable _TYPE_ on the output data set.

ACTUAL	the actual values from the input data set
PREDICT	the values predicted by the model
RESIDUAL	residual = actual - predicted
U95M	upper 95% confidence limit for mean
L95M	lower 95% confidence limit for mean
U95	upper 95% confidence limit for prediction
L95	lower 95% confidence limit for prediction
D	Cook's *D* influence statistic
BYOUT	requests that only the first BY group be used to estimate the model. Subsequent BY groups have scoring statistics computed on the output data set only. BYOUT is used only when a BY statement is specified.

WEIGHT Statement

WEIGHT *variable*;

The WEIGHT statement names a numeric variable on the input data set whose values are to be used as relative weights on the observations to produce weighted least-squares estimates.

ID Statement

ID *variables*;

The ID statement names variables that you want to transfer to the output data set in addition to the variables mentioned in other statements.

BY Statement

BY *variables*;

A BY statement may be used with PROC RSREG to obtain separate analyses on observations in groups defined by the BY variables. When a BY statement appears, the procedure expects the input data set to be sorted in order of the BY variables. If your input data set is not sorted in ascending order, use the SORT procedure with a similar BY statement to sort the data, or, if appropriate, use the BY statement options NOTSORTED or DESCENDING. For more information, see the discussion of the BY statement in Chapter 8, "Statements Used in the PROC Step," in *SAS User's Guide: Basics, 1982 Edition*.

DETAILS

Missing Values

If an observation has missing data for any of the variables used by the procedure, then that observation is not used in the estimation process. If one or more response variables are missing, but no factor or covariate variables are missing, then predicted values and confidence limits are computed for the output data set, but the residual and *D* statistic are missing.

Output Data Set

An output data set is created whenever OUT= is specified on the PROC RSREG statement. The data set contains the following variables:

- the BY variables
- the ID variables
- the WEIGHT variable
- the independent variables in the MODEL statement
- the variable _TYPE_ which identifies the observation type on the output data set. _TYPE_ is a character variable of length 8 and takes on the values ACTUAL, PREDICT, RESIDUAL, U95M, L95M, U95, L95, and D.
- the response variables containing special output values identified by the _TYPE_ variable.

All confidence limits use the two-tailed Student's *t* value.

Lack-of-Fit Test

If the LACKFIT option is specified, the data should be sorted so that repeated observations appear together. If the data are not sorted, the procedure cannot find these repeats. Since all other test statistics for the model are tested by total error rather than pure error, you may want to hand-calculate the tests with respect to pure error if the lack-of-fit is significant.

Plotting the Surface

You can generate predicted values for a grid of points to be plotted with the PREDICT option. (See example below.) Contour plots are possible for only two factor variables at a time, with other factor variables held to constant values. First, form a grid of points in the input data set using DO loops after the end-of-file. Response variables should be set to missing so that the grid data do not affect the estimates. The OUT= feature and PREDICT option create the output data set from PROC RSREG. The data set is then subset and used with the PLOT procedure to plot the contours in the grid. See the **EXAMPLE** later in this chapter.

Searching for Multiple Response Conditions

Suppose you want to find the factor setting that produces responses in a certain region. For example, you want to find the values of x_1 and x_2 that maximize y_1 subject to $y_2<2$ and $y_3<y_2+y_1$. The exact answer is not easy to obtain analytically, but brute force can be applied. Approach the problem by checking conditions across a grid of values in the range of interest.

```
DATA B;  SET A END=EOF;  OUTPUT;
  IF EOF THEN DO;  Y1=.;  Y2=.;  Y3=.;
    DO X1=1 TO 5 BY .1;
      DO X2=1 TO 5 BY .1;
        OUTPUT;
        END;
      END;
    END;
PROC RSREG DATA=B OUT=C;
  MODEL Y1 Y2 Y3=X1 X2 / PREDICT;
DATA D;  SET C;
  IF Y2<2;  IF Y3<Y2+Y1;
PROC SORT DATA=D;  BY DESCENDING Y1;
PROC PRINT;
```

Computational Method

The model can be written in the form:

$$y_i = x_i' \mathbf{A} x_i + \mathbf{b}' x_i + \mathbf{c}' z_i + \varepsilon_i$$

where

- y_i is the ith observation on the response variable
- $x_i = (x_{i1}, x_{i2}, \ldots, x_{ik})$ are the k factor variables for the ith observation
- $z_i = (z_{i1}, z_{i2}, \ldots, z_{iL})$ are the L covariate variables including the intercept term
- $\mathbf{A}$ is the k by k matrix of quadratic parameters
- $\mathbf{b}$ is the k by 1 vector of linear parameters
- $\mathbf{c}$ is the L by 1 vector of covariate parameters.

The parameters in **A**, **b**, and **c** are estimated by least squares. To optimize y with respect to x, take partial derivatives, set them to zero, and solve. Making **A** symmetric, this is written:

$$\partial y / \partial x = 2x' \mathbf{A} + \mathbf{b}' = 0$$
$$x = -.5\mathbf{A}^{-1}\mathbf{b}$$

To determine if the solution of critical values is a maximum or minimum, find out if **A** is negative or positive definite by looking at the eigenvalues of **A**.

if eigenvalues	then solution is
are all negative	maximum
are all positive	minimum
have mixed signs	saddle-point
contain zeros	in a flat area

The eigenvectors are also printed. The eigenvector for the largest eigenvalue gives the direction of steepest ascent, if positive, or steepest descent, if negative. The eigenvectors corresponding to small or zero eigenvalues point in directions of relative flatness.

Printed Output

All estimates and hypothesis tests depend on the correctness of the model and the error distributed according to classical statistical assumptions.

The individual items in the output from RSREG are:

1. RESPONSE MEAN is the mean of the response variable in the sample.
2. ROOT MSE estimates the standard deviation of the response variable by the square root of the TOTAL ERROR mean square.
3. R-SQUARE is R^2, or the coefficient of determination. R^2 measures the portion of the variation in the response that is attributed to the model rather than to random error.
4. COEF OF VARIATION is the coefficient of variation, which is equal to 100*rootmse/mean for the response variable.
5. Terms are brought into the regression in four steps: (1) INTERCEPT and COVARIATES (not shown), (2) LINEAR terms like X1 and X2, (3) QUADRATIC terms like X1*X1 or X2*X2, and (4) CROSSPRODUCT terms like X1*X2.
6. DF indicates degrees of freedom and should be the same as the number of parameters unless one or more of the parameters is not estimable.
7. TYPE I SS, also called the sequential sums of squares, measure the reduction in the error sum of squares as terms are added to the model individually (in the order given in the MODEL statement).
8. These R-SQUAREs measure the portion of total R^2 contributed as each set of terms (LINEAR, QUADRATIC, etc.) is added to the model.
9. Each F-RATIO tests the hypothesis that all parameters in the term are zero using the TOTAL ERROR mean square as the denominator. This is a test of a TYPE I hypothesis, which is the usual *F* test numerator conditional on the effects below it not being in the model.
10. PROB is the significance value or probability of obtaining at least as great an *F* ratio given that the hypothesis is true. When PROB<.05, the effect is usually termed "significant."
11. The TOTAL ERROR sum of squares can be partitioned into LACK OF FIT and PURE ERROR. When LACK OF FIT is significantly different from PURE ERROR, then there is variation in the model not accounted for by random error.
12. The TOTAL ERROR MEAN SQUARE estimates σ^2, the variance.
13. If an effect is a linear combination of previous effects, the parameter for it is not estimable. When this happens, the DF is zero, the parameter estimate is set to zero, and the estimates and tests on other parameters are conditional on this parameter being zero (not shown).
14. The ESTIMATE column contains the parameter estimates.
15. The STD DEV column contains the estimated standard deviations of the parameter estimates.
16. The T-RATIO column contains *t* values of a test of the hypothesis that the true parameter is zero.
17. PROB gives the significance value or probability of a greater absolute *t* ratio given that the hypothesis is true.
18. The test on a factor, say X1, is a joint test on all the parameters involving that factor. For example, the test for X1 tests the hypothesis that the

parameters for X1, X1*X1, and X1*X2 are all zero.

19. The CRITICAL VALUEs for the factor variables are solved to find the factor combinations that yield the optimum response. The critical values may be at a minimum, maximum, or saddle point.
20. The EIGENVALUES and EIGENVECTORS are from the matrix of quadratic parameter estimates that determine the curvature of the response surface.

EXAMPLE

The following example uses the three-factor quadratic model discussed in John (1971). The objective is to minimize the unpleasant odor of a chemical.

```
* ----------------------------RESPONSE SURFACE EXPERIMENT---------------------------- *
|  SCHNEIDER AND STOCKETT (1963) PERFORMED AN EXPERIMENT AIMED AT                     |
|  REDUCING THE UNPLEASANT ODOR OF A CHEMICAL PRODUCT WITH SEVERAL                    |
|  FACTORS. FROM PETER W. M. JOHN, STATISTICAL DESIGN AND ANALYSIS                    |
|  OF EXPERIMENTS, MACMILLAN 1971.                                                    |
* ----------------------------------------------------------------------------------- * ;
 DATA A;
   INPUT Y X1-X3 @@;
     LABEL Y=ODOR
           X1=TEMPERATURE
           X2=GAS-LIQUID RATIO
           X3=PACKING HEIGHT;
   CARDS;
 66 -1 -1  0      39  1 -1  0      43 -1  1  0      49  1  1  0
 58 -1  0 -1      17  1  0 -1      -5 -1  0  1     -40  1  0  1
 65  0 -1 -1       7  0  1 -1      43  0 -1  1     -22  0  1  1
-31  0  0  0     -35  0  0  0     -26  0  0  0
 ;
 PROC SORT;  BY X1-X3;
 PROC RSREG;
   MODEL Y=X1-X3 / LACKFIT;
```

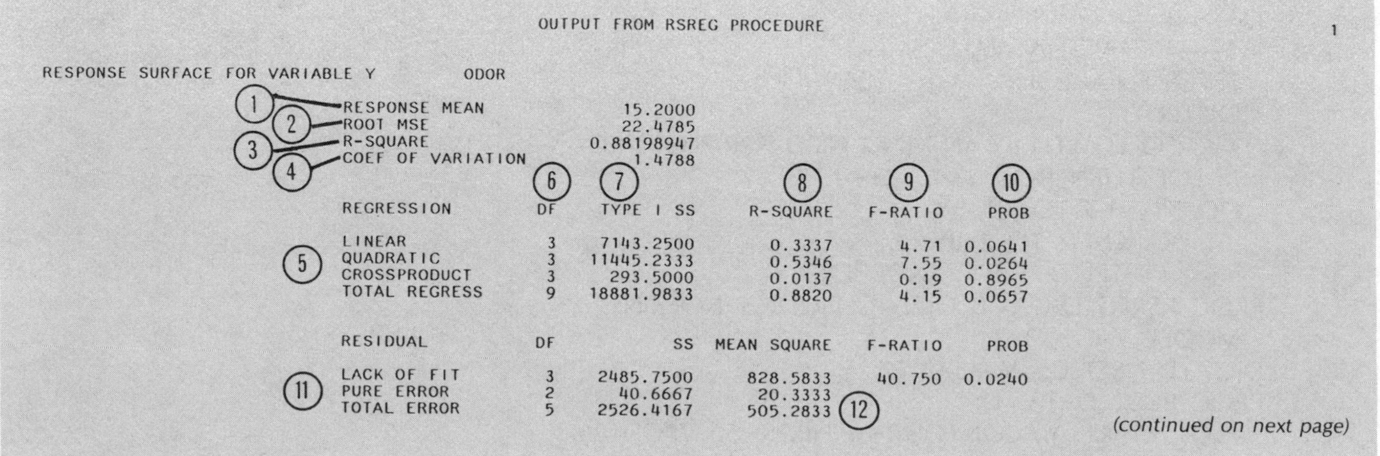

```
                                OUTPUT FROM RSREG PROCEDURE                                  1

RESPONSE SURFACE FOR VARIABLE Y          ODOR

             (1) RESPONSE MEAN                 15.2000
             (2) ROOT MSE                      22.4785
             (3) R-SQUARE                   0.88198947
             (4) COEF OF VARIATION              1.4788
                                      (6)      (7)            (8)          (9)        (10)
                 REGRESSION           DF    TYPE I SS      R-SQUARE      F-RATIO      PROB

                 LINEAR                3    7143.2500       0.3337         4.71     0.0641
             (5) QUADRATIC             3   11445.2333       0.5346         7.55     0.0264
                 CROSSPRODUCT          3     293.5000       0.0137         0.19     0.8965
                 TOTAL REGRESS         9   18881.9833       0.8820         4.15     0.0657

                 RESIDUAL             DF           SS   MEAN SQUARE      F-RATIO      PROB

                 LACK OF FIT           3    2485.7500     828.5833       40.750     0.0240
            (11) PURE ERROR            2      40.6667      20.3333
                 TOTAL ERROR           5    2526.4167     505.2833 (12)
```

(continued on next page)

(continued from previous page)

PARAMETER	DF	(14) ESTIMATE	(15) STD DEV	(16) T-RATIO	(17) PROB
INTERCEPT	1	-30.6667	12.9780	-2.36	0.0645
X1	1	-12.1250	7.9474	-1.53	0.1876
X2	1	-17.0000	7.9474	-2.14	0.0854
X3	1	-21.3750	7.9474	-2.69	0.0433
X1*X1	1	32.0833	11.6982	2.74	0.0407
X1*X2	1	8.2500	11.2393	0.73	0.4959
X2*X2	1	47.8333	11.6982	4.09	0.0095
X1*X3	1	1.5000	11.2393	0.13	0.8990
X2*X3	1	-1.7500	11.2393	-0.16	0.8824
X3*X3	1	6.0833	11.6982	0.52	0.6252

(18)

FACTOR	DF	SS	MEAN SQUARE	F-RATIO	PROB	
X1	4	5258.0160	1314.5040	2.60	0.1613	TEMPERATURE
X2	4	11044.6026	2761.1506	5.46	0.0454	GAS-LIQUID RATIO
X3	4	3813.0160	953.2540	1.89	0.2510	PACKING HEIGHT

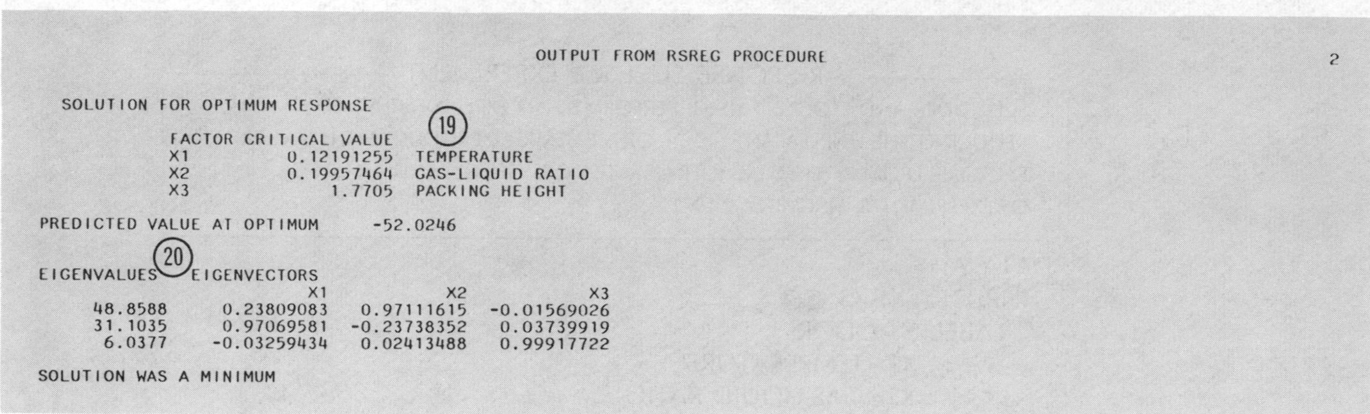

```
                              OUTPUT FROM RSREG PROCEDURE                              2

  SOLUTION FOR OPTIMUM RESPONSE
                                  (19)
      FACTOR CRITICAL VALUE
      X1          0.12191255   TEMPERATURE
      X2          0.19957464   GAS-LIQUID RATIO
      X3              1.7705   PACKING HEIGHT

PREDICTED VALUE AT OPTIMUM       -52.0246

              (20)
EIGENVALUES      EIGENVECTORS
                          X1             X2             X3
    48.8588      0.23809083     0.97111615    -0.01569026
    31.1035      0.97069581    -0.23738352     0.03739919
     6.0377     -0.03259434     0.02413488     0.99917722

SOLUTION WAS A MINIMUM
```

To plot the response surface with respect to two of the dimensions, we fix X3 at the optimum and generate a grid of points for X1 and X2. PROC RSREG computes the predicted values, which are output and plotted using the CONTOUR feature of PROC PLOT.

```
DATA B;
  *——THE ACTUAL VALUES——;
  SET A END=EOF;
  OUTPUT;
  *——FOLLOWED BY AN X1*X2 GRID FOR PLOTTING——;
  IF EOF THEN DO; Y=.; X3=1.77;
  DO X1=-1.5 TO 1.5 BY .1;
      DO X2=-2 TO 2 BY .1;
        OUTPUT; END; END; END;
PROC RSREG DATA=B OUT=C PREDICT NOPRINT;
  MODEL Y=X1-X3;
DATA D; SET C; IF X3=1.77;
PROC PLOT DATA=D;
  PLOT X1*X2=Y / CONTOUR=6 HPOS=50 VPOS=36;
```

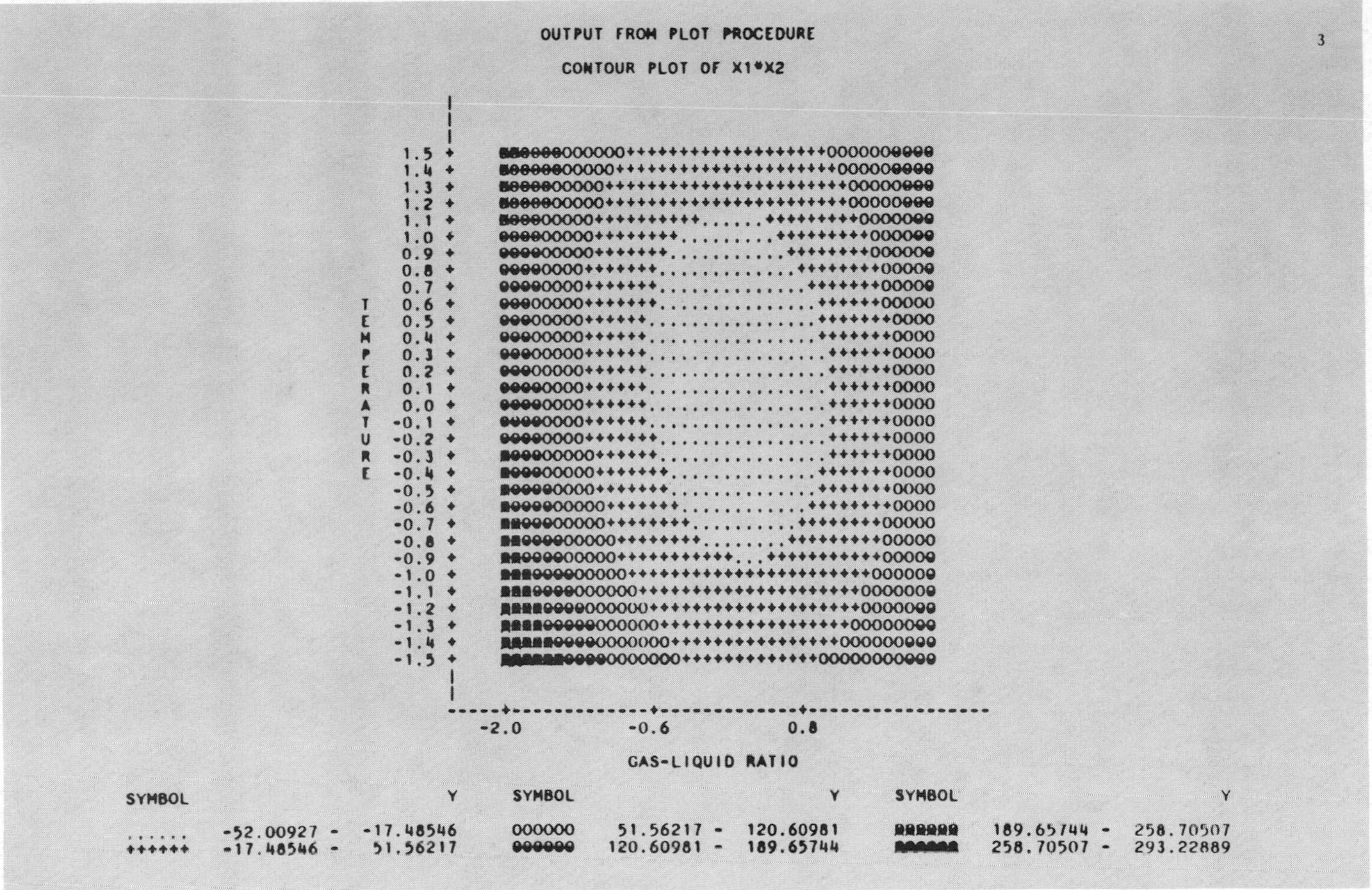

REFERENCES

Box, G.E.P. and Hunter, J.S. (1957), "Multifactor Experimental Designs for Exploring Response Surfaces," *Annuals of Mathematical Statistics*, 28, 195-242.

Box, G.E.P. and Wilson, K.J. (1951), "On the Experimental Attainment of Optimum Conditions," *Journal of the Royal Statistical Society*, Ser. B, 13, 1-45.

Cochran, W.G. and Cox, G.M. (1957), *Experimental Designs*, (2nd ed.), New York: John Wiley & Sons.

John, P.W.M. (1971), *Statistical Design and Analysis of Experiments*, New York: Macmillan.

Myers, Raymond H. (1976), *Response Surface Methodology*, Blacksburg, Virginia: Virginia Polytechnic Institute and State University.

The STEPWISE Procedure

ABSTRACT

The STEPWISE procedure provides five methods for stepwise regression. STEPWISE is useful when you have many independent variables and want to find which of the variables should be included in a regression model.

INTRODUCTION

STEPWISE is most helpful for exploratory analysis, because it can give you insight into the relationships between the independent variables and the dependent or response variable. However, STEPWISE is not guaranteed to give you the "best" model for your data, or even the model with the largest R^2. And no model developed by these means can be guaranteed to represent real-world processes accurately.

STEPWISE and Other Model-Building Procedures

STEPWISE differs from RSQUARE, another procedure used for exploratory model analysis. RSQUARE finds the R^2 value for all possible combinations of the independent variables. Therefore, RSQUARE always identifies the model with the largest R^2 for each number of variables considered. STEPWISE uses the selection strategies described below in choosing the variables for the models it considers. Also, when STEPWISE evaluates a model, it prints a complete report on the regression, while RSQUARE prints only the R^2 value for each model. RSQUARE requires much more computer time than STEPWISE.

Model-Selection Methods

The five methods of model selection implemented in PROC STEPWISE are:

FORWARD	forward selection
BACKWARD	backward elimination
STEPWISE	stepwise regression, forward and backward
MAXR	forward selection with pair switching
MINR	forward selection with pair searching

A survey article by Hocking (1976) describes these and other variable-selection methods. The five methods are described below with the keyword value of the METHOD= option to request each method.

Forward selection (FORWARD) The forward-selection technique begins with no variables in the model. For each of the independent variables, FORWARD calculates *F* statistics reflecting the variable's contribution to the model if it is included. These *F* statistics are compared to the SLENTRY= value that is specified in the MODEL statement (or to .50 if SLENTRY= is omitted). If no *F* statistic has a significance level greater than the SLENTRY= value, FORWARD stops. Otherwise, FORWARD adds the variable that has the largest *F* statistic to the model. FORWARD then calculates *F* statistics again for the variables still remaining outside the model, and the evaluation process is repeated. Variables are thus added one by one to the model until no remaining variable produces a significant *F* statistic. Once a variable is in the model, it stays.

Backward elimination (BACKWARD) The backward elimination technique begins by calculating statistics for a model including all of the independent variables. Then the variables are deleted from the model one by one until all the variables remaining in the model produce *F* statistics significant at the SLSTAY= level specified in the MODEL statement (or at the .10 level, if SLSTAY= is omitted). At each step, the variable showing the smallest contribution to the model is deleted.

Stepwise (STEPWISE) The stepwise method is a modification of the forward-selection technique and differs in that variables already in the model do not necessarily stay there. As in the forward-selection method, variables are added one by one to the model, and the *F* statistic for a variable to be added must be significant at the SLENTRY= level. After a variable is added, however, the stepwise method looks at all the variables already included in the model and deletes any variable that does not produce an *F* statistic significant at the SLSTAY= level. Only after this check is made and the necessary deletions accomplished can another variable be added to the model. The stepwise process ends when none of the variables outside the model has an *F* statistic significant at the SLENTRY= level and every variable in the model is significant at the SLSTAY= level, or when the variable to be added to the model is one just deleted from it.

Maximum R^2 improvement (MAXR) The maximum R^2 improvement technique developed by James Goodnight is considered superior to the stepwise technique and almost as good as all possible regressions. Unlike the three techniques above, this method does not settle on a single model. Instead, it tries to find the best one-variable model, the best two-variable model, and so forth, although it is not guaranteed to find the model with the largest R^2 for each size.

The MAXR method begins by finding the one-variable model producing the highest R^2. Then another variable, the one that yields the greatest increase in R^2, is added. Once the two-variable model is obtained, each of the variables in the model is compared to each variable not in the model. For each comparison, MAXR determines if removing one variable and replacing it with the other variable increases R^2. After comparing all possible switches, the one that produces the largest increase in R^2 is made. Comparisons begin again, and the process continues until MAXR finds that no switch could increase R^2. The two-variable model thus achieved is considered the "best" two-variable model the technique can find. Another variable is then added to the model, and the comparing-and-switching process is repeated to find the "best" three-variable model, and so forth.

The difference between the stepwise technique and the maximum R^2 improvement method is that all switches are evaluated before any switch is made in the MAXR method. In the stepwise method, the "worst" variable may be removed without considering what adding the "best" remaining variable might accomplish. The MAXR method may require much more computer time than the STEPWISE method.

Minimum R^2 improvement (MINR) The MINR method closely resembles MAXR, but the switch chosen is the one that produces the smallest increase in R^2. For a given number of variables in the model, MAXR and MINR usually produce the same "best" model, but MINR considers more models of each size.

Significance Levels

When many significance tests are performed, each at a level of, say 5%, the overall probability of rejecting at least one true null hypothesis is much larger than 5%. If you want to guard against including any variables that do not contribute to the predictive power of the model in the population, you should specify a very small significance level. In most applications many variables considered have some predictive power, however small. If you want to choose the model that provides the best prediction using the sample estimates, you need only guard against estimating more parameters than can be reliably estimated with the given sample size, so you should use a moderate significance level, perhaps in the range of 10% to 25%.

C_p Statistic

C_p was proposed by Mallows as a criterion for selecting a model. C_p is a measure of total squared error defined:

$$C_p = (SSE_p/s^2) - (N - 2p) ,$$

where

s^2 is the MSE for the full model,

SSE_p is the sum of squares error for a model with p parameters plus the intercept, if any.

If C_p is graphed with p, Mallows recommends the model where C_p first approaches p. When the right model is chosen, the parameter estimates are unbiased, and this reflects in C_p near p. For further discussion, see Daniel and Wood, 1980.

SPECIFICATIONS

The statements used to control PROC STEPWISE are:

PROC STEPWISE *options;*
  MODEL *dependents = independents / options;*
  WEIGHT *variable;*
  BY *variables;*

STEPWISE needs at least one MODEL statement. The BY and WEIGHT statements can be placed anywhere.

PROC STEPWISE Statement

PROC STEPWISE *options;*

Only one option is used on the PROC statement:

DATA=*SASdataset*	names the SAS data set containing the data for the regression. If it is omitted, the most recently created data set is used.

MODEL Statement

MODEL *dependents=independents / options;*

In the MODEL statement, list the dependent variables on the left side of the equal sign and the independent variables on the right side of the equal sign.

For each dependent variable given, STEPWISE goes through the model-building process using the independent variables listed. Any number of MODEL statements may be included. The options below may be specified in the MODEL statement after a slash (/).

NOINT	prevents the procedure from automatically including an intercept term in the model.
FORWARD F	requests the forward-selection technique.
BACKWARD B	requests the backward-elimination technique.
STEPWISE	requests the stepwise technique, the default.
MAXR	requests the maximum R^2 improvement technique.
MINR	requests the minimum R^2 improvement technique.
SLENTRY=*value* SLE=*value*	specifies the significance level for entry into the model used in the forward-selection and stepwise techniques. If SLENTRY= is omitted, STEPWISE uses the SLENTRY= value .50 for forward selection, .15 for stepwise.
SLSTAY=*value* SLS=*value*	specifies the significance level for staying in the model for the backward elimination and stepwise techniques. If it is omitted, STEPWISE uses the SLSTAY= value .10 for backward elimination, .15 for stepwise.
INCLUDE=*n*	forces the first *n* independent variables always to be included in the model. The selection techniques are performed on the other variables in the MODEL statement.
START=*s*	is used to begin the comparing-and-switching process for a model containing the first *s* independent variables in the MODEL statement, where *s* is the START value. Consequently, no model is evaluated that contains fewer than *s* variables. This applies only to the MAXR or MINR methods.
STOP=*s*	causes STEPWISE to stop when it has found the "best" *s*-variable model, where *s* is the STOP value. This applies only to the MAXR or MINR methods.

WEIGHT Statement

WEIGHT *variable;*

The WEIGHT statement is used to specify a variable on the data set containing weights for the observations. Only observations with positive values of the WEIGHT variable are used in the analysis.

BY Statement

BY *variables;*

A BY statement may be used with PROC STEPWISE to obtain separate analyses on observations in groups defined by the BY variables. When a BY statement appears, the procedure expects the input data set to be sorted in order of the BY variables. If your input data set is not sorted in ascending order, use the SORT procedure with a similar BY statement to sort the data, or, if appropriate, use the BY statement options NOTSORTED or DESCENDING. For more information, see the discussion of the BY statement in Chapter 8, "Statements Used in the PROC Step,"in *SAS User's Guide: Basics, 1982 Edition*.

DETAILS

Missing Values

STEPWISE omits observations from the calculations for a given model if the observation has missing values for any of the variables in the model. The observation is included for any models that do not include the variables with missing values.

Limitations

Any number of dependent variables may be included in a MODEL statement. Although there is no built-in limit on the number of independent variables, the calculations for a model with many variables are lengthy. For the MAXR or MINR technique, a reasonable maximum for the number of independent variables in a single MODEL statement is about 20.

Printed Output

For each model of a given size, STEPWISE prints an analysis-of-variance table, the regression coefficients, and related statistics.

The analysis-of-variance table includes:

1. the source of variation REGRESSION, which is the variation that is attributed to the independent variables in the model
2. the source of variation ERROR, which is the residual variation that is not accounted for by the model
3. the source of variation TOTAL, which is corrected for the mean of *y* if an intercept is included in the model, uncorrected if an intercept is not included
4. DF, degrees of freedom
5. SUMS OF SQUARES for REGRESSION, ERROR, and TOTAL
6. MEAN SQUARES for REGRESSION and ERROR

7. the F value, which is the ratio of the REGRESSION mean square to the ERROR mean square
8. PROB > F, the significance probability of the *F* value
9. R SQUARE or R^2, the square of the multiple correlation coefficient
10. C(P) statistic proposed by Mallows.

Below the analysis-of-variance table are printed:

11. the names of the independent variables included in the model
12. B VALUES, the corresponding estimated regression coefficients
13. STD ERROR of the estimates
14. TYPE II SS (sum of squares) for each variable, which is the SS that is added to the error SS if that one variable is removed from the model
15. F values and PROB > F associated with the Type II sums of squares.

EXAMPLE

The example below asks for the FORWARD, BACKWARD, and MAXR methods.

```
* ---------------------------------DATA ON PHYSICAL FITNESS--------------------------------- *
| THESE MEASUREMENTS WERE MADE ON MEN INVOLVED IN A PHYSICAL FITNESS                        |
| COURSE AT N.C. STATE UNIV. THE VARIABLES ARE AGE(YEARS), WEIGHT(KG),                      |
| OXYGEN UPTAKE RATE(ML PER KG BODY WEIGHT PER MINUTE), TIME TO RUN                         |
| 1.5 MILES(MINUTES), HEART RATE WHILE RESTING, HEART RATE WHILE                            |
| RUNNING (SAME TIME OXYGEN RATE MEASURED), AND MAXIMUM HEART RATE                          |
| RECORDED WHILE RUNNING. CERTAIN VALUES OF MAXPULSE WERE MODIFIED                          |
| FOR CONSISTENCY. DATA COURTESY DR. A. C. LINNERUD                                         |
* ------------------------------------------------------------------------------------------ * ;
DATA FITNESS;
 INPUT AGE WEIGHT OXY RUNTIME RSTPULSE
       RUNPULSE MAXPULSE @@;
 CARDS;
44 89.47 44.609 11.37 62 178 182   40 75.07 45.313 10.07 62 185 185
44 85.84 54.297  8.65 45 156 168   42 68.15 59.571  8.17 40 166 172
38 89.02 49.874  9.22 55 178 180   47 77.45 44.811 11.63 58 176 176
40 75.98 45.681 11.95 70 176 180   43 81.19 49.091 10.85 64 162 170
44 81.42 39.442 13.08 63 174 176   38 81.87 60.055  8.63 48 170 186
44 73.03 50.541 10.13 45 168 168   45 87.66 37.388 14.03 56 186 192
45 66.45 44.754 11.12 51 176 176   47 79.15 47.273 10.60 47 162 164
54 83.12 51.855 10.33 50 166 170   49 81.42 49.156  8.95 44 180 185
51 69.63 40.836 10.95 57 168 172   51 77.91 46.672 10.00 48 162 168
48 91.63 46.774 10.25 48 162 164   49 73.37 50.388 10.08 67 168 168
57 73.37 39.407 12.63 58 174 176   54 79.38 46.080 11.17 62 156 165
52 76.32 45.441  9.63 48 164 166   50 70.87 54.625  8.92 48 146 155
51 67.25 45.118 11.08 48 172 172   54 91.63 39.203 12.88 44 168 172
51 73.71 45.790 10.47 59 186 188   57 59.08 50.545  9.93 49 148 155
49 76.32 48.673  9.40 56 186 188   48 61.24 47.920 11.50 52 170 176
52 82.78 47.467 10.50 53 170 172
;
PROC STEPWISE;
 MODEL OXY=AGE WEIGHT RUNTIME RUNPULSE RSTPULSE
           /FORWARD BACKWARD MAXR;
```

```
                              S T A T I S T I C A L   A N A L Y S I S   S Y S T E M                                    1

                              FORWARD SELECTION PROCEDURE FOR DEPENDENT VARIABLE OXY
STEP 1    VARIABLE RUNTIME ENTERED      R SQUARE = 0.74338010 (9)    C(P) =      13.69884048 (10)

                             (4) DF          (5) SUM OF SQUARES       (6) MEAN SQUARE            (7) F   (8) PROB>F
            (1)
    (2) ---- REGRESSION          1            632.90009985          632.90009985            84.01       0.0001
             ERROR              29            218.48144499            7.53384293
         (3) TOTAL              30            851.38154484

                            (12) B VALUE     (13) STD ERROR          (14) TYPE II SS                F  (15) PROB>F

       (11) INTERCEPT      82.42177268
            RUNTIME        -3.31055536          0.36119485          632.90009985            84.01       0.0001
-----------------------------------------------------------------------------------------------------------------

STEP 2    VARIABLE AGE ENTERED          R SQUARE = 0.76424693        C(P) =      12.38944895

                               DF              SUM OF SQUARES           MEAN SQUARE                F       PROB>F

             REGRESSION         2             650.66573237          325.33286618            45.38       0.0001
             ERROR             28             200.71581247            7.16842187
             TOTAL             30             851.38154484

                                B VALUE            STD ERROR            TYPE II SS                F       PROB>F

             INTERCEPT     88.46228749
             AGE           -0.15036567          0.09551468           17.76563252             2.48       0.1267
             RUNTIME       -3.20395056          0.35877488          571.67750579            79.75       0.0001
-----------------------------------------------------------------------------------------------------------------

STEP 3    VARIABLE RUNPULSE ENTERED     R SQUARE = 0.81109446        C(P) =       6.95962673

                               DF              SUM OF SQUARES           MEAN SQUARE                F       PROB>F

             REGRESSION         3             690.55085627          230.18361876            38.64       0.0001
             ERROR             27             160.83068857            5.95669217
             TOTAL             30             851.38154484

                                B VALUE            STD ERROR            TYPE II SS                F       PROB>F

             INTERCEPT    111.71806443
             AGE           -0.25639826          0.09622892           42.28867438             7.10       0.0129
             RUNTIME       -2.82537867          0.35828041          370.43528607            62.19       0.0001
             RUNPULSE      -0.13090870          0.05059011           39.88512390             6.70       0.0154
-----------------------------------------------------------------------------------------------------------------
```

```
                              S T A T I S T I C A L   A N A L Y S I S   S Y S T E M                                    2

                              FORWARD SELECTION PROCEDURE FOR DEPENDENT VARIABLE OXY

STEP 4    VARIABLE MAXPULSE ENTERED     R SQUARE = 0.83681815        C(P) =       4.87995808

                               DF              SUM OF SQUARES           MEAN SQUARE                F       PROB>F

             REGRESSION         4             712.45152692          178.11288173            33.33       0.0001
             ERROR             26             138.93001792            5.34346223
             TOTAL             30             851.38154484

                                B VALUE            STD ERROR            TYPE II SS                F       PROB>F

             INTERCEPT     98.14788797
             AGE           -0.19773470          0.09563662           22.84231496             4.27       0.0488
             RUNTIME       -2.76757879          0.34053643          352.93569605            66.05       0.0001
             RUNPULSE      -0.34810795          0.11749917           46.90088674             8.78       0.0064
             MAXPULSE       0.27051297          0.13361978           21.90067065             4.10       0.0533
-----------------------------------------------------------------------------------------------------------------

STEP 5    VARIABLE WEIGHT ENTERED       R SQUARE = 0.84800181        C(P) =       5.10627546

                               DF              SUM OF SQUARES           MEAN SQUARE                F       PROB>F

             REGRESSION         5             721.97309402          144.39461880            27.90       0.0001
             ERROR             25             129.40845082            5.17633803
             TOTAL             30             851.38154484

                                B VALUE            STD ERROR            TYPE II SS                F       PROB>F

             INTERCEPT    102.20427520
             AGE           -0.21962138          0.09550245           27.37429100             5.29       0.0301
             WEIGHT        -0.07230234          0.05331009            9.52156710             1.84       0.1871
             RUNTIME       -2.68252297          0.34098544          320.35967836            61.89       0.0001
             RUNPULSE      -0.37340085          0.11714109           52.59623720            10.16       0.0038
             MAXPULSE       0.30490783          0.13393642           26.82640270             5.18       0.0316
-----------------------------------------------------------------------------------------------------------------

NO OTHER VARIABLES MET THE 0.5000 SIGNIFICANCE LEVEL FOR ENTRY INTO THE MODEL.
```

```
                         S T A T I S T I C A L   A N A L Y S I S   S Y S T E M                                          3

                          BACKWARD ELIMINATION PROCEDURE FOR DEPENDENT VARIABLE OXY

STEP 0    ALL VARIABLES ENTERED         R SQUARE = 0.84867192         C(P) =       7.00000000

                         DF              SUM OF SQUARES           MEAN SQUARE                F        PROB>F

         REGRESSION       6                722.54360701          120.42393450            22.43        0.0001
         ERROR           24                128.83793783            5.36824741
         TOTAL           30                851.38154484

                          B VALUE             STD ERROR           TYPE II SS                 F        PROB>F

         INTERCEPT    102.93447948
         AGE           -0.22697380           0.09983747          27.74577148              5.17        0.0322
         WEIGHT        -0.07417741           0.05459316           9.91058836              1.85        0.1869
         RUNTIME       -2.62865282           0.38456220         250.82210090             46.72        0.0001
         RUNPULSE      -0.36962776           0.11985294          51.05805832              9.51        0.0051
         RSTPULSE      -0.02153364           0.06605428           0.57051299              0.11        0.7473
         MAXPULSE       0.30321713           0.13649519          26.49142405              4.93        0.0360
--------------------------------------------------------------------------------------------------------------

STEP 1    VARIABLE RSTPULSE REMOVED     R SQUARE = 0.84800181         C(P) =       5.10627546

                         DF              SUM OF SQUARES           MEAN SQUARE                F        PROB>F

         REGRESSION       5                721.97309402          144.39461880            27.90        0.0001
         ERROR           25                129.40845082            5.17633803
         TOTAL           30                851.38154484

                          B VALUE             STD ERROR           TYPE II SS                 F        PROB>F

         INTERCEPT    102.20427520
         AGE           -0.21962138           0.09550245          27.37429100              5.29        0.0301
         WEIGHT        -0.07230234           0.05331009           9.52156710              1.84        0.1871
         RUNTIME       -2.68252297           0.34098544         320.35967836             61.89        0.0001
         RUNPULSE      -0.37340085           0.11714109          52.59623720             10.16        0.0038
         MAXPULSE       0.30490783           0.13393642          26.82640270              5.18        0.0316
--------------------------------------------------------------------------------------------------------------

STEP 2    VARIABLE WEIGHT REMOVED       R SQUARE = 0.83681815         C(P) =       4.87995808

                         DF              SUM OF SQUARES           MEAN SQUARE                F        PROB>F

         REGRESSION       4                712.45152692          178.11288173            33.33        0.0001
         ERROR           26                138.93001792            5.34346223
         TOTAL           30                851.38154484

                          B VALUE             STD ERROR           TYPE II SS                 F        PROB>F

         INTERCEPT     98.14788797
         AGE           -0.19773470           0.09563662          22.84231496              4.27        0.0488
         RUNTIME       -2.76757879           0.34053643         352.93569605             66.05        0.0001
         RUNPULSE      -0.34810795           0.11749917          46.90088674              8.78        0.0064
         MAXPULSE       0.27051297           0.13361978          21.90067065              4.10        0.0533
--------------------------------------------------------------------------------------------------------------

ALL VARIABLES IN THE MODEL ARE SIGNIFICANT AT THE 0.1000 LEVEL.
```

```
                         S T A T I S T I C A L   A N A L Y S I S   S Y S T E M                                          4

                          MAXIMUM R-SQUARE IMPROVEMENT FOR DEPENDENT VARIABLE OXY

STEP 1    VARIABLE RUNTIME ENTERED          R SQUARE = 0.74338010         C(P) =      13.69884048

                         DF              SUM OF SQUARES           MEAN SQUARE                F        PROB>F

         REGRESSION       1                632.90009985          632.90009985            84.01        0.0001
         ERROR           29                218.48144499            7.53384293
         TOTAL           30                851.38154484

                          B VALUE             STD ERROR           TYPE II SS                 F        PROB>F

         INTERCEPT     82.42177268
         RUNTIME       -3.31055536           0.36119485         632.90009985             84.01        0.0001
--------------------------------------------------------------------------------------------------------------

THE ABOVE MODEL IS THE BEST  1 VARIABLE MODEL FOUND.

STEP 2    VARIABLE AGE ENTERED              R SQUARE = 0.76424693         C(P) =      12.38944895

                         DF              SUM OF SQUARES           MEAN SQUARE                F        PROB>F

         REGRESSION       2                650.66573237          325.33286618            45.38        0.0001
         ERROR           28                200.71581247            7.16842187
         TOTAL           30                851.38154484

                          B VALUE             STD ERROR           TYPE II SS                 F        PROB>F

         INTERCEPT     88.46228749
         AGE           -0.15036567           0.09551468          17.76563252              2.48        0.1267
         RUNTIME       -3.20395056           0.35877488         571.67750579             79.75        0.0001
--------------------------------------------------------------------------------------------------------------

THE ABOVE MODEL IS THE BEST  2 VARIABLE MODEL FOUND.
```

(continued on next page)

(continued from previous page)

```
STEP 3    VARIABLE RUNPULSE ENTERED        R SQUARE = 0.81109446        C(P) =      6.95962673

                          DF               SUM OF SQUARES        MEAN SQUARE              F       PROB>F

      REGRESSION           3                 690.55085627       230.18361876          38.64       0.0001
      ERROR               27                 160.83068857         5.95669217
      TOTAL               30                 851.38154484

                              B VALUE           STD ERROR        TYPE II SS               F       PROB>F

      INTERCEPT       111.71806443
      AGE              -0.25639826          0.09622892        42.28867438            7.10       0.0129
      RUNTIME          -2.82537867          0.35828041       370.43528607           62.19       0.0001
      RUNPULSE         -0.13090870          0.05059011        39.88512390            6.70       0.0154
-----------------------------------------------------------------------------------------------------

THE ABOVE MODEL IS THE BEST  3 VARIABLE MODEL FOUND.
```

```
                       S T A T I S T I C A L   A N A L Y S I S   S Y S T E M                              5

                          MAXIMUM R-SQUARE IMPROVEMENT FOR DEPENDENT VARIABLE OXY

STEP 4    VARIABLE MAXPULSE ENTERED        R SQUARE = 0.83681815        C(P) =      4.87995808

                          DF               SUM OF SQUARES        MEAN SQUARE              F       PROB>F

      REGRESSION           4                 712.45152692       178.11288173          33.33       0.0001
      ERROR               26                 138.93001792         5.34346223
      TOTAL               30                 851.38154484

                              B VALUE           STD ERROR        TYPE II SS               F       PROB>F

      INTERCEPT        98.14788797
      AGE              -0.19773470          0.09563662        22.84231496            4.27       0.0488
      RUNTIME          -2.76757879          0.34053643       352.93569605           66.05       0.0001
      RUNPULSE         -0.34810795          0.11749917        46.90088674            8.78       0.0064
      MAXPULSE          0.27051297          0.13361978        21.90067065            4.10       0.0533
-----------------------------------------------------------------------------------------------------

THE ABOVE MODEL IS THE BEST  4 VARIABLE MODEL FOUND.

STEP 5    VARIABLE WEIGHT ENTERED          R SQUARE = 0.84800181        C(P) =      5.10627546

                          DF               SUM OF SQUARES        MEAN SQUARE              F       PROB>F

      REGRESSION           5                 721.97309402       144.39461880          27.90       0.0001
      FRROR               25                 129.40845082         5.17633803
      TOTAL               30                 851.38154484

                              B VALUE           STD ERROR        TYPE II SS               F       PROB>F

      INTERCEPT       102.20427520
      AGE              -0.21962138          0.09550245        27.37429100            5.29       0.0301
      WEIGHT           -0.07230234          0.05331009         9.52156710            1.84       0.1871
      RUNTIME          -2.68252297          0.34098544       320.35967836           61.89       0.0001
      RUNPULSE         -0.37340085          0.11714109        52.59623720           10.16       0.0038
      MAXPULSE          0.30490783          0.13393642        26.82640270            5.18       0.0316
-----------------------------------------------------------------------------------------------------

THE ABOVE MODEL IS THE BEST  5 VARIABLE MODEL FOUND.

STEP 6    VARIABLE RSTPULSE ENTERED        R SQUARE = 0.84867192        C(P) =      7.00000000

                          DF               SUM OF SQUARES        MEAN SQUARE              F       PROB>F

      REGRESSION           6                 722.54360701       120.42393450          22.43       0.0001
      ERROR               24                 128.83793783         5.36824741
      TOTAL               30                 851.38154484

                              B VALUE           STD ERROR        TYPE II SS               F       PROB>F

      INTERCEPT       102.93447948
      AGE              -0.22697380          0.09983747        27.74577148            5.17       0.0322
      WEIGHT           -0.07417741          0.05459316         9.91058836            1.85       0.1869
      RUNTIME          -2.62865282          0.38456220       250.82210090           46.72       0.0001
      RUNPULSE         -0.36962776          0.11985294        51.05805832            9.51       0.0051
      RSTPULSE         -0.02153364          0.06605428         0.57051299            0.11       0.7473
      MAXPULSE          0.30321713          0.13649519        26.49142405            4.93       0.0360
-----------------------------------------------------------------------------------------------------

THE ABOVE MODEL IS THE BEST  6 VARIABLE MODEL FOUND.
```

REFERENCES

Daniel, Cuthbert and Wood, Fred S. (1980), *Fitting Equations to Data*, Second Edition, New York: John Wiley and Sons, Inc.

Draper, N.R. and Smith, H. (1981), *Applied Regression Analysis,* Second Edition, New York: John Wiley and Sons, Inc.

Hocking, R.R. (1976), "The Analysis and Selection of Variables in Linear Regression," *Biometrics*, 32, 1-50.

Mallows, C.L. (1964), "Some Comments on C_p," *Technometrics*, 15, 661-675.

Sall, J. (1981), *SAS Regression Applications*, Technical Report A-102, Cary, N.C.: SAS Institute.

ANALYSIS OF VARIANCE

Introduction

ANOVA

GLM

NESTED

NPAR1WAY

PLAN

TTEST

VARCOMP

The Four Types of Estimable Functions

Introduction to SAS Analysis-of-Variance Procedures

This chapter reviews the SAS procedures that are used for analysis of variance: GLM, ANOVA, NESTED, VARCOMP, NPAR1WAY, TTEST, and PLAN.

The most general analysis-of-variance procedure is GLM, which can handle most problems. Other procedures are used for special cases as described below:

GLM	performs analysis of variance, regression, analysis of covariance, and multivariate analysis of variance
ANOVA	handles analysis of variance for balanced designs
NESTED	performs analysis of variance for purely nested random models
VARCOMP	estimates variance components
NPAR1WAY	performs nonparametric one-way analysis of rank scores
TTEST	compares the means of two groups of observations
PLAN	generates random permutations for experimental plans.

Introduction

These procedures perform analysis of variance, which is a technique for analyzing experimental data. A continuous response variable is measured under various experimental conditions identified by classification variables. The variation in the response is "explained" as due to effects in the classification with random error accounting for the remaining variation.

For each observation, the *ANOVA* model predicts the response, often by a sample mean. The difference between the actual and the predicted response is the residual error. Analysis-of-variance procedures fit parameters to minimize the sum of squares of residual errors. Thus the method is called *least squares*. The variance of the random error, σ^2, is estimated by the mean squared error (MSE or s^2).

Analysis of variance was pioneered by R.A. Fisher (1925). For a general introduction to analysis of variance, see an intermediate statistical methods textbook such as Steel and Torrie (1980), Snedecor and Cochran (1980), Mendenhall (1968), John (1971), Ott (1977), or Kirk (1968). A classic source is Scheffe (1959). Freund and Littell (1981) bring together a treatment of these statistical methods and SAS procedures. Linear models texts include Searle (1971), Graybill (1961), and Bock (1975). Kennedy and Gentle (1980) survey the computing aspects.

ANOVA for Balanced Designs

One of the factors that determines which procedure to use is whether your data are balanced or unbalanced. When you design an experiment, you choose how many experimental units to assign to each combination of levels (or cells) in the classification. In order to achieve good statistical properties and simplify the statistical arithmetic, you typically attempt to assign the same number of units to every cell in the design. These designs are called *balanced*.

If you have balanced data, the arithmetic for calculating sums of squares can be greatly simplified. In SAS, you can use the ANOVA procedure rather than the more expensive GLM procedure for balanced data. Generalizations of the balanced concept can be made to use the arithmetic for balanced designs even though the design does not contain an equal number of observations per cell. You can use balanced arithmetic for all one-way models regardless of how unbalanced the cell counts are. You can even use the balanced arithmetic for Latin squares that do not always have data in all cells.

However, if you use the ANOVA procedure to analyze a design that is not balanced, you may get incorrect results, including negative values reported for the sums of squares.

Analysis-of-variance procedures construct ANOVA tests by comparing mean squares relative to their expected values under the null hypothesis. Each mean square in a fixed analysis-of-variance model has an expected value that is composed of two components: quadratic functions of fixed parameters and random variation. For a fixed effect called A, the expected value of its mean square is written:

$$E(MS(A)) = Q(\beta) + \sigma_e^2 \ .$$

The mean square is constructed so that under the hypothesis to be tested (null hypothesis) the fixed portion $Q(\beta)$ of the expected value is zero. This mean square is then compared to another mean square, say MS(E), that is independent of the first, yet has the expected value σ_e^2. The ratio of the two mean squares is an *F* statistic that has the *F* distribution under the null hypothesis:

$$F = MS(A)/MS(E) \ .$$

When the null hypothesis is false, the numerator term has a larger expected value, but the expected value of the denominator remains the same. Thus large *F* values lead to rejection of the null hypothesis. The test decides an outcome by controlling for the Type 1 error rate, the probability of rejecting a true null hypothesis. You look at the significance probability, the probability of getting an even larger *F* value if the null hypothesis is true. If this probability is small, say below .05 or .01, you are wrong in rejecting less than .05 or .01 of the time respectively. If you are unable to reject the hypothesis, you conclude that either the null hypothesis was true or that you do not have enough data to detect the small differences to be tested.

General Linear Models

If your data do not fit into a balanced design, then you probably need the framework of linear models in the GLM procedure.

An analysis-of-variance model can be written as a linear model, an equation to predict the response as a linear function of parameters and design variables. In general we write:

$$y_i = \beta_0 x_{0i} + \beta_1 x_{1i} + \ldots + \beta_k x_{ki} + \varepsilon_i \quad , \qquad i = 1 \ldots n$$

where y_i is the response for the ith observation, β_k are unknown parameters to be estimated, and x_{ij} are design variables. Design variables for analysis of variance are indicator variables, that is, they are always either 0 or 1.

The simplest model is to fit a single mean to all observations. In this case there is only one parameter, β_0, and one design variable, x_{0i}, which always has the value 1:

$$y_i = \beta_0 x_{0i} + \varepsilon_i$$
$$= \beta_0 + \varepsilon_i \quad .$$

The least-squares estimator of β_0 is the mean of the y_i. This simple model underlies all more complex models, and all larger models are compared to this simple mean model.

A one-way model is written by introducing an indicator variable for each level of the classification variable. Suppose that a variable A has four levels, with two observations per level. The indicators are created as shown below:

intercept	a1	a2	a3	a4
1	1	0	0	0
1	1	0	0	0
1	0	1	0	0
1	0	1	0	0
1	0	0	1	0
1	0	0	1	0
1	0	0	0	1
1	0	0	0	1

The linear model can be written:

$$y_i = \beta_0 + a1_i\beta_1 + a2_i\beta_2 + a3_i\beta_3 + a4_i\beta_4$$

To construct crossed and nested effects, you can simply multiply out all combinations of the main-effect columns. This is described in detail in the section **Parameterization** in the GLM procedure.

Linear hypotheses When models are expressed in the framework of linear models, hypothesis tests are expressed in terms of a linear function of the parameters. For example, you may want to test that $\beta_2 - \beta_3 = 0$. In general, the coefficients for linear hypotheses are some set of Ls:

$$H_0\colon L_0\beta_0 + L_1\beta_1 + \ldots + L_k\beta_k = 0$$

Several of these linear functions can be combined to make one joint test. Tests can also be expressed in one matrix equation:

$$H_0\colon L\beta = 0 \ .$$

For each linear hypothesis, a sum of squares due to that hypothesis can be constructed. This sums of squares can be calculated either as a quadratic form of the estimates:

$$SS(L\beta = 0) = (Lb)'(L(X'X)^{-}L')^{-1}(Lb)$$

or as the increase in SSE for the model constrained by the hypothesis

$$SS(L\beta=0)= SSE(constrained) - SSE(full).$$

This SS is then divided by degrees of freedom and used as a numerator of an *F* statistic.

Random effects To estimate the variances of random effects, use the VARCOMP or NESTED procedures; PROC GLM does not estimate variance components but can produce expected mean squares.

A *random effect* is an effect whose parameters are drawn from a normally distributed random process with mean zero and common variance. Effects are declared random when the levels are randomly selected from a large population of possible levels. The inferences concern fixed effects but can be generalized across the whole population of random effects levels rather than only those levels in your sample.

In agricultural experiments, it is common to declare location or plot as random since these levels are chosen randomly from a large population and you assume fertility to vary normally across locations. In repeated-measures experiments with people or animals as subjects, subjects are declared random since they are selected from the larger population to which you want to generalize.

When effects are declared random in GLM, the expected mean square of each effect is calculated. Each expected mean square is a function of variances of random effects and quadratic functions of parameters of fixed effects. To test a given effect, you must search for a term that has the same expectation as your numerator term, except for the portion of the expectation that you want to test to be zero. If the two mean squares are independent, then the *F* test is valid. Sometimes, however, you will not be able to find a proper denominator term.

Comparison of means When you have more than two means to compare, an *ANOVA F* test tells you if the means are significantly different from each other, but it does not tell you which means differ from which other means.

If you have specific comparisons in mind, you can use the CONTRAST statement in GLM to make these comparisons. However, if you make many comparisons using some alpha level to judge significance, you are more likely to make a Type 1 error (rejecting incorrectly a hypothesis that means are equal) simply because you have more chances to make the error.

Multiple comparison methods give you more detailed information about the differences among the means and allow you to control error rates for a multitude of comparisons. A variety of multiple comparison methods are available with the MEANS statement in the ANOVA and GLM procedures. These are described in detail in the section **Comparison of Means** in GLM.

Nonparametric analysis Analysis of variance is sensitive to the distribution of the error term. If the error term is not normally distributed, the statistics based on normality may be misleading. The traditional test statistics are called *parametric tests* because they depend on the specification of a certain probability distribution up to a set of free parameters. Nonparametric methods make the tests without making distributional assumptions. If the data are distributed normally, often nonparametric methods are almost as powerful as parametric methods.

Most nonparametric methods are based on taking the ranks of a variable and analyzing these ranks (or transformations of them) instead of the original values. The NPAR1WAY procedure is available to perform a nonparametric one-way analysis of variance. Other nonparametric tests can be performed by taking ranks of the data (using PROC RANK) and using a regular parametric procedure to perform the analysis. Some of these techniques are outlined in the description of PROC RANK and Conover and Iman (1981) cited below.

REFERENCES

Bock, M.E. (1975), "Minimax Estimators of the Mean of a Multivariate Normal Distribution," *Annals of Statistics*, 3, 209-218.

Conover, W.J. and Iman, R.L. (1981), "Rank Transformations as a Bridge Between Parametric and Nonparametric Statistics," *The American Statistician*, 35, 124-129.

Fisher, R.A. (1925), *Statistical Methods for Research Workers*, Edinburgh: Oliver & Boyd.

Freund, R.J. and Littell, R.C. (1981), *SAS for Linear Models*, Cary, NC: SAS Institute.

Graybill, F.A. (1961), *An Introduction to Linear Statistical Models*, New York: McGraw-Hill.

John, P. (1971), *Statistical Design and Analysis of Experiments*, New York: Macmillan.

Kennedy, W.J., Jr. and Gentle, J.E. (1980), *Statistical Computing*, New York: Marcel Dekker.

Kirk, R.E. (1968), *Experimental Design: Procedures for the Behavioral Sciences*, Monterey, CA: Brooks/Cole.

Mendenhall, W. (1968), *Introduction to Linear Models and The Design and Analysis of Experiments*, Belmont, CA: Duxbury.

Ott, L. (1977), *An Introduction of Statistical Methods and Data Analysis*; Belmont, CA: Duxbury.

Scheffe, H. (1959), *The Analysis of Variance*, New York: John Wiley & Sons.

Searle, S.R. (1971), *Linear Models*, New York: John Wiley & Sons.

Snedecor, G.W. and Cochran, W.G. (1980), *Statistical Methods*, Seventh Edition, Ames, Iowa: The Iowa State University Press.

Steel, R.G.D and Torrie, J.H. (1980), *Principles and Procedures of Statistics*, Second Edition, New York: McGraw-Hill.

REFERENCES

The ANOVA Procedure

ABSTRACT

The ANOVA procedure performs analysis of variance for balanced data from a wide variety of experimental designs.

INTRODUCTION

ANOVA is one of several procedures in SAS for *analysis of variance*, which is a technique for analyzing experimental data. A continuous response variable is measured under experimental conditions identified by classification variables. The variation in the response is "explained" as due to effects in the classification with random error accounting for the remaining variation. Fisher (1942) is the pioneering work.

The ANOVA procedure is designed to handle balanced data, while the GLM procedure can analyze both balanced and unbalanced data. But ANOVA is faster and uses less storage than GLM for balanced data because ANOVA takes into account the special structure of a balanced design.

Use ANOVA for the analysis of balanced data only, with the exception of Latin-square designs, certain balanced incomplete blocks designs, completely nested (hierarchical) designs, and designs whose cell frequencies are proportional to each other and are also proportional to the background population. For further discussion, see Searle (1971, p. 138).

ANOVA does not check to see if your design is balanced or is one of the special cases described above. **If you use ANOVA for analysis of unbalanced data, you must assume responsibility for the validity of the output.**

Specification of Effects

In SAS analysis-of-variance procedures, the variables that identify levels of the classifications are called *classification variables* and are declared in the CLASS statement. Classification variables may also be called *categorical*, *qualitative*, *discrete*, or *nominal* variables. The values of a class variable are called *levels*. Class variables can be either numeric or character. This is in contrast to the *response* (or *dependent*) variables, which are continuous; response variables must be numeric.

The analysis-of-variance model specifies *effects*, which are combinations of classification variables to which the dependent variables are fit, in the following manner:

- main effects are specified by writing the variables by themselves: A B C
- crossed effects (interactions) are specified by joining the class variables with asterisks: A*B A*C A*B*C
- nested effects are specified by placing a parenthetical field after a class variable or interaction indicating the variables within which the effect is nested: B(A) C*D(A B).

The general form of an effect can be illustrated using the class variables A, B, C, D, E, and F:

```
A*B*C(D E F) .
```

The crossed list should come first, followed by the nested list in parentheses. Note that the nested list does not contain an asterisk between class variable names or before the first parenthesis.

Main-effects models For a three-factor main-effects model with A, B, and C as the factors and Y as the dependent variable, the necessary statements are:

```
PROC ANOVA;
  CLASS A B C;
  MODEL Y=A B C;
```

Models with crossed factors You can join variables with an asterisk (*) to denote crossed factors (interactions) in a model. For example, these statements specify a complete factorial model, which includes all the interactions:

```
PROC ANOVA;
  CLASS A B C;
  MODEL Y=A B C A*B A*C B*C A*B*C;
```

Bar notation You can shorten the specifications of a full factorial model using bar notation. For example, the statements above can also be written:

```
PROC ANOVA;
  CLASS A B C;
  MODEL Y=A|B|C;
```

When the bar (|) is used, the expression on the right side of the equal sign is expanded from left to right using the equivalents of rules 2-4 given in Searle (1971, p. 390). Other examples of the bar notation:

```
A|C(B)         is equivalent to A C(B) A*C(B)
A(B)|C(B)      is equivalent to A(B) C(B) A*C(B)
A(B)|B(D E)    is equivalent to A(B) B(D E)
A|B(A)|C       is equivalent to A B(A) C A*C B*C(A) .
```

Consult the description of the GLM procedure for further details on the bar notation.

Nested models Write the effect that is nested within another effect first, followed by the other effect in parentheses. For example, if B is nested within A, and C is nested within A and B, the statements for PROC ANOVA are:

```
PROC ANOVA;
  CLASS A B C;
  MODEL Y=A B(A) C(A B);
```

The identity of a level is viewed within the context of the level of the containing effects. For example, CITY is nested within STATE.

The distinguishing feature of a nested specification is that nested effects never appear as main effects also. Another way of viewing nested effects is that they are effects that pool the main effect with the interaction of the nesting variable. See **Automatic pooling** below.

Models involving nested, crossed, and main effects Asterisks and parentheses may be combined in the MODEL statement for models involving nested and crossed effects:

```
PROC ANOVA;
  CLASS A B C;
  MODEL Y=A B(A) C(A) B*C(A);
```

Automatic pooling In line with the general philosophy of the GLM procedure, there is no difference between the statements

```
MODEL Y=A B(A);
```

and

```
MODEL Y=A A*B;
```

The effect B becomes a nested effect by virtue of the fact that it does not occur as a main effect. If B is not written as a main effect in addition to participating in A*B, then the sum of squares that would be associated with B is pooled into A*B.

This feature allows the automatic pooling of sums of squares. If an effect is omitted from the model, it is automatically pooled with all the higher-level effects containing the class variables in the omitted effect (or within-error). This feature is most useful in split-plot designs.

SPECIFICATIONS

These statements are available in ANOVA:

PROC ANOVA *option*;
  CLASS *variables*;
  MODEL *dependents=effects / option*;
  MEANS *effects / options*;
  ABSORB *variables*;
  FREQ *variable*;
  TEST H=*effects* **E=***effect*;
  MANOVA H=*effects* **E=***effect / options*;
  BY *variables*;

The CLASS statement must precede the MODEL statement. TEST and MANOVA statements, if used, should follow the MODEL statement. More than one MEANS, TEST, or MANOVA statement may appear; the other statements may be used only once with a PROC ANOVA statement.

PROC ANOVA Statement

PROC ANOVA DATA=*SASdataset*;

DATA=*SASdataset* names the SAS data set to be analyzed by PROC ANOVA. If no DATA= option is specified, ANOVA uses the most recently created SAS data set.

CLASS Statement

CLASS *variables*;

Any variables used as classification variables in ANOVA must be declared first in the CLASS (or CLASSES) statement to identify the groups for the analysis. Typical classification variables are TRT, SEX, RACE, GROUP, and REP. They may be either numeric or character, but a character variable used in a CLASS statement cannot have a length greater than 16.

MODEL Statement

MODEL *dependents*=*effects* / *option*;

The MODEL statement names the dependent variables and independent effects. The syntax of effects is described in the introductory section **Specification of Effects**. If no effects are specified, ANOVA fits the response to a constant mean.

The option listed below may be specified in the MODEL statement and must be separated from the list of independent effects by a slash (/).

NOUNI requests that ANOVA not print the univariate descriptive statistics that are produced by default. Use NOUNI when you want only the multivariate statistics produced by a MANOVA statement.

MEANS Statement

MEANS *effects* / *options*;

Means can be computed for any effect involving class variables, whether or not the effect is specified in the MODEL statement. Any number of MEANS statements may be used either before or after the MODEL statement.

For example:

```
PROC ANOVA;
  CLASS A B C;
  MODEL Y=A B C;
  MEANS A B C A*B;
```

Means are printed for each level of the variables A, B, and C and for the combined levels of A and B.

The options below may appear in the MEANS statement after a slash (/). For a further discussion of these options, see **Comparison of Means** in the GLM procedure description.

BON performs Bonferroni *t* tests of differences between means for all main-effect means in the MEANS statement.

DUNCAN performs Duncan's multiple-range test on all main-effect means given in the MEANS statement.

GABRIEL performs Gabriel's multiple-comparison procedure on all main-effect means in the MEANS statement.

REGWF performs the Ryan-Einot-Gabriel-Welsch multiple *F* test on all main-effect means in the MEANS statement.

REGWQ performs the Ryan-Einot-Gabriel-Welsch multiple range test on all main-effect means in the MEANS statement.

SCHEFFE performs Scheffe's multiple-comparison procedure on all main-effect means in the MEANS statement.

SIDAK performs pairwise *t* tests on differences between means with levels adjusted according to Sidak's inequality for all main-effect means in the MEANS statement.

SMM
GT2 performs pairwise comparisons based on the studentized maximum modulus and Sidak's uncorrelated-*t* inequality, yielding Hochberg's GT2 method when sample sizes are unequal, for all main-effect means in the MEANS statement.

SNK performs the Student-Newman-Keuls multiple range test on all main-effect means in the MEANS statement.

T
LSD performs pairwise *t* tests, equivalent to Fisher's least-significant-difference test in the case of equal cell sizes, for all main-effect means in the MEANS statement.

TUKEY performs Tukey's studentized range test (HSD) on all main-effects means in the MEANS statement.

ALPHA=*p* gives the level of significance for comparisons among the means. The default ALPHA value is .05. With the DUNCAN option, you may specify only values of .01, .05, or .1.

WALLER requests that the Waller-Duncan k-ratio *t* test be performed on all main-effect means in the MEANS statement.

KRATIO=*value* gives the type1/type2 error seriousness ratio for the Waller-Duncan test. Reasonable values for KRATIO are 50, 100, 500, which roughly correspond for the two-level case to ALPHA levels of .1, .05, and .01. If KRATIO is omitted, the procedure uses the default value of 100.

LINES requests that the results of the BON, DUNCAN, GABRIEL, REGWF, REGWQ, SCHEFFE, SIDAK, SMM, GT2, SNK, T, LSD, TUKEY, and WALLER options be presented by listing the means in descending order and indicating nonsignificant subsets by line segments beside the corresponding means. LINES is appropriate for equal cell sizes, for which it is the default. LINES is also the default if DUNCAN, REGWF, REGWQ, SNK, or WALLER is specified. If the cell sizes are unequal, the harmonic means is used, which may lead to somewhat liberal tests if the cell sizes are highly disparate.

CLDIFF requests that the results of the BON, GABRIEL, SCHEFFE, SIDAK, SMM, GT2, T, LSD, and TUKEY options be presented as confidence intervals for all pairwise differences between means. CLDIFF is the default for unequal cell sizes unless DUNCAN, REGWF, REGWQ, SNK, or WALLER is specified.

E=*effect* specifies the error mean square to use in the multiple comparisons. If E= is omitted, the residual MS is used. The effect specified with the E= option must be a term in the model; otherwise, the residual MS is used. See **Comparison of Means** in the GLM procedure description for details on multiple comparison methods.

NOSORT prevents the means from being sorted into descending order when CLDIFF is specified.

ABSORB Statement

ABSORB *variables;*

The technique of absorption, requested by the ABSORB statement, saves time and reduces storage requirements for certain types of models. The analysis of variance is adjusted for the absorbed effects. See **Absorption** in the GLM procedure description for more information.

Restrictions: with the ABSORB statement, the data set (or each BY group, if a BY statement appears) must be sorted by the variables in the ABSORB statement. Including an absorbed variable in the CLASS list or in the MODEL statement produces erroneous sums of squares.

FREQ Statement

FREQ *variable;*

When a FREQ (or FREQUENCY) statement appears, each observation in the input data set is assumed to represent *n* observations in the experiment, where *n* is the value of the FREQ variable. If the value of the FREQ variable is less than 1, the observation is not used in the analysis. If the value is not an integer, only the integer portion is used.

The analysis produced using a FREQ statement is identical to an analysis produced using a data set that contains *n* observations (where *n* is the value of the FREQ variable) in place of each observation of the input data set. Therefore, means and total degrees of freedom reflect the expanded number of observations.

TEST Statement

TEST H=*effects* E=*effect;*

Although an F value is computed for all SS in the analysis using the residual MS as an error term, you may request additional *F* tests using other effects as error terms. You need this feature when a non-Iσ^2 error structure (as in a split plot) exists.

These terms are specified on the TEST statement:

H=*effects* specifies the effects in the preceding model to be used as hypothesis (numerator) effects.

E=*effect* specifies one, and only one, effect to be used as the error (denominator) term. If you use a TEST statement, you must specify an error term with E=.

For example,

```
PROC ANOVA;
  CLASS A B C;
  MODEL Y=A | B(A) | C;
  TEST H=A E=B(A);
  TEST H=C A*C E=B*C(A);
```

MANOVA Statement

MANOVA H=*effects* E=*effect / options*;

If the MODEL statement includes more than one dependent variable, multivariate statistics may be requested with the MANOVA statement.

When a MANOVA statement appears, ANOVA enters a multivariate mode for handling missing values: observations with missing independent or dependent variables are excluded from the analysis. Even when you do not want multivariate statistics, you can use the statement

```
MANOVA;
```

to request this method of treating missing values. See **Missing Values** below for more information.

These terms are specified on the MANOVA statement:

H=*effects* specifies the effects in the preceding model to be used for hypotheses. For each **H** matrix (the SSCP matrix associated with that effect), the characteristic roots and vectors of $\mathbf{E}^{-1}\mathbf{H}$ are printed (where **E** is the matrix associated with the error effect). The Hotelling-Lawley trace, Pillai's trace, Wilk's criterion, and Roy's maximum-root criterion are printed with approximate *F* statistics. The statistics test the joint hypothesis that the effect is not significant for any of the dependent variables.

E=*effect* specifies the error effect. If E= is not specified, the error SSCP (residual) matrix from the analysis is used.

The options below may appear in the MANOVA statement after a slash (/).

PRINTH requests that the **H** matrix (the SSCP matrix) associated with each effect specified by the H= option be printed.

PRINTE requests printing of the **E** matrix. The partial correlations are also printed if **E** contains the residuals. For example, the statement

```
MANOVA / PRINTE;
```

prints the error SSCP matrix and the partial correlation matrix computed from the error SSCP matrix.

Examples of the MANOVA statement:

```
PROC ANOVA;
  CLASS A B;
  MODEL Y1-Y5=A B(A);
  MANOVA H=A E=B(A) / PRINTH PRINTE;
  MANOVA H=B(A) / PRINTE;
```

In this example, the first MANOVA statement specifies A as the hypothesis effect and B(A) as the error effect. The PRINTH option requests that the **H** matrix associated with the A effect be printed, and the PRINTE option requests that the **E** matrix associated with the B(A) effect be printed.

The second MANOVA statement specifies B(A) as the hypothesis effect. Since no error effect is specified, the error SSCP matrix from the analysis is used as the **E** matrix. The PRINTE option requests that this **E** matrix be printed. Since the **E** matrix is the error SSCP matrix from the analysis, the partial correlation matrix computed from this matrix is also printed.

BY Statement

BY *variables*;

A BY statement may be used with PROC ANOVA to obtain separate analyses on observations in groups defined by the BY variables. When a BY statement appears, the procedure expects the input data set to be sorted in order of the BY variables. If your input data set is not sorted in ascending order, use the SORT procedure with a similar BY statement to sort the data, or, if appropriate, use the BY statement options NOTSORTED or DESCENDING. For more information, see the discussion of the BY statement in Chapter 8, "Statements Used in the PROC Step," in *SAS User's Guide: Basics, 1982 Edition*.

DETAILS

Missing Values

The dependent variables are grouped based on the similarity of their missing values. This feature is similar to the way GLM treats missing values.

Computational Method

Let **X** represent the $n \times p$ design matrix. The columns of **X** contain only 0s and 1s. Let **Y** represent the $n \times 1$ vector of dependent variables.

In the GLM procedure, **X'X**, **X'Y**, and **Y'Y** are formed in main storage. However, in the ANOVA procedure only the diagonals of **X'X** are computed, along with **X'Y** and **Y'Y**. Thus, ANOVA saves a considerable amount of storage as well as time.

The elements of **X'Y** are cell totals, and the diagonal elements of **X'X** are cell frequencies. Since ANOVA automatically pools omitted effects into the next higher-level effect containing the names of the omitted effect (or within-error), a slight modification to the rules given by Searle (1971, p. 389) is used.

> Step 1: The sum of squares for each effect is computed as if it were a main effect. In other words, for each effect, square each cell total and divide by its cell frequency. Add these quantities together and then subtract the correction factor for the mean (total squared over N).
>
> Step 2: For each effect involving two class names, subtract the SS for any main effect whose name is contained in the two-factor effect.
>
> Step 3: For each effect involving three class names, subtract the SS for all main effects and two-factor effects whose names are contained in the three-factor effect. If effects involving four or more class names are present, continue this process.

Printed Output

ANOVA first prints a table that includes:

1. the name of each variable in the CLASS statement
2. the number of different values or LEVELS of the CLASS variables
3. the VALUES of the CLASS variables
4. the number of observations in the data set and the number of observations excluded from the analysis because of missing values, if any.

ANOVA then prints an analysis-of-variance table for each dependent variable in the MODEL statement. This table breaks down:

5. the CORRECTED TOTAL sum of squares for the dependent variable
6. into the portion attributed to the MODEL
7. and the portion attributed to ERROR.
8. the MEAN SQUARE term is the
9. SUM OF SQUARES divided by the
10. DEGREES OF FREEDOM (DF).
11. The MEAN SQUARE for ERROR is an estimate of σ^2, the variance of the true errors.
12. the F VALUE is the ratio produced by dividing MS(MODEL), the mean square for the model, by the MS(ERROR), the mean square for error. It tests how well the model as a whole (adjusted for the mean) accounts for the dependent variable's behavior. This *F* test is a test that all parameters except the intercept are zero.
13. The significance probability associated with the *F* statistic, labeled PR>F.
14. R-SQUARE, R^2, measures how much variation in the dependent variable can be accounted for by the model. R^2, which can range from 0 to 1, is the ratio of the sum of squares for the model divided by the sum of squares for the corrected total. In general, the larger the R^2 value, the better the model fits the data.
15. C.V., the coefficient of variation, is often used to describe the amount of variation in the population. The C.V. is 100 times the standard deviation of the dependent variable, STD DEV, divided by the MEAN. The coefficient of variation is often a preferred measure because it is unitless.
16. ROOT MSE estimates the standard deviation of the dependent variable and is computed as the square root of MS(ERROR), the mean square of the error term
17. the MEAN of the dependent variable.

For each effect (or source of variation) in the model, ANOVA then prints:

18. DF, degrees of freedom
19. ANOVA SS, the sum of squares
20. the F VALUE for testing the hypothesis that the group means for that effect are equal
21. PR>F, the significance probability value associated with the F VALUE.

When a TEST statement is used, ANOVA prints the results of the tests requested in the TEST statement. When a MANOVA statement is used and the model includes more than one dependent variable, ANOVA prints these additional statistics (not shown in example output):

22. for each **H** matrix, the characteristic roots and vectors of $\mathbf{E}^{-1}\mathbf{H}$
23. the Hotelling-Lawley trace
24. Pillai's trace

25. Wilks' criterion
26. Roy's maximum root criterion.

These MANOVA tests are discussed in Chapter 1, "Introduction to SAS Regression Procedures" and in the GLM procedure description.

EXAMPLES

One-Way Layout with Means Comparisons: Example 1

The following data are derived from an experiment by Erdman (1946) and analyzed in Chapters 7 and 8 of Steel and Torrie (1980). The measurements are the nitrogen content of red clover plants inoculated with cultures of Rhizobium trifolii strains of bacteria and a composite of five Rhizobium meliloti strains. Several different means comparisons methods are requested.

```
DATA CLOVER;
  INPUT STRAIN $ NITROGEN @@;
  CARDS;
3DOK1  19.4  3DOK1  32.6  3DOK1  27.0  3DOK1  32.1  3DOK1  33.0
3DOK5  17.7  3DOK5  24.8  3DOK5  27.9  3DOK5  25.2  3DOK5  24.3
3DOK4  17.0  3DOK4  19.4  3DOK4   9.1  3DOK4  11.9  3DOK4  15.8
3DOK7  20.7  3DOK7  21.0  3DOK7  20.5  3DOK7  18.8  3DOK7  18.6
3DOK13 14.3  3DOK13 14.4  3DOK13 11.8  3DOK13 11.6  3DOK13 14.2
COMPOS 17.3  COMPOS 19.4  COMPOS 19.1  COMPOS 16.9  COMPOS 20.8
;
PROC ANOVA;
  CLASS STRAIN;
  MODEL NITROGEN=STRAIN;
  MEANS STRAIN / DUNCAN WALLER;
  MEANS STRAIN / LSD TUKEY CLDIFF;
```

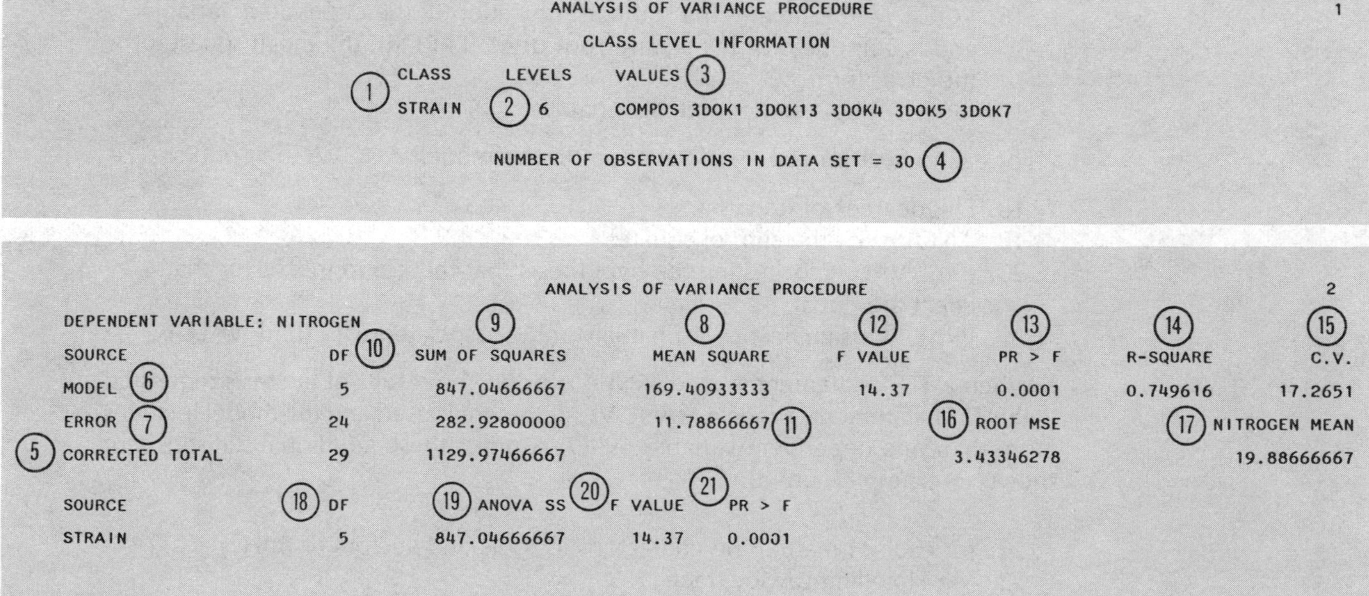

```
                              ANALYSIS OF VARIANCE PROCEDURE                                         1

                                 CLASS LEVEL INFORMATION

                (1) CLASS     LEVELS      VALUES (3)
                    STRAIN   (2) 6        COMPOS 3DOK1 3DOK13 3DOK4 3DOK5 3DOK7

                          NUMBER OF OBSERVATIONS IN DATA SET = 30 (4)
```

```
                              ANALYSIS OF VARIANCE PROCEDURE                                         2
DEPENDENT VARIABLE: NITROGEN          (9)                (8)            (12)        (13)        (14)          (15)
SOURCE                   DF (10)  SUM OF SQUARES      MEAN SQUARE      F VALUE     PR > F     R-SQUARE        C.V.
MODEL (6)                 5       847.04666667        169.40933333       14.37      0.0001     0.749616      17.2651
ERROR (7)                24       282.92800000         11.78866667 (11)       (16) ROOT MSE   (17) NITROGEN MEAN
(5) CORRECTED TOTAL      29      1129.97466667                                    3.43346278       19.88666667

SOURCE              (18) DF     (19) ANOVA SS (20) F VALUE (21) PR > F
STRAIN                    5       847.04666667      14.37      0.0001
```

```
                         ANALYSIS OF VARIANCE PROCEDURE                              3

WALLER-DUNCAN K-RATIO T TEST FOR VARIABLE: NITROGEN
NOTE: THIS TEST MINIMIZES THE BAYES RISK UNDER ADDITIVE LOSS
      AND CERTAIN OTHER ASSUMPTIONS.

KRATIO=100  DF=24  MSE=11.7887  F=14.3705
CRITICAL VALUE OF T=1.91853
MINIMUM SIGNIFICANT DIFFERENCE=4.16612

MEANS WITH THE SAME LETTER ARE NOT SIGNIFICANTLY DIFFERENT.

WALLER   GROUPING            MEAN      N  STRAIN

                A          28.820      5  3DOK1

                B          23.980      5  3DOK5
                B
         C      B          19.920      5  3DOK7
         C
         C      D          18.700      5  COMPOS
                D
         E      D          14.640      5  3DOK4
         E
         E                 13.260      5  3DOK13
```

```
                         ANALYSIS OF VARIANCE PROCEDURE                              4

DUNCAN'S MULTIPLE RANGE TEST FOR VARIABLE: NITROGEN
NOTE: THIS TEST CONTROLS THE TYPE I COMPARISONWISE ERROR RATE,
      NOT THE EXPERIMENTWISE ERROR RATE.

ALPHA=0.05  DF=24  MSE=11.7887

MEANS WITH THE SAME LETTER ARE NOT SIGNIFICANTLY DIFFERENT.

DUNCAN   GROUPING            MEAN      N  STRAIN

                A          28.820      5  3DOK1

                B          23.980      5  3DOK5
                B
         C      B          19.920      5  3DOK7
         C
         C      D          18.700      5  COMPOS
                D
         E      D          14.640      5  3DOK4
         E
         E                 13.260      5  3DOK13
```

```
                         ANALYSIS OF VARIANCE PROCEDURE                              5

T TESTS (LSD) FOR VARIABLE: NITROGEN
NOTE: THIS TEST CONTROLS THE TYPE I COMPARISONWISE ERROR RATE,
      NOT THE EXPERIMENTWISE ERROR RATE.

ALPHA=0.05  CONFIDENCE=0.95  DF=24  MSE=11.7887
CRITICAL VALUE OF T=2.0639
LEAST SIGNIFICANT DIFFERENCE=4.48178

COMPARISONS SIGNIFICANT AT THE 0.05 LEVEL ARE INDICATED BY '***'

                       LOWER    DIFFERENCE     UPPER
    STRAIN          CONFIDENCE   BETWEEN    CONFIDENCE
  COMPARISON           LIMIT      MEANS        LIMIT


3DOK1  - 3DOK5         0.358      4.840        9.322    ***
3DOK1  - 3DOK7         4.418      8.900       13.382    ***
3DOK1  - COMPOS        5.638     10.120       14.602    ***
3DOK1  - 3DOK4         9.698     14.180       18.662    ***
3DOK1  - 3DOK13       11.078     15.560       20.042    ***

3DOK5  - 3DOK1        -9.322     -4.840       -0.358    ***
3DOK5  - 3DOK7        -0.422      4.060        8.542
3DOK5  - COMPOS        0.798      5.280        9.762    ***
3DOK5  - 3DOK4         4.858      9.340       13.822    ***
3DOK5  - 3DOK13        6.238     10.720       15.202    ***

3DOK7  - 3DOK1       -13.382     -8.900       -4.418    ***
3DOK7  - 3DOK5        -8.542     -4.060        0.422
3DOK7  - COMPOS       -3.262      1.220        5.702
3DOK7  - 3DOK4         0.798      5.280        9.762    ***
3DOK7  - 3DOK13        2.178      6.660       11.142    ***

COMPOS - 3DOK1       -14.602    -10.120       -5.638    ***
COMPOS - 3DOK5        -9.762     -5.280       -0.798    ***
COMPOS - 3DOK7        -5.702     -1.220        3.262
COMPOS - 3DOK4        -0.422      4.060        8.542
COMPOS - 3DOK13        0.958      5.440        9.922    ***
```

(continued on next page)

(continued from previous page)

```
3DOK4  - 3DOK1    -18.662    -14.180    -9.698   ***
3DOK4  - 3DOK5    -13.822     -9.340    -4.858   ***
3DOK4  - 3DOK7     -9.762     -5.280    -0.798   ***
3DOK4  - COMPOS    -8.542     -4.060     0.422
3DOK4  - 3DOK13    -3.102      1.380     5.862

3DOK13 - 3DOK1    -20.042    -15.560   -11.078   ***
3DOK13 - 3DOK5    -15.202    -10.720    -6.238   ***
3DOK13 - 3DOK7    -11.142     -6.660    -2.178   ***
3DOK13 - COMPOS    -9.922     -5.440    -0.958   ***
3DOK13 - 3DOK4     -5.862     -1.380     3.102
```

```
                               ANALYSIS OF VARIANCE PROCEDURE                              6

TUKEY'S STUDENTIZED RANGE (HSD) TEST FOR VARIABLE: NITROGEN
NOTE: THIS TEST CONTROLS THE TYPE I EXPERIMENTWISE ERROR RATE

ALPHA=0.05  CONFIDENCE=0.95  DF=24  MSE=11.7887
CRITICAL VALUE OF STUDENTIZED RANGE=4.37266
MINIMUM SIGNIFICANT DIFFERENCE=6.71419

COMPARISONS SIGNIFICANT AT THE 0.05 LEVEL ARE INDICATED BY '***'

                   SIMULTANEOUS               SIMULTANEOUS
                      LOWER      DIFFERENCE      UPPER
     STRAIN        CONFIDENCE     BETWEEN     CONFIDENCE
   COMPARISON         LIMIT        MEANS         LIMIT

3DOK1  - 3DOK5     -1.874      4.840     11.554
3DOK1  - 3DOK7      2.186      8.900     15.614   ***
3DOK1  - COMPOS     3.406     10.120     16.834   ***
3DOK1  - 3DOK4      7.466     14.180     20.894   ***
3DOK1  - 3DOK13     8.846     15.560     22.274   ***

3DOK5  - 3DOK1    -11.554     -4.840      1.874
3DOK5  - 3DOK7     -2.654      4.060     10.774
3DOK5  - COMPOS    -1.434      5.280     11.994
3DOK5  - 3DOK4      2.626      9.340     16.054   ***
3DOK5  - 3DOK13     4.006     10.720     17.434   ***

3DOK7  - 3DOK1    -15.614     -8.900     -2.186   ***
3DOK7  - 3DOK5    -10.774     -4.060      2.654
3DOK7  - COMPOS    -5.494      1.220      7.934
3DOK7  - 3DOK4     -1.434      5.280     11.994
3DOK7  - 3DOK13    -0.054      6.660     13.374

COMPOS - 3DOK1    -16.834    -10.120     -3.406   ***
COMPOS - 3DOK5    -11.994     -5.280      1.434
COMPOS - 3DOK7     -7.934     -1.220      5.494
COMPOS - 3DOK4     -2.654      4.060     10.774
COMPOS - 3DOK13    -1.274      5.440     12.154

3DOK4  - 3DOK1    -20.894    -14.180     -7.466   ***
3DOK4  - 3DOK5    -16.054     -9.340     -2.626   ***
3DOK4  - 3DOK7    -11.994     -5.280      1.434
3DOK4  - COMPOS   -10.774     -4.060      2.654
3DOK4  - 3DOK13    -5.334      1.380      8.094

3DOK13 - 3DOK1    -22.274    -15.560     -8.846   ***
3DOK13 - 3DOK5    -17.434    -10.720     -4.006   ***
3DOK13 - 3DOK7    -13.374     -6.660      0.054
3DOK13 - COMPOS   -12.154     -5.440      1.274
3DOK13 - 3DOK4     -8.094     -1.380      5.334
```

Randomized Complete Block: Example 2

The example below shows statements for the analysis of a randomized block. Since the data for the analysis are balanced, we use PROC ANOVA. The blocking variable BLOCK and the treatment variable TRTMENT appear in the CLASS statement, and the MODEL statement requests an analysis for each of the two dependent variables YIELD and WORTH.

```
TITLE RANDOMIZED COMPLETE BLOCK;
DATA RCB;
  INPUT BLOCK TRTMENT $ YIELD WORTH;
  CARDS;
```

```
1 A 32.6 112
1 B 36.4 130
1 C 29.5 106
2 A 42.7 139
2 B 47.1 143
2 C 32.9 112
3 A 35.3 124
3 B 40.1 134
3 C 33.6 116
PROC ANOVA;
  CLASS BLOCK TRTMENT;
  MODEL YIELD WORTH = BLOCK TRTMENT;
```

```
                    RANDOMIZED COMPLETE BLOCK                                   1
                  ANALYSIS OF VARIANCE PROCEDURE
                     CLASS LEVEL INFORMATION
                    CLASS      LEVELS     VALUES
                    BLOCK           3     1 2 3
                    TRTMENT         3     A B C

              NUMBER OF OBSERVATIONS IN DATA SET = 9
```

```
                    RANDOMIZED COMPLETE BLOCK                                   2
                  ANALYSIS OF VARIANCE PROCEDURE
DEPENDENT VARIABLE: YIELD
SOURCE              DF     SUM OF SQUARES      MEAN SQUARE     F VALUE       PR > F     R-SQUARE         C.V.
MODEL                4       225.27777778      56.31944444        8.94       0.0283     0.899424       6.8400
ERROR                4        25.19111111       6.29777778                 ROOT MSE                YIELD MEAN
CORRECTED TOTAL      8       250.46888889                                2.50953736               36.68888889

SOURCE              DF           ANOVA SS    F VALUE    PR > F
BLOCK                2        98.17555556       7.79    0.0417
TRTMENT              2       127.10222222      10.09    0.0274
```

```
                    RANDOMIZED COMPLETE BLOCK                                   3
                  ANALYSIS OF VARIANCE PROCEDURE
DEPENDENT VARIABLE: WORTH
SOURCE              DF     SUM OF SQUARES      MEAN SQUARE     F VALUE       PR > F     R-SQUARE         C.V.
MODEL                4      1247.33333333     311.83333333        8.28       0.0323     0.892227       4.9494
ERROR                4       150.66666667      37.66666667                 ROOT MSE                WORTH MEAN
CORRECTED TOTAL      8      1398.00000000                                6.13731755              124.00000000

SOURCE              DF           ANOVA SS    F VALUE    PR > F
BLOCK                2       354.66666667       4.71    0.0889
TRTMENT              2       892.66666667      11.85    0.0209
```

Split Plot: Example 3

The statements below produce an analysis for a split-plot design. The CLASS statement includes the variables BLOCK, A, and B. The MODEL statement includes the independent effects BLOCK, A, BLOCK*A, B, and A*B. The TEST statement asks for an *F* test using the BLOCK*A effect as the error term and the A effect as the hypothesis effect.

```
* -------------------------------------------SPLIT  PLOT---------------------------------------- *
|  B DEFINES SUBPLOTS WITHIN A*BLOCK WHOLE PLOTS.                                                  |
|  THE WHOLE PLOT EFFECTS MUST BE TESTED WITH A                                                    |
|  TEST STATEMENT AGAINST BLOCK*A. THE SUBPLOT                                                     |
|  EFFECTS CAN BE TESTED AGAINST THE RESIDUAL.                                                     |
* ---------------------------------------------------------------------------------------------- * ;
 DATA SPLIT;
   INPUT BLOCK 1 A 2 B 3 RESPONSE ;
   CARDS;
 142 40.0
 141 39.5
 112 37.9
 111 35.4
 121 36.7
 122 38.2
 132 36.4
 131 34.8
 221 42.7
 222 41.6
 212 40.3
 211 41.6
 241 44.5
 242 47.6
 231 43.6
 232 42.8
 PROC ANOVA;
   CLASS BLOCK A B;
   MODEL RESPONSE = BLOCK A BLOCK*A B A*B ;
   TEST H=A E=BLOCK*A;
   TITLE SPLIT PLOT DESIGN;
```

```
                              SPLIT PLOT DESIGN                                   1
                        ANALYSIS OF VARIANCE PROCEDURE
                           CLASS LEVEL INFORMATION
                        CLASS       LEVELS     VALUES
                        BLOCK            2     1 2
                        A                4     1 2 3 4
                        B                2     1 2
                  NUMBER OF OBSERVATIONS IN DATA SET = 16
```

```
                              SPLIT PLOT DESIGN                                   2
                        ANALYSIS OF VARIANCE PROCEDURE
DEPENDENT VARIABLE: RESPONSE
SOURCE           DF     SUM OF SQUARES      MEAN SQUARE    F VALUE     PR > F     R-SQUARE       C.V.
MODEL            11       182.02000000      16.54727273       7.85     0.0306     0.955736     3.6090
```

(continued on next page)

(continued from previous page)

ERROR	4	8.43000000	2.10750000	ROOT MSE	RESPONSE MEAN
CORRECTED TOTAL	15	190.45000000		1.45172311	40.22500000

SOURCE	DF	ANOVA SS	F VALUE	PR > F
BLOCK	1	131.10250000	62.21	0.0014
A	3	40.19000000	6.36	0.0530
BLOCK*A	3	6.92750000	1.10	0.4476
B	1	2.25000000	1.07	0.3599
A*B	3	1.55000000	0.25	0.8612

TESTS OF HYPOTHESES USING THE ANOVA MS FOR BLOCK*A AS AN ERROR TERM

SOURCE	DF	ANOVA SS	F VALUE	PR > F
A	3	40.19000000	5.80	0.0914

Latin-Square Split Plot: Example 4

The Latin-square design below using data from W.G. Smith (1951) is used to evaluate 6 different sugar beet varieties arranged in a 6-row (REP) by 6-column (COL) square. Then the data are recollected for a second harvest. Then HARVEST becomes a split plot on the original Latin-square design for whole plots.

```
DATA BEETS;
  DO HARVEST=1 TO 2;
    DO REP=1 TO 6;
      DO COL=1 TO 6;
        INPUT VARIETY Y @; OUTPUT;
        END;
      END;
    END;
CARDS;
3 19.1 6 18.3 5 19.6 1 18.6 2 18.2 4 18.5
6 18.1 2 19.5 4 17.6 3 18.7 1 18.7 5 19.9
1 18.1 5 20.2 6 18.5 4 20.1 3 18.6 2 19.2
2 19.1 3 18.8 1 18.7 5 20.2 4 18.6 6 18.5
4 17.5 1 18.1 2 18.7 6 18.2 5 20.4 3 18.5
5 17.7 4 17.8 3 17.4 2 17.0 6 17.6 1 17.6
3 16.2 6 17.0 5 18.1 1 16.6 2 17.7 4 16.3
6 16.0 2 15.3 4 16.0 3 17.1 1 16.5 5 17.6
1 16.5 5 18.1 6 16.7 4 16.2 3 16.7 2 17.3
2 17.5 3 16.0 1 16.4 5 18.0 4 16.6 6 16.1
4 15.7 1 16.1 2 16.7 6 16.3 5 17.8 3 16.2
5 18.3 4 16.6 3 16.4 2 17.6 6 17.1 1 16.5
;
PROC ANOVA;
  CLASS COL REP VARIETY HARVEST;
  MODEL Y=REP COL VARIETY REP*COL*VARIETY
              HARVEST HARVEST*REP
              HARVEST*VARIETY;

  TEST H=REP COL VARIETY E=REP*COL*VARIETY;
  TEST H=HARVEST E=HARVEST*REP;
```

ANALYSIS OF VARIANCE PROCEDURE 1

CLASS LEVEL INFORMATION

CLASS	LEVELS	VALUES
COL	6	1 2 3 4 5 6
REP	6	1 2 3 4 5 6
VARIETY	6	1 2 3 4 5 6
HARVEST	2	1 2

NUMBER OF OBSERVATIONS IN DATA SET = 72

ANALYSIS OF VARIANCE PROCEDURE 2

DEPENDENT VARIABLE: Y

SOURCE	DF	SUM OF SQUARES	MEAN SQUARE	F VALUE	PR > F	R-SQUARE	C.V.
MODEL	46	98.91472222	2.15032005	7.22	0.0001	0.929971	3.0855
ERROR	25	7.44847222	0.29793889		ROOT MSE		Y MEAN
CORRECTED TOTAL	71	106.36319444			0.54583779		17.69027778

SOURCE	DF	ANOVA SS	F VALUE	PR > F
REP	5	4.32069444	2.90	0.0337
COL	5	1.57402778	1.06	0.4075
VARIETY	5	20.61902778	13.84	0.0001
COL*REP*VARIETY	20	3.25444444	0.55	0.9144
HARVEST	1	60.68347222	203.68	0.0001
REP*HARVEST	5	7.71736111	5.18	0.0021
VARIETY*HARVEST	5	0.74569444	0.50	0.7729

TESTS OF HYPOTHESES USING THE ANOVA MS FOR COL*REP*VARIETY AS AN ERROR TERM

SOURCE	DF	ANOVA SS	F VALUE	PR > F
REP	5	4.32069444	5.31	0.0029
COL	5	1.57402778	1.93	0.1333
VARIETY	5	20.61902778	25.34	0.0001

TESTS OF HYPOTHESES USING THE ANOVA MS FOR REP*HARVEST AS AN ERROR TERM

SOURCE	DF	ANOVA SS	F VALUE	PR > F
HARVEST	1	60.68347222	39.32	0.0015

Strip-Split Plot: Example 5

In this example, the fertilizer treatments are laid out in vertical strips, which are then split into calcium-effect subplots. Soil type is stripped across the split-plot experiment, and the entire experiment is then replicated three times.

The input data are the 96 values of Y, arranged so that the calcium value (CA) changes most rapidly, then the fertilizer value (FERTIL), then the SOIL value, and finally the REP value. Values are shown for CA (0 and 1); FERTIL (0, 1, 2, 3); SOIL (1, 2, 3); and REP (1, 2, 3, 4).

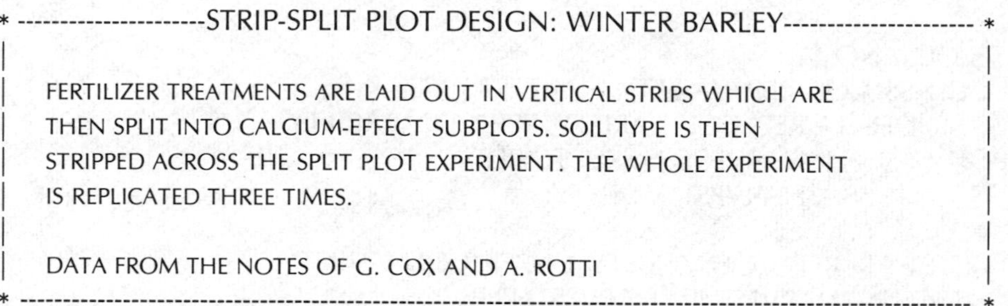

```
* ---------------------STRIP-SPLIT PLOT DESIGN: WINTER BARLEY--------------------- *
|                                                                                  |
|  FERTILIZER TREATMENTS ARE LAID OUT IN VERTICAL STRIPS WHICH ARE                 |
|  THEN SPLIT INTO CALCIUM-EFFECT SUBPLOTS. SOIL TYPE IS THEN                      |
|  STRIPPED ACROSS THE SPLIT PLOT EXPERIMENT. THE WHOLE EXPERIMENT                 |
|  IS REPLICATED THREE TIMES.                                                      |
|                                                                                  |
|  DATA FROM THE NOTES OF G. COX AND A. ROTTI                                      |
* -------------------------------------------------------------------------------- * ;
```

```
DATA BARLEY;
  DO REP=1 TO 4;
    DO SOIL=1 TO 3;      * 1=D 2=H 3=P;
      DO FERTIL=0 TO 3;
        DO CA=0,1;
          INPUT Y @;  OUTPUT;
          END;
        END;
      END;
    END;
  CARDS;
4.91 4.63 4.76 5.04 5.38 6.21 5.60 5.08
4.94 3.98 4.64 5.26 5.28 5.01 5.45 5.62
5.20 4.45 5.05 5.03 5.01 4.63 5.80 5.90
6.00 5.39 4.95 5.39 6.18 5.94 6.58 6.25
5.86 5.41 5.54 5.41 5.28 6.67 6.65 5.94
5.45 5.12 4.73 4.62 5.06 5.75 6.39 5.62
4.96 5.63 5.47 5.31 6.18 6.31 5.95 6.14
5.71 5.37 6.21 5.83 6.28 6.55 6.39 5.57
4.60 4.90 4.88 4.73 5.89 6.20 5.68 5.72
5.79 5.33 5.13 5.18 5.86 5.98 5.55 4.32
5.61 5.15 4.82 5.06 5.67 5.54 5.19 4.46
5.13 4.90 4.88 5.18 5.45 5.80 5.12 4.42
;

* NOTE THAT SINCE THE MODEL IS COMPLETELY SPECIFIED AND SEVERAL
* ERROR TERMS ARE PRESENT, THE TEST STATEMENT MUST BE USED TO
* OBTAIN THE PROPER TEST STATISTICS. THE TOP PORTION OF THE OUTPUT
* SHOULD BE IGNORED, SINCE THE RESIDUAL ERROR TERM IS NOT
* MEANINGFUL HERE;

PROC ANOVA;
  CLASS REP SOIL CA FERTIL;
  MEANS F CA S CA*F;
  MODEL Y=REP
            FERTIL FERTIL*REP
            CA CA*FERTIL CA*REP(FERTIL)
            SOIL SOIL*REP
            SOIL*FERTIL SOIL*REP*FERTIL
            SOIL*CA SOIL*FERTIL*CA
            SOIL*CA*REP(FERTIL);
  TEST H=FERTIL E=FERTIL*REP;
  TEST H=CA CA*FERTIL E=CA*REP(FERTIL);
  TEST H=SOIL E=SOIL*REP;
  TEST H=SOIL*FERTIL E=SOIL*REP*FERTIL;
  TEST H=SOIL*CA
           SOIL*FERTIL*CA E=SOIL*CA*REP(FERTIL);
  TITLE STRIP-SPLIT PLOT;
```

```
                              STRIP-SPLIT PLOT                                  1
                       ANALYSIS OF VARIANCE PROCEDURE
                          CLASS LEVEL INFORMATION
                        CLASS      LEVELS     VALUES
                        REP             4     1 2 3 4
                        SOIL            3     1 2 3
                        CA              2     0 1
                        FERTIL          4     0 1 2 3
                 NUMBER OF OBSERVATIONS IN DATA SET = 96
```

```
                              STRIP-SPLIT PLOT                                  2
                       ANALYSIS OF VARIANCE PROCEDURE
DEPENDENT VARIABLE: Y
SOURCE               DF    SUM OF SQUARES     MEAN SQUARE    F VALUE      PR > F     R-SQUARE          C.V.
MODEL                95       31.89149583      0.33569996   99999.99      0.0000     1.000000        0.0000
ERROR                 0        0.00000000      0.00000000               ROOT MSE                     Y MEAN
CORRECTED TOTAL      95       31.89149583                               0.00000000               5.42729167

SOURCE               DF          ANOVA SS    F VALUE    PR > F

REP                   3        6.27974583          .         .
FERTIL                3        7.22127083          .         .
REP*FERTIL            9        6.08211250          .         .
CA                    1        0.27735000          .         .
CA*FERTIL             3        1.96395833          .         .
REP*CA(FERTIL)       12        1.76705833          .         .
SOIL                  2        1.92658958          .         .
REP*SOIL              6        1.66761042          .         .
SOIL*FERTIL           6        0.68828542          .         .
REP*SOIL*FERTIL      18        1.58698125          .         .
SOIL*CA               2        0.04493125          .         .
SOIL*CA*FERTIL        6        0.18936042          .         .
REP*SOIL*CA(FERTIL)  24        2.19624167          .         .

TESTS OF HYPOTHESES USING THE ANOVA MS FOR REP*FERTIL AS AN ERROR TERM
SOURCE               DF          ANOVA SS    F VALUE    PR > F
FERTIL                3        7.22127083       3.56    0.0604

TESTS OF HYPOTHESES USING THE ANOVA MS FOR REP*CA(FERTIL) AS AN ERROR TERM
SOURCE               DF          ANOVA SS    F VALUE    PR > F
CA                    1        0.27735000       1.88    0.1950
CA*FERTIL             3        1.96395833       4.45    0.0255

TESTS OF HYPOTHESES USING THE ANOVA MS FOR REP*SOIL AS AN ERROR TERM
SOURCE               DF          ANOVA SS    F VALUE    PR > F
SOIL                  2        1.92658958       3.47    0.0999
```

```
                              STRIP-SPLIT PLOT                                  3
                       ANALYSIS OF VARIANCE PROCEDURE
DEPENDENT VARIABLE: Y

TESTS OF HYPOTHESES USING THE ANOVA MS FOR REP*SOIL*FERTIL AS AN ERROR TERM
SOURCE               DF          ANOVA SS    F VALUE    PR > F
SOIL*FERTIL           6        0.68828542       1.30    0.3063

TESTS OF HYPOTHESES USING THE ANOVA MS FOR REP*SOIL*CA(FERTIL) AS AN ERROR TERM
SOURCE               DF          ANOVA SS    F VALUE    PR > F
SOIL*CA               2        0.04493125       0.25    0.7843
SOIL*CA*FERTIL        6        0.18936042       0.34    0.9059
```

```
                         STRIP-SPLIT PLOT                              4
                  ANALYSIS OF VARIANCE PROCEDURE
                              MEANS

                  FERTIL            N             Y

                  0                24     5.18416667
                  1                24     5.12916667
                  2                24     5.75458333
                  3                24     5.64125000

                    CA            N             Y

                    0            48     5.48104167
                    1            48     5.37354167

                    SOIL            N             Y

                    1              32     5.54312500
                    2              32     5.51093750
                    3              32     5.22781250

              CA    FERTIL            N             Y

              0     0                12     5.34666667
              0     1                12     5.08833333
              0     2                12     5.62666667
              0     3                12     5.86250000
              1     0                12     5.02166667
              1     1                12     5.17000000
              1     2                12     5.88250000
              1     3                12     5.42000000
```

REFERENCES

Erdman, L.W. (1946), "Studies to Determine If Antibiosis Occurs Among Rhizobia," *Journal of the American Society of Agronomy*, 38, 251-258.

Fisher, R.A. (1942), *The Design of Experiments,* Third Edition, Edinburgh: Oliver & Boyd.

Freund, R.J. and Littell, R.C. (1981), *SAS for Linear Models: A Guide to the ANOVA and GLM Procedures*, Cary, NC: SAS Institute, Inc.

Graybill, F.A. (1961), *An Introduction to Linear Statistical Models,* Vol. I, New York: McGraw-Hill.

Henderson, C.R. (1953), "Estimation of Variance and Covariance Components," *Biometrics*, 9, 226-252.

Remington, R.D. and Schork, M.A. (1970), *Statistics with Applications to the Biological and Health Sciences*, Englewood Cliffs, New Jersey: Prentice-Hall, Inc.

Scheffe, H. (1959), *The Analysis of Variance*, New York: John Wiley & Sons.

Searle, S.R. (1971), *Linear Models*, New York: John Wiley & Sons.

Snedecor, G.W. and Cochran, W.G. (1967), *Statistical Methods,* Sixth Edition, Ames, Iowa: The Iowa State University Press.

Steel, R.G.D. and Torrie, J.H. (1980), *Principles and Procedures of Statistics*, New York: McGraw-Hill.

The GLM Procedure

ABSTRACT

The GLM procedure uses the method of least squares to fit general linear models. Among the statistical methods available in GLM are regression, analysis of variance, analysis of covariance, multivariate analysis of variance, and partial correlation.

INTRODUCTION

PROC GLM analyzes data within the framework of **G**eneral **L**inear **M**odels, hence the name GLM. GLM handles classification variables, which have discrete levels, as well as continuous variables, which measure quantities. Thus GLM can be used for many different analyses including:

- simple regression
- multiple regression
- analysis of variance (*ANOVA*), especially for unbalanced data
- analysis of covariance
- response-surface models
- weighted regression
- polynomial regression
- partial correlation
- multivariate analysis of variance (*MANOVA*).

Specification of Effects

Each term in an analysis-of-variance model is an effect that is specified with a special notation. Effects are constructed with variable names and operators. There are two kinds of variables: *classification* or *class variables* and *continuous variables*. There are two kinds of operators: *crossing* and *nesting*. A third operator, the *bar operator*, is used to construct other effects.

Analysis-of-variance models require variables that identify classification levels. In SAS these are called *class variables* and are declared in the CLASS statement. (They may also be called *categorical*, *qualitative*, *discrete*, or *nominal variables*.) Class variables may be either *numeric* or *character*. The values of a class variable are called *levels*.

Any variable used in a model that is not declared in the CLASS statement is assumed to be continuous. Continuous variables, which must be numeric, are used for response variables and covariates.

There are seven different types of effects used in GLM. In the following list assume that A, B, C, D, and E are class variables and X1, X2 and Y are continuous variables:

- regressor effects are specified by writing continuous variables by themselves: X1 X2
- polynomial effects are specified by joining two or more continuous variables with asterisks: X1*X1 X1*X2
- main effects are specified by writing class variables by themselves: A B C

- crossed effects (interactions) are specified by joining class variables with asterisks: A*B B*C A*B*C
- nested effects are specified by placing a parenthetical field after a variable or interaction indicating the class variable within which the effect is nested: B(A) C(B A) D*E(C B A)
  Note: B(A) is read "B nested within A."
- continuous-by-class effects are written by joining continuous variables with class variables: X1*A
- continuous-nesting effects consist of continuous variables followed by a parenthetical field of class variables: X1(A) X1*X2(A B).

One example of the general form of an effect involving seven variables is:

```
X1*X2*A*B*C(D E)  .
```

This contains polynomial terms by crossed terms nested within multiple class variables. The continuous list comes first, followed by the crossed list, followed by the nested list in parentheses. Note that no asterisks appear within the nested list or before the left parenthesis. For details on how the design matrix and parameters are defined with respect to the effects specified in this section, see **Parameterization** below.

The MODEL statement and several other statements use these effects. Some examples of MODEL statements using various kinds of effects are shown below. A, B, and C represent class variables and X1-X3 represent continuous variables.

specification	*kind of model*
MODEL Y=X1;	simple regression
MODEL Y=X1 X2;	multiple regression
MODEL Y=X1 X1*X1;	polynomial regression
MODEL Y1 Y2=X1 X2;	multivariate regression
MODEL Y=A;	one-way layout
MODEL Y=A B C;	main-effects model
MODEL Y=A B A*B;	factorial model (with interaction)
MODEL Y=A B(A) C(B A);	nested model
MODEL Y1 Y2=A B;	multivariate analysis of variance
MODEL Y=A X1;	analysis-of-covariance model
MODEL Y=A X1(A);	separate-slopes model
MODEL Y=A X1 X1*A;	homogeneity-of-slopes model

You can shorten the specification of a full factorial model using bar notation. For example, two ways of writing a full three-way factorial are:

```
PROC GLM;
  CLASS A B C;
  MODEL Y=A B C A*B A*C B*C A*B*C;

PROC GLM;
  CLASS A B C;
  MODEL Y=A|B|C;
```

When the bar (|) is used, the right and left sides become effects, and the cross of them becomes an effect. Multiple bars are permitted. The expressions are expanded from left to right, using rules 2-4 given in Searle (1971, p. 390):

- multiple bars are evaluated left to right. For instance, A|B|C is [A|B]|C, which is [A B A*B]|C, which is

  A B A*B C A*C B*C A*B*C.

- crossed and nested groups of variables are combined. For example, A(B)|C(D) generates A*C(B D), among other terms.
- duplicate variables are removed. For example, A(C)|B(C) generates A*B(C), among other terms, and the extra C is removed.
- effects are discarded if a variable occurs on both the crossed and nested sides of an effect. For instance, A(B)|B(D E) generates A*B(B D E), but this effect is eliminated immediately.

Other examples of the bar notation:

A|C(B)	is equivalent to A C(B) A*C(B)
A(B)|C(B)	is equivalent to A(B) C(B) A*C(B)
A(B)|B(D E)	is equivalent to A(B) B(D E)
A|B(A)|C	is equivalent to A B(A) C A*C B*C(A).

GLM for Multiple Regression

In multiple regression, the values of a dependent variable (or response variable) are described or predicted in terms of one or more independent or explanatory variables. The statements

```
PROC GLM;
  MODEL dependent=independents;
```

can be used to describe a multiple regression model in GLM.

GLM for Unbalanced ANOVA

The ANOVA procedure should be used whenever possible for analysis of variance, since ANOVA processes data more efficiently than GLM. However, GLM is used in most unbalanced situations.

Here is an example of a 2×2 factorial model. The data are shown in a table and then read into a SAS data set:

		A = 1	A = 2
B	1	12 14	20 18
B	2	11 9	17

```
DATA EXP;
  INPUT A $ B $ Y @@;
  CARDS;
A1 B1 12 A1 B1 14 A1 B2 11 A1 B2 9
A2 B1 20 A2 B1 18 A2 B2 17
;
```

Note that for the second levels of A and B there is only one value. Since one cell contains a different number of values from the other cells in the table, this is an unbalanced design, and GLM should be used. The statements needed for this two-way factorial model are:

```
PROC GLM;
  CLASS A B;
  MODEL Y=A B A*B;
```

The results from GLM are shown below.

```
                                                                                         1
                         GENERAL LINEAR MODELS PROCEDURE

                            CLASS LEVEL INFORMATION

                          CLASS       LEVELS      VALUES

                          A                2      A1 A2

                          B                2      B1 B2

                    NUMBER OF OBSERVATIONS IN DATA SET = 7
```

```
                                                                                                  2
                         GENERAL LINEAR MODELS PROCEDURE

DEPENDENT VARIABLE: Y

SOURCE             DF      SUM OF SQUARES        MEAN SQUARE     F VALUE        PR > F       R-SQUARE          C.V.

MODEL               3         91.71428571        30.57142857       15.29        0.0253       0.938596        9.8015

ERROR               3          6.00000000         2.00000000                   ROOT MSE                      Y MEAN

CORRECTED TOTAL     6         97.71428571                                    1.41421356                 14.42857143

SOURCE             DF           TYPE I SS    F VALUE     PR > F      DF       TYPE III SS    F VALUE      PR > F

A                   1         80.04761905      40.02     0.0080       1       67.60000000      33.80      0.0101
B                   1         11.26666667       5.63     0.0982       1       10.00000000       5.00      0.1114
A*B                 1          0.40000000       0.20     0.6850       1        0.40000000       0.20      0.6850
```

Four types of estimable functions of parameters are available for testing hypotheses in GLM. For data with no missing cells, the TYPE III and TYPE IV estimable functions are the same and test the same hypotheses that would be tested if the data were balanced.

The TYPE III results on this printout indicate no A*B interaction and a significant A effect. Further investigation of factor B could be made, however, since the significance level of the B main effect is $p = .1114$.

GLM Features

The following list summarizes the features in PROC GLM.

- When more than one dependent variable is specified, GLM automatically groups together the dependent variables with similar missing value

structures within the data set or within a BY group. This insures that the analysis for each dependent variable brings into use all possible observations.
- GLM allows the specification of any degree of interaction (crossed effects) and nested effects. It also provides for continuous-by-continuous, continuous-by-class, and continuous-nesting effects.
- Through the concept of estimability, GLM can provide tests of hypotheses for the effects of a linear model regardless of the number of missing cells or the extent of confounding. GLM prints not only the SS associated with each hypothesis tested, but also, upon request, the form of the estimable functions employed in the test. GLM can produce the general form of all estimable functions.
- GLM can create an output data set containing predicted and residual values from the analysis and all of the original variables.
- The MANOVA statement allows you to specify both the hypothesis effects and the error effect to use for a multivariate analysis of variance.
- The RANDOM statement allows you to specify random effects in the model; expected mean squares are printed for each TYPE I, TYPE II, TYPE III, TYPE IV, and contrast mean square used in the analysis.
- You can use the ESTIMATE statement to specify an **L** vector for estimating a linear function of the parameters $\mathbf{L}\boldsymbol{\beta}$.
- You can use the CONTRAST statement to specify a contrast vector or matrix for testing the hypothesis that $\mathbf{L}\boldsymbol{\beta}=0$.

SPECIFICATIONS

Although there are numerous statements and options available in GLM, many applications use only a few of them. Often you can find the features you need by looking at an example or by quickly scanning through this section. The statements available in GLM are:

PROC GLM *options*;

CLASS *variables*;	must precede **MODEL** and **MEANS** statements
MODEL *dependents = independents / options*;	required statement
CONTRAST *'label' [effect values]... / options*; **ESTIMATE** *'label' effect values... / options*; **LSMEANS** *effects / options*; **MANOVA H**=*effects* **E**=*effect / options*; **OUTPUT OUT**=*SASdataset keyword=names...*; **RANDOM** *effects / options*; **TEST H**=*effects* **E**=*effect / options*;	must follow **MODEL** statement
ABSORB *variables*; **BY** *variables*; **FREQ** *variable*; **ID** *variables*; **MEANS** *effects / options*; **WEIGHT** *variable*;	may be placed anywhere among statements

The PROC GLM and MODEL statements are required. If classification effects are used, the class variables must be declared in a CLASS statement.

The statements used with PROC GLM in addition to the PROC statement are (in alphabetical order):

ABSORB	absorbs classification effects in a model
BY	processes BY-groups
CLASS	declares classification variables
CONTRAST	constructs and tests linear functions of the parameters
ESTIMATE	also constructs and tests linear functions of the parameters
FREQ	specifies a frequency variable (similar to the WEIGHT statement)
ID	identifies observations on printed output
LSMEANS	computes least-squares (marginal) means
MANOVA	performs a multivariate analysis of variance
MEANS	requests that means be printed and compared
MODEL	defines the model to be fit
OUTPUT	requests an output data set containing predicted values and residuals
RANDOM	declares certain effects to be random and computes expected mean squares
TEST	constructs tests using the sums of squares for effects and the error term you specify
WEIGHT	specifies a variable for weighting observations.

PROC GLM Statement

PROC GLM *options;*

Only one option can be used on the PROC GLM statement:

DATA=*SASdataset*	names the SAS data set to be used by GLM. If DATA= is omitted, GLM uses the most recently created SAS data set.

ABSORB Statement

ABSORB *variables;*

Absorption is a computational technique that provides a large reduction in time and storage requirements for certain types of models.

For a main-effect variable with a large number of levels that does not participate in interactions, you can absorb the effect by naming it in an ABSORB statement. This means that the effect can be adjusted out prior to the construction and solution of the rest of the model.

Several variables can be specified, in which case each one is assumed to be nested in the preceding one in the ABSORB statement.

Restrictions: when the ABSORB statement is used, the data set (or each BY group) must be sorted by the variables in the ABSORB statement. GLM cannot produce predicted values or create an output data set if ABSORB is used.

See the detail section **Absorption** below for more information.

BY Statement

BY *variables;*

A BY statement may be used with PROC GLM to obtain separate analyses on observations in groups defined by the BY variables. When a BY statement appears, the procedure expects the input data set to be sorted in order of the BY variables. If your input data set is not sorted in ascending order, use the SORT procedure with a similar BY statement to sort the data, or, if appropriate, use the BY statement options NOTSORTED or DESCENDING. For more information, see the discussion of the BY statement in Chapter 8, "Statements Used in the PROC Step," in *SAS User's Guide: Basics, 1982 Edition*.

CLASS Statement

CLASS *variables;*

The CLASS or CLASSES statement names the classification variables to be used in the analysis. Typical class variables are TRTMENT, SEX, RACE, GROUP, and REP.

Classification variables may be either character or numeric. Only the first 16 characters of a character variable are used.

Class levels are determined from the formatted values of the CLASS variables. Thus you can use formats to group values into levels. See the discussion of the FORMAT procedure, the FORMAT statement, and Chapter 13, "SAS Formats," in *SAS User's Guide: Basics, 1982 Edition*.

CONTRAST Statement

CONTRAST *'label' [effect values,...] / options;*

The CONTRAST statement provides a mechanism for specifying an **L** vector or matrix for testing the hypothesis $\mathbf{L}\beta=0$. If the hypothesis is testable, the $SS(H_0\text{: }\mathbf{L}\beta=0)$ is computed as

$$(\mathbf{Lb})'(\mathbf{L}(\mathbf{X'X})^{-}\mathbf{L'})^{-1}(\mathbf{Lb}) \quad ,$$

where $\mathbf{b}=(\mathbf{X'X})^{-}\mathbf{X'y}$. This is the SS printed on the analysis-of-variance table.

You may specify the effect to be used as a denominator in the *F* test. If you use a RANDOM statement, the expected mean square of the contrast is printed. There is no limit to the number of CONTRAST statements, but they must come after the MODEL statement.

In the CONTRAST statement, *'label'* is 20 characters or less in single quotes and is used on the printout to identify the contrast. *Effect* is the name of an effect that appears in the MODEL statement; the keyword INTERCEPT may be used as an effect when an intercept is fitted in the model. The constants are elements of the **L** vector associated with the preceding effect. Not all effects in the MODEL statement need to be included.

Multiple-degree-of-freedom hypotheses may be specified by separating the rows of the **L** matrix with commas, as shown above.

For example, for the model

```
MODEL Y=A B;
```

with A at 5 levels and B at 2 levels, the parameter vector is

$$(\mu\ \alpha_1\ \alpha_2\ \alpha_3\ \alpha_4\ \alpha_5\ \beta_1\ \beta_2) \quad .$$

To test the hypothesis that the pooled A linear and A quadratic effect is zero, you may use the following **L** matrix:

```
L =  0 -2 -1  0  1 2 0 0
     0  2 -1 -2 -1 2 0 0
```

The corresponding CONTRAST statement is:

```
CONTRAST 'A LINEAR & QUADRATIC'
            A -2 -1     0 1 2,
            A  2 -1 -2 -1 2;
```

If the first level of A is a control level and you want a test of control versus others, you can use this statement:

```
CONTRAST 'CONTROL VS OTHERS' A -1 .25 .25 .25 .25;
```

The **L** matrix should be of full row rank. However, if it is not, the degrees of freedom associated with the hypotheses are reduced to the row rank of **L**. The SS computed in this situation are equivalent to the SS computed using an **L** matrix with any row deleted, that is, a linear combination of previous rows.

These three options are available in the CONTRAST statement and are specified after a slash (/):

E	requests that the entire **L** vector be printed.
E=*effect*	specifies an effect in the model to use as an error term. If none is specified, the error MS is used.
ETYPE=*n*	specifies the type (1,2,3,4) of the E= effect. If E= is specified and ETYPE= is not, the highest type computed in the analysis is used.

See the discussion of the ESTIMATE statement below and **Specification of ESTIMATE Expressions** for rules on specification, construction, distribution, and estimability in the CONTRAST statement.

ESTIMATE Statement

ESTIMATE *'label' [effect values...] / options;*

The ESTIMATE statement may be used to estimate linear functions of the parameters by multiplying the vector **L** by the parameter estimate vector **b** resulting in **Lb**. All of the elements of the **L** vector may be given, or if only certain portions of the **L** vector are given, the remaining elements are constructed by GLM from the context (in a manner similar to rule 4 discussed in the section **Least-Squares Means**).

The linear function is checked for estimability. The estimate **Lb**, where $\mathbf{b}=(\mathbf{X'X})^{-}\mathbf{X'y}$, is printed along with its associated standard error, $\sqrt{(\mathbf{L}(\mathbf{X'X})^{-}\mathbf{L'}s^2)}$, and *t* test. There is no limit to the number of ESTIMATE statements, but they must come after the MODEL statement.

In the ESTIMATE statement, *'label'* is 20 characters or less in single quotes and is used on the printout to identify the estimate. *Effect* is the name of an effect that appears in the MODEL statement; the keyword INTERCEPT may be used as an effect when an intercept is fitted in the model. The constants are the elements of the **L** vector associated with the preceding effect. For example, with no options,

```
ESTIMATE 'A1 VS A2' A 1 -1;
```

Not all effects in the MODEL statement need to be included.

The options below can appear in the ESTIMATE statement after a slash (/):

E — requests that the entire **L** vector be printed.

DIVISOR= *number* — specifies a value by which to divide all coefficients so that fractional coefficients may be entered as integer numerators. For example:

```
ESTIMATE '1/3(A1+A2)-2/3A3' A 1 1 -2 /
  DIVISOR=3;
```

instead of

```
ESTIMATE '1/3(A1+A2)-2/3A3' A .33333
  .33333 -.66667;
```

See also **Specification of ESTIMATE Expressions** below.

FREQ Statement

FREQ *variable*;

When a FREQ statement appears, each observation in the input data set is assumed to represent *n* observations in the experiment, where *n* is the value of the FREQ variable.

If the value of the FREQ variable is less than 1, the observation is not used in the analysis. If the value is not an integer, only the integer portion is used.

The analysis produced using a FREQ statement is identical to an analysis produced using a data set that contains *n* observations in place of each observation of the input data set, where *n* is the value of the FREQ variable. Therefore, means and total degrees of freedom reflect the expanded number of observations.

ID Statement

ID *variables*;

When predicted values are requested as a MODEL statement option, values of the variables given in the ID statement are printed beside each observed, predicted, and residual value for identification. Although there are no restrictions on the number or length of ID variables, GLM may truncate the number of values printed in order to print on one line.

LSMEANS Statement

LSMEANS *effects* / *options*;

Least-squares means are computed for each effect listed in the LSMEANS statement.

Least-squares estimates of marginal means (LSMs) are to unbalanced designs as class and subclass arithmetic means are to balanced designs. LSMs are simply estimators of the class or subclass marginal means that would be expected had the design been balanced. For further information see the detail section, **Least-Squares Means**.

Least-squares means (LSMs) can be computed for any effect involving class variables as long as the effect is in the model. Any number of LSMEANS statements can be used. They must be given after the MODEL statement.

Here is an example:

```
PROC GLM;
  CLASS A B;
  MODEL Y=A B A*B;
  LSMEANS A B A*B;
```

Least-squares means are printed for each level of the A, B, and A*B effects.

The options below may appear in the LSMEANS statement after a slash (/):

E prints the estimable functions used to compute the LSM.

STDERR prints the standard error of the LSM and the probability level for the hypothesis H_0:LSM=0.

PDIFF requests that all possible probability values for the hypothesis H_0: LSM(*i*) = LSM(*j*) be printed.

EPSILON= *number* tunes the estimability checking. If ABS(**L**− **LH**)<C*EPSILON for any row, then the **L** is declared non-estimable. **H** is the **(X′X)⁻X′X** matrix, and **C** is ABS(**L**) except for rows where **L** is zero, in which case it is 1. The default is 1E-4.

MANOVA Statement

MANOVA H=*effects* E=*effect* / *options*;

If the MODEL statement includes more than one dependent variable, additional multivariate statistics may be requested with the MANOVA statement.

When a MANOVA statement appears, GLM enters a multivariate mode with respect to the handling of missing values: observations with missing independent or dependent variables are excluded from the analysis. Even when you do not want multivariate statistics, the statement

MANOVA;

can be used to request the multivariate mode of handling missing values. The terms below are specified on the MANOVA statement:

H=*effects* specifies which effects in the preceding model are to be used as hypothesis matrices. For each **H** matrix (the SSCP matrix associated with that effect), the characteristic roots and vectors of $\mathbf{E}^{-1}\mathbf{H}$ are printed (where **E** is the matrix associated with the error effect). The Hotelling-Lawley trace, Pillai's trace, Wilk's criterion, and Roy's maximum root criterion are printed with approximate *F* statistics. For background and further details, see the detail section **Multivariate Analysis of Variance**.

E=*effect* specifies the error effect. If E= is omitted, the error SSCP (residual) matrix from the analysis is used.

The options below can appear in the MANOVA statement after a slash (/):

PRINTH requests that the **H** matrix (the SSCP matrix) associated with each effect specified by the H parameter be printed.

PRINTE requests printing of the **E** matrix. If the **E** matrix is the error SSCP (residual) matrix from the analysis, the partial correlations of the dependent variables given the

independent variables are also printed.
For example, the statement

```
MANOVA / PRINTE;
```

prints the error SSCP matrix and the partial correlation matrix computed from the error SSCP matrix.

HTYPE=*n* specifies the TYPE (I, II, III, or IV) of the **H** matrix, where HTYPE= 1 for TYPE I, and so on. If an HTYPE= value *n* is given, the corresponding test must have been performed in the MODEL statement, either by options SS*n*, E*n*, or the default Type I and Type III. If no HTYPE= option appears in the MANOVA statement, the HTYPE= value defaults to the highest type (largest *n*) used in the analysis.

ETYPE=*n* specifies the TYPE (I, II, III, or IV) of the **E** matrix. You need this option if you use E= (rather than residual error) and you want to specify the type of SS used for the effect. See HTYPE= above for further rules.

Here is another example of the MANOVA statement:

```
PROC GLM;
  CLASS A B;
  MODEL Y1-Y5=A B(A);
  MANOVA H=A E=B(A) / PRINTH PRINTE HTYPE=1 ETYPE=1;
  MANOVA H=B(A) / PRINTE;
```

Since this MODEL statement requests no options for type, GLM uses TYPE I and TYPE III. The first MANOVA statement specifies A as the hypothesis effect and B(A) as the error effect. The PRINTH option requests that the **H** matrix associated with the A effect be printed, and the PRINTE option requests that the **E** matrix associated with the B(A) effect be printed. The HTYPE= 1 option specifies that the **H** matrix be TYPE I; the ETYPE= 1 option specifies that the **E** matrix be TYPE I.

The second MANOVA statement specifies B(A) as the hypothesis effect. Since no error effect is specified, GLM uses the error SSCP matrix from the analysis as the **E** matrix. The PRINTE option requests that this **E** matrix be printed. Since the **E** matrix is the error SSCP matrix from the analysis, the partial correlation matrix computed from this matrix is also printed.

MEANS Statement

MEANS *effects / options*;

GLM can compute means for any effect involving CLASS variables whether or not the effect is specified in the MODEL statement. You can use any number of MEANS statements either before or after the MODEL statement.

For example:

```
PROC GLM;
  CLASS A B C;
  MODEL Y=A B C;
  MEANS A B C A*B;
```

Means are printed for each level of the variables A, B, and C and for the combined levels of A and B.

The options below may appear in the MEANS statement after a slash (/):

DEPONLY indicates that only the dependent variable means are to be printed. By default, GLM prints means for all continuous variables, including independent variables.

BON performs Bonferroni *t* tests of differences between means for all main-effect means in the MEANS statement.

DUNCAN performs Duncan's multiple-range test on all main--effect means given in the MEANS statement.

GABRIEL performs Gabriel's multiple-comparison procedure on all main-effect means in the MEANS statement.

REGWF performs the Ryan-Einot-Gabriel-Welsch multiple *F* test on all main-effect means in the MEANS statement.

REGWQ performs the Ryan-Einot-Gabriel-Welsch multiple-range test on all main-effect means in the MEANS statement.

SCHEFFE performs Scheffe's multiple-comparison procedure on all main-effect means in the MEANS statement.

SIDAK performs pairwise *t* tests on differences between means with levels adjusted according to Sidak's inequality for all main-effect means in the MEANS statement.

SMM
GT2 performs pairwise comparisons based on the studentized maximum modulus and Sidak's uncorrelated-*t* inequality, yielding Hochberg's GT2 method when sample sizes are unequal, for all main-effect means in the MEANS statement.

SNK performs the Student-Newman-Keuls multiple range test on all main-effect means in the MEANS statement.

T
LSD performs pairwise *t* tests, equivalent to Fisher's least-significant-difference test in the case of equal cell sizes, for all main-effect means in the MEANS statement.

TUKEY performs Tukey's studentized range test (HSD) on all main-effects means in the MEANS statement.

ALPHA=*p* gives the level of significance for comparisons among the means. The default ALPHA value is .05. With the DUNCAN option, you may specify only values of .01, .05, or .1.

WALLER requests that the Waller-Duncan k-ratio *t* test be performed on all main-effect means in the MEANS statement.

KRATIO=*value* gives the type1/type2 error seriousness ratio for the Waller-Duncan test. Reasonable values for KRATIO are 50, 100, 500, which roughly correspond for the two-level case to ALPHA levels of .1, .05, and .01. If KRATIO is omitted, the procedure uses the default value of 100.

LINES requests that the results of the BON, DUNCAN, GABRIEL, REGWF, REGWQ, SCHEFFE, SIDAK, SMM, GT2, SNK, T, LSD, TUKEY, and WALLER options be presented by listing the means in descending order and indicating nonsignificant subsets by line segments beside the corresponding means. LINES is appropriate for equal cell sizes, for which it is the default. LINES is also the default if DUNCAN, REGWF, REGWQ, SNK, or WALLER is specified. If the cell sizes are unequal, the harmonic mean is used, which may lead to somewhat liberal tests if the cell sizes are highly disparate.

CLDIFF
: requests that the results of the BON, GABRIEL, SCHEFFE, SIDAK, SMM, GT2, T, LSD, and TUKEY options be presented as confidence intervals for all pairwise differences between means. CLDIFF is the default for unequal cell sizes unless DUNCAN, REGWF, REGWQ, SNK, or WALLER is specified.

NOSORT
: prevents the means from being sorted into descending order when CLDIFF is specified.

E=*effect*
: specifies the error mean square to use in the multiple comparisons. If E= is omitted, GLM uses the residual MS. The effect specified with the E= option must be a term in the model; otherwise, the procedure uses the residual MS.

ETYPE=*n*
: specifies the type of mean square for the error effect. When E= effect is specified, it is sometimes necessary to indicate which type (1, 2, 3, or 4) MS is to be used. The *n* value must be one of the types specified or implied by the MODEL statement. The default MS type is the highest type used in the analysis.

HTYPE=*n*
: gives the MS type for the hypothesis MS. The HTYPE= option is needed only when the WALLER option is specified. The default HTYPE value is the highest type used in the model.

MODEL Statement

MODEL *dependents*=*independents* / *options*;

The MODEL statement names the dependent variables and independent effects. The syntax of effects is described in the introductory section **Specification of Effects.** If no independent effects are specified, only an intercept term is fit.

The options listed below may be specified in the MODEL statement after a slash (/):

Option for the intercept

NOINT
: requests that the intercept parameter not be included in the model.

Options to request printouts

NOUNI
: requests that no univariate statistics be printed. You typically use the NOUNI option with a multivariate analysis of variance when you do not need any univariate statistics printed.

SOLUTION
: asks GLM to print a solution to the normal equations (parameter estimates). GLM always prints a solution when no CLASS statement appears.

TOLERANCE requests that the tolerances used in the SWEEP routine be printed. The tolerances are of the form D/UCSS or D/CSS, as described in the discussion of EPSILON (under **Tuning options** below). The tolerance value for the intercept is not divided by its uncorrected SS.

Options to control standard hypothesis tests

E asks GLM to print the general form of all estimable functions.

E1 requests that the TYPE I estimable functions for each effect in the model be printed.

E2 requests that the TYPE II estimable functions for each effect in the model be printed.

E3 requests that the TYPE III estimable functions for each effect in the model be printed.

E4 requests that the TYPE IV estimable functions for each effect in the model be printed.

SS1 asks GLM to print the SS associated with TYPE I estimable functions for each effect.

SS2 asks GLM to print the SS associated with TYPE II estimable functions for each effect.

SS3 asks GLM to print the SS associated with TYPE III estimable functions for each effect.

SS4 asks GLM to print the SS associated with TYPE IV estimable functions for each effect.

Note: if E1, E2, E3, or E4 is specified, the corresponding SS for each effect are printed. By default, the procedure prints the TYPE I and TYPE III SS for each effect.

Options for predicted values and residuals

P asks GLM to print observed, predicted, and residual values for each observation that does not contain missing values for independent variables. The Durbin-Watson statistic is also printed when P is specified. The PRESS statistic is also printed if either CLM or CLI is specified.

CLM prints confidence limits for a mean predicted value for each observation. The P option must also appear.

CLI prints confidence limits for individual predicted values for each observation. The P option must also appear. CLI should not be used with CLM; it is ignored if CLM is also specified.

ALPHA=*p* specifies the alpha level for confidence intervals. The only acceptable values for ALPHA are .01, .05, and .10. If no ALPHA level is given, GLM uses .05.

Options to print intermediate calculations

XPX prints the **X′X** crossproducts matrix.

INVERSE | I prints the inverse or the generalized inverse of the **X′X** matrix.

Tuning options

EPSILON=*value* tunes the sensitivity of the regression routine to linear dependencies in the design. If a diagonal pivot element is less than C*EPSILON as GLM sweeps the **X'X** matrix, the associated design column is declared to be linearly dependent with previous columns, and the associated parameter is zeroed.

The C value adjusts the check to the relative scale of the variable C, the corrected SS for the variable unless the corrected SS is 0, in which case C is 1. If NOINT is specified but the ABSORB option is not, GLM uses the uncorrected SS instead.

Note: the default value of EPSILON, 1E-8, is perhaps too small but is necessary in order to handle the high-degree polynomials used in the literature to compare regression routines.

ZETA=*value* tunes the sensitivity of the check for estimability for Type III and Type IV functions. Any element in the estimable function basis with an absolute value less than ZETA is set to zero. The default value for ZETA is 1E-8, which suffices for all *ANOVA*-type models.

Note: although it is possible to generate data for which this absolute check can be defeated, it suffices in most practical examples. Additional research needs to be performed to make this check relative rather than absolute.

OUTPUT Statement

OUTPUT OUT=*SASdataset* PREDICTED=*variables* RESIDUAL=*variables;*

or

OUTPUT OUT=*SASdataset* P=*variables* R=*variables;*

The OUTPUT statement asks GLM to create a new SAS data set. Predicted and residual values as well as all the variables in the original data set are included in the new data set. If you want to create a permanent SAS data set, you must specify a two-level name (see Chapter 12, "SAS Data Sets," in *SAS User's Guide: Basics, 1982 Edition*, for more information on permanent SAS data sets).

The options below are given in the OUTPUT statement:

OUT= gives the name of the new data set. If OUT= is omitted, SAS names the new data set using the DATA*n* convention.

PREDICTED=
P=
RESIDUAL=
R= specifies new variable names. The names correspond to the dependent variable or variables given in the MODEL statement.

For example, the statements

```
PROC GLM;
  CLASS A B;
  MODEL Y=A B A*B;
  OUTPUT OUT=NEW P=YHAT R=RESID;
```

request an output data set named NEW. In addition to all the variables from the original data set, NEW contains the variable YHAT, whose values are predicted values of the dependent variable Y. NEW also contains the variable RESID, whose values are the residual values of Y.

Another example:

```
PROC GLM;
  BY GROUP;
  CLASS A;
  MODEL Y1-Y5=A X(A);
  OUTPUT OUT=POUT PREDICTED=PY1-PY5;
```

Data set POUT contains five new variables, PY1-PY5. PY1's values are the predicted values of Y1; PY2's values are the predicted values of Y2; and so on.

The predicted value is missing for any observation with one or more missing independent variables. If an observation from the original data set is not used in the analysis, its residual value is also missing.

RANDOM Statement

RANDOM *effects* / *options*;

The RANDOM statement specifies which effects in the model are random. When you use a RANDOM statement, GLM prints the expected value of each TYPE I, TYPE II, TYPE III, TYPE IV, or contrast MS used in the analysis but does not make use of the information pertaining to expected mean squares in any way. Since other features in GLM assume that all effects are fixed, all tests and estimability checks are based on a fixed-effects model, even when you use a RANDOM statement.

You may use only one RANDOM statement, and it must come after the MODEL statement.

The list of effects in the RANDOM statement should contain one or more of the pure classification effects (main effects, crossed effects, or nested effects) specified in the MODEL statement. The levels of each effect specified are assumed to be normally and independently distributed with common variance. Levels in different effects are assumed independent.

The option below can appear in the RANDOM statement after a slash (/):

Q requests a complete printout of all quadratic forms in the fixed effects that appear in the expected mean squares.

See **Expected Mean Squares for Random Effects** below for more information on the calculation of expected mean squares.

TEST Statement

TEST H=*effects* E=*effect* / options;

Although an *F* value is computed for all SS in the analysis using the residual MS as an error term, you may request additional *F* tests using other effects as error terms. You need a TEST statement when a non-Iσ^2 error structure (as in a split-plot) exists. However, in most unbalanced models with non-Iσ^2 error structures, most MSs are not independent and do not have equal expectations under the null hypothesis.

If you use a TEST statement, E= is required.

GLM does not check any of the assumptions underlying the *F* statistic. **When you**

specify a TEST statement, you assume sole responsibility for the validity of the F statistic produced. To help validate a test, you may use the RANDOM statement and inspect the expected mean squares.

These terms are specified on the TEST statement:

H=*effects* specifies which effects in the preceding model are to be used as hypothesis (numerator) effects.

E=*effect* specifies one, and only one, effect to use as the error (denominator) term.

By default, the SS type for all hypothesis SS and error SS is the highest type computed in the model. If the hypothesis type or error type is to be another type that was computed in the model, you should specify one or both of these options after a slash (/):

HTYPE=*n* specifies the type of SS to use for the hypothesis. The type must be a type computed in the model (*n*= 1, 2, 3, or 4).

ETYPE=*n* specifies the type of SS to use for the error term. The type must be a type computed in the model (*n*= 1, 2, 3, or 4).

This example illustrates the TEST statement with a balanced split-plot model:

```
PROC GLM;
  CLASS A B C;
  MODEL Y=A B(A) C A*C B*C(A);
  TEST H=A E=B(A) / HTYPE=1 ETYPE=1;
  TEST H=C A*C E=B*C (A) / HTYPE=1 ETYPE=1;
```

WEIGHT Statement

WEIGHT *variable*;

When a WEIGHT statement is used, a weighted residual sum of squares

$$\Sigma w(y - \hat{y})^2$$

is minimized, where *w* is the value of the WEIGHT variable.

The observation is used in the analysis only if the value of the WEIGHT variable is greater than zero.

Means and total degrees of freedom are unaffected by the presence of a WEIGHT statement. The normal equations used when a WEIGHT statement is present are:

$$\beta = (\mathbf{X'WX})^{-}\mathbf{X'W\,Y}$$

where **W** is a diagonal matrix consisting of the values of the WEIGHT variable.

If the weights for the observations are proportional to the reciprocals of the error variances, then the weighted least-squares estimates are B.L.U.E (best linear unbiased estimators).

DETAILS

Missing Values

For an analysis involving one dependent variable, GLM uses an observation if values are present for that dependent variable and all the variables used in independent effects.

For an analysis involving multiple dependent variables without the MANOVA statement, a missing value in one dependent variable does not eliminate the observation from the analysis of other nonmissing dependent variables. For an analysis with the MANOVA statement, GLM requires values for all dependent variables to be present for an observation to be used for any.

During processing, GLM groups the dependent variables on their missing values across observations so that sums and crossproducts can be collected in the most efficient manner.

Output Data Set

The OUTPUT statement produces an output data set that contains:

- all original data from the SAS data set input to GLM
- the PREDICT= variables named in the OUTPUT statement to contain predicted values
- the RESIDUAL= variables named in the OUTPUT statement to contain the residual values.

With multiple dependent variables, a name may be specified for predicted and residual values for each of the dependent variables in the order they occur in the MODEL statement.

For example, suppose the input data set A contains the variables Y1, Y2, Y3, X1, and X2. Then you can code:

```
PROC GLM DATA=A;
  MODEL Y1 Y2 Y3=X1;
  OUTPUT P=Y1HAT Y2HAT Y3HAT R=Y1RESID;
```

The output data set contains Y1, Y2, Y3, X1, X2, Y1HAT, Y2HAT, Y3HAT, and Y1RESID. X2 is output even though it was not used by GLM. Although predicted variables are generated for all three dependent variables, residuals are output for only the first dependent variable.

On the output data set the predicted values are missing when any independent variable in the analysis is missing. The residuals are missing if either an independent variable in the analysis or the dependent variable is missing.

Computer Resources

Memory For large problems, most of the memory resources are required for holding the **X'X** matrix of the sums and crossproducts. The section on **Parameterization of GLM Models** describes how columns of the **X** matrix are allocated for various types of effects. For each level that occurs in the data for a combination of class variables in a given effect, a row and column for **X'X** is needed.

An example illustrates the calculation. Suppose A has 20 levels, B has 4, and C has 3. Then consider the model:

```
PROC GLM;
  CLASS A B C;
  MODEL Y1 Y2 Y3=A B A*B C A*C B*C A*B*C X1 X2;
```

The **X'X** matrix (bordered by **X'Y** and **Y'Y**) could have as many as 425 rows and columns:

1 for the intercept term
20 for A
4 for B
80 for A*B
3 for C
60 for A*C
12 for B*C
240 for A*B*C
2 for X1 and X2 (continuous variables)
3 for Y1, Y2, and Y3 (dependent variables).

The matrix only has 425 rows and columns if all combinations of levels occurred for each effect in the model. For m rows and columns, $8 \times m^2/2$ bytes are needed for crossproducts. In this case, $8 \times 425^2/2$ is 722500 bytes. To convert to K units, divide by 1024.

For this example, the analysis requires 706K of memory for **X'X** and 200K or more memory for SAS and the GLM program; so at least 900K should be requested.

The required memory grows as the square of the number of columns of **X** and **X'X**; most is for the A*B*C interaction. Without A*B*C, we have 185 columns and need only 134K for **X'X**. Without A*B we only need 43K. If A is recoded to have ten levels, then even the full model has only 220 columns and requires only 189K.

If you have a very large model that will not fit the region or would be too expensive to run, these are your options:

- cut out terms, especially high-level interactions
- cut down the number of levels for variables with many levels
- use ABSORB for parts of the model that are large
- use PROC ANOVA rather than PROC GLM, if your design allows.

Cputime For large problems, two operations consume a lot of cputime: the collection of sums and crossproducts and the solution of the normal equations.

The time required for collecting sums and crossproducts is difficult to calculate since it is a complicated function of the model. For a model with m columns and n rows (observations) in X, the worst case occurs if all columns are continuous variables, involving $n \times m^2/2$ multiplications and additions. If the columns are levels of a classification, then only m sums may be needed, but a significant amount of time may be spent in lookup operations. Solving the normal equations requires time for approximately $m^3/2$ multiplications and additions.

Parameterization of GLM Models

GLM constructs a linear model according to the specifications on the MODEL statement. Each effect generates one or more columns in a design matrix **X**. This section shows precisely how **X** is built.

Intercept All models automatically include a column of 1s to estimate an intercept parameter μ. You can use the NOINT option to suppress the intercept.

Regression effects Regression effects (covariables) have the values of the variables copied into the design matrix directly. Polynomial terms are multiplied out and then installed in **X**.

Main effects If a class variable has m levels, GLM generates m columns in the design matrix for its main effect. Each column is an indicator variable for a given level. The order of the columns is the sort order of the formatted values of their levels. For example:

data		int	A		B		
A	B	μ	A1	A2	B1	B2	B3
1	1	1	1	0	1	0	0
1	2	1	1	0	0	1	0
1	3	1	1	0	0	0	1
2	1	1	0	1	1	0	0
2	2	1	0	1	0	1	0
2	3	1	0	1	0	0	1

There are more columns for these effects than there are degrees of freedom for them; in other words, GLM is using an over-parameterized model.

Crossed effects First, GLM reorders the terms to correspond to the order of the variables in the CLASS statement; thus B*A becomes A*B if A precedes B in the CLASS statement. Then GLM generates columns for all combinations of levels that occur in the data. The order of the columns is such that the rightmost variables in the cross index faster than the leftmost variables. Empty columns (that would contain all zeros) are not generated.

data		int	A		B			A*B					
A	B	μ	A1	A2	B1	B2	B3	A1B1	A1B2	A1B3	A2B1	A2B2	A2B3
1	1	1	1	0	1	0	0	1	0	0	0	0	0
1	2	1	1	0	0	1	0	0	1	0	0	0	0
1	3	1	1	0	0	0	1	0	0	1	0	0	0
2	1	1	0	1	1	0	0	0	0	0	1	0	0
2	2	1	0	1	0	1	0	0	0	0	0	1	0
2	3	1	0	1	0	0	1	0	0	0	0	0	1

In the above matrix, main-effects columns are not linearly independent of crossed-effect columns; in fact, the column space for the crossed effects contains the space of the main effect.

Nested effects Nested effects are generated in the same manner as crossed effects. Hence the design columns generated by the following statements are the same (but the ordering of the columns is different):

```
MODEL Y=A B(A);        (B nested within A)
```

and

```
MODEL =A A*B;          (omitted main effect for B).
```

The nesting operator in GLM is more a notational convenience than an operation distinct from crossing. Nested effects are characterized by the property that the nested variables never appear as main effects. The order of the variables within nesting parentheses is made to correspond to the order of these variables in the CLASS statement. The order of the columns is such that crossed variables index faster than nested ones, and the rightmost nested variables index faster than the leftmost ones.

data		int	A		B(A)					
A	B	μ	A1	A2	A1B1	A2B1	A1B2	A2B2	A1B3	A2B3
1	1	1	1	0	1	0	0	0	0	0
1	2	1	1	0	0	0	1	0	0	0
1	3	1	1	0	0	0	0	0	1	0
2	1	1	0	1	0	1	0	0	0	0
2	2	1	0	1	0	0	0	1	0	0
2	3	1	0	1	0	0	0	0	0	1

Continuous-nesting-class effects When a continuous variable nests with a class variable, the design columns are constructed by multiplying the continuous values into the design columns for the class effect.

data		int	A		X(A)	
X	A	μ	A1	A2	X(A1)	X(A2)
21	1	1	1	1	21	0
24	1	1	1	1	24	0
22	1	1	1	1	22	0
28	2	1	0	0	0	28
19	2	1	0	0	0	19
23	2	1	0	0	0	23

This model estimates a separate slope for **X** within each level of A.

Continuous-by-class effects Continuous-by-class effects generate the same design columns as continuous-nesting class effects. The two models are made different by the presence of the continuous variable as a regressor by itself as well as a contributor to a compound effect.

data		int	X	A		X*A	
X	A	μ	X	A1	A2	X*A1	X*A2
21	1	1	21	1	1	21	0
24	1	1	24	1	1	24	0
22	1	1	22	1	1	22	0
28	2	1	28	0	0	0	28
19	2	1	19	0	0	0	19
23	2	1	23	0	0	0	23

Continuous-by-class effects are used to test the homogeneity of slopes. If the continuous-by-class effect is nonsignificant, the effect can be removed so that the response with respect to **X** is the same for the levels of the class variables.

General effects An example which combines all the effects is:

X1*X2*A*B*C(D E)

The continuous list comes first, followed by the crossed list, followed by the nested list in parentheses.

The sequencing of parameters is not important to learn unless you contemplate using the CONTRAST or ESTIMATE statements to compute some function of the parameter estimates.

Effects may be retitled by GLM to correspond to ordering rules. For example, B*A(E D) might be retitled A*B(D E) to satisfy the following:

- class variables that occur outside parentheses (crossed effects) are sorted in the order they appear in the CLASS statement
- variables within parentheses (nested effects) are sorted in the order they appear in a CLASS statement.

The sequencing of the parameters generated by an effect can be described by which variables have their levels indexed faster:

- variables in the crossed part index faster than variables in the nested list
- within a crossed or nested list, variables to the right index faster than variables to the left.

For example, suppose a model includes four effects—A, B, C, and D—each having two levels, 1 and 2. If the CLASS statement is

```
CLASS A B C D;
```

then the order of the parameters for the effect B*A(C D), which is retitled A*B(C D), is:

$A_1B_1C_1D_1 \rightarrow A_1B_2C_1D_1 \rightarrow A_2B_1C_1D_1 \rightarrow A_2B_2C_1D_1 \rightarrow A_1B_1C_1D_2 \rightarrow$
$A_1B_2C_1D_2 \rightarrow A_2B_1C_1D_2 \rightarrow A_2B_2C_1D_2 \rightarrow A_1B_1C_2D_1 \rightarrow A_1B_2C_2D_1 \rightarrow$
$A_2B_1C_1D_2 \rightarrow A_2B_2C_2D_1 \rightarrow A_1B_1C_2D_2 \rightarrow A_1B_2C_2D_2 \rightarrow A_2B_1C_2D_2 \rightarrow$
$A_2B_2C_2D_2$.

Note that first the crossed effects B and A are sorted in the order that they appear in the CLASS statement so that A precedes B in the parameter list. Then, for each combination of the nested effects in turn, combinations of A and B appear. B moves fastest since it is rightmost in the cross list. Then A moves next fastest. D moves next fastest. C is the slowest, since it is leftmost in the nested list.

When noninteger numeric levels are used, levels are sorted by their character format, which may not correspond to their numeric sort sequence. Therefore, it is advisable to include a format for noninteger numeric levels.

Degrees of freedom For models with class variables, there are more design columns constructed than there are degrees of freedom for the effect. There are thus linear dependencies among the columns. In this event, the parameters are not estimable; there will be an infinite number of least-squares solutions. GLM uses a generalized (G2) inverse to obtain values for the estimates. The solution values are not even printed unless the SOLUTION option is specified. The solution has the characteristic that estimates are zero whenever the design column for that parameter is a linear combination of previous columns. (Strictly termed, the solution values should not even be called "estimates.") With this full parameterization, hypothesis tests are constructed to test linear functions of the parameters that are estimable.

Other programs (such as FUNCAT) reparameterize models to full rank using certain restrictions on the parameters. GLM does not reparameterize, making the hypotheses that are commonly tested more understandable. See Goodnight (1978) for additional reasons for not reparameterizing.

GLM does not actually construct the design matrix **X** rather, the procedure constructs directly the crossproduct matrix, **X'X**, which is made up of counts, sums, and crossproducts.

Hypothesis Testing in GLM

A complete discussion of the four standard types of hypothesis tests is located in Chapter 15, "The Four Types of Estimable Functions."

Example To illustrate the four types of tests and the principles upon which they are based, consider a two-way design with interaction based on these data:

		B 1	B 2
A	1	23.5 23.7	28.7
A	2	8.9	5.6 8.9
A	3	10.3 12.5	13.6 14.6

Invoke GLM and ask for all the estimable functions options to examine what GLM can test. The code below is followed by the summary *ANOVA* table from the printout.

```
DATA EXAMPLE;
  INPUT A B Y @@;
  CARDS;
1 1 23.5   1 1 23.7   1 2 28.7   2 1  8.9   2 2  5.6
2 2  8.9   3 1 10.3   3 1 12.5   3 2 13.6   3 2 14.6
;
PROC GLM;
  CLASS A B;
  MODEL Y=A B A*B / E E1 E2 E3 E4;
```

```
                                                                                                   1
                                    GENERAL LINEAR MODELS PROCEDURE

DEPENDENT VARIABLE: Y

SOURCE                  DF     SUM OF SQUARES       MEAN SQUARE      F VALUE      PR > F     R-SQUARE        C.V.

MODEL                    5       520.47600000      104.09520000        49.66      0.0011     0.984145      9.6330

ERROR                    4         8.38500000        2.09625000                  ROOT MSE                 Y MEAN

CORRECTED TOTAL          9       528.86100000                                  1.44784322             15.03000000
```

The following sections show the general form of estimable functions and discuss the four standard tests, their properties, and abbreviated printouts for our two-way crossed example.

Estimability The first printout is the general form of estimable functions. In order to be testable, a hypothesis must be able to fit within the framework printed here.

```
                                                                                      2
                              GENERAL LINEAR MODELS PROCEDURE

DEPENDENT VARIABLE: Y

GENERAL FORM OF ESTIMABLE FUNCTIONS

EFFECT                COEFFICIENTS

INTERCEPT             L1

A                1    L2
                 2    L3
                 3    L1-L2-L3

B                1    L5
                 2    L1-L5

A*B              1 1  L7
                 1 2  L2-L7
                 2 1  L9
                 2 2  L3-L9
                 3 1  L5-L7-L9
                 3 2  L1-L2-L3-L5+L7+L9
```

If a hypothesis is estimable, the Ls in the above scheme can be set to values that match the hypothesis. All the standard tests in GLM can be shown in the format above, with some of the Ls zeroed, some set to functions of other Ls.

The following sections show how many of the hypotheses can be tested by comparing the model sum-of-squares regression from one model to a submodel. The notation used is:

$$SS(Beffects|Aeffects) = SS(Beffects, Aeffects) - SS(Aeffects)$$

where SS(*Aeffects*) denotes the regression model sum of squares for the model consisting of *Aeffects*. This notation is equivalent to the "reduction" notation defined by Searle (1971).

Type I tests Type I sums of squares, also called *sequential sums of squares*, are the incremental improvement in error SS as each effect is added to the model. They can be computed by fitting the model in steps, and recording the difference in SSE at each step.

Source	*Type I SS*
A	$SS(A\|\mu)$
B	$SS(B\|\mu,A)$
A*B	$SS(A*B\|\mu,A,B)$

Type I SS are printed out by default since they are easy to obtain and can be used in various hand-calculations to produce SS values for a series of different models.

The Type I hypotheses have these properties:

- Type I SS for all effects add up to the model SS. None of the other SS types have this property, except in special cases.
- Type I hypotheses can be derived from rows of the forward Dolittle transformation of **X'X**.
- Type I SS are statistically independent from each other if the residual errors are independent and identically distributed normal.
- Type I hypotheses depend on the order in which effects are specified in the MODEL.
- Type I hypotheses are uncontaminated by effects preceding the effect

being tested; however, the hypotheses usually involve parameters for effects following the tested effect in the model. For example, in the model

Y=A B;

the Type I hypothesis for B does not involve A parameters, but the Type I hypothesis for A does involve B parameters.

- Type I hypotheses are functions of the cell counts for unbalanced data; the hypotheses are not usually the same hypotheses that are tested if the data are balanced.
- Type I SS are useful for polynomial models where you want to know the contribution of a term as though it had been made orthogonal to preceding effects. Thus, Type I SS correspond to tests of the orthogonalized polynomials.

```
                                      GENERAL LINEAR MODELS PROCEDURE                          3

DEPENDENT VARIABLE: Y

TYPE I ESTIMABLE FUNCTIONS FOR: A         FUNCTIONS FOR: B             FUNCTIONS FOR: A*B

EFFECT          COEFFICIENTS              COEFFICIENTS                 COEFFICIENTS

INTERCEPT       0                         0                            0

A          1    L2                        0                            0
           2    L3                        0                            0
           3    -L2-L3                    0                            0

B          1    0.1667*L2-0.1667*L3       L5                           0
           2    -0.1667*L2+0.1667*L3      -L5                          0

A*B        1 1  0.6667*L2                 0.2857*L5                    L7
           1 2  0.3333*L2                 -0.2857*L5                   -L7
           2 1  0.3333*L3                 0.2857*L5                    L9
           2 2  0.6667*L3                 -0.2857*L5                   -L9
           3 1  -0.5*L2-0.5*L3            0.4286*L5                    -L7-L9
           3 2  -0.5*L2-0.5*L3            -0.4286*L5                   L7+L9


SOURCE                DF          TYPE I SS      F VALUE      PR > F

A                      2       494.03100000       117.84      0.0003
B                      1        10.71428571         5.11      0.0866
A*B                    2        15.73071429         3.75      0.1209
```

Type II tests The Type II tests can also be calculated by comparing the error SS for subset models. The Type II SS are the reduction in error SS due to adding the term after all other terms have been added to the model except terms that contain the effect being tested. An effect is contained in another effect if it can be derived by deleting terms in the effect. For example, A and B are both contained in A*B. For our model:

Source	*Type II SS*
A	SS(A|μ,B)
B	SS(B|μ,A)
A*B	SS(A*B|μ,A,B)

Type II SS have these properties:

- Type II SS do not necessarily add to the model SS.
- The hypothesis for an effect does not involve parameters of other effects except for containing effects (which it must involve to be estimable).

- Type II SS are invariant to the ordering of effects in the model.
- For unbalanced designs, Type II hypotheses for effects that are contained in other effects are not usually the same hypotheses that are tested if the data are balanced. The hypotheses are generally functions of the cell counts.

```
                                    GENERAL LINEAR MODELS PROCEDURE                    4

DEPENDENT VARIABLE: Y

TYPE II ESTIMABLE FUNCTIONS FOR: A          FUNCTIONS FOR: B           FUNCTIONS FOR: A*B

EFFECT              COEFFICIENTS            COEFFICIENTS               COEFFICIENTS

INTERCEPT           0                       0                          0

A            1      L2                      0                          0
             2      L3                      0                          0
             3      -L2-L3                  0                          0

B            1      0                       L5                         0
             2      0                       -L5                        0

A*B          1 1    0.619*L2+0.0476*L3      0.2857*L5                  L7
             1 2    0.381*L2-0.0476*L3      -0.2857*L5                 -L7
             2 1    -0.0476*L2+0.381*L3     0.2857*L5                  L9
             2 2    0.0476*L2+0.619*L3      -0.2857*L5                 -L9
             3 1    -0.5714*L2-0.4286*L3    0.4286*L5                  -L7-L9
             3 2    -0.4286*L2-0.5714*L3    -0.4286*L5                 L7+L9


DF          TYPE II SS      F VALUE    PR > F

2        499.12028571        119.05    0.0003
1         10.71428571          5.11    0.0866
2         15.73071429          3.75    0.1209
```

Type III and Type IV tests Type III and Type IV SS, sometimes referred to as *partial sums of squares*, are considered by many to be the most desirable. These SS cannot in general be computed by comparing model SS from several models using GLM's parameterization. (However, they can sometimes be computed by "reduction" for methods that reparameterize to full rank.) In GLM they are computed by constructing an estimated hypothesis matrix **L** and then computing the SS associated with the hypothesis $\mathbf{L}\beta=0$. As long as there are no missing cells in the design, Type III and Type IV SS are the same.

These are properties of Type III and Type IV SS:

- The hypothesis for an effect does not involve parameters of other effects except for containing effects (which it must involve to be estimable).
- The hypotheses to be tested are invariant to the ordering of effects in the model.
- The hypotheses are the same hypotheses that are tested if there are no missing cells. They are not functions of cell counts.
- The SS do not normally add up to the model SS.

They are constructed from the general form of estimable functions. Type III and Type IV tests are only different if the design has missing cells. In this case the Type III tests have an orthogonality property, while the Type IV tests have a balancing property. These properties are discussed in Chapter 15, "The Four Types of Estimable Functions." For this example, Type IV tests are identical to the Type III tests that are shown.

```
                                          GENERAL LINEAR MODELS PROCEDURE                                             5
DEPENDENT VARIABLE: Y

TYPE III ESTIMABLE FUNCTIONS FOR: A            FUNCTIONS FOR: B             FUNCTIONS FOR: A*B

EFFECT              COEFFICIENTS               COEFFICIENTS                 COEFFICIENTS

INTERCEPT           0                          0                            0

A            1      L2                         0                            0
             2      L3                         0                            0
             3      -L2-L3                     0                            0

B            1      0                          L5                           0
             2      0                          -L5                          0

A*B          1 1    0.5*L2                     0.3333*L5                    L7
             1 2    0.5*L2                     -0.3333*L5                   -L7
             2 1    0.5*L3                     0.3333*L5                    L9
             2 2    0.5*L3                     -0.3333*L5                   -L9
             3 1    -0.5*L2-0.5*L3             0.3333*L5                    -L7-L9
             3 2    -0.5*L2-0.5*L3             -0.3333*L5                   L7+L9


SOURCE                  DF        TYPE III SS      F VALUE     PR > F        DF        TYPE IV SS      F VALUE     PR > F

A                        2       479.10785714       114.28     0.0003         2      479.10785714       114.28     0.0003
B                        1         9.45562500         4.51     0.1009         1        9.45562500         4.51     0.1009
A*B                      2        15.73071429         3.75     0.1209         2       15.73071429         3.75     0.1209
```

Computational Method

Let **X** represent the $n \times p$ design matrix. (When effects containing only class variables are involved, the columns of **X** corresponding to these effects contain only 0s and 1s. No reparameterization is made.) Let **Y** represent the $n \times 1$ vector of dependent variables.

The normal equations $\mathbf{X'X}\beta = \mathbf{X'Y}$ are solved using a modified sweep routine that produces a generalized (g2) inverse $(\mathbf{X'X})^-$ and a solution $\mathbf{b} = (\mathbf{X'X})^- \mathbf{X'y}$ (Pringle and Raynor, 1971).

For each effect in the model, a matrix **L** is computed such that the rows of **L** are estimable. Tests of the hypothesis $\mathbf{L}\beta = 0$ are then made by first computing

$$SS(\mathbf{L}\beta = 0) = (\mathbf{Lb})'(\mathbf{L}(\mathbf{X'X})^-\mathbf{L'})^{-1}(\mathbf{Lb}) \quad ,$$

then computing the associated F value using the mean squared error.

Absorption

Absorption is a computational technique used to reduce computing resource needs in certain cases. The classic use of absorption occurs when a blocking factor with a large number of levels is a term in the model.

For example, the statements

```
PROC GLM;
  ABSORB HERD;
  CLASS A B;
  MODEL Y=A B A*B;
```

are equivalent to

```
PROC GLM;
  CLASS HERD A B;
  MODEL Y=HERD A B A*B;
```

with the exception that the TYPE II, TYPE III, or TYPE IV SS for HERD are not computed when HERD is absorbed.

Several effects may be absorbed at one time. For example, these statements

```
PROC GLM;
  ABSORB HERD COW;
  CLASS A B;
  MODEL Y=A B A*B;
```

are equivalent to

```
PROC GLM;
  CLASS HERD COW A B;
  MODEL Y=HERD COW(HERD) A B A*B;
```

When you use absorption, the size of the **X'X** matrix is a function only of the effects in the MODEL statement. The effects being absorbed do not contribute to the size of the **X'X** matrix.

For the example above, A and B could be absorbed:

```
PROC GLM;
  ABSORB A B;
  CLASS HERD COW;
  MODEL Y=HERD COW(HERD);
```

Although the sources of variation in the results are listed as

```
A B(A) HERD COW(HERD) ,
```

all types of estimable functions for HERD and COW(HERD) are free of A, B, and A*B parameters.

To illustrate the savings in computing using ABSORB, we ran GLM on general data with 1276 degrees of freedom in the model with these statements:

```
DATA A;
  LENGTH HERD COW TRTMENT 2;
  DO HERD=1 TO 40;
    N=1+UNIFORM(1234567)*60;
    DO COW=1 TO N;
      DO TRTMENT=1 TO 3;
        DO REP=1 TO 2;
          Y=HERD/5+COW/10+TRTMENT+NORMAL(1234567);
          OUTPUT;
          END;
        END;
      END;
    END;
  DROP N;
```

This analysis would have required over 6 megabytes of memory for the **X'X** matrix had GLM solved it directly. However, GLM only needed a 4×4 matrix for intercept and treatment, since the other effects were absorbed.

```
PROC GLM;
  ABSORB HERD COW;
  CLASS TRTMENT;
  MODEL Y=TRTMENT;
```

```
                                                                                          1
                         GENERAL LINEAR MODELS PROCEDURE

                             CLASS LEVEL INFORMATION

                           CLASS       LEVELS     VALUES

                           TRTMENT        3       1 2 3

                  NUMBER OF OBSERVATIONS IN DATA SET = 7650
```

```
                                                                                          2
                         GENERAL LINEAR MODELS PROCEDURE

DEPENDENT VARIABLE: Y  (6)          (5)                (4)            (8)        (9)          (10)          (11)
SOURCE                 DF    SUM OF SQUARES     MEAN SQUARE      F VALUE     PR > F      R-SQUARE       C.V.

(2) MODEL            1276    62208.68496838     48.75288791        48.14     0.0001      0.905997     12.4013

(3) ERROR            6373     6454.52506857  (7) 1.01279226            (12) ROOT MSE               (13) Y MEAN

(1) CORRECTED TOTAL  7649    68663.21003694                             1.00637580                8.11507114
                                  (14)                                          (15)          (16)
SOURCE                 DF       TYPE I SS     F VALUE    PR > F     DF     TYPE III SS     F VALUE    PR > F

HERD                   39  44824.52687244     1134.83    0.0001
COW(HERD)            1235  12224.34481401        9.77    0.0001
TRTMENT                 2   5159.81328193     2547.32    0.0001      2   5159.81328193     2547.32    0.0001
```

Expected Mean Squares for Random Effects

The RANDOM statement in GLM declares one or more effects in the model to be random rather than fixed components. GLM does not estimate variance components, nor does it determine what tests are appropriate. However, GLM does print out the coefficients of the expected mean squares. From this printout you can form variance component estimates and determine proper tests by yourself.

The expected mean squares are computed as follows. Consider the model

$$Y = X_0\beta_0 + X_1\beta_1 + X_2\beta_2 \ldots + X_k\beta_k + \varepsilon$$

where β_0 represents the fixed effects, and β_1, β_2, ..., ε represent the random effects normally and independently distributed. For any **L** in the row space of $\mathbf{X} = (X_0|X_1|X_2|\ldots|X_k)$, then

$$\begin{aligned} E(SS_L) &= \beta_0'C_0'C_0\beta_0 + SSQ(C_1)\sigma_1^2 + SSQ(C_2)\sigma_2^2 \ldots \\ &\quad + SSQ(C_k)\sigma_k^2 + \text{rank}(\mathbf{L})\sigma_\varepsilon^2 \ , \end{aligned}$$

where **C** is of the same dimensions as **L** and partitioned as the **X** matrix. In other words,

$$\mathbf{C} = (C_0|C_1|\ldots|C_k) \ .$$

Furthermore, **C**=**ML** where **M** is the inverse of the lower triangular Cholesky decomposition matrix of **L(X'X)⁻L'**. SSQ(**A**) is defined as tr(**A'A**).

For the model in this MODEL statement

```
MODEL Y=A B(A) C A*C;
```

with B(A) declared as random, the expected mean square of each effect is printed as

```
VAR(ERROR) + constant*VAR(B(A)) + Q(A,C,A*C).
```

If any fixed effects appear in the expected mean square of an effect, the letter Q followed by the list of fixed effects in the expected value is printed. The actual numeric values of the quadratic form (**Q** matrix) may be printed using the Q option.

An example at the end of this chapter illustrates the use of the RANDOM statement for mixed models. See Goodnight and Speed (1978) for further theoretical discussion.

Comparisons of Means

When comparing more than two means, an *ANOVA F* test tells you if the means are significantly different from each other, but it does not tell you which means differ from which other means. Multiple comparison methods (also called *mean separation tests*) give you more detailed information about the differences among the means. A variety of multiple comparison methods are available with the MEANS statement in the ANOVA and GLM procedures.

BY *multiple comparisons* we mean **more than one comparison among three or more means**. There is a serious lack of standardized terminology in the literature on comparison of means. Einot and Gabriel (1975), for example, use the term *multiple comparison procedure* to mean what we define below as a *step-down multiple-stage test*. Some methods for multiple comparisons have not yet been given names, such as those referred to below as REGWQ and REGWF. When reading the literature, you may need to determine what methods are being discussed based on the formulas and references given.

When you interpret multiple comparisons, it is important to remember that failure to reject the hypothesis that two or more means are equal should not lead to the conclusion that the population means are in fact equal. Failure to reject the null hypothesis implies only that the difference between population means, if any, is not large enough to be detected with the given sample size. A related point is that nonsignificance is nontransitive: given three sample means, the largest and smallest may be significantly different from each other, while neither is significantly different from the middle one. Nontransitive results of this type occur frequently in multiple comparisons.

Multiple comparisons can also lead to counter-intuitive results when the cell sizes are unequal. Consider four cells labeled A, B, C, and D, with sample means in the order A>B>C>D. If A and D each have two observations, and B and C each have 10,000 observations, then the difference between B and C may be significant while the difference between A and D is not.

Confidence intervals may be more useful than significance tests in multiple comparisons. Confidence intervals show the degree of uncertainty in each comparison in an easily interpretable way, make it easier to assess the practical significance of a difference as well as the statistical significance, and are less likely to lead nonstatisticians to the invalid conclusion that nonsignificantly different sample means imply equal population means.

The simplest approach to multiple comparisons is to do a *t* test on every pair of means (the T option on the MEANS statement). For the ith and j^{th} means you can reject the null hypothesis that the population means are equal if

$$|\bar{y}_i - \bar{y}_j|/s(1/n_i + 1/n_j)^{1/2} >= t(\alpha;\nu) \ ,$$

where $\bar{y}_i$ and $\bar{y}_j$ are the means, n_i and n_j are the number of observations in the two cells, s is the root mean square error based on ν degrees of freedom, α is the significance level, and $t(\alpha;\nu)$ is the two-tailed critical value from a Student's *t* distribution. If the cell sizes are all equal to, say, n, the above formula can be rearranged to give

$$|\bar{y}_i - \bar{y}_j| >= t(\alpha;\nu)s(2/n)^{1/2} \ ,$$

the value of the right-hand side being Fisher's least significant difference (LSD).

There is a problem with repeated *t* tests, however. Suppose there are ten means and each *t* test is performed at the .05 level. There are $10(10-1)/2=45$ pairs of means to compare, each with a .05 probability of a type 1 error (a false rejection of the null hypothesis). The chance of making at least one type 1 error is much higher than .05. It is difficult to calculate the exact probability, but you can derive a pessimistic approximation by assuming the comparisons are independent, giving an upper bound to the probability of making at least one type 1 error (the experimentwise error rate) of

$$1 - (1 - .05)^{45} = .90 \ .$$

The actual probability is somewhat less than .90, but as the number of means increases, the chance of making at least one type 1 error approaches 1.

It is up to you to decide whether to control the comparisonwise error rate or the experimentwise error rate, but there are many situations in which the experimentwise error rate should be held to a small value. Statistical methods for making two or more inferences while controlling the probability of making at least one type 1 error are called *simultaneous inference methods* (Miller, 1981), although Einot and Gabriel (1975) use the term *simultaneous test procedure* in a much more restrictive sense.

It has been suggested that the experimentwise error rate can be held to the α level by performing the overall *ANOVA F* test at the α level and making further comparisons only if the *F* test is significant, as in Fisher's protected LSD. This assertion is false if there are more than three means (Einot and Gabriel, 1975). Consider again the situation with ten means. Suppose that one population mean differs from the others by a sufficiently large amount that the power (probability of correctly rejecting the null hypothesis) of the *F* test is near 1, but that all the other population means are equal to each other. There will be $9(9-1)/2=36$ *t* tests of true null hypotheses, with an upper limit of .84 on the probability of at least one type 1 error. Thus you must distinguish between the experimentwise error rate under the complete null hypothesis, in which all population means are equal, and the experimentwise error rate under a partial null hypothesis, in which some means are equal but others differ. The following abbreviations are used in the discussion below:

CER	comparisonwise error rate
EERC	experimentwise error rate under the complete null hypothesis

EERP experimentwise error rate under a partial null hypothesis

MEER maximum experimentwise error rate under any complete or partial null hypothesis.

A preliminary *F* test controls the EERC but not the EERP or the MEER.

The MEER can be controlled at the α level by setting the CER to a sufficiently small value. The Bonferroni inequality (Miller, 1981) has been widely used for this purpose. If

$$\mathrm{CER} = \alpha/c \quad ,$$

where c is the total number of comparisons, then the MEER is less than α. Bonferroni *t* tests (the BON option) with MEER<α declare two means to be significantly different if

$$|\bar{y}_i - \bar{y}_j|/s(1/n_i + 1/n_j)^{1/2} >= t(\varepsilon;\nu) \quad ,$$

where $\varepsilon = \alpha/(k(k-1)/2)$ for comparison of *k* means. If the cell sizes are equal, the test simplifies to

$$|\bar{y}_i - \bar{y}_j| >= t(\varepsilon;\nu)s(2/n)^{1/2} \quad .$$

Sidak (1967) has provided a tighter bound, showing that

$$\mathrm{CER} = 1 - (1-\alpha)^{1/c}$$

also insures MEER<=α for any set of *c* comparisons. A Sidak *t* test (Games, 1977), provided by the SIDAK option, is thus given by

$$|\bar{y}_i - \bar{y}_j|/s(1/n_i + 1/n_j)^{1/2} >= t(\varepsilon;\nu) \quad ,$$

where $\varepsilon = 1-(1-\alpha)^{1/(k(k-1)/2)}$ for comparison of *k* means. If the sample sizes are equal, the test simplifies to

$$|\bar{y}_i - \bar{y}_j| >= t(\varepsilon;\nu)s(2/n)^{1/2} \quad .$$

The Bonferroni additive inequality and the Sidak multiplicative inequality can be used to control the MEER for any set of contrasts or other hypothesis tests, not just pairwise comparisons. The Bonferroni inequality can provide simultaneous inferences in any statistical application requiring tests of more than one hypothesis. Other methods discussed below for pairwise comparisons can also be adapted for general contrasts (Miller, 1981).

Scheffe (1953, 1959) proposed another method to control the MEER for any set of contrasts or other linear hypotheses in the analysis of linear models, including pairwise comparisons, obtained with the SCHEFFE option. Two means are declared significantly different if

$$|\bar{y}_i - \bar{y}_j|/s(1/n_i + 1/n_j)^{1/2} >= ((k-1)F(\alpha;k-1,\nu))^{1/2} \quad ,$$

or, for equal cell sizes,

$$|\bar{y}_i - \bar{y}_j| >= ((k-1)F(\alpha;k-1,\nu)s(2/n))^{1/2} \quad ,$$

where $F(\alpha;k-1,\nu)$ is the α-level critical value of an F distribution with $k-1$ numerator degrees of freedom and ν denominator degrees of freedom.

Scheffe's test is compatible with the overall *ANOVA F* test in that Scheffe's method never declares a contrast significant if the overall F test is nonsignificant. Most other multiple comparison methods are capable of finding significant contrasts when the overall F is nonsignificant and will therefore suffer a loss of power when used with a preliminary F test.

Scheffe's method may be more powerful than the Bonferroni or Sidak methods if the number of comparisons is large relative to the number of means. For pairwise comparisons, Sidak t tests are generally more powerful.

Tukey (1952, 1953) proposed a test designed specifically for pairwise comparisons based on the studentized range, sometimes called the "honestly significant difference" test, that controls the MEER when the sample sizes are equal. Tukey (1953) and Kramer (1956) independently proposed a modification for unequal cell sizes. The Tukey or Tukey-Kramer method is provided by the TUKEY option. There is not yet a general proof that the Tukey-Kramer procedure controls the MEER but the method has fared extremely well in Monte Carlo studies (Dunnett, 1980). The Tukey-Kramer method is more powerful than the Bonferroni, Sidak, or Scheffe methods for pairwise comparisons. Two means are considered significantly different by the Tukey-Kramer criterion if

$$|\bar{y}_i - \bar{y}_j|/s((1/n_i + 1/n_j)/2)^{1/2} >= q(\alpha;k,\nu) \quad ,$$

where $q(\alpha;k,\nu)$ is the α-level critical value of a studentized range distribution of k independent normal random variables with ν degrees of freedom. For equal cell sizes, Tukey's method rejects the null hypothesis of equal population means if

$$|\bar{y}_i - \bar{y}_j| >= q(\alpha;k,\nu)s/n^{1/2} \quad .$$

Hochberg (1974) devised a method (the GT2 or SMM option) similar to Tukey's but using the studentized maximum modulus instead of the studentized range and employing Sidak's (1967) uncorrelated-t inequality. It was proved to hold the MEER at a level not exceeding α with unequal sample sizes. It is generally less powerful than the Tukey-Kramer method and always less powerful than Tukey's test for equal cell sizes. Two means are declared significantly different if

$$|\bar{y}_i - \bar{y}_j|/s(1/n_i + 1/n_j)^{1/2} >= m(\alpha;c,\nu) \quad ,$$

where $m(\alpha;c,\nu)$ is the α-level critical value of the studentized maximum modulus distribution of c independent normal random variables with ν degrees of freedom and $c=k(k-1)/2$. For equal cell sizes, the test simplifies to

$$|\bar{y}_i - \bar{y}_j| >= m(\alpha;c,\nu)s(2/n)^{1/2} \quad .$$

Gabriel (1978) proposed another method (the GABRIEL option) based on the studentized maximum modulus for unequal cell sizes that rejects if

$$|\bar{y}_i - \bar{y}_j|/s((2n_i)^{-1/2} + (2n_j)^{-1/2}) >= m(\alpha;k,\nu) \quad .$$

For equal cell sizes, Gabriel's test is equivalent to Hochberg's GT2 method. For unequal cell sizes, Gabriel's method is more powerful than GT2 but may become liberal with highly disparate cell sizes (see also Dunnett, 1980). Gabriel's test is the only method for unequal sample sizes that lends itself to a convenient graphical representation. Assuming $\bar{y}_i > \bar{y}_j$, the above inequality can be rewritten as

$$\bar{y}_i - m(\alpha;k,\nu)s/(2n_i)^{1/2} >= \bar{y}_j + m(\alpha;k,\nu)s/(2n_j)^{1/2} \quad .$$

The expression on the left does not depend on *j*, nor does the expression on the right depend on *i*. Hence one can form what Gabriel calls an (*l*,*u*)-interval around each sample mean and declare two means to be significantly different if their (*l*,*u*)-intervals do not overlap.

All of the methods discussed so far can be used to obtain simultaneous confidence intervals (Miller, 1981). By sacrificing the facility for simultaneous estimation, it is possible to obtain simultaneous tests with greater power using multiple-stage tests (MSTs). MSTs come in both step-up and step-down varieties (Welsch, 1977). The step-down methods, which have been more widely used, are available in SAS.

Step-down MSTs first test the homogeneity of all of the means at a level γ_k. If the test results in a rejection, then each subset of $k-1$ means is tested at level γ_{k-1}; otherwise, the procedure stops. In general, if the hypothesis of homogeneity of a set of *p* means is rejected at the γ_p level, then each subset of $p-1$ means is tested at the γ_{p-1} level; otherwise, the set of *p* means is considered not to differ significantly and none of its subsets are tested. The many varieties of MSTs that have been proposed differ in the levels γ_p and the statistics on which the subset tests are based. Clearly, the EERC of a step-down MST is not greater than γ_k, and the CER is not greater than γ_2, but the MEER is a complicated function of γ_p, $p=2,\ldots,k$.

MSTs can be used with unequal cell sizes but the resulting operating characteristics are undesirable, so only the balanced case will be considered here. With equal sample sizes, the means can be arranged in ascending or descending order, and only contiguous subsets need be tested. It is common practice to report the results of an MST by writing the means in such an order and drawing lines parallel to the list of means spanning the homogeneous subsets. This form of presentation is also convenient for pairwise comparisons with equal cell sizes.

The best known MSTs are the Duncan (the DUNCAN option) and Student-Newman-Keuls (the SNK option) methods (Miller, 1981). Both use the studentized range statistic and hence are called *multiple range tests*. Duncan's method is often called the "new" multiple range test despite the fact that it is one of the oldest MSTs in current use. The Duncan and SNK methods differ in the γ_p values used. For Duncan's method they are

$$\gamma_p = 1 - (1 - \alpha)^{p-1} \quad ,$$

while the SNK method uses

$$\gamma_p = \alpha \quad .$$

Duncan's method controls the CER at the α level. Its operating characteristics appear similar to those of Fisher's unprotected LSD or repeated *t* tests at level α (Petrinovich and Hardyck, 1969). Since repeated *t* tests are easier to compute, easier to explain, and applicable to unequal sample sizes, Duncan's method is not recommended. Several published studies (for example, Carmer and Swanson, 1973) have claimed that Duncan's method is superior to Tukey's because of greater power without considering that the greater power of Duncan's method is due to its higher type 1 error rate (Einot and Gabriel, 1975).

The SNK method holds the EERC to the α level but does not control the EERP (Einot and Gabriel, 1975). Consider ten population means that occur in five pairs such that means within a pair are equal, but there are large differences between pairs. Making the usual sampling assumptions and also assuming that the sample sizes are very large, all subset homogeneity hypotheses for three or more means are rejected. The SNK method then comes down to five independent tests, one for each pair, each at the α level. Letting α be .05, the probability of at least one false rejection is

$$1 - (1 - .05)^5 = .23 \quad .$$

As the number of means increases, the MEER approaches 1. Therefore, the SNK method cannot be recommended.

A variety of MSTs that control the MEER have been proposed, but these methods are not as well known as those of Duncan and SNK. An approach developed by Ryan (1959, 1960), Einot and Gabriel (1975), and Welsch (1977) sets

$$\begin{aligned} \gamma_p &= 1 - (1 - \alpha)^{p/k} && \text{for } p<k-1 \quad , \\ &= \alpha && \text{for } p >= k-1 \quad . \end{aligned}$$

Either range or F statistics may be used, leading to what we call the REGWQ and REGWF methods, respectively, after the authors' initials. Assuming the sample means have been arranged in descending order from $\bar{y}_1$ through $\bar{y}_k$, the homogeneity of means $\bar{y}_i, \ldots, \bar{y}_j$, $i<j$, is rejected by REGWQ if

$$\bar{y}_i - \bar{y}_j >= q(\gamma_p;p,\nu)s/n^{1/2} \quad ,$$

or by REGWF if

$$n(\Sigma\bar{y}_u^2 - (\Sigma\bar{y}_u)^2/k)/(p-1)s^2 >= F(\gamma_p;p-1,\nu) \quad ,$$

where $p=j-i+1$ and the summations are over $u=i,\ldots,j$ (Einot and Gabriel, 1975).

REGWQ and REGWF appear to be the most powerful step-down MSTs in the current literature (for example, Ramsey, 1978). REGWF has the advantage of being compatible with the overall *ANOVA F* test in that REGWF rejects the complete null hypothesis if and only if the overall F test does so, since the latter is identical to the first step in REGWF. Use of a preliminary F test decreases the power of all the other multiple comparison methods discussed above except for Scheffe's test.

Other multiple comparison methods proposed by Peritz (Marcus, Peritz, and Gabriel, 1976; Begun and Gabriel, 1981) and Welsch (1977) are still more powerful than the REGW procedures. These methods have not yet been implemented in SAS.

Waller and Duncan (1969) and Duncan (1975) take an approach to multiple comparisons that differs from all the methods discussed above in minimizing the Bayes risk under additive loss rather than controlling type 1 error rates. For each pair of population means μ_i and μ_j, null (H_0^{ij}) and alternative (H_a^{ij}) hypotheses are defined:

$$\begin{aligned} H_0^{ij}&: \mu_i - \mu_j <= 0 \\ H_a^{ij}&: \mu_i - \mu_j > 0 \quad . \end{aligned}$$

For any i,j pair let d_0 indicate a decision in favor of H_0^{ij} and d_a indicate a decision in favor of H_a^{ij}, and let $\delta=\mu_i - \mu_j$. The loss function for the decision on the i,j pair is

$$\begin{aligned} L(d_0|\delta) &= 0 && \text{if } \delta<= 0 \\ &= \delta && \text{if } \delta > 0 \\ L(d_a|\delta) &= -k\delta && \text{if } \delta<= 0 \\ &= 0 && \text{if } \delta>0 \quad , \end{aligned}$$

where k represents a constant that you specify rather than the number of means. The loss for the joint decision involving all pairs of means is the sum of the losses for each individual decision. The population means are assumed to have a normal prior distribution with unknown variance, the logarithm of the variance of the means having a uniform prior distribution. For the i,j pair, the null hypothesis is rejected if

$$\bar{y}_i - \bar{y}_j >= t_B s(2/n)^{1/2}$$

where t_B is the Bayesian t value (Waller and Kemp, 1975) depending on k, the F statistic for the one-way *ANOVA*, and the degrees of freedom for F. The value of t_B is a decreasing function of F, so the Waller-Duncan test becomes more liberal as F increases.

In summary, if you wish to control the CER, the recommended methods are repeated t tests or Fisher's unprotected LSD (the T or LSD option). If you wish to control the MEER, do not need confidence intervals, and have equal cell sizes, then the REGWF and REGWQ methods are recommended. If you wish to control the MEER and need confidence intervals or have unequal cell sizes, then use the Tukey or Tukey-Kramer methods (the TUKEY option). If you agree with the Bayesian approach and Waller and Duncan's assumptions, you should use the Waller-Duncan test (the WALLER option).

Multivariate Analysis of Variance

If you fit several dependent variables to the same effects, you may want to make tests jointly involving parameters of several dependent variables. Suppose you have p dependent variables, k parameters for each dependent variable, and n observations. The models can be collected into one equation:

$$\mathbf{Y} = \mathbf{X}\beta + \varepsilon$$

where $\mathbf{Y}$ is $n \times p$, $\mathbf{X}$ is $n \times k$, β is $k \times p$, and ε is $n \times p$. Each of the p models can be estimated and tested separately. However, you may also want to consider the joint distribution. With p dependent variables, there are $n \times p$ errors that are independent across observations, but not across dependent variables. Assume:

$$\text{vec}(\varepsilon) \sim \text{N}(0, \text{I}_n \otimes \Sigma)$$

where vec(ε) strings ε out by rows, and Σ is $p \times p$. Σ can be estimated by:

$$\mathbf{S} = (e'e)/n = (\mathbf{Y} - \mathbf{XB})'(\mathbf{Y} - \mathbf{XB})/n$$

where $\mathbf{B} = (\mathbf{X'X})^{-}\mathbf{X'Y}$.

If $\mathbf{S}$ is scaled to unit diagonals, the values in $\mathbf{S}$ are called *partial correlations of the Ys adjusting for the Xs*. This matrix can be printed by GLM if PRINTE is specified as a MANOVA option.

You can form hypotheses for linear combinations across columns as well as across rows of β. The multivariate general linear hypothesis is written:

$$\mathbf{L}\beta\mathbf{M} = 0 \quad .$$

The GLM procedure can test special cases where $\mathbf{L}$ is for Type I, Type II, Type III, or Type IV tests and $\mathbf{M}$ is the $p \times p$ identity matrix. In other words, GLM makes joint tests that the Type I, Type II, Type III, or Type IV hypothesis holds for all dependent variables in the model.

Multivariate tests first construct the matrices **H** and **E** that correspond to the numerator and denominator of a univariate *F* test.

$$\mathbf{H} = \mathbf{M}'(\mathbf{LB})'(\mathbf{L}(\mathbf{X}'\mathbf{X})^{-}\mathbf{L}')^{-1}(\mathbf{LB})\mathbf{M}$$

$$\mathbf{E} = \mathbf{M}'(\mathbf{Y}'\mathbf{Y} - \mathbf{B}'(\mathbf{X}'\mathbf{X})\mathbf{B})\mathbf{M} \quad .$$

With GLM, the diagonal elements of **H** and **E** correspond to the hypothesis and error SS for the univariate tests.

Four test statistics, all functions of the eigenvalues of $\mathbf{E}^{-1}\mathbf{H}$ (or $(\mathbf{E}+\mathbf{H})^{-1}\mathbf{H}$) are constructed:

- Wilks' lambda = $\det(\mathbf{E})/\det(\mathbf{H}+\mathbf{E})$
- Pillai's trace = $\mathrm{trace}(\mathbf{H}(\mathbf{H}+\mathbf{E})^{-1})$
- Hotelling-Lawley trace = $\mathrm{trace}(\mathbf{E}^{-1}\mathbf{H})$
- Roy's maximum root = λ, largest eigenvalue of $\mathbf{E}^{-1}\mathbf{H}$.

All four are reported with *F* approximations. For further details on these four statistics, see the section **Multivariate Tests** in "Introduction to SAS Regression Procedures."

GLM does not have a feature to compare parameters in different columns of β; in other words, GLM does not support an **M** other than the identity. However, it is possible to construct a test for any multivariate hypothesis by suitably transforming the dependent variables:

$$\mathbf{Y} = \mathbf{X}\beta + \varepsilon$$
$$H_0\colon \mathbf{L}\beta\mathbf{M} = 0$$

Consider transforming the dependents:

$$\mathbf{Y}^* = \mathbf{YM} = \mathbf{X}\beta\mathbf{M} + \varepsilon\mathbf{M} = \mathbf{X}\delta + u$$

$$H_0\colon \mathbf{L}\delta = 0, \quad \text{which is } \mathbf{L}\beta\mathbf{M} = 0 \quad .$$

To get the tests you want, apply the **M** to the dependent variables, fit the model, and apply the tests to the transformed model.

Examples Suppose you have three dependent variables fit to an effect:

```
PROC GLM DATA=A;
  CLASS SEX;
  MODEL Y1 Y2 Y3=SEX;
  MANOVA H=SEX;
```

GLM tests that SEX does not affect any of the three dependent variables:

$$\mathbf{L} = (0\ 1\ -1)\ , \quad \mathbf{M} = \mathbf{I}_3 \quad .$$

You may also want to test the weaker hypothesis that SEX does not affect the average across the three responses. Then:

$$\mathbf{L} = (0\ 1\ -1)\ , \quad \mathbf{M} = (1\ 1\ \ 1)$$

and you code these statements:

```
DATA B;
  SET A;
  SUM=(Y1+Y2+Y3);
PROC GLM DATA=A;
  CLASS SEX;
  MODEL SUM=SEX;
```

Another common test is that the parameters for SEX are the same across the three responses:

$$\mathbf{L} = (0\ 1\ -1) \qquad \mathbf{M} = \begin{bmatrix} 1 & 0 \\ 0 & 1 \\ -1 & -1 \end{bmatrix}$$

```
DATA C;
  SET A;
  Y13=Y1-Y3;
  Y23=Y2-Y3;
PROC GLM DATA=C;
  CLASS SEX;
  MODEL Y13 Y23=SEX;
  MANOVA H=SEX;
```

Least-Squares Means

Simply put, least-squares means, or *population marginal means*, are the expected value of class or subclass means that you would expect for a balanced design involving the class variable with all covariates at their mean value. This informal concept is explained further in Searle, Speed, and Milliken (1980).

To construct a least-squares mean (LSM) for a given level of a given effect, construct a set of Xs according to the following rules and use them in the linear model with the parameter estimates to yield the value of the LSM.

1. Hold all covariates (continuous variables) to their mean value.
2. Consider effects contained by the given effect. Give the Xs for levels associated with the given level a value of 1. Make the other Xs equal to 0.
3. Consider the given effect. Make the X associated with the given level equal to 1. Set the Xs for the other levels to 0.
4. Consider the effects that contain the given effect. For the columns associated with the given level, use $1/k$ where k is the number of such columns. For the other columns use 0.
5. Consider the other effects not yet considered. Use $1/j$ where j is the number of levels in the effect.

The consequence of these rules is that the sum of the Xs within any classification effect is 1. This set of Xs forms a linear combination of the parameters that is checked for estimability before it is evaluated.

For example, consider the model:

```
PROC GLM;
  CLASS A B C;
  MODEL Y=A B A*B C X;
  LSMEANS A B A*B C;
```

Assume A has 3 levels, B has 2, and C has 2 and assume that every combination of levels of A and B exists in the data. Assume also that the average of X is 12.5. Then the least-squares means are computed by the following linear combinations of the parameter estimates:

	A			B		A*B						C		X
	1	2	3	1	2	11	21	31	12	22	32	1	2	.
LSM( )	1/3	1/3	1/3	1/2	1/2	1/6	1/6	1/6	1/6	1/6	1/6	1/2	1/2	12.5
LSM(A1)	1	0	0	1/2	1/2	1/2	0	0	1/2	0	0	1/2	1/2	12.5
LSM(A2)	0	1	0	1/2	1/2	0	1/2	0	0	1/2	0	1/2	1/2	12.5
LSM(A3)	0	0	1	1/2	1/2	0	0	1/2	0	0	1/2	1/2	1/2	12.5
LSM(B1)	1/3	1/3	1/3	1	0	1/3	1/3	1/3	0	0	0	1/2	1/2	12.5
LSM(B2)	1/3	1/3	1/3	0	1	0	0	0	1/3	1/3	1/3	1/2	1/2	12.5
LSM(AB11)	1	0	0	1	0	1	0	0	0	0	0	1/2	1/2	12.5
LSM(AB21)	0	1	0	1	0	0	1	0	0	0	0	1/2	1/2	12.5
LSM(AB31)	0	0	1	1	0	0	0	1	0	0	0	1/2	1/2	12.5
LSM(AB12)	1	0	0	0	1	0	0	0	1	0	0	1/2	1/2	12.5
LSM(AB22)	0	1	0	0	1	0	0	0	0	1	0	1/2	1/2	12.5
LSM(AB32)	0	0	1	0	1	0	0	0	0	0	1	1/2	1/2	12.5
LSM(C1)	1/3	1/3	1/3	1/2	1/2	1/6	1/6	1/6	1/6	1/6	1/6	1	0	12.5
LSM(C2)	1/3	1/3	1/3	1/2	1/2	1/6	1/6	1/6	1/6	1/6	1/6	0	1	12.5

Specification of ESTIMATE Expressions

For this example of the regression model:

```
MODEL Y=X1 X2 X3;
```

the associated parameters are β_0, β_1, β_2, and β_3 (where β_0 represents the intercept). To estimate $3\beta_1+2\beta_2$, you need the following **L** vector:

$$L=(0\ 3\ 2\ 0)\ .$$

The corresponding ESTIMATE statement is:

```
ESTIMATE '3B1+2B2' X1 3 X2 2;
```

To estimate $\beta_0+\beta-2\beta_3$ you need this **L** vector:

$$L=(1\ 1\ 0\ -2)\ .$$

The corresponding ESTIMATE statement is:

```
ESTIMATE 'B0+B1-2B3' INTERCEPT 1 X1 1 X3 -2;
```

Now consider models involving class variables such as

```
MODEL Y=A B A*B;
```

with the associated parameters:

$(\mu\ \alpha_1\ \alpha_2\ \alpha_3\ \beta_1\ \beta_2\ \alpha\beta_{11}\ \alpha\beta_{12}\ \alpha\beta_{21}\ \alpha\beta_{22}\ \alpha\beta_{31}\ \alpha\beta_{32})$.

To estimate the least-squares mean for α_1, you need following **L** vector:

L = (1|1 0 0|.5 .5|.5 .5 0 0 0 0)

and you could use this ESTIMATE statement:

```
ESTIMATE 'LSM(A1)' INTERCEPT 1 A 1 B .5. 5. A*B .5 .5;
```

Note in the above statement that only one element of **L** was specified following the A effect, even though A has three levels. Whenever the list of constants following an effect name is shorter than the effect's number of levels, zeros are used as the remaining constants. In the event that the list of constants is longer than the number of levels for the effect, the extra constants are ignored.

To estimate the A linear effect in the model above, assuming equally spaced levels for A, the following **L** could be used

L = (0|−1 0 1|0 0|−.5 −.5 0 0 .5 .5) .

The ESTIMATE statement for the above **L** is written as

```
ESTIMATE 'A LINEAR' A -1 0 1;
```

If the elements of **L** are not specified for an effect which "contains" a specified effect, then the elements of the specified effect are equitably distributed over the levels of the higher-order effect. The distribution of lower-order coefficients to higher-order effect coefficients follows the same general rules as in the LSMEANS statement and is similar to that used to construct TYPE IV **L**s. In the previous example, the −1 associated with α_1 is divided by the number of $\alpha\beta_{1j}$ parameters; then each $\alpha\beta_{1j}$ coefficient is set to −1/(number of $\alpha\beta_{1j}$). The 1 associated with α_3 is distributed among the $\alpha\beta_{3j}$ parameters in a similar fashion. In the event that an unspecified effect contains several specified effects, distribution of coefficients to the higher-order effect is additive.

Note: numerous syntactical expressions for the ESTIMATE statement were considered, including many that involved specifying the effect and level information associated with each coefficient. For models involving higher-level effects, the requirement of specifying level information would lead to very bulky specifications. Consequently, the simpler form of the ESTIMATE statement described above was implemented. The syntax of this ESTIMATE statement puts a burden on you to know *a priori* the order of the parameter list associated with each effect. When you first begin to use this statement, use the E option to make sure that the actual **L** constructed is the one you envisioned.

A note on estimability Each **L** is checked for estimability using the relationship: **L** = **LH** where **H** = $(\mathbf{X'X})^{-}\mathbf{X'X}$. The **L** vector is declared non-estimable, if for any *i*

$$\mathrm{ABS}(\mathbf{L}_i - (\mathbf{LH})_i) > \begin{cases} \text{1E-4} & \text{if } \mathbf{L}_i = 0 \text{ or} \\ \text{1E-4*ABS}(\mathbf{L}_i) & \text{otherwise} \end{cases} .$$

Continued fractions (like 1/3) should be specified to at least six decimal places, or the DIVISOR parameter should be used.

Printed Output

The GLM procedure produces the following printed output by default:

1. The overall analysis-of-variance table breaks down the CORRECTED TOTAL sum of squares for the dependent variable
2. into the portion attributed to the MODEL,
3. and the portion attributed to ERROR.
4. The MEAN SQUARE term is the
5. SUM OF SQUARES divided by the
6. DEGREES OF FREEDOM (DF).
7. The MEAN SQUARE for ERROR, (MS(ERROR)), is an estimate of ς^2, the variance of the true errors.
8. The F VALUE is the ratio produced by dividing MS(MODEL) by MS(ERROR). It tests how well the model as a whole (adjusted for the mean) accounts for the dependent variable's behavior. An *F* test is a joint test that all parameters except the intercept are zero.
9. A small significance probability, PR>F, indicates that some linear function of the parameters is significantly different from zero.
10. R-SQUARE, R^2, measures how much variation in the dependent variable can be accounted for by the model. R^2, which can range from 0 to 1, is the ratio of the sum of squares for the model divided by the sum of squares for the corrected total. In general, the larger the value of R^2, the better the model's fit.
11. C.V., the coefficient of variation, which describes the amount of variation in the population, is 100 times the standard deviation estimate of the dependent variable, ROOT MSE, divided by the MEAN. The coefficient of variation is often a preferred measure because it is unitless.
12. ROOT MSE estimates the standard deviation of the dependent variable (or equivalently, the error term) and equals the square root of MS(ERROR).
13. MEAN is the sample mean of the dependent variable.

These tests are used primarily in analysis-of-variance applications:

14. The TYPE I SS measures incremental sums of squares for the model as each variable is added.
15. The TYPE III SS is the sum of squares that results when that variable is added last to the model.
16. The F VALUE and PR>F values for TYPE III tests in this section of the output.
17. This section of the output gives the ESTIMATES for the model PARAMETERs—the intercept and the coefficients.
18. T FOR HO: PARAMETER=0 is the Student's *t* value for testing the null hypothesis that the parameter (if it is estimable) equals zero.
19. The significance level, PR>|T|, is the probability of getting a larger value of *t* if the parameter is truly equal to zero. A very small value for this probability leads to the conclusion that the independent variable contributes significantly to the model.
20. The STD ERROR OF ESTIMATE is the standard error of the estimate of the true value of the parameter.

EXAMPLES

Balanced Data from Randomized Complete Block with Means Comparisons and Contrasts: Example 1

Since these data are balanced, you could obtain the same answer more efficiently using the ANOVA procedure; however, GLM presents the results in a slightly different way. Notice that since the data are balanced, the Type I and Type III SS are the same and will equal the ANOVA SS.

First, the standard analysis is shown followed by a run that uses the SOLUTION option and includes MEANS and CONTRAST statements. The SOLUTION option requests a printout of the parameter estimates, which are only printed by default if there are no CLASS variables. A MEANS statement is used to request a printout of the means with two of the multiple comparisons procedures requested. In experiments with well understood treatment levels, CONTRAST statements are preferable to a blanket means comparison method.

```
* -------------------------------SNAPDRAGON  EXPERIMENT-------------------------------- *
|   AS REPORTED BY STENSTROM, 1940, AN EXPERIMENT WAS                                   |
|   UNDERTAKEN TO INVESTIGATE HOW SNAPDRAGONS GREW IN                                   |
|   VARIOUS SOILS. EACH SOIL TYPE WAS USED IN THREE                                     |
|   BLOCKS.                                                                             |
* ------------------------------------------------------------------------------------- *  ;

DATA PLANTS;
  INPUT TYPE $ @;
  DO BLOCK=1 TO 3;
    INPUT STEMLENG @;
    OUTPUT;
    END;
  CARDS;
CLARION  32.7 32.3 31.5
CLINTON  32.1 29.7 29.1
KNOX     35.7 35.9 33.1
O'NEILL  36.0 34.2 31.2
COMPOST  31.8 28.0 29.2
WABASH   38.2 37.8 31.9
WEBSTER  32.5 31.1 29.7
;

PROC GLM;
  CLASS TYPE BLOCK;
  MODEL STEMLENG=TYPE BLOCK;
PROC GLM;
  CLASS TYPE BLOCK;
  MODEL STEMLENG=TYPE BLOCK / SOLUTION;
  MEANS TYPE / WALLER REGWQ;

*-TYPE-ORDER ---------------------------------------------CLRN-CLTN-KNOX-ONEL-CPST-WBSH-WSTR;
CONTRAST 'COMPOST VS OTHERS'      TYPE  -1   -1   -1   -1    6   -1   -1;
CONTRAST 'RIVER SOILS VS NON'     TYPE  -1   -1   -1   -1    0    5   -1,
                                  TYPE  -1    4   -1   -1    0    0   -1;
CONTRAST 'GLACIAL VS DRIFT'       TYPE  -1    0    1    1    0    0   -1;
CONTRAST 'CLARION VS WEBSTER'     TYPE  -1    0    0    0    0    0    1;
CONTRAST 'KNOX VS ONEILL'         TYPE   0    0    1   -1    0    0    0;
```

```
                                                                                                1
                                GENERAL LINEAR MODELS PROCEDURE
                                   CLASS LEVEL INFORMATION
                 CLASS      LEVELS      VALUES
                 TYPE            7      CLARION CLINTON COMPOST KNOX O'NEILL WABASH WEBSTER
                 BLOCK           3      1 2 3

                              NUMBER OF OBSERVATIONS IN DATA SET = 21
```

```
                                                                                                2
                                GENERAL LINEAR MODELS PROCEDURE
DEPENDENT VARIABLE: STEMLENG
SOURCE                  DF      SUM OF SQUARES        MEAN SQUARE      F VALUE         PR > F       R-SQUARE          C.V.
MODEL                    8        142.18857143        17.77357143        10.80         0.0002       0.878079        3.9397
ERROR                   12         19.74285714         1.64523810                    ROOT MSE                 STEMLENG MEAN
CORRECTED TOTAL         20        161.93142857                                     1.28266835                   32.55714286

SOURCE                  DF         TYPE I SS     F VALUE    PR > F        DF        TYPE III SS     F VALUE     PR > F
TYPE                     6      103.15142857       10.45    0.0004         6       103.15142857       10.45     0.0004
BLOCK                    2       39.03714286       11.86    0.0014         2        39.03714286       11.86     0.0014
```

```
                                                                                                3
                                GENERAL LINEAR MODELS PROCEDURE
                                   CLASS LEVEL INFORMATION
                 CLASS      LEVELS      VALUES
                 TYPE            7      CLARION CLINTON COMPOST KNOX O'NEILL WABASH WEBSTER
                 BLOCK           3      1 2 3

                              NUMBER OF OBSERVATIONS IN DATA SET = 21
```

```
                                                                                                4
                                GENERAL LINEAR MODELS PROCEDURE
DEPENDENT VARIABLE: STEMLENG
SOURCE                  DF      SUM OF SQUARES        MEAN SQUARE      F VALUE         PR > F       R-SQUARE          C.V.
MODEL                    8        142.18857143        17.77357143        10.80         0.0002       0.878079        3.9397
ERROR                   12         19.74285714         1.64523810                    ROOT MSE                 STEMLENG MEAN
CORRECTED TOTAL         20        161.93142857                                     1.28266835                   32.55714286

SOURCE                  DF         TYPE I SS     F VALUE    PR > F        DF        TYPE III SS     F VALUE     PR > F
TYPE                     6      103.15142857       10.45    0.0004         6       103.15142857       10.45     0.0004
BLOCK                    2       39.03714286       11.86    0.0014         2        39.03714286       11.86     0.0014

CONTRAST                DF                SS     F VALUE    PR > F
COMPOST VS OTHERS        1        5.40642857        3.29    0.0949
RIVER SOILS VS.NON       2       53.68916667       16.32    0.0004
GLACIAL VS DRIFT         1        1.26750000        0.77    0.3973
CLARION VS WEBSTER       1        1.70666667        1.04    0.3285
KNOX VS ONEILL           1       41.08166667       24.97    0.0003
```

(continued on next page)

(continued from previous page)

```
                                              (18)            (19)               (20)
                                           T FOR HO:       PR > |T|          STD ERROR OF
PARAMETER                    ESTIMATE    PARAMETER=0                           ESTIMATE
              (17)
INTERCEPT              29.35714286 B        34.96          0.0001            0.83970354
TYPE        CLARION     1.06666667 B         1.02          0.3285            1.04729432
            CLINTON    -0.80000000 B        -0.76          0.4597            1.04729432
            COMPOST    -1.43333333 B        -1.37          0.1962            1.04729432
            KNOX        3.80000000 B         3.63          0.0035            1.04729432
            O'NEILL     2.70000000 B         2.58          0.0242            1.04729432
            WABASH      4.86666667 B         4.65          0.0006            1.04729432
            WEBSTER     0.00000000 B          .              .                 .
BLOCK       1           3.32857143 B         4.85          0.0004            0.68561507
            2           1.90000000 B         2.77          0.0169            0.68561507
            3           0.00000000 B          .              .                 .

NOTE: THE X'X MATRIX HAS BEEN DEEMED SINGULAR AND A GENERALIZED INVERSE HAS BEEN EMPLOYED TO SOLVE THE NORMAL EQUATIONS.
      THE ABOVE ESTIMATES REPRESENT ONLY ONE OF MANY POSSIBLE SOLUTIONS TO THE NORMAL EQUATIONS. ESTIMATES FOLLOWED BY
      THE LETTER B ARE BIASED AND DO NOT ESTIMATE THE PARAMETER BUT ARE BLUE FOR SOME LINEAR COMBINATION OF PARAMETERS
      (OR ARE ZERO). THE EXPECTED VALUE OF THE BIASED ESTIMATORS MAY BE OBTAINED FROM THE GENERAL FORM OF ESTIMABLE
      FUNCTIONS. FOR THE BIASED ESTIMATORS, THE STD ERR IS THAT OF THE BIASED ESTIMATOR AND THE T VALUE TESTS
      HO: E(BIASED ESTIMATOR) = 0. ESTIMATES NOT FOLLOWED BY THE LETTER B ARE BLUE FOR THE PARAMETER.
```

```
                                                                                                    5
                                   GENERAL LINEAR MODELS PROCEDURE

WALLER-DUNCAN K-RATIO T TEST FOR VARIABLE: STEMLENG
NOTE: THIS TEST MINIMIZES THE BAYES RISK UNDER ADDITIVE LOSS
      AND CERTAIN OTHER ASSUMPTIONS.

KRATIO=100  DF=12  MSE=1.64524  F=10.4495
CRITICAL VALUE OF T=2.12034
MINIMUM SIGNIFICANT DIFFERENCE=2.22063

MEANS WITH THE SAME LETTER ARE NOT SIGNIFICANTLY DIFFERENT.

WALLER   GROUPING           MEAN      N   TYPE

               A          35.967      3   WABASH
               A
               A          34.900      3   KNOX
               A
         B     A          33.800      3   O'NEILL
         B
         B     C          32.167      3   CLARION
               C
         D     C          31.100      3   WEBSTER
         D     C
         D     C          30.300      3   CLINTON
         D
         D                29.667      3   COMPOST
```

```
                                                                                                    6
                                   GENERAL LINEAR MODELS PROCEDURE

RYAN-EINOT-GABRIEL-WELSCH MULTIPLE RANGE TEST FOR VARIABLE: STEMLENG
NOTE: THIS TEST CONTROLS THE TYPE I EXPERIMENTWISE ERROR RATE.

ALPHA=0.05  DF=12  MSE=1.64524

MEANS WITH THE SAME LETTER ARE NOT SIGNIFICANTLY DIFFERENT.

REGWQ    GROUPING           MEAN      N   TYPE

               A          35.967      3   WABASH
               A
         B     A          34.900      3   KNOX
         B     A
         B     A    C     33.800      3   O'NEILL
         B          C
         B     D    C     32.167      3   CLARION
               D    C
               D    C     31.100      3   WEBSTER
               D
               D          30.300      3   CLINTON
               D
               D          29.667      3   COMPOST
```

Regression with Mileage Data: Example 2

A car is tested for gas mileage at various speeds to determine at what speed the car achieves the greatest gas mileage. A quadratic response surface is fit to the experimental data.

```
*------GASOLINE MILEAGE EXPERIMENT-------;

DATA MILEAGE;
  INPUT MPH MPG @@;
  CARDS;
20 15.4 30 20.2 40 25.7 50 26.2 50 26.6 50 27.4 55   .  60 24.8
;
PROC GLM;
  MODEL MPG=MPH MPH*MPH / P CLM;
  OUTPUT OUT=PP P=MPGPRED R=RESID;
PROC PLOT DATA=PP;
  PLOT  MPG*MPH='A' MPGPRED*MPH='P' / OVERLAY;
```

```
                                                                                                              1

                                       GENERAL LINEAR MODELS PROCEDURE

                                        DEPENDENT VARIABLE INFORMATION

                                   NUMBER OF OBSERVATIONS IN DATA SET = 8

NOTE: ALL DEPENDENT VARIABLES ARE CONSISTENT WITH RESPECT TO THE PRESENCE OR ABSENCE OF MISSING VALUES. HOWEVER, ONLY  7
      OBSERVATIONS IN DATA SET CAN BE USED IN THIS ANALYSIS.
```

```
                                                                                                              2

                                       GENERAL LINEAR MODELS PROCEDURE

DEPENDENT VARIABLE: MPG

SOURCE                  DF       SUM OF SQUARES        MEAN SQUARE       F VALUE        PR > F     R-SQUARE          C.V.

MODEL                    2         111.80861827        55.90430913         77.96        0.0006     0.974986        3.5646

ERROR                    4           2.86852459         0.71713115                    ROOT MSE                   MPG MEAN

CORRECTED TOTAL          6         114.67714286                                     0.84683596                23.75714286

SOURCE                  DF          TYPE I SS    F VALUE     PR > F          DF        TYPE III SS     F VALUE      PR > F

MPH                      1        85.64464286     119.43     0.0004           1        41.01171219       57.19      0.0016
MPH*MPH                  1        26.16397541      36.48     0.0038           1        26.16397541       36.48      0.0038

                                  T FOR HO:       PR > |T|           STD ERROR OF
PARAMETER               ESTIMATE     PARAMETER=0                      ESTIMATE

INTERCEPT            -5.98524590          -1.88      0.1334            3.18522249
MPH                   1.30524590           7.56      0.0016            0.17259876
MPH*MPH              -0.01309836          -6.04      0.0038            0.00216852

OBSERVATION          OBSERVED            PREDICTED              RESIDUAL         LOWER 95% CL          UPPER 95% CL
                        VALUE                VALUE                                   FOR MEAN              FOR MEAN

       1          15.40000000          14.88032787            0.51967213          12.69704364           17.06361209
       2          20.20000000          21.38360656           -1.18360656          20.01729100           22.74992212
       3          25.70000000          25.26721311            0.43278689          23.87461985           26.65980638
       4          26.20000000          26.53114754           -0.33114754          25.44574938           27.61654570
       5          26.60000000          26.53114754            0.06885246          25.44574938           27.61654570
       6          27.40000000          26.53114754            0.86885246          25.44574938           27.61654570
       7 *                  .          26.18073770                     .          24.88681114           27.47466427
       8          24.80000000          25.17540984           -0.37540984          23.05957931           27.29124036

* OBSERVATION WAS NOT USED IN THIS ANALYSIS

          SUM OF RESIDUALS                                    0.00000000
          SUM OF SQUARED RESIDUALS                            2.86852459
          SUM OF SQUARED RESIDUALS - ERROR SS                -0.00000000
          PRESS STATISTIC                                    23.18107335
          FIRST ORDER AUTOCORRELATION                        -0.54376613
          DURBIN-WATSON D                                     2.94425592
```

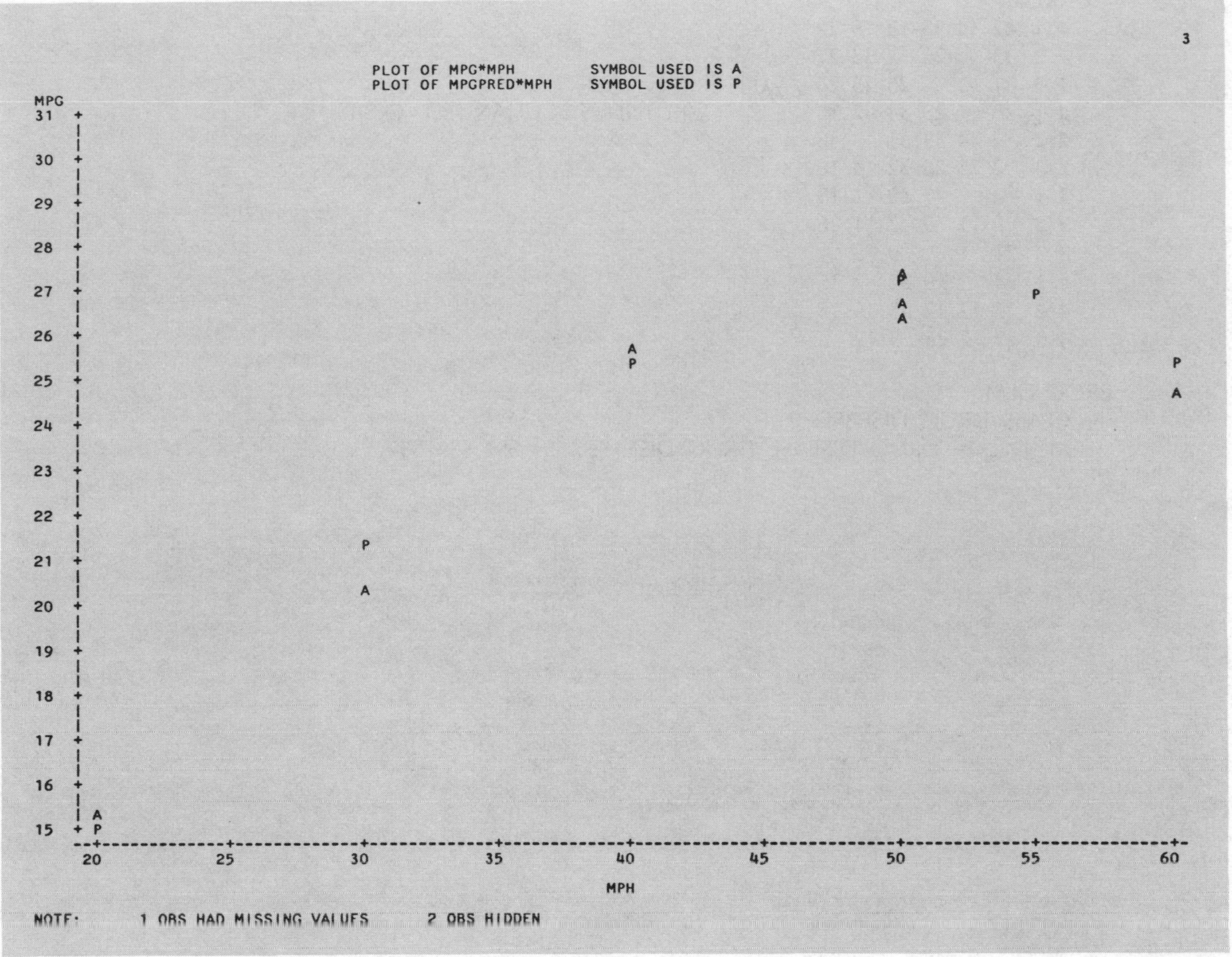

Unbalanced ANOVA for Two-Way Design with Interaction: Example 3

This example uses data from Kutner (1974) to illustrate a two-way analysis of variance. The original data source is Afifi and Azen (1972).

```
*------------------------------------------------------------------------*
|  A TWO-WAY ANALYSIS OF VARIANCE EXAMPLE USING THE DATA FROM:           |
|  KUTNER, MICHAEL H., THE AMERICAN STATISTICIAN, AUG. 74, P.98.         |
|  ORIGINAL DATA SOURCE: AFIFI AND AZEN (1972), STATISTICAL ANALYSIS:    |
|  A COMPUTER-ORIENTED APPROACH. ACADEMIC PRESS, NY, P.166.              |
*------------------------------------------------------------------------*  ;

 DATA A;
   INPUT DRUG DISEASE @;
   DO I=1 TO 6;
     INPUT Y @;
     OUTPUT;
     END;
```

```
CARDS;
  1 1  42 44 36 13 19 22
  1 2  33  . 26  . 33 21
  1 3  31 -3  . 25 25 24
  2 1  28  . 23 34 42 13          KUTNER'S 24 CHANGED TO 34
  2 2   . 34 33 31  . 36
  2 3   3 26 28 32  4 16
  3 1   .  .  1 29  . 19
  3 2   . 11  9  7  1 -6
  3 3  21  1  .  9  3  .
  4 1  24  .  9 22 -2 15
  4 2  27 12 12 -5 16 15
  4 3  22  7 25  5 12  .
;
PROC GLM;
  CLASS DRUG DISEASE;
  MODEL Y=DRUG DISEASE DRUG*DISEASE / SS1 SS2 SS3 SS4;
```

```
                                                                                                  1
                              GENERAL LINEAR MODELS PROCEDURE
                                 CLASS LEVEL INFORMATION
                               CLASS       LEVELS      VALUES
                               DRUG             4      1 2 3 4
                               DISEASE          3      1 2 3

                          NUMBER OF OBSERVATIONS IN DATA SET = 72

NOTE: ALL DEPENDENT VARIABLES ARE CONSISTENT WITH RESPECT TO THE PRESENCE OR ABSENCE OF MISSING VALUES. HOWEVER, ONLY 58
      OBSERVATIONS IN DATA SET CAN BE USED IN THIS ANALYSIS.
```

```
                                                                                                  2
                              GENERAL LINEAR MODELS PROCEDURE
DEPENDENT VARIABLE: Y

SOURCE               DF     SUM OF SQUARES      MEAN SQUARE    F VALUE      PR > F     R-SQUARE        C.V.
MODEL                11     4259.33850575       387.21259143      3.51      0.0013     0.456024     55.6675
ERROR                46     5080.81666667       110.45253623               ROOT MSE                  Y MEAN
CORRECTED TOTAL      57     9340.15517241                                 10.50964016             18.87931034

SOURCE               DF        TYPE I SS    F VALUE    PR > F       DF        TYPE II SS   F VALUE    PR > F
DRUG                  3    3133.23850575       9.46    0.0001        3     3063.43286350      9.25    0.0001
DISEASE               2     418.83374069       1.90    0.1617        2      418.83374069      1.90    0.1617
DRUG*DISEASE          6     707.26625931       1.07    0.3958        6      707.26625931      1.07    0.3958

SOURCE               DF      TYPE III SS    F VALUE    PR > F       DF        TYPE IV SS   F VALUE    PR > F
DRUG                  3    2997.47186048       9.05    0.0001        3     2997.47186048      9.05    0.0001
DISEASE               2     415.87304632       1.88    0.1637        2      415.87304632      1.88    0.1637
DRUG*DISEASE          6     707.26625931       1.07    0.3958        6      707.26625931      1.07    0.3958
```

Analysis of Covariance: Example 4

Analysis of covariance combines some of the features of regression and analysis of variance. Typically, a continuous variable (the covariate) is introduced into the model of an analysis-of-variance experiment.

Data in the following example were selected from a larger experiment on the use of drugs in the treatment of leprosy (Snedecor and Cochran, 1967, p. 422).

Variables in the study are:

DRUG	two antibiotics (A and D) and a control (F)
X	a pre-treatment score of leprosy bacilli
Y	a post-treatment score of leprosy bacilli.

Ten patients were selected for each treatment (DRUG), and six sites on each patient were measured for leprosy bacilli.

The covariate (a pre-treatment score) is included in the model for increased precision in determining the effect of drug treatments on the post-treatment count of bacilli.

The code for creating the data set and invoking GLM is:

```
*FROM SNEDECOR AND COCHRAN, (1967), STATISTICAL METHODS, P. 422.;

DATA DRUGTEST;
  INPUT DRUG $ X Y @@;
  CARDS;
A 11  6 A  8  0 A  5  2 A 14  8 A 19 11
A  6  4 A 10 13 A  6  1 A 11  8 A  3  0
D  6  0 D  6  2 D  7  3 D  8  1 D 18 18
D  8  4 D 19 14 D  8  9 D  5  1 D 15  9
F 16 13 F 13 10 F 11 18 F  9  5 F 21 23
F 16 12 F 12  5 F 12 16 F  7  1 F 12 20
;
PROC GLM;
  CLASS DRUG;
  MODEL Y=DRUG X / SOLUTION;
  LSMEANS DRUG / STDERR PDIFF;
```

```
                                                                                 1

                         GENERAL LINEAR MODELS PROCEDURE

                            CLASS LEVEL INFORMATION

                         CLASS       LEVELS      VALUES

                         DRUG           3        A D F

                    NUMBER OF OBSERVATIONS IN DATA SET = 30
```

```
                                                                                                  2

                                   GENERAL LINEAR MODELS PROCEDURE

DEPENDENT VARIABLE: Y

SOURCE                DF      SUM OF SQUARES        MEAN SQUARE      F VALUE        PR > F      R-SQUARE           C.V.

MODEL                  3        871.49740304       290.49913435        18.10        0.0001      0.676261        50.7060

ERROR                 26        417.20259696        16.04625373                    STD DEV                      Y MEAN

CORRECTED TOTAL       29       1288.70000000                                    4.00577754                  7.90000000

SOURCE                DF      (1) TYPE I SS     F VALUE    PR > F         DF     (2) TYPE III SS     F VALUE     PR > F

DRUG                   2        293.60000000        9.15    0.0010          2        68.55371060         2.14     0.1384
X                      1        577.89740304       36.01    0.0001          1       577.89740304        36.01     0.0001
```

(continued on next page)

(continued from previous page)

```
                                         T FOR H0:           PR > |T|            STD ERROR OF
PARAMETER              ESTIMATE         PARAMETER=0                                ESTIMATE

INTERCEPT          -0.43467116 B           -0.18             0.8617              2.47135356
DRUG          A    -3.44613828 B           -1.83             0.0793              1.88678065
              D    -3.33716695 B           -1.80             0.0835              1.85386642
              F     0.00000000 B             .                 .                  .
X                   0.98718381              6.00             0.0001              0.16449757

NOTE: THE X'X MATRIX HAS BEEN DEEMED SINGULAR AND A GENERALIZED INVERSE HAS BEEN EMPLOYED TO SOLVE THE NORMAL EQUATIONS.
      THE ABOVE ESTIMATES REPRESENT ONLY ONE OF MANY POSSIBLE SOLUTIONS TO THE NORMAL EQUATIONS. ESTIMATES FOLLOWED BY
      THE LETTER B ARE BIASED AND DO NOT ESTIMATE THE PARAMETER BUT ARE BLUE FOR SOME LINEAR COMBINATION OF PARAMETERS
      (OR ARE ZERO). THE EXPECTED VALUE OF THE BIASED ESTIMATORS MAY BE OBTAINED FROM THE GENERAL FORM OF ESTIMABLE
      FUNCTIONS. FOR THE BIASED ESTIMATORS, THE STD ERR IS THAT OF THE BIASED ESTIMATOR AND THE T VALUE TESTS
      H0: E(BIASED ESTIMATOR) = 0. ESTIMATES NOT FOLLOWED BY THE LETTER B ARE BLUE FOR THE PARAMETER.
```

```
                                                                                                                        3
                                        GENERAL LINEAR MODELS PROCEDURE

        (3)                                (4) LEAST SQUARES MEANS          (5)

DRUG                    Y           STD ERR      PROB > |T|       PROB > |T| H0: LSMEAN(I)=LSMEAN(J)
                   LSMEAN            LSMEAN    H0:LSMEAN=0        I/J       1          2          3

A               6.7149635         1.2884943         0.0001        1         .     0.9521     0.0793
D               6.8239348         1.2724690         0.0001        2    0.9521          .     0.0835
F              10.1611017         1.3159234         0.0001        3    0.0793     0.0835          .

NOTE: TO ENSURE OVERALL PROTECTION LEVEL, ONLY PROBABILITIES ASSOCIATED WITH PRE-PLANNED COMPARISONS SHOULD BE USED.
```

The numbers on the printout correspond to the numbered descriptions that follow:

1. The TYPE I SS for DRUG gives the between-drug sums of squares that would be obtained for the analysis-of-variance model Y = DRUG.
2. TYPE III SS for DRUG gives the DRUG SS "adjusted" for the covariate.
3. The LSMEANS printed are the same as adjusted means (means adjusted for the covariate).
4. The STDERR option on the LSMEANS statement causes the standard error of the least-squares means and the probability of getting a larger $|t|$ value under the hypothesis H_0:LSM = 0 to be printed.
5. Specifying the PDIFF option causes all probability values for the hypothesis H_0: LSM(I) = LSM(J) to be printed.

Random Effects in Repeated-Measures Designs: Example 5

The following repeated-measures example from Winer (1971) is a good example of a mixed model. The subjects factor and all its interactions (for example, subjects within whole plot factors) are random effects. The expected mean squares are examined to determine what tests are valid. A fixed effect is tested by comparing its mean square with another mean square that has the same expectation under the hypothesis that the fixed effect is zero.

Below is the code to read in the data and perform the standard analysis.

```
* -------------------------------------------------------------------- *
| THIS EXAMPLE IS A THREE-FACTOR EXPERIMENT WITH REPEATED              |
| MEASURES ON ONE FACTOR FROM WINER (1971, P. 564) ON DATA             |
| FROM A STUDY BY MEYER AND NOBLE (1958). THE EXPERIMENT               |
| EVALUATED ERRORS ON FOUR LEARNING TRIALS AS A FUNCTION OF            |
| HIGH AND LOW LEVELS OF MANIFEST ANXIETY AND TWO LEVELS OF            |
| MUSCULAR TENSION (DEFINED BY PRESSURE ON A DYNAMOMETER).             |
| AS IN WINER, A IS ANXIETY, B IS TENSION, SUBJ IS SUBJECT,            |
| AND TRIAL IS THE REPEATED MEASUREMENTS.                              |
|                                                                      |
| THE REPEATED-MEASURES EXPERIMENT CAN BE THOUGHT OF AS A              |
| SPLIT-PLOT EXPERIMENT WHERE SUBJECTS FORM THE WHOLE PLOTS.           |
* -------------------------------------------------------------------- * ;

DATA B;
  INPUT SUBJ A B @;
  DO C=1 TO 4;
    INPUT RESPONSE @;
    OUTPUT;
    END;
  LABEL A=ANXIETY
        B=TENSION
        SUBJ=SUBJECT
        C=EXPERIMENTAL TRIAL;
  CARDS;
 1 1 1 18 14 12 6
 2 1 1 19 12  8 4
 3 1 1 14 10  6 2
 4 1 2 16 12 10 4
 5 1 2 12  8  6 2
 6 1 2 18 10  5 1
 7 2 1 16 10  8 4
 8 2 1 18  8  4 1
 9 2 1 16 12  6 2
10 2 2 19 16 10 8
11 2 2 16 14 10 9
12 2 2 16 12  8 8
;
PROC GLM;
   CLASS A B SUBJ C;
   MODEL RESPONSE=A B A*B SUBJ(A B)   C A*C B*C A*B*C;
   RANDOM SUBJ(A B);
   TEST H=A B A*B E=SUBJ(A B);
   TITLE COMPLETELY BALANCED DATA;
```

Since the data are balanced, all four of the standard tests are the same. Below is the printout of the expected mean squares.

```
                         COMPLETELY BALANCED DATA                                    2
                      GENERAL LINEAR MODELS PROCEDURE
DEPENDENT VARIABLE: RESPONSE

SOURCE              TYPE I EXPECTED MEAN SQUARE
A                   VAR(ERROR) + 4 VAR(SUBJ(A*B)) + Q(A,A*B,A*C,A*B*C)
B                   VAR(ERROR) + 4 VAR(SUBJ(A*B)) + Q(B,A*B,B*C,A*B*C)
A*B                 VAR(ERROR) + 4 VAR(SUBJ(A*B)) + Q(A*B,A*B*C)
SUBJ(A*B)           VAR(ERROR) + 4 VAR(SUBJ(A*B))
C                   VAR(ERROR) + Q(C,A*C,B*C,A*B*C)
A*C                 VAR(ERROR) + Q(A*C,A*B*C)
B*C                 VAR(ERROR) + Q(B*C,A*B*C)
A*B*C               VAR(ERROR) + Q(A*B*C)

SOURCE              TYPE III EXPECTED MEAN SQUARE
A                   VAR(ERROR) + 4 VAR(SUBJ(A*B)) + Q(A,A*B,A*C,A*B*C)
B                   VAR(ERROR) + 4 VAR(SUBJ(A*B)) + Q(B,A*B,B*C,A*B*C)
A*B                 VAR(ERROR) + 4 VAR(SUBJ(A*B)) + Q(A*B,A*B*C)
SUBJ(A*B)           VAR(ERROR) + 4 VAR(SUBJ(A*B))
C                   VAR(ERROR) + Q(C,A*C,B*C,A*B*C)
A*C                 VAR(ERROR) + Q(A*C,A*B*C)
B*C                 VAR(ERROR) + Q(B*C,A*B*C)
A*B*C               VAR(ERROR) + Q(A*B*C)
```

To test the whole-plot factors A, B, and A*B, you need a mean square with expectation: VAR(ERROR)+4*VAR(SUBJ(A*B)). The SUBJ(A B) effect fits our needs for a denominator term; thus we justify the TEST statement:

TEST H=A B A*B E=SUBJ(A B);

The GLM report:

```
                         COMPLETELY BALANCED DATA                                    3
                      GENERAL LINEAR MODELS PROCEDURE
DEPENDENT VARIABLE: RESPONSE

SOURCE            DF    SUM OF SQUARES      MEAN SQUARE     F VALUE      PR > F      R-SQUARE        C.V.
MODEL             23     1205.83333333      52.42753623       24.12      0.0001      0.958532     14.7432
ERROR             24       52.16666667       2.17361111                  ROOT MSE              RESPONSE MEAN
CORRECTED TOTAL   47     1258.00000000                                1.47431717                10.00000000

SOURCE            DF       TYPE I SS    F VALUE    PR > F      DF     TYPE III SS    F VALUE    PR > F
A                  1     10.08333333       4.64    0.0415       1     10.08333333       4.64    0.0415
B                  1      8.33333333       3.83    0.0619       1      8.33333333       3.83    0.0619
A*B                1     80.08333333      36.84    0.0001       1     80.08333333      36.84    0.0001
SUBJ(A*B)          8     82.50000000       4.74    0.0014       8     82.50000000       4.74    0.0014
C                  3    991.50000000     152.05    0.0001       3    991.50000000     152.05    0.0001
A*C                3      8.41666667       1.29    0.3003       3      8.41666667       1.29    0.3003
B*C                3     12.16666667       1.87    0.1624       3     12.16666667       1.87    0.1624
A*B*C              3     12.75000000       1.96    0.1477       3     12.75000000       1.96    0.1477

TESTS OF HYPOTHESES USING THE TYPE III MS FOR SUBJ(A*B) AS AN ERROR TERM

SOURCE            DF     TYPE III SS    F VALUE    PR > F
A                  1     10.08333333       0.98    0.3517
B                  1      8.33333333       0.81    0.3949
A*B                1     80.08333333       7.77    0.0237
```

Unbalanced in whole plots To see whether the same mean squares are valid if the data are unbalanced in whole plots, we delete a subject in the data set. Each subject still has the same number of trials, but there are a different number of subjects for one A*B combination. Since the design is unbalanced, look at all four types of mean squares to find the ones you want.

```
DATA C; SET B; IF SUBJ=1 THEN DELETE;
PROC GLM;
  CLASS A B SUBJ C;
  MODEL RESPONSE=A B A*B SUBJ(A B) C A*C B*C A*B*C
          / SS1 SS2 SS3 SS4;
  RANDOM SUBJ(A B);
  TEST H=A B A*B E=SUBJ(A B) / HTYPE=1 ETYPE=1;
  TITLE AMONG SUBJECTS (WHOLE PLOT) IS UNBALANCED;
```

```
                         AMONG SUBJECTS (WHOLE PLOT) IS UNBALANCED                          5
                              GENERAL LINEAR MODELS PROCEDURE

DEPENDENT VARIABLE: RESPONSE

SOURCE              TYPE I EXPECTED MEAN SQUARE

A                   VAR(ERROR) + 4 VAR(SUBJ(A*B)) + Q(A,B,A*B,A*C,B*C,A*B*C)

B                   VAR(ERROR) + 4 VAR(SUBJ(A*B)) + Q(B,A*B,B*C,A*B*C)

A*B                 VAR(ERROR) + 4 VAR(SUBJ(A*B)) + Q(A*B,A*B*C)

SUBJ(A*B)           VAR(ERROR) + 4 VAR(SUBJ(A*B))

C                   VAR(ERROR) + Q(C,A*C,B*C,A*B*C)

A*C                 VAR(ERROR) + Q(A*C,B*C,A*B*C)

B*C                 VAR(ERROR) + Q(B*C,A*B*C)

A*B*C               VAR(ERROR) + Q(A*B*C)

SOURCE              TYPE II EXPECTED MEAN SQUARE

A                   VAR(ERROR) + 4 VAR(SUBJ(A*B)) + Q(A,A*B,A*C,A*B*C)

B                   VAR(ERROR) + 4 VAR(SUBJ(A*B)) + Q(B,A*B,B*C,A*B*C)

A*B                 VAR(ERROR) + 4 VAR(SUBJ(A*B)) + Q(A*B,A*B*C)

SUBJ(A*B)           VAR(ERROR) + 4 VAR(SUBJ(A*B))

C                   VAR(ERROR) + Q(C,A*C,B*C,A*B*C)

A*C                 VAR(ERROR) + Q(A*C,A*B*C)

B*C                 VAR(ERROR) + Q(B*C,A*B*C)

A*B*C               VAR(ERROR) + Q(A*B*C)
```

```
                         AMONG SUBJECTS (WHOLE PLOT) IS UNBALANCED                          6
                              GENERAL LINEAR MODELS PROCEDURE

DEPENDENT VARIABLE: RESPONSE

SOURCE              TYPE III EXPECTED MEAN SQUARE

A                   VAR(ERROR) + 4 VAR(SUBJ(A*B)) + Q(A,A*B,A*C,A*B*C)

B                   VAR(ERROR) + 4 VAR(SUBJ(A*B)) + Q(B,A*B,B*C,A*B*C)

A*B                 VAR(ERROR) + 4 VAR(SUBJ(A*B)) + Q(A*B,A*B*C)

SUBJ(A*B)           VAR(ERROR) + 4 VAR(SUBJ(A*B))

C                   VAR(ERROR) + Q(C,A*C,B*C,A*B*C)

A*C                 VAR(ERROR) + Q(A*C,A*B*C)

B*C                 VAR(ERROR) + Q(B*C,A*B*C)

A*B*C               VAR(ERROR) + Q(A*B*C)
```

(continued on next page)

(continued from previous page)

SOURCE	TYPE IV EXPECTED MEAN SQUARE
A	VAR(ERROR) + 3.58301158 VAR(SUBJ(A*B)) + Q(A,A*B,A*C,A*B*C)
B	VAR(ERROR) + 3.58301158 VAR(SUBJ(A*B)) + Q(B,A*B,B*C,A*B*C)
A*B	VAR(ERROR) + 4 VAR(SUBJ(A*B)) + Q(A*B,A*B*C)
SUBJ(A*B)	VAR(ERROR) + 4 VAR(SUBJ(A*B))
C	VAR(ERROR) + Q(C,A*C,B*C,A*B*C)
A*C	VAR(ERROR) + Q(A*C,A*B*C)
B*C	VAR(ERROR) + Q(B*C,A*B*C)
A*B*C	VAR(ERROR) + Q(A*B*C)

AMONG SUBJECTS (WHOLE PLOT) IS UNBALANCED 7

GENERAL LINEAR MODELS PROCEDURE

DEPENDENT VARIABLE: RESPONSE

SOURCE	DF	SUM OF SQUARES	MEAN SQUARE	F VALUE	PR > F	R-SQUARE	C.V.
MODEL	22	1107.68560606	50.34934573	22.01	0.0001	0.958432	15.4769
ERROR	21	48.04166667	2.28769841		ROOT MSE		RESPONSE MEAN
CORRECTED TOTAL	43	1155.72727273			1.51251394		9.77272727

SOURCE	DF	TYPE I SS	F VALUE	PR > F	DF	TYPE II SS	F VALUE	PR > F
A	1	24.81893939	10.85	0.0035	1	30.00000000	13.11	0.0016
B	1	27.07500000	11.84	0.0025	1	27.07500000	11.84	0.0025
A*B	1	45.37500000	19.83	0.0002	1	45.37500000	19.83	0.0002
SUBJ(A*B)	7	56.45833333	3.53	0.0116	7	56.45833333	3.53	0.0116
C	3	921.54545455	134.28	0.0001	3	921.54545455	134.28	0.0001
A*C	3	6.66287879	0.97	0.4251	3	8.09629630	1.18	0.3413
B*C	3	15.92129630	2.32	0.1046	3	15.92129630	2.32	0.1046
A*B*C	3	9.82870370	1.43	0.2616	3	9.82870370	1.43	0.2616

SOURCE	DF	TYPE III SS	F VALUE	PR > F	DF	TYPE IV SS	F VALUE	PR > F
A	1	23.81332599	10.41	0.0041	1	35.68870656	15.60	0.0007
B	1	21.22477974	9.28	0.0062	1	34.18935006	14.94	0.0009
A*B	1	45.37500000	19.83	0.0002	1	45.37500000	19.83	0.0002
SUBJ(A*B)	7	56.45833333	3.53	0.0116	7	56.45833333	3.53	0.0116
C	3	907.12500000	132.17	0.0001	3	907.12500000	132.17	0.0001
A*C	3	6.34722222	0.92	0.4460	3	6.34722222	0.92	0.4460
B*C	3	13.27314815	1.93	0.1550	3	13.27314815	1.93	0.1550
A*B*C	3	9.82870370	1.43	0.2616	3	9.82870370	1.43	0.2616

TESTS OF HYPOTHESES USING THE TYPE I MS FOR SUBJ(A*B) AS AN ERROR TERM

SOURCE	DF	TYPE I SS	F VALUE	PR > F
A	1	24.81893939	3.08	0.1228
B	1	27.07500000	3.36	0.1096
A*B	1	45.37500000	5.63	0.0495

UNBALANCED WITHIN REPEATED MEASURES (SUBPLOT) 8

GENERAL LINEAR MODELS PROCEDURE

CLASS LEVEL INFORMATION

CLASS	LEVELS	VALUES
A	2	1 2
B	2	1 2
SUBJ	12	1 2 3 4 5 6 7 8 9 10 11 12
C	4	1 2 3 4

NUMBER OF OBSERVATIONS IN DATA SET = 47

TYPE I mean squares have the right expectation with respect to the random terms, but they do not test the right hypothesis with respect to the fixed effects. TYPE II and TYPE III are appropriate. TYPE IV mean squares are not appropriate because the expected mean squares for whole-plot main effects do not have the proper coefficients for the variances for the random terms to use SUBJ(A B) as a denominator of an *F* test.

Unbalanced in subplots To find some tests to use if the data are unbalanced within subjects (the subplots), go back to the original data and delete only one observation. The first subject now has one less observation than the others.

```
DATA D;  SET B;  IF _N_=1 THEN DELETE;
PROC GLM;
  CLASS A B SUBJ C;
  MODEL RESPONSE=A B A*B SUBJ(A B) C A*C B*C A*B*C
        / SS1 SS2 SS3 SS4;
  RANDOM SUBJ(A B);
  TEST H=A B A*B E=SUBJ(A B);
  TITLE UNBALANCED WITHIN REPEATED MEASURES (SUBPLOT);
```

```
                    UNBALANCED WITHIN REPEATED MEASURES (SUBPLOT)                    9
                           GENERAL LINEAR MODELS PROCEDURE

DEPENDENT VARIABLE: RESPONSE

SOURCE              TYPE I EXPECTED MEAN SQUARE

A                   VAR(ERROR) + 3.933395 VAR(SUBJ(A*B)) + Q(A,B,A*B,C,A*C,B*C,A*B*C)
B                   VAR(ERROR) + 3.93043478 VAR(SUBJ(A*B)) + Q(B,A*B,C,A*C,B*C,A*B*C)
A*B                 VAR(ERROR) + 3.92727273 VAR(SUBJ(A*B)) + Q(A*B,C,A*C,B*C,A*B*C)
SUBJ(A*B)           VAR(ERROR) + 3.90909091 VAR(SUBJ(A*B)) + Q(C,A*C,B*C,A*B*C)
C                   VAR(ERROR) + Q(C,A*C,B*C,A*B*C)
A*C                 VAR(ERROR) + Q(A*C,B*C,A*B*C)
B*C                 VAR(ERROR) + Q(B*C,A*B*C)
A*B*C               VAR(ERROR) + Q(A*B*C)

SOURCE              TYPE II EXPECTED MEAN SQUARE

A                   VAR(ERROR) + 3.92307692 VAR(SUBJ(A*B)) + Q(A,A*B,A*C,A*B*C)
B                   VAR(ERROR) + 3.92307692 VAR(SUBJ(A*B)) + Q(B,A*B,B*C,A*B*C)
A*B                 VAR(ERROR) + 3.91428571 VAR(SUBJ(A*B)) + Q(A*B,A*B*C)
SUBJ(A*B)           VAR(ERROR) + 3.875 VAR(SUBJ(A*B))
C                   VAR(ERROR) + Q(C,A*C,B*C,A*B*C)
A*C                 VAR(ERROR) + Q(A*C,A*B*C)
B*C                 VAR(ERROR) + Q(B*C,A*B*C)
A*B*C               VAR(ERROR) + Q(A*B*C)
```

```
                         UNBALANCED WITHIN REPEATED MEASURES (SUBPLOT)                          10

                              GENERAL LINEAR MODELS PROCEDURE

DEPENDENT VARIABLE: RESPONSE

SOURCE              TYPE III EXPECTED MEAN SQUARE

A                   VAR(ERROR) + 3.84 VAR(SUBJ(A*B)) + Q(A,A*B,A*C,A*B*C)

B                   VAR(ERROR) + 3.84 VAR(SUBJ(A*B)) + Q(B,A*B,B*C,A*B*C)

A*B                 VAR(ERROR) + 3.84 VAR(SUBJ(A*B)) + Q(A*B,A*B*C)

SUBJ(A*B)           VAR(ERROR) + 3.875 VAR(SUBJ(A*B))

C                   VAR(ERROR) + Q(C,A*C,B*C,A*B*C)

A*C                 VAR(ERROR) + Q(A*C,A*B*C)

B*C                 VAR(ERROR) + Q(B*C,A*B*C)

A*B*C               VAR(ERROR) + Q(A*B*C)

SOURCE              TYPE IV EXPECTED MEAN SQUARE

A                   VAR(ERROR) + 3.84 VAR(SUBJ(A*B)) + Q(A,A*B,A*C,A*B*C)

B                   VAR(ERROR) + 3.84 VAR(SUBJ(A*B)) + Q(B,A*B,B*C,A*B*C)

A*B                 VAR(ERROR) + 3.84 VAR(SUBJ(A*B)) + Q(A*B,A*B*C)

SUBJ(A*B)           VAR(ERROR) + 3.875 VAR(SUBJ(A*B))

C                   VAR(ERROR) + Q(C,A*C,B*C,A*B*C)

A*C                 VAR(ERROR) + Q(A*C,A*B*C)

B*C                 VAR(ERROR) + Q(B*C,A*B*C)

A*B*C               VAR(ERROR) + Q(A*B*C)
```

No mean squares have the correct expectation to use as a denominator to the whole-plot effects. A synthetic effect could be constructed:

MS = MS(SUBJ(A B))*3.84/3.875 + MS(ERROR)*(1 – 3.84/3.875)

which would have the right expectation if the mean squares were independent. Satterthwaite's approximation could be used to derive a degrees of freedom for the approximate test.

Multivariate Analysis of Variance: Example 6

Using data from A. Anderson, Oregon State University, this example performs a multivariate analysis of variance.

```
* -----------------------MULTIVARIATE ANALYSIS OF VARIANCE----------------------- *
|  DATA FROM A. ANDERSON, OREGON STATE UNIVERSITY.                                 |
|  FOUR DIFFERENT RESPONSE VARIABLES ARE MEASURED.                                 |
|  THE HYPOTHESIS THAT WE WANT TO TEST IS THAT SEX                                 |
|  DOES NOT AFFECT ANY OF THE FOUR RESPONSES.                                      |
* -------------------------------------------------------------------------------- * ;

DATA SKULL;
  INPUT SEX $ LENGTH BASILAR ZYGOMAT POSTORB @@;
  CARDS;
```

```
M 6460 4962 3286 1100 M 6252 4773 3239 1061 M 5772 4480 3200 1097
M 6264 4806 3179 1054 M 6622 5113 3365 1071 M 6656 5100 3326 1012
M 6441 4918 3153 1061 M 6281 4821 3133 1071 M 6606 5060 3227 1064
M 6573 4977 3392 1110 M 6563 5025 3234 1090 M 6552 5086 3292 1010
M 6535 4939 3261 1065 M 6573 4962 3320 1091 M 6537 4990 3309 1059
M 6302 4761 3204 1135 M 6449 4921 3256 1068 M 6481 4887 3233 1124
M 6368 4824 3258 1130 M 6372 4844 3306 1137 M 6592 5007 3284 1148
M 6229 4746 3257 1153 M 6391 4834 3244 1169 M 6560 4981 3341 1038
M 6787 5181 3334 1104 M 6384 4834 3195 1064 M 6282 4757 3180 1179
M 6340 4791 3300 1110 M 6394 4879 3272 1241 M 6153 4557 3214 1039
M 6348 4886 3160  991 M 6534 4990 3310 1028 M 6509 4951 3282 1104
F 6287 4845 3218  996 F 6583 4992 3300 1107 F 6518 5023 3246 1035
F 6432 4790 3249 1117 F 6450 4888 3259 1060 F 6379 4844 3266 1115
F 6424 4855 3322 1065 F 6615 5088 3280 1179 F 6760 5206 3337 1219
F 6521 5011 3208  989 F 6416 4889 3200 1001 F 6511 4910 3230 1100
F 6540 4997 3320 1078 F 6780 5259 3358 1174 F 6336 4781 3165 1126
F 6472 4954 3125 1178 F 6476 4896 3148 1066 F 6276 4709 3150 1134
F 6693 5177 3236 1131 F 6328 4792 3214 1018 F 6661 5104 3395 1141
F 6266 4721 3257 1031 F 6660 5146 3374 1069 F 6624 5032 3384 1154
F 6331 4819 3278 1008 F 6298 4683 3270 1150
;
PROC GLM;
  CLASS SEX;
  MODEL LENGTH BASILAR ZYGOMAT POSTORB=SEX;
  MANOVA H=SEX / PRINTE PRINTH;
  TITLE MULTIVARIATE ANALYSIS OF VARIANCE;
```

```
                    MULTIVARIATE ANALYSIS OF VARIANCE                    1
                     GENERAL LINEAR MODELS PROCEDURE
                         CLASS LEVEL INFORMATION
                         CLASS     LEVELS    VALUES
                         SEX          2      F M

                 NUMBER OF OBSERVATIONS IN DATA SET = 59
```

```
                         MULTIVARIATE ANALYSIS OF VARIANCE                                        2
                          GENERAL LINEAR MODELS PROCEDURE
DEPENDENT VARIABLE: LENGTH
SOURCE              DF     SUM OF SQUARES      MEAN SQUARE     F VALUE       PR > F    R-SQUARE         C.V.
MODEL                1     47060.93163052   47060.93163052        1.59       0.2119    0.027201       2.6624
ERROR               57   1683039.20396301   29527.00357830                 ROOT MSE              LENGTH MEAN
CORRECTED TOTAL     58   1730100.13559353                              171.83423285            6454.22033898

SOURCE              DF        TYPE I SS   F VALUE    PR > F       DF       TYPE III SS    F VALUE    PR > F
SEX                  1   47060.93163052      1.59    0.2119        1    47060.93163052       1.59    0.2119
```

MULTIVARIATE ANALYSIS OF VARIANCE 3

GENERAL LINEAR MODELS PROCEDURE

DEPENDENT VARIABLE: BASILAR

SOURCE	DF	SUM OF SQUARES	MEAN SQUARE	F VALUE	PR > F	R-SQUARE	C.V.
MODEL	1	23985.10578405	23985.10578405	1.02	0.3174	0.017539	3.1229
ERROR	57	1343555.19930095	23571.14384739		ROOT MSE		BASILAR MEAN
CORRECTED TOTAL	58	1367540.30508500			153.52896745		4916.16949153

SOURCE	DF	TYPE I SS	F VALUE	PR > F	DF	TYPE III SS	F VALUE	PR > F
SEX	1	23985.10578405	1.02	0.3174	1	23985.10578405	1.02	0.3174

MULTIVARIATE ANALYSIS OF VARIANCE 4

GENERAL LINEAR MODELS PROCEDURE

DEPENDENT VARIABLE: ZYGOMAT

SOURCE	DF	SUM OF SQUARES	MEAN SQUARE	F VALUE	PR > F	R-SQUARE	C.V.
MODEL	1	66.95272806	66.95272806	0.01	0.9045	0.000255	2.0823
ERROR	57	262647.62354317	4607.85304462		ROOT MSE		ZYGOMAT MEAN
CORRECTED TOTAL	58	262714.57627124			67.88116856		3259.91525424

SOURCE	DF	TYPE I SS	F VALUE	PR > F	DF	TYPE III SS	F VALUE	PR > F
SEX	1	66.95272806	0.01	0.9045	1	66.95272806	0.01	0.9045

MULTIVARIATE ANALYSIS OF VARIANCE 5

GENERAL LINEAR MODELS PROCEDURE

DEPENDENT VARIABLE: POSTORB

SOURCE	DF	SUM OF SQUARES	MEAN SQUARE	F VALUE	PR > F	R-SQUARE	C.V.
MODEL	1	192.91266643	192.91266643	0.06	0.8132	0.000987	5.3592
ERROR	57	195162.71445222	3423.90727109		ROOT MSE		POSTORB MEAN
CORRECTED TOTAL	58	195355.62711865			58.51416300		1091.84745763

SOURCE	DF	TYPE I SS	F VALUE	PR > F	DF	TYPE III SS	F VALUE	PR > F
SEX	1	192.91266643	0.06	0.8132	1	192.91266643	0.06	0.8132

MULTIVARIATE ANALYSIS OF VARIANCE 6

GENERAL LINEAR MODELS PROCEDURE

E = ERROR SS&CP MATRIX

DF=57	LENGTH	BASILAR	ZYGOMAT	POSTORB
LENGTH	1683039.20396301	1430839.75174861	386107.03613056	74382.90326344
BASILAR	1430839.75174861	1343555.19930095	324249.61888114	38106.47202801
ZYGOMAT	386107.03613056	324249.61888114	262647.62354317	33070.58857810
POSTORB	74382.90326344	38106.47202801	33070.58857810	195162.71445222

PARTIAL CORRELATION COEFFICIENTS FROM THE ERROR SS&CP MATRIX / PROB > |R|

DF=56	LENGTH	BASILAR	ZYGOMAT	POSTORB
LENGTH	1.000000 0.0000	0.951516 0.0001	0.580729 0.0001	0.129786 0.3315
BASILAR	0.951516 0.0001	1.000000 0.0000	0.545840 0.0001	0.074417 0.5788
ZYGOMAT	0.580729 0.0001	0.545840 0.0001	1.000000 0.0000	0.146069 0.2739
POSTORB	0.129786 0.3315	0.074417 0.5788	0.146069 0.2739	1.000000 0.0000

```
                    MULTIVARIATE ANALYSIS OF VARIANCE                                        7

                      GENERAL LINEAR MODELS PROCEDURE

                   H = TYPE III SS&CP MATRIX FOR: SEX

   DF=1              LENGTH                 BASILAR                ZYGOMAT               POSTORB

   LENGTH        47060.93163052          33597.04486192          1775.06556438         3013.07978744
   BASILAR       33597.04486192          23985.10578405          1267.22857651         2151.05339576
   ZYGOMAT        1775.06556438           1267.22857651            66.95272806          113.64871005
   POSTORB        3013.07978744           2151.05339576           113.64871005          192.91266643

CHARACTERISTIC ROOTS AND VECTORS OF: E INVERSE * H, WHERE  H = TYPE III SS&CP MATRIX FOR: SEX     E = ERROR SS&CP MATRIX

         CHARACTERISTIC    PERCENT      CHARACTERISTIC VECTOR   V'EV=1
              ROOT
                                         LENGTH          BASILAR          ZYGOMAT          POSTORB

           0.04525085      100.00      0.00181900      -0.00111840      -0.00115382      0.00005517

           0.00000000        0.00      0.00029966      -0.00053092       0.00210429      0.00000000

           0.00000000        0.00     -0.00048407       0.00047922      -0.00017569      0.00232068

           0.00000000        0.00     -0.00179395       0.00251286       0.00000000      0.00000000
```

```
                    MULTIVARIATE ANALYSIS OF VARIANCE                                        8

                      GENERAL LINEAR MODELS PROCEDURE

        MANOVA TEST CRITERIA FOR THE HYPOTHESIS OF NO OVERALL SEX EFFECT

                   H = TYPE III SS&CP MATRIX FOR: SEX
                   E = ERROR SS&CP MATRIX
                   P = DEP. VARIABLES =      4
                   Q = HYPOTHESIS DF  =      1
                   NE= DF OF E        =     57
                   S = MIN(P,Q)       =      1
                   M = .5(ABS(P-Q)-1) =    1.0
                   N = .5(NE-P-1)     =   26.0

-----------------------------------------------------------------------------------------------

HOTELLING-LAWLEY TRACE = TR(E**-1*H) =          0.04525085      (SEE PILLAI'S TABLE #3)

      F APPROXIMATION = 2(S*N+1)*TR(E**-1*H)/(S*S*(2M+S+1))    WITH S(2M+S+1) AND 2(S*N+1) DF

            F(4,54) =      0.61      PROB > F = 0.6566

-----------------------------------------------------------------------------------------------

PILLAI'S TRACE        V = TR(H*INV(H+E)) =      0.04329185      (SEE PILLAI'S TABLE #2)

      F APPROXIMATION = (2N+S+1)/(2M+S+1) * V/(S-V)            WITH S(2M+S+1) AND S(2N+S+1) DF

            F(4,54) =      0.61      PROB > F = 0.6566

-----------------------------------------------------------------------------------------------

WILKS' CRITERION       L = DET(E)/DET(H+E) =    0.95670815      (SEE RAO 1973 P 555)

      EXACT F = (1-L)/L*(NE+Q-P)/P                             WITH P AND NE+Q-P DF

            F(4,54) =      0.61      PROB > F = 0.6566

-----------------------------------------------------------------------------------------------

ROY'S MAXIMUM ROOT CRITERION =                  0.04525085      (SEE AMS VOL 31 P 625)

      FIRST CANONICAL VARIABLE YIELDS AN F UPPER BOUND

            F(1,57) =      2.58      (UPPER BOUND)

-----------------------------------------------------------------------------------------------
```

REFERENCES

Afifi, A.A. and Azen, S.P. (1972), *Statistical Analysis: A Computer-Oriented Approach*, New York: Academic Press.

Anderson, T.W. (1952), *An Introduction to Multivariate Statistical Analysis*, New York: John Wiley & Sons.

Begun, J.M. and Gabriel, K.R. (1981), "Closure of the Newman-Keuls Multiple Comparisons Procedure," *Journal of the American Statistical Association*, 76, 374.

Carmer, S.G. and Swanson, M.R. (1973), "Evaluation of Ten Pairwise Multiple Comparison Procedures by Monte-Carlo Methods," *Journal of the American Statistical Association*, 68, 66–74.

Draper, N.R. and Smith, H. (1966), *Applied Regression Analysis*, New York: John Wiley & Sons.

Duncan, D.B. (1975), "*t*-Tests and Intervals for Comparisons Suggested by the Data," *Biometrics*, 31, 339–359.

Dunnett, C.W. (1980), "Pairwise Multiple Comparisons in the Homogeneous Variance, Unequal Sample Size Case," *Journal of the American Statistical Association*, 75, 372.

Einot, I. and Gabriel, K.R. (1975), "A Study of the Powers of Several Methods of Multiple Comparisons," *Journal of the American Statistical Association*, 70, 351.

Freund, R.J. and Littell, R. (1981), *SAS For Linear Models*, Cary, NC: SAS Institute.

Gabriel, K.R. (1978), "A Simple Method of Multiple Comparisons of Means," *Journal of the American Statistical Association*, 73, 364.

Games, P.A. (1977), "An Improved *t* Table for Simultaneous Control on g Contrasts," *Journal of the American Statistical Association*, 72, 359.

Goodnight, J.H. (1976), "The New General Linear Models Procedure," *Proceedings of the First International SAS Users' Meeting*, Cary, NC: SAS Institute.

Goodnight, J.H. (1978), *Tests of Hypothesis in Fixed Effects Linear Models*, SAS Technical Report R-101, Cary, NC: SAS Institute.

Goodnight, J.H. (1979), "A Tutorial on the Sweep Operator," *American Statistician*, 33, 149–158. (Also available as SAS Technical Report R-106.)

Goodnight, J.H. and Harvey, W.R. (1978), *Least Squares Means in the Fixed Effects General Linear Model*, SAS Technical Report R-103, Cary, NC: SAS Institute.

Goodnight, J.H. and Speed, F.M. (1978), *Computing Expected Mean Squares*, SAS Technical Report R-102, Cary, NC: SAS Institute.

Graybill, F.A. (1961), *An Introduction to Linear Statistical Models, Volume I*, New York: McGraw-Hill.

Harvey, Walter R. (1975), *Least-Squares Analysis of Data With Unequal Subclass Numbers*, USDA Report ARS H-4.

Heck, D.L. (1960), "Charts of Some Upper Percentage Points of the Distribution of the Largest Characteristic Root," *Annals of Mathematical Statistics*, 31, 625–642.

Hochburg, Y. (1974), "Some Conservative Generalizations of the T-Method in Simultaneous Inference," *Journal of Multivariate Analysis*, 4, 224–234.

Hocking, R.R. (1976), "The Analysis and Selection of Variables in a Linear Regression," *Biometrics*, 32, 1–50.

Kennedy, W.J., Jr. and Gentle, J.E. (1980), *Statistical Computing*, Chapter 9, New York: Marcel Dekker.

Kramer, C.Y. (1956), "Extension of Multiple Range Tests to Group Means with Unequal Numbers of Replications," *Biometrics*, 12, 307–310.

Kutner, M.H. (1974), "Hypothesis Testing in Linear Models (Eisenhart Model)," *American Statistician*, 28, 98–100.

Marcus, R., Peritz, E., and Gabriel, K.R. (1976), "On Closed Testing Procedures With Special Reference to Ordered Analysis of Variance," *Biometrika*, 63, 655–660.

Miller, R.G., Jr. (1981), *Simultaneous Statistical Inference*, New York: Springer-Verlag.

Morrison, D.F. (1976), *Multivariate Statistical Methods,* Second Edition, New York: McGraw-Hill.

Petrinovich, L.F. and Hardyck, C.D. (1969), "Error Rates for Multiple Comparison Methods: Some Evidence Concerning the Frequency of Erroneous Conclusions," *Psychological Bulletin*, 71, 43–54.

Pillai, K.C.S. (1960), *Statistical Tables for Tests of Multivariate Hypotheses*, Manila: The Statistical Center, University of the Philippines.

Pringle, R.M. and Raynor, A.A. (1971), *Generalized Inverse Matrices with Applications to Statistics*, New York: Hafner Publishing Company.

Ramsey, P.H. (1978), "Power Differences Between Pairwise Multiple Comparisons," *Journal of the American Statistical Association*, 73, 363.

Rao, C.R. (1965), *Linear Statistical Inference and Its Applications*, New York: John Wiley & Sons.

Ryan, T.A. (1959), "Multiple Comparisons in Psychological Research," *Psychological Bulletin*, 56, 26–47.

Ryan, T.A. (1960), "Significance Tests for Multiple Comparison of Proportions, Variances, and Other Statistics," *Psychological Bulletin*, 57, 318–328.

Schatzoff, M. (1966), "Exact Distributions of Wilk's Likelihood Ratio Criterion," *Biometrika*, 53, 347–358.

Scheffe, H. (1953), "A Method for Judging All Contrasts in the Analysis of Variance," *Biometrika*, 40, 87–104.

Scheffe, H. (1959), *The Analysis of Variance*, New York: John Wiley & Sons.

Searle, S.R. (1971), *Linear Models*, New York: John Wiley & Sons.

Searle, S.R., Speed, F.M., and Milliken, G.A. (1980), "Populations Marginal Means in the Linear Model: An Alternative to Least Squares Means," *The American Statistican*, 34, 216–221.

Sidak, Z. (1967), "Rectangular Confidence Regions for the Means of Multivariate Normal Distributions," *Journal of the American Statistical Association*, 62, 626–633.

Steel, R.G.D. and Torrie, J.H. (1960), *Principles and Procedures of Statistics*, New York: McGraw-Hill.

Tukey, J.W. (1952), "Allowances for Various Types of Error Rates," Unpublished IMS address, Chicago, Illinois.

Tukey, J.W. (1953), "The Problem of Multiple Comparisons," Unpublished manuscript.

Waller, R.A. and Duncan, D.B. (1969), "A Bayes Rule for the Symmetric Multiple Comparison Problem," *Journal of the American Statistical Association*, 64, 1484–1499, and (1972) Corrigenda, 67, 253–255.

Waller, R.A. and Kemp, K.E. (1976), "Computations of Bayesian t-Values for Multiple Comparisons," *Journal of Statistical Computation and Simulation*, 75, 169–172.

Welsch, R.E. (1977), "Stepwise Multiple Comparison Procedures," *Journal of the American Statistical Association*, 72, 359.

The NESTED Procedure

ABSTRACT

The NESTED procedure performs analysis of variance and analysis of covariance for data from an experiment with a nested (hierarchical) structure. Each effect is assumed to be a random effect.

INTRODUCTION

Although both the GLM and VARCOMP procedures provide similar analyses, NESTED is more efficient for this special type of design, especially for designs involving large numbers of levels and observations.

The data set that NESTED uses must first be sorted by the classification or CLASS variables defining the effects.

The CLASS variables in PROC NESTED are assumed to form a nested set of effects. For example, these statements for PROC NESTED:

```
CLASS A B C;
  VAR Y;
```

form a design specification that is specified in the GLM, ANOVA, or VARCOMP procedures as:

```
CLASS A B C;
  MODEL Y=A B(A) C(A B);
```

Note: NESTED is modeled after the General Purpose Nested Analysis of Variance program of the Dairy Cattle Research Branch of the United States Department of Agriculture. That program was originally written by Merrill R. Swanson, Statistical Reporting Service, USDA.

SPECIFICATIONS

The NESTED procedure is specified by the following statements:

PROC NESTED *option*;
  CLASS *variables*;
  VAR *variables*;
  BY *variables*;

The PROC and CLASS statements are required.

PROC NESTED Statement

PROC NESTED *options;*

The options below may appear in the PROC NESTED statement.

DATA=*SASdataset* names the SAS data set to be used by PROC NESTED. If DATA= is omitted, the most recently created SAS data set is used.

AOV suppresses the analysis of covariance statistics if you only want statistics for the analysis of variance.

CLASS Statement

CLASS *variables;*

A CLASS statement specifying the classification variables for the analysis **must** be included. The data set must be sorted by the classification variables in the order that they are given in the CLASS statement. Use PROC SORT to sort the data if they are not already sorted.

Values of a variable in the CLASS statement denote the levels of an effect. The name of that variable is also the name of the corresponding effect.

The second effect is assumed to be nested within the first effect, the third effect is assumed to be nested within the second effect, and so on.

VAR Statement

VAR *variables;*

List the dependent variables for the analysis in the VAR statement. If the VAR statement is omitted, NESTED performs an analysis of variance for all numeric variables in the data set, except those in the CLASS statement.

BY Statement

BY *variables;*

A BY statement may be used with PROC NESTED to obtain separate analyses on observations in groups defined by the BY variables. The input data set must be sorted by the BY variables.

DETAILS

Missing Values

An observation with missing values on any of the variables used by NESTED is omitted from the analysis. Blank values of CLASS variables are treated as missing values.

Printed Output

For each effect in the model, NESTED prints:

1. COEFFICIENTS OF EXPECTED MEAN SQUARES, the coefficients of the variance components making up the expected mean square.

For every dependent variable, NESTED prints an analysis-of-variance table containing:

2. VARIANCE SOURCE, sources of variation
3. D.F., degrees of freedom
4. SUM OF SQUARES
5. MEAN SQUARES
6. VARIANCE COMPONENT, estimates of variance components
7. PERCENT, the percentage associated with a source of variance is 100 times the ratio of that source's estimated variance component to the total variance component.

Below each analysis-of-variance table is printed:

8. MEAN, the overall mean
9. STANDARD DEVIATION
10. COEFFICIENT OF VARIATION of the response variable, based on the error mean square.

For each pair of dependent variables, NESTED prints an analysis-of-covariance table (unless AOV is specified). For each source of variation, this table includes:

11. D.F., the degrees of freedom
12. SUM OF PRODUCTS
13. MEAN PRODUCTS
14. COVARIANCE COMPONENT, the estimate of the covariance component
15. VARIANCE COMPONENT CORRELATION, the covariance component correlation
16. MEAN SQUARE CORRELATION.

EXAMPLE

In the following example from Snedecor and Cochran (1967), an experiment is conducted to determine the variability of calcium concentration in turnip greens. Four plants are selected at random, then three leaves are randomly selected from each plant. Two 100-mg samples are taken from each leaf. The amount of calcium is determined by microchemical methods.

Since the data are read in sorted order, it is not necessary to use PROC SORT on the class variables. LEAF is nested in PLANT; SAMPLE is nested in LEAF and is left for the residual term. All the effects are random effects.

```
TITLE CALCIUM CONCENTRATION IN TURNIP LEAVES — NESTED RANDOM
 MODEL;
TITLE2 SNEDECOR AND COCHRAN, STATISTICAL METHODS, 1967, P. 286;

DATA TURNIP;
  DO PLANT=1 TO 4;
    DO LEAF=1 TO 3;
      DO SAMPLE=1 TO 2;  INPUT CALCIUM @@;  OUTPUT;
        END;
      END;
    END;
  CARDS;
```

```
3.28 3.09 3.52 3.48 2.88 2.80
2.46 2.44 1.87 1.92 2.19 2.19
2.77 2.66 3.74 3.44 2.55 2.55
3.78 3.87 4.07 4.12 3.31 3.31
;
PROC NESTED;
  CLASSES PLANT LEAF;
  VAR CALCIUM;
```

① CALCIUM CONCENTRATION IN TURNIP LEAVES -- NESTED RANDOM MODEL 1
SNEDECOR AND COCHRAN, STATISTICAL METHODS, 1967, P. 286

COEFFICIENTS OF EXPECTED MEAN SQUARES

SOURCE	PLANT	LEAF	ERROR
PLANT	6.00000	2.00000	1.00000
LEAF	0.0	2.00000	1.00000
ERROR	0.0	0.0	1.00000

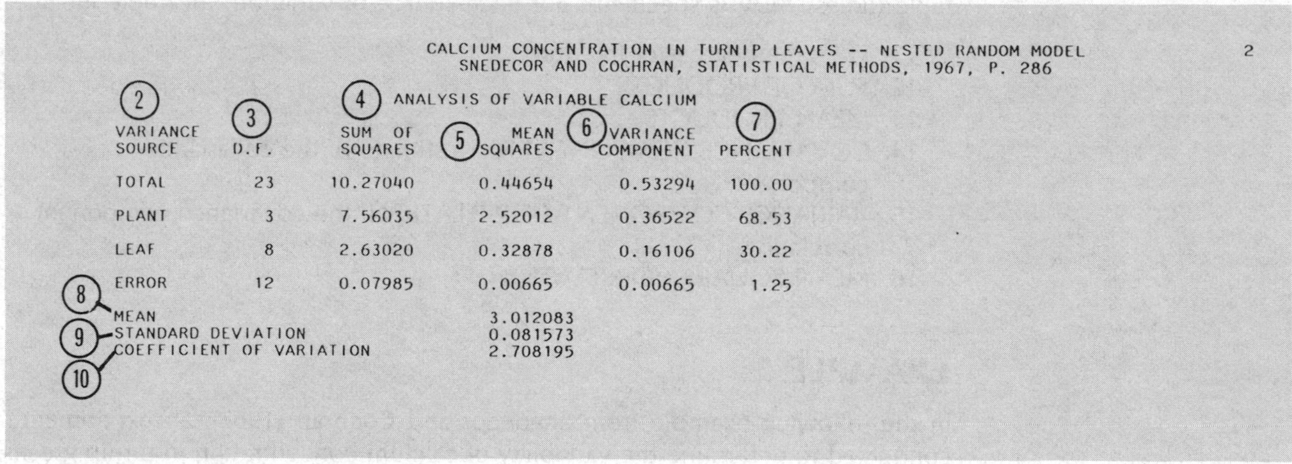

CALCIUM CONCENTRATION IN TURNIP LEAVES -- NESTED RANDOM MODEL 2
SNEDECOR AND COCHRAN, STATISTICAL METHODS, 1967, P. 286

④ ANALYSIS OF VARIABLE CALCIUM

② VARIANCE SOURCE	③ D.F.	SUM OF SQUARES	⑤ MEAN SQUARES	⑥ VARIANCE COMPONENT	⑦ PERCENT
TOTAL	23	10.27040	0.44654	0.53294	100.00
PLANT	3	7.56035	2.52012	0.36522	68.53
LEAF	8	2.63020	0.32878	0.16106	30.22
ERROR	12	0.07985	0.00665	0.00665	1.25

⑧ MEAN	3.012083
⑨ STANDARD DEVIATION	0.081573
⑩ COEFFICIENT OF VARIATION	2.708195

REFERENCES

Snedecor, George W. and Cochran, William G. (1967), *Statistical Methods*, Ames, Iowa: The Iowa State University Press.

Steel, Robert G.D. and Torrie, James H. (1980), *Principles and Procedures of Statistics*, New York: McGraw-Hill Book Company, Inc.

The NPAR1WAY Procedure

ABSTRACT

NPAR1WAY performs analysis of variance on ranks and certain rank scores of a response across a one-way classification. NPAR1WAY is a nonparametric procedure for testing that the distribution of a variable has the same location parameter across different groups.

INTRODUCTION

Most nonparametric tests are derived by examining the distribution of rank scores of the response variable. The rank scores are simply functions of the ranks of the response, where the values are ranked from low to high. Statistics defined as linear combinations of these rank scores are called *linear rank statistics*. The NPAR1WAY procedure calculates these four scores:

- **Wilcoxon scores** are the ranks:

  $$z_i = R_i$$

  and are locally most powerful for location shifts of a logistic distribution.
- **Median scores** are 1 for points above the median, 0 otherwise:

  $$z_i = (R_i > (n + 1)/2)$$

  and are locally most powerful for double exponential distributions.
- **Van der Waerden** scores are approximations to the expected values of the order statistics for a normal distribution:

  $$z_i = \Phi^{-1}(R_i/(n + 1))$$

  where Φ is the distribution function for the normal distribution. Van der Waerden scores are powerful for normal distributions.
- **Savage scores** are expected values minus 1 of order statistics for the exponential distribution:

  $$z_i = \Sigma_{j=1}^{R_i} 1/(n - j + 1) - 1$$

  and are powerful for comparing scale differences in exponential distributions (Hajek, 1969, p.83). NPAR1WAY subtracts 1 to center the scores around 0.

The statistics computed by PROC NPAR1WAY can also be computed by calculating the rank scores using PROC RANK and analyzing these rank scores with PROC ANOVA.

NPAR1WAY scores...	*correspond to these tests if data are classified in two levels...*[1]	*correspond to these tests for a one-way layout or k-sample location test...*[2]
Wilcoxon	Wilcoxon rank sum test Mann-Whitney U test	Kruskal-Wallis test
Median	median test for two samples	k-sample median test (Brown-Mood)
van der Waerden	van der Waerden test	k-sample van der Waerden test
Savage	Savage test	k-sample Savage test

[1] The tests are two-tailed. For a one-tailed test transform the significance probability by $p/2$ or $(1 - p/2)$.

[2] NPAR1WAY provides a chi-square approximate test.

SPECIFICATIONS

The statements used to control NPAR1WAY are:

PROC NPAR1WAY *options;*
  VAR *variables;*
  CLASS *variable;*
  BY *variables;*

The PROC and CLASS statements are required.

PROC NPAR1WAY Statement

PROC NPAR1WAY *options;*

The options available are:

DATA=*SASdataset* names the SAS data set containing the data to be analyzed. If DATA= is omitted, the most recently created SAS data set is used.

These options may be specified on the PROC NPAR1WAY statement. If none of these options are specified, then all five are performed by default.

ANOVA requests a standard analysis of variance in addition to the nonparametric *ANOVA* on the ranks.

WILCOXON requests an analysis of the ranks of the data, or the Wilcoxon scores. For two levels, this is the same as a Wilcoxon rank-sum test. For any number of levels, this is a Kruskal-Wallis test.

MEDIAN requests an analysis of the median scores. The median score is 1 for points above the median, 0 otherwise. For two samples, this produces a median test; for any number of levels, this is the Brown-Mood test.

VW requests that van der Waerden scores be analyzed. These are approximate normal scores derived by applying the inverse distribution function of the normal to the fractional ranks:

$$\Phi^{-1}(R_i/(n + 1)).$$

For two levels, this is the standard van der Waerden test.

SAVAGE requests that Savage scores be analyzed. These are expected order statistics for the exponential minus 1. This test is appropriate for comparing groups of data with exponential distributions.

VAR Statement

VAR *variables;*

This statement names the response or dependent variables to be analyzed. If the VAR statement is omitted, all numeric variables in the data set are analyzed.

CLASS Statement

CLASS *variable;*

The CLASS statement, which is required, names one and only one classification variable.

BY Statement

BY *variables;*

A BY statement may be used with PROC NPAR1WAY to obtain separate analyses on observations in groups defined by the BY variables. When a BY statement appears, the procedure expects the input data set to be sorted in order of the BY variables. If your input data set is not sorted in ascending order, use the SORT procedure with a similar BY statement to sort the data, or, if appropriate, use the BY statement options NOTSORTED or DESCENDING. For more information, see the discussion of the BY statement in Chapter 8, "Statements Used in the PROC Step," in *SAS User's Guide: Basics, 1982 Edition*.

DETAILS

Missing Values

If an observation has a missing value for a response variable or the classification variable, then that observation is excluded from the analysis.

Limitations

No more than 32767 = (2^{15} − 1) observations can be analyzed at one time. The procedure must have 20**n* bytes of memory available to store the data, where *n* is the number of nonmissing observations.

Resolution of Tied Values

Although the nonparametric tests were developed for continuous distributions, tied values do occur in practice. Ties are handled in all methods by assigning the average score for the different ranks corresponding to the tied values. Adjustments to variance estimates are performed in the manner described by Hajek (1969, Chapter 7).

Printed Output

NPAR1WAY produces the printed output described below.

If the ANOVA option is specified, NPAR1WAY prints:

1. the traditional ANALYSIS OF VARIANCE table
2. the effect mean square is reported as AMONG MS
3. the error mean square is reported as WITHIN MS.

(These values are the same that would result from using a procedure such as ANOVA or GLM.)

NPAR1WAY produces a table for each rank score, including for each level in the classification:

4. the LEVEL
5. the number of observations in the level (N)
6. the SUM OF SCORES
7. the EXPECTED sum of scores UNDER H0, the null hypothesis
8. STD DEV, the standard deviation estimate of the sum of scores, and
9. the MEAN SCORE.

For two or more levels, NPAR1WAY prints:

10. a chi-square statistic (CHISQ)
11. its degrees of freedom (DF)
12. PROB>CHISQ, the significance probability.

If there are only two levels, NPAR1WAY reports:

13. the smallest mean score as S
14. the ratio (S-expected)/std as Z, which is approximately normally distributed under the null hypothesis
15. PROB>|Z|, the probability of a greater observed Z value
16. T-TEST APPROX, the significance level for the *t*-test approximation.

EXAMPLE

The data are read in with a variable number of observations per record. In this example, NPAR1WAY first performs all five analyses on five levels of the class variable DOSE. Then the two lowest levels are output to a second data set to illustrate the two-sample tests.

```
TITLE WEIGHT GAINS WITH GOSSYPOL ADDITIVE;
TITLE3 HALVERSON AND SHERWOOD - 1932;
  DATA G;
    INPUT DOSE N;
    DO I=1 TO N;
      INPUT GAIN @@;
      OUTPUT;
      END;
```

```
  CARDS;
  0 16
   228 229 218 216 224 208 235 229 233 219 224 220 232 200 208 232
.04 11
   186 229 220 208 228 198 222 273 216 198 213
.07 12
   179 193 183 180 143 204 114 188 178 134 208 196
.10 17
   130 87 135 116 118 165 151 59 126 64 78 94 150 160 122 110 178
.13 11
   154 130 130 118 118 104 112 134 98 100 104
;
PROC NPAR1WAY;
  CLASS DOSE;
  VAR GAIN;
DATA G2;
  SET G;
  IF DOSE<=.04;
PROC NPAR1WAY;
  CLASS DOSE;
  VAR GAIN;
  TITLE4 'DOSES<=.04';
```

```
                 WEIGHT GAINS WITH GOSSYPOL ADDITIVE                        1

                    HALVERSON AND SHERWOOD - 1932

       ANALYSIS FOR VARIABLE GAIN CLASSIFIED BY VARIABLE DOSE

                  AVERAGE SCORES WERE USED FOR TIES

                       ANALYSIS OF VARIANCE        (2)        (3)
                  (1)
  LEVEL                 N        MEAN              AMONG MS   WITHIN MS
                                                   35020.7    627.452
               0       16       222.19
            0.04       11       217.36             F VALUE     PROB>F
            0.07       12       175.00               55.81     0.0001
             0.1       17       120.18
            0.13       11       118.36


                    WILCOXON SCORES (RANK SUMS)
  (4)                                (7)         (8)
                            (6) SUM OF   EXPECTED    STD DEV   (9)   MEAN
  LEVEL            (5)   N      SCORES   UNDER H0    UNDER H0        SCORE

               0       16       890.50     544.00       67.98        55.66
            0.04       11       555.00     374.00       59.06        50.45
            0.07       12       395.50     408.00       61.14        32.96
             0.1       17       275.50     578.00       69.38        16.21
            0.13       11       161.50     374.00       59.06        14.68

            KRUSKAL-WALLIS TEST (CHI-SQUARE APPROXIMATION)
            CHISQ=  52.67     DF=   4     PROB > CHISQ=0.0001
              (10)             (11)              (12)
              MEDIAN SCORES (NUMBER POINTS ABOVE MEDIAN)

                                SUM OF   EXPECTED    STD DEV         MEAN
  LEVEL                  N      SCORES   UNDER H0    UNDER H0        SCORE

               0       16        16.00       7.88        1.76         1.00
            0.04       11        11.00       5.42        1.53         1.00
            0.07       12         6.00       5.91        1.58         0.50
             0.1       17         0.00       8.37        1.79         0.00
            0.13       11         0.00       5.42        1.53         0.00

            MEDIAN 1-WAY ANALYSIS (CHI-SQUARE APPROXIMATION)
            CHISQ= 54.176      DF=   4      PROB > CHISQ=0.0001
```

```
                    WEIGHT GAINS WITH GOSSYPOL ADDITIVE                         2

                       HALVERSON AND SHERWOOD - 1932

         ANALYSIS FOR VARIABLE GAIN CLASSIFIED BY VARIABLE DOSE


                     VAN DER WAERDEN SCORES (NORMAL)

                               SUM OF    EXPECTED     STD DEV        MEAN
LEVEL                N         SCORES    UNDER HO     UNDER HO       SCORE

                0   16          16.12        0.00         3.33        1.01
             0.04   11           8.34        0.00         2.89        0.76
             0.07   12          -0.58        0.00         2.99       -0.05
              0.1   17         -14.69        0.00         3.39       -0.86
             0.13   11          -9.19        0.00         2.89       -0.84

          VAN DER WAERDEN 1-WAY (CHI-SQUARE APPROXIMATION)
          CHISQ=  47.30     DF=  4      PROB > CHISQ=0.0001

                     SAVAGE SCORES (EXPONENTIAL)

                               SUM OF    EXPECTED     STD DEV        MEAN
LEVEL                N         SCORES    UNDER HO     UNDER HO       SCORE

                0   16          16.07        0.00         3.39        1.00
             0.04   11           7.69        0.00         2.94        0.70
             0.07   12          -3.58        0.00         3.04       -0.30
              0.1   17         -11.98        0.00         3.46       -0.70
             0.13   11          -8.20        0.00         2.94       -0.75

          SAVAGE 1-WAY (CHI-SQUARE APPROXIMATION)
          CHISQ=  39.49     DF=  4      PROB > CHISQ=0.0001
```

```
                    WEIGHT GAINS WITH GOSSYPOL ADDITIVE                         3

                       HALVERSON AND SHERWOOD - 1932
                                DOSES<=.04

         ANALYSIS FOR VARIABLE GAIN CLASSIFIED BY VARIABLE DOSE

                       AVERAGE SCORES WERE USED FOR TIES

                            ANALYSIS OF VARIANCE

LEVEL                N          MEAN                      AMONG MS   WITHIN MS
                                                          151.684     271.479
                0   16        222.19
             0.04   11        217.36                      F VALUE      PROB>F
                                                             0.56      0.4617


                      WILCOXON SCORES (RANK SUMS)

                               SUM OF    EXPECTED     STD DEV        MEAN
LEVEL                N         SCORES    UNDER HO     UNDER HO       SCORE

                0   16         253.50      224.00        20.22       15.84
             0.04   11         124.50      154.00        20.22       11.32

          WILCOXON 2-SAMPLE TEST (NORMAL APPROXIMATION)
    (13)  (WITH CONTINUITY CORRECTION OF .5)
          S=  124.50        Z=-1.4341       PROB >|Z|=0.1515
    (15)  T-TEST APPROX.  SIGNIFICANCE=0.1635

    (14)  KRUSKAL-WALLIS TEST (CHI-SQUARE APPROXIMATION)
          CHISQ=    2.13     DF=  1    PROB > CHISQ=0.1446

               MEDIAN SCORES (NUMBER POINTS ABOVE MEDIAN)

                               SUM OF    EXPECTED     STD DEV        MEAN
LEVEL                N         SCORES    UNDER HO     UNDER HO       SCORE

                0   16           9.00        7.70         1.30        0.56
             0.04   11           4.00        5.30         1.30        0.36

          MEDIAN 2-SAMPLE TEST (NORMAL APPROXIMATION)
          S=    4.00      Z=-0.9965     PROB >|Z|=0.3190

          MEDIAN 1-WAY ANALYSIS (CHI-SQUARE APPROXIMATION)
          CHISQ=  0.994      DF=  1     PROB > CHISQ=0.3187
```

```
                                                                          4
                  WEIGHT GAINS WITH GOSSYPOL ADDITIVE

                     HALVERSON AND SHERWOOD - 1932
                              DOSES<=.04

   ANALYSIS FOR VARIABLE GAIN CLASSIFIED BY VARIABLE DOSE


                  VAN DER WAERDEN SCORES (NORMAL)

                            SUM OF     EXPECTED     STD DEV         MEAN
 LEVEL               N      SCORES     UNDER HO    UNDER HO         SCORE

           0        16        3.35        0.00        2.32          0.21
        0.04        11       -3.35        0.00        2.32         -0.30

         VAN DER WAERDEN 2-SAMPLE TEST (NORMAL APPROXIMATION)
         S=   -3.35       Z= -1.442       PROB >|Z|=0.1492

         VAN DER WAERDEN 1-WAY (CHI-SQUARE APPROXIMATION)
         CHISQ=    2.08      DF=  1       PROB > CHISQ=0.1492


                  SAVAGE SCORES (EXPONENTIAL)

                            SUM OF     EXPECTED     STD DEV         MEAN
 LEVEL               N      SCORES     UNDER HO    UNDER HO         SCORE

           0        16        1.83        0.00        2.40          0.11
        0.04        11       -1.83        0.00        2.40         -0.17

         SAVAGE 2-SAMPLE TEST (NORMAL APPROXIMATION)
         S=   -1.83       Z= -0.764       PROB >|Z|=0.4450

         SAVAGE 1-WAY (CHI-SQUARE APPROXIMATION)
         CHISQ=    0.58      DF=  1       PROB > CHISQ=0.4450
```

REFERENCES

Conover, W.J. (1980), *Practical Nonparametric Statistics*, Second Edition, New York: John Wiley & Sons.

Hajek, J. (1969), *A Course in Nonparametric Statistics*, San Francisco: Holden-Day.

Lehmann, E.L. (1975), *Nonparametrics: Statistical Methods Based on Ranks*, San Francisco: Holden-Day.

Quade, D. (1966), "On Analysis of Variance for the k-Sample Problem," *Annals of Mathematical Statistics*, 37, 1747-1758.

The PLAN Procedure

ABSTRACT

The PLAN procedure generates randomized plans for experiments.

INTRODUCTION

These plans are represented as groups of random permutations of positive integers. One or more random permutations can be generated for each item in another random permutation; there is no limit to the depth to which the random permutations may be nested. Any number of randomized plans may be generated.

The random permutations are selected from uniform pseudo-random variates generated as in the UNIFORM function (see Chapter 5, "SAS Functions," in the *SAS User's Guide: Basics, 1982 Edition*).

SPECIFICATIONS

The PLAN procedure is controlled by two statements:

PROC PLAN *options;*
FACTORS *requests;*

You include a FACTORS statement for each plan you want. Several FACTORS statements are permitted.

PROC PLAN Statement

PROC PLAN *option;*

Since the PLAN procedure needs no input data, the DATA= option is never used.

The option below may appear in the PROC PLAN statement:

SEED= specifies a 5-, 6-, or 7-digit odd integer for PLAN to use to generate the first random permutation. If the SEED parameter is omitted, a reading of the time of day from the computer's clock is used to generate the first random permutation.

FACTORS Statement

FACTORS *requests;*

request specifies the randomized plan to be provided by PLAN. The form of a *request* is:

```
name=<m OF >n <ORDERED> ...
```

where brackets (< >) denote an optional specification. More than one request can appear in the same FACTORS statement. The names in a request must be valid SAS names. *N* or *m* values must be positive integers.

A positive integer *n* appearing alone after an equal sign produces a random permutation of the integers 1, 2, ..., *n*.

A positive integer *n* followed by the word ORDERED generates simply the list of integers 1, 2, ..., *n*, in that order.

The specification *m* OF *n* tells PLAN to pick a random sample of *m* integers (without replacement) from the set of integers 1, 2, ..., *n* and to arrange the sample randomly.

For every integer generated for the first name specified, a permutation is generated for the second name according to the specifications following the second equal sign; for each of the integers generated, a permutation is generated for the second name, the third name, and so forth. For example,

```
PROC PLAN;
  FACTORS ONE=4 TWO=3;
```

You can think of a factor TWO as being nested within factor ONE, where the levels of factor ONE are to be randomly assigned to 4 units.

Six random permutations, say, of the numbers 1, 2, 3 can be generated simply by specifying

```
FACTORS A=6 ORDERED B=3;
```

DETAILS

Printed Output

The PLAN procedure prints for each factor:

1. the initial random number
2. the number of levels of nesting in the plan
3. the random permutations making up each plan.

EXAMPLES

This first plan is appropriate for a completely random design with 12 experimental units and several treatments. The FACTORS statement requests a permutation of the integers 1, 2, ..., 12. If there are two treatments, the experimenter might then assign treatment 1 to the units corresponding to the first 6 integers in the permutations and treatment 2 to the other units.

```
PROC PLAN SEED=27371;
  TITLE COMPLETELY RANDOMIZED DESIGN;
  FACTORS U=12;
```

```
                    COMPLETELY RANDOMIZED DESIGN                              1

PROCEDURE PLAN.     RANDOM NUMBER SEED=      27371

FACTOR     SELECT  LEVELS  RANDOMIZED?
------     ------  ------  -----------
U              12      12       RANDOM

        U
----+----+----+----+----+----+----+----+----+----+----+----+

    5   10   12   11    3    7    2    1    9    4    6    8
```

The second plan is appropriate for a split-plot design with main plots forming a randomized complete blocks design. Say that there are 3 blocks, 4 main plots per block, and 2 subplots per main plot. Three random permutations (one for each of the blocks) of the integers 1, 2, 3, and 4 are produced. The four integers correspond to the four levels of factor A; the permutation determines how the levels of A are assigned to the main plots within a block. For each of those twelve numbers, a random permutation of the integers 1 and 2 is produced. Each two-integer permutation determines the assignment of the two levels of factor B to the subplots within a main plot.

```
PROC PLAN SEED=37277;
  TITLE SPLIT PLOT DESIGN;
  FACTORS BLOCK=3 ORDERED A=4 B=2;
```

```
                    SPLIT PLOT DESIGN                              2

PROCEDURE PLAN.     RANDOM NUMBER SEED=      37277

FACTOR     SELECT  LEVELS  RANDOMIZED?
------     ------  ------  -----------
BLOCK          3       3       ORDERED
A              4       4        RANDOM
B              2       2        RANDOM

     BLOCK         A         B
  --------  --------  ----+----+

         1         1    2    1
                   4    2    1
                   3    1    2
                   2    1    2
         2         1    1    2
                   4    2    1
                   2    2    1
                   3    1    2
         3         4    1    2
                   3    1    2
                   2    1    2
                   1    2    1
```

The third plan is appropriate for a hierarchical design. The FACTORS statement requests a random permutation of the numbers 1, 2, and 3; a random permutation of the numbers 1, 2, 3 and 4 for each of those first three numbers; and a random permutation of 1, 2, and 3 for each of the twelve integers in the second set of permutations.

```
PROC PLAN SEED=17431;
  TITLE HIERARCHICAL DESIGN;
  FACTORS HOUSES=3 POTS=4 PLANTS=3;
```

```
                         HIERARCHICAL DESIGN                                  3

PROCEDURE PLAN.     RANDOM NUMBER SEED=        17431

FACTOR      SELECT  LEVELS  RANDOMIZED?
------      ------  ------  -----------
HOUSES           3       3       RANDOM
POTS             4       4       RANDOM
PLANTS           3       3       RANDOM

  HOUSES      POTS    PLANTS
--------  --------  ----+----+----+

       3         1     1    2    3

                 3     2    1    3

                 2     3    1    2

                 4     1    3    2

       2         2     1    2    3

                 1     2    3    1

                 4     1    2    3

                 3     1    3    2

       1         2     2    1    3

                 1     3    2    1

                 4     2    3    1

                 3     2    3    1
```

REFERENCES

Cochran, William G. and Cox, G.M. (1957), *Experimental Designs*, Second Edition, New York: John Wiley and Sons, Inc.

Lewis, P.A.W., Goodman, A.S., and Miller, J.M. (1969), "A Pseudo-Random Number Generator for the System/360," *IBM Systems Journal*, 8(2).

The TTEST Procedure

ABSTRACT

The TTEST procedure computes a t statistic for testing the hypothesis that the means of two groups of observations in a SAS data set are equal.

INTRODUCTION

Means for a variable are computed for each of the two groups of observations identified by values of a classification or CLASS variable. The t test tests the hypothesis that the true means are the same. This can be considered as a special case of a one-way analysis of variance with two levels of classification.

TTEST computes the t statistic based on the assumption that the variances of the two groups are equal and also computes an approximate t based on the assumption that the variances are unequal. For each t, the degrees of freedom and probability level are given; Satterthwaite's approximation (1946) is used to compute the degrees of freedom associated with the approximate t. An F (folded) statistic is computed to test for equality of the two variances. (Steel and Torrie, 1980)

The TTEST procedure was not designed for paired comparisons. See **EXAMPLES** below for a method of using the MEANS procedure to get a paired-comparisons t test.

Note that the underlying assumption of the t test computed by this procedure is that the variables are normally and independently distributed within each group.

SPECIFICATIONS

The statements used to control the procedure are:

PROC TTEST *option*;
CLASS *variable*;
VAR *variables*;
BY *variables*;

Each statement may be given only once. There is no restriction on the order of the statements after the PROC statement. The CLASS statement is required.

PROC TTEST Statement

PROC TTEST *option*;

The option is:

DATA=*SASdataset* names the SAS data set for the procedure to use. If DATA= is not given, TTEST uses the most recently created SAS data set.

CLASS Statement

CLASS *variable;*

A CLASS statement giving the name of the grouping variable must accompany the PROC TTEST statement. The grouping variable may have two, and only two, values. TTEST divides the observations into the two groups for the *t* test using the values of this variable.

Either a numeric or character variable may be used in the CLASS statement. If a character variable longer than 16 characters is used, the value is truncated and a warning message is issued.

VAR Statement

VAR *variables;*

The VAR statement gives the names of the variables whose means are to be compared. If the VAR statement is omitted, all numeric variables in the input data set (except a numeric variable appearing in the CLASS statement) are included in the analysis.

BY Statement

BY *variables;*

A BY statement may be used with PROC TTEST to obtain separate analyses on observations in groups defined by the BY variables. When a BY statement appears, the procedure expects the input data set to be sorted in order of the BY variables. If your input data set is not sorted in ascending order, use the SORT procedure with a similar BY statement to sort the data, or, if appropriate, use the BY statement options NOTSORTED or DESCENDING. For more information, see the discussion of the BY statement in Chapter 8, "Statements Used in the PROC Step."

DETAILS

Missing Values

An observation with a missing value for either the CLASS variable or for the variable to be tested is never included in the calculations.

The F′ (Folded) Statistic

The usual *t* statistic for testing the equality of means $\bar{x}_1$ and $\bar{x}_2$ from two independent samples

$$t = (\bar{x}_1 - \bar{x}_2)/\sqrt{s^2(1/n_1 + 1/n_2)}$$

for n_1 and n_2 observations includes a term for the pooled variance s^2.

The pooled variance

$$s^2 = [(n_1 - 1)s_1^2 + (n_2 - 1)s_2^2]/(n_1 + n_2 - 2),$$

where s_1^2 and s_2^2 are the variances of the two groups, depends on the assumption that $\sigma_1^2 = \sigma_2^2$.

You can use the folded form of the F statistic, F′, to test the assumption that the variances are equal, where

$$F' = (\text{larger of } s_1^2, s_2^2)/(\text{smaller of } s_1^2, s_2^2).$$

A test of F′ is a two-tailed F test, since we do not specify which s^2 we expect to be larger. PROB>F gives the probability of a greater F value under the null hypothesis that $\sigma_1^2 = \sigma_2^2$.

Under the assumption of EQUAL variances, the t statistic is computed with the formula given above using the pooled estimate of s^2.

Under the assumption of UNEQUAL variances, the approximate t is computed as

$$t = (\bar{x}_1 - \bar{x}_2)/\sqrt{(s_1^2/n_1 + s_2^2/n_2)}.$$

The formula for Satterthwaite's (1946) approximation for df is:

$$df = \frac{(s_1^2/n_1 + s_2^2/n_2)^2}{(s_1^2/n_1)^2/(n_1 - 1) + (s_2^2/n_2)^2/(n_2 - 1)}$$

Refer to Steel and Torrie (1980) and *SAS for Linear Models* (Freund and Littell, 1981) for more information.

Printed Output

For each variable included in the analysis, TTEST prints the following statistics for each group:

1. the name of the variable
2. the levels of the classification variable
3. N, the number of nonmissing values
4. the MEAN or average
5. STD DEV, or the standard deviation
6. STD ERROR, or the standard error
7. the MINIMUM value
8. the MAXIMUM value.

Under the assumption of UNEQUAL variances, TTEST prints:

9. T, an approximate t statistic
10. DF, Satterthwaite's approximation for the degrees of freedom
11. PROB > |T|, the probability of a greater absolute value of t.

Under the assumption of EQUAL variances, TTEST prints:

12. T, the t statistic
13. DF, the degrees of freedom
14. PROB > |T|, the probability of a greater absolute value of t.

TTEST then gives the results of the test of equality of variances:

15. the F' (folded) statistic (see the **DETAIL** section above)
16. the degrees of freedom, DF, in each group
17. PROB > F′, the probability of a greater F value.

EXAMPLES

Comparing Group Means: Example 1

The data for this example consist of golf scores for a physical education class. We want to use a *t* test to determine if the mean golf score for the males in the class differs significantly from the mean score for the females.

The grouping variable is SEX, and it appears in the CLASS statement.

The numbers on the sample output correspond to the statistics described above.

```
DATA SCORES;
  INPUT SEX $ SCORE @@;
  CARDS;
F 75  F 76  F 80  F 77  F 80  F 77  F 73
M 82  M 80  M 85  M 85  M 78  M 87  M 82
;
PROC TTEST;
  CLASS SEX;
  VAR SCORE;
  TITLE GOLF SCORES;
```

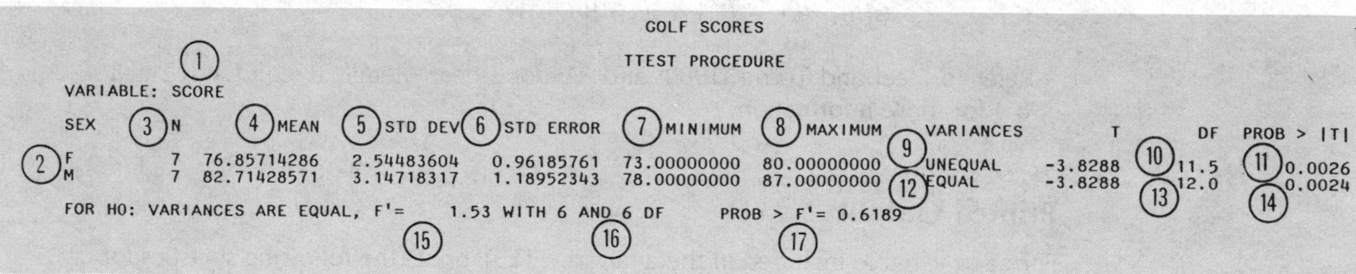

```
                                        GOLF SCORES                                                          1

                                      TTEST PROCEDURE
VARIABLE: SCORE

SEX      N       MEAN       STD DEV     STD ERROR     MINIMUM      MAXIMUM     VARIANCES        T        DF    PROB > |T|

F        7   76.85714286   2.54483604   0.96185761   73.00000000  80.00000000  UNEQUAL     -3.8288    11.5     0.0026
M        7   82.71428571   3.14718317   1.18952343   78.00000000  87.00000000  EQUAL       -3.8288    12.0     0.0024

FOR H0: VARIANCES ARE EQUAL, F'=     1.53 WITH 6 AND 6 DF      PROB > F'= 0.6189
```

Paired Comparisons Using PROC MEANS: Example 2

For paired comparisons, use PROC MEANS rather than PROC TTEST. You can create a new variable containing the differences between the paired variables and use the T and PRT options of PROC MEANS to test whether the mean difference is significantly different from zero.

For example, say you have a PRETEST and POSTTEST value for each observation in a data set. You want to test whether there is a significant difference between the two sets of scores.

Following the INPUT statement in the DATA step is an assignment statement to create a new variable DIFF by subtracting PRETEST from POSTTEST. Then, you use PROC MEANS with the T and PRT options to get a *t* statistic and a probability value for the hypothesis that DIFF's value is equal to zero.

```
DATA A;
  INPUT ID PRETEST POSTTEST;
  DIFF=POSTTEST-PRETEST;
  CARDS;
```

```
1  80  82
2  73  71
3  70  95
4  60  69
5  88 100
6  84  71
7  65  75
8  37  60
9  91  95
10 98  99
11 52  65
12 78  83
13 40  60
14 79  86
15 59  62
;
PROC MEANS MEAN STDERR T PRT;
  VAR DIFF;
  TITLE PAIRED-COMPARISONS T TEST;
```

```
                    PAIRED-COMPARISONS T TEST                                  1

VARIABLE          MEAN          STD FRROR          T        PR>|T|
                                 OF MEAN

DIFF          7.93333333       2.56434651        3.09      0.0079
```

REFERENCES

Freund, Rudolf J. and Littell, Ramon C. (1981), *SAS for Linear Models: A Guide to the ANOVA and GLM Procedures,* Cary, NC: SAS Institute Inc.

Satterthwaite, F. W. (1946), "An Approximate Distribution of Estimates of Variance Components," *Biometrics Bulletin*, 2, 110-114.

Steel, R.G.D. and Torrie, J.H. (1980), *Principles and Procedures of Statistics*, 2nd edition, New York: McGraw-Hill Book Company.

The VARCOMP Procedure

ABSTRACT

The VARCOMP procedure computes estimates of the variance components in a general linear model.

INTRODUCTION

VARCOMP is designed to handle models that have random effects. Random effects are classification effects where the levels of the effect are assumed to be randomly selected from an infinite population of normally distributed levels. The goal of VARCOMP is to estimate the variances of each one of these populations.

A single MODEL statement specifies the dependent variables and the effects: main effects, interactions, and nested effects. The effects must be composed of class variables; no continuous variables are allowed on the right-hand side of the equal sign.

You may specify certain effects as fixed (non-random) by putting them first in the MODEL statement and indicating the number of fixed effects with the FIXED option. An intercept is always fitted and assumed fixed. Except the effects specified as fixed, all other effects are assumed to be normally and independently distributed.

The dependent variables are grouped based on the similarity of their missing values. Each group of dependent variables is then analyzed separately. The columns of the design matrix **X** are formed in the same order as the effects are specified in the MODEL statement. No reparameterization is done. Thus, the columns of **X** contain only 0s and 1s.

Three methods of estimation are available:

The Type I method This method (METHOD=TYPE1) computes the Type I sum of squares for each effect, equates each mean square involving only random effects to its expected value, and solves the resulting system of equations (Gaylor, Lucas, and Anderson, 1970). The **X'X|X'Y** matrix is computed and adjusted in segments whenever memory is not sufficient to hold the entire matrix.

The MIVQUE0 method Based on the technique suggested by Hartley, Rao, and LaMotte (1978), the MIVQUE0 method produces estimates that are invariant with respect to the fixed effects of the model and are locally best quadratic unbiased estimates given that the true ratio of each component to the residual error component is zero. The technique is similar to TYPE1 except that the random effects are adjusted only for the fixed effects. This affords a considerable timing advantage over the TYPE1 method; thus MIVQUE0 is the default method used in VARCOMP.

The **X'X|X'Y** matrix is computed and adjusted in segments whenever memory is not sufficient to hold the entire matrix. For more information, refer to Rao (1971, 1972).

The maximum-likelihood method The ML method (METHOD=ML) computes maximum-likelihood estimates of the variance components using the W-transformation developed by Hemmerle and Hartley (1973). Initial estimates of the components are computed using MIVQUE0. The procedure then iterates until the log-likelihood objective function converges.

SPECIFICATIONS

The statements used in VARCOMP are:

PROC VARCOMP *options;*
CLASS *variables;*
MODEL *dependents = effects / options;*
BY *variables;*

PROC VARCOMP Statement

PROC VARCOMP *options;*

The options below may appear in the PROC VARCOMP statement:

METHOD=TYPE1 METHOD=MIVQUE0 METHOD=ML	specifies which of the three methods (TYPE1, MIVQUE0, or ML) should use the VARCOMP procedure. If METHOD= is omitted, MIVQUE0 is the default.
MAXITER=*number*	specifies the maximum number of iterations for METHOD=ML. If a value for MAXITER is omitted, its value is set to 50.
EPSILON=*number*	specifies the convergence value of the objective function for METHOD=ML. If EPSILON= is omitted, its value is 1E-8.
DATA=*SASdataset*	names the SAS data set to be used by VARCOMP. If DATA= is omitted, VARCOMP uses the most recently created SAS data set.

CLASS Statement

CLASS *variables;*

The CLASS statement specifies the classification variables to be used in the analysis. Class variables may be either numeric or character; if character, their lengths must be 16 or less.

Numeric class variables are not restricted to integers, since a variable's format determines the levels. For more information, see the FORMAT statement in *SAS User's Guide: Basics, 1982 Edition*.

MODEL Statement

MODEL *dependents = effects / options;*

The MODEL statement gives the dependent variables and independent effects. If more than one dependent variable is specified, a separate analysis is performed for each one. The independent effects are limited to main effects, interactions, and nested effects; no continuous effects are allowed. Effects are specified in VARCOMP in the same way as described for the ANOVA procedure. Only one MODEL statement is allowed.

Only one option is available on the MODEL statement:

FIXED=*n* specifies to VARCOMP that the first *n* effects in the MODEL statement are fixed effects; the remaining effects are assumed to be random.

BY Statement

BY *variables;*

A BY statement may be used with PROC VARCOMP to obtain separate analyses on observations in groups defined by the BY variables. When a BY statement appears, the procedure expects the input data set to be sorted in order of the BY variables. If your input data set is not sorted in ascending order, use the SORT procedure with a similar BY statement to sort the data, or, if appropriate, use the BY statement options NOTSORTED or DESCENDING. For more information, see the discussion of the BY statement in Chapter 8, "Statements Used in the PROC Step," *SAS User's Guide: Basics, 1982 Edition*.

DETAILS

Missing Values

The dependent variables are grouped based on the similarity of their missing values. This feature is similar to the way GLM treats missing values.

Printed Output

VARCOMP prints the following items:

1. CLASS LEVEL INFORMATION, for verifying the levels and number of observations in your data
2. for METHOD=TYPE1, an analysis-of-variance table with SOURCE, DF, TYPE I SS, TYPE I MS, and EXPECTED MEAN SQUARE
3. for METHOD=MIVQUE0, the SSQ MATRIX containing sums of squares of partitions of the **X'X** crossproducts matrix adjusted for the fixed effects. Each element (i,j) of this matrix is computed:

   $$\mathrm{SSQ}(\mathbf{X}_i'\mathbf{M}\mathbf{X}_j)$$

   where

   $$\mathbf{M} = \mathrm{I} - \mathbf{X}_0(\mathbf{X}_0'\mathbf{X}_0)^{-1}\mathbf{X}_0' \quad ,$$

   $\mathbf{X}_0$ is part of the design matrix for the fixed effects,
   $\mathbf{X}_i$ is part of the design matrix for one of the random effects, and
   SSQ is an operator that takes the sum of squares of the elements.
4. for METHOD=ML, the iteration history, including the OBJECTIVE function, as well as variance component estimates.

EXAMPLE

In this example A and B are classification variables and Y is the dependent variable. A is declared fixed, and B and A*B are random. VARCOMP is invoked three times, once for each of the estimation methods. The data are from Hemmerle and Hartley (1978, p. 829).

```
DATA A;
  INPUT A B Y @@;
  CARDS;
1 1 237 1 1 254 1 1 246 1 2 178 1 2 179 2 1 208
2 1 178 2 1 187 2 2 146 2 2 145 2 2 141 3 1 186
3 1 183 3 2 142 3 2 125 3 2 136
;
PROC VARCOMP METHOD=TYPE1;
  CLASS A B;
  MODEL Y=A|B / FIXED=1;
PROC VARCOMP METHOD=MIVQUE0;
  CLASS A B;
  MODEL Y=A|B / FIXED=1;
PROC VARCOMP METHOD=ML;
  CLASS A B;
  MODEL Y=A|B / FIXED=1;
```

```
                                                                              1
                 VARIANCE COMPONENT ESTIMATION PROCEDURE
             (1)     CLASS LEVEL INFORMATION
                   CLASS       LEVELS     VALUES
                   A                3     1 2 3
                   B                2     1 2

                 NUMBER OF OBSERVATIONS IN DATA SET = 16
```

```
                                                                              2
(2)              VARIANCE COMPONENT ESTIMATION PROCEDURE
DEPENDENT VARIABLE: Y
SOURCE              DF        TYPE I SS        TYPE I MS     EXPECTED MEAN SQUARE
A                    2   11736.43750000    5868.21875000     VAR(ERROR) + 2.725 VAR(A*B) + 0.1 VAR(B) + Q(A)
B                    1   11448.12564103   11448.12564103     VAR(ERROR) + 2.63076923 VAR(A*B) + 7.8 VAR(B)
A*B                  2     299.04102564     149.52051282     VAR(ERROR) + 2.58461538 VAR(A*B)
ERROR               10     786.33333333      78.63333333     VAR(ERROR)
CORRECTED TOTAL     15   24269.93750000

VARIANCE COMPONENT                ESTIMATE
VAR(B)                       1448.37683150
VAR(A*B)                       27.42658730
VAR(ERROR)                     78.63333333
```

```
                                                                          3
                    VARIANCE COMPONENT ESTIMATION PROCEDURE

                           CLASS LEVEL INFORMATION

                          CLASS     LEVELS     VALUES

                          A            3       1 2 3

                          B            2       1 2

                    NUMBER OF OBSERVATIONS IN DATA SET = 16
```

```
                                                                          4
          (3)  MIVQUE(0) VARIANCE COMPONENT ESTIMATION PROCEDURE

                                  SSQ MATRIX

SOURCE                          B            A*B           ERROR                Y

B                     60.84000000    20.52000000      7.80000000   89295.38000000
A*B                   20.52000000    20.52000000      7.80000000   30181.30000000
ERROR                  7.80000000     7.80000000     13.00000000   12533.50000000

VARIANCE COMPONENT       ESTIMATE
                                Y

VAR(B)              1466.12301587
VAR(A*B)             -35.49170274
VAR(ERROR)           105.73659674
```

```
                                                                          5
                    VARIANCE COMPONENT ESTIMATION PROCEDURE

                           CLASS LEVEL INFORMATION

                          CLASS     LEVELS     VALUES

                          A            3       1 2 3

                          B            2       1 2

                    NUMBER OF OBSERVATIONS IN DATA SET = 16
```

```
                                                                          6
       (4)  MAXIMUM LIKELIHOOD VARIANCE COMPONENT ESTIMATION PROCEDURE

DEPENDENT VARIABLE: Y

        ITERATION       OBJECTIVE           VAR(B)       VAR(A*B)       VAR(ERROR)

            0         78.38503712    1031.49069751     0.00000074      74.39097179
            1         79.11104067     345.15271871     0.00000091      91.30657141
            2         78.61867933     435.77212280     0.00000085      85.36540947
            3         78.37684582     535.96918079     0.00000082      81.52153518
            4         78.28681411     628.62366953     0.00000079      79.21098160
            5         78.26590307     691.34441528     0.00000078      78.04344579
            6         78.26364080     717.05370759     0.00000078      77.63119991
            7         78.26354899     722.72439338     0.00000078      77.54470514
            8         78.26354715     723.54524926     0.00000078      77.53231081
            9         78.26354712     723.65063133     0.00000078      77.53072190
           10         78.26354712     723.66391194     0.00000078      77.53052170

        CONVERGENCE CRITERION MET
```

REFERENCES

Gaylor, D.W., Lucas, H.L., and Anderson, R.L. (1970), "Calculations of Expected Mean Squares by the Abbreviated Doolittle and Square Root Method," *Biometrics*, 26, 641-655.

Goodnight, J.H. (1978), "Computing MIVQUE0 Estimates of Variance Components," SAS Technical Report R-105. Cary, NC: SAS Institute.

Goodnight, J.H. and Hemmerle, W.J. (1979), "A Simplified Algorithm for the W-Transformation in Variance Component Estimation," *Technometrics*, 21, 265-268.

Hartley, H.O., Rao, J.N.K., and LaMotte, Lynn (1978), "A Simple Synthesis-Based Method of Variance Component Estimation," *Biometrics*, 34, 233-242.

Hemmerle, W.J. and Hartley, H.O. (1973), "Computing Maximum Likelihood Estimates for the Mixed AOV Model Using the W-Transformation," *Technometrics*, 15, 819-831.

Rao, C.R. (1971), "Minimum Variance Quadratic Unbiased Estimation of Variance Components," *Journal of Multivariate Analysis*, 1, 445-456.

Rao, C.R. (1972), "Estimation of Variance and Covariance Components in Linear Models," *Journal of the American Statistical Association*, 57, 112-15.

The Four Types of Estimable Functions

INTRODUCTION

GLM, VARCOMP, and other SAS procedures label the sums of squares associated with the various effects in the model as TYPE I, TYPE II, TYPE III, and TYPE IV. The four types of hypotheses available in GLM may not always be sufficient for a statistician to perform all desired hypothesis tests, but should suffice for the vast majority of analyses. The purpose of this chapter is to explain the hypotheses tested by each of the four types of SS (sums of squares). For additional discussion, see *SAS for Linear Models* (Freund and Littell, 1981).

ESTIMABILITY

For linear models, such as

$$\mathbf{Y} = \mathbf{X}\beta + \varepsilon$$

which have $E(\mathbf{Y}) = \mathbf{X}\beta$, a primary analytical goal is to estimate or test (where possible) the elements of β or certain linear combinations of the elements of β. This is accomplished by computing linear combinations of the observed **Y**s. To estimate a specific linear function of the βs, say $\mathbf{L}\beta$, we must be able to find a linear combination of the **Y**s that has an expected value of $\mathbf{L}\beta$. Hence the following definition:

> $\mathbf{L}\beta$ is estimable if and only if a linear combination of the **Y**s exists that has an expected value of $\mathbf{L}\beta$.

Any linear combination of the **Y**s that is computed, say KY, will have $E(K\mathbf{Y}) = K\mathbf{X}\beta$. Thus the expected value of any linear combination of the **Y**s is equal to that same linear combination of the rows of **X** multiplied by β. Therefore, $\mathbf{L}\beta$ is estimable if and only if we can find a linear combination of the rows of **X** that is equal to **L**.

Thus the rows of **X** form a generating set from which an **L**, such that $\mathbf{L}\beta$ is estimable, can be constructed. Since **X** can be reconstructed from the rows of $\mathbf{X}'\mathbf{X}$ (that is, $\mathbf{X} = [\mathbf{X}(\mathbf{X}'\mathbf{X})\mathbf{1}(\mathbf{X}'\mathbf{X})]$, therefore the rows of $\mathbf{X}'\mathbf{X}$ also form a generating set from which all **L**s, such that $\mathbf{L}\beta$ is estimable, can be constructed. Similarly, the rows of $(\mathbf{X}'\mathbf{X})^{-}\mathbf{X}'\mathbf{X}$ also form a generating set for **L**.

Therefore, if **L** is generated as a linear combination of the rows of **X**, $\mathbf{X}'\mathbf{X}$, or $(\mathbf{X}'\mathbf{X})^{-}\mathbf{X}'\mathbf{X}$, $\mathbf{L}\beta$ is estimable. Furthermore, any number of row operations that do not destroy the row rank can be performed on **X**, $\mathbf{X}'\mathbf{Y}$, or $(\mathbf{X}'\mathbf{X})^{-}\mathbf{X}'\mathbf{X}$. The rows of the resulting matrices also form a generating set for **L**.

Once an **L** of full row rank has been formed from a generating set, we can estimate $\mathbf{L}\beta$ by computing $\mathbf{Lb}$, where $\mathbf{b} = (\mathbf{X}'\mathbf{X})^{-}\mathbf{X}'\mathbf{Y}$. From the general theory of linear models, **Lb** will be the best linear unbiased estimator of $\mathbf{L}\beta$. To test the hypothesis that $\mathbf{L}\beta = 0$, we compute SS(H0: $\mathbf{L}\beta = 0$) = $(\mathbf{Lb})'(\mathbf{L}(\mathbf{X}'\mathbf{X})^{-}\mathbf{L}')^{-1}\mathbf{Lb}$ and form an *F* test using the appropriate error term.

General Form of an Estimable Function

Although any generating set for **L**, such as **X**, **X′ X**, or $(\mathbf{X}'\mathbf{X})^{-}\mathbf{X}'\mathbf{X}$ could be printed to inform the user of what could be estimated, the volume of output would usually defeat the purpose. A rather simple shorthand technique for printing any generating set is demonstrated below.

Suppose

$$\mathbf{X} = \begin{bmatrix} 1 & 1 & 0 & 0 \\ 1 & 1 & 0 & 0 \\ 1 & 0 & 1 & 0 \\ 1 & 0 & 1 & 0 \\ 1 & 0 & 0 & 1 \\ 1 & 0 & 0 & 1 \end{bmatrix} \quad \text{and} \quad \beta = \begin{bmatrix} \mu \\ A1 \\ A2 \\ A3 \end{bmatrix}$$

Although **X** is a generating set for **L**, so also is

$$\mathbf{X}^* = \begin{bmatrix} 1 & 1 & 0 & 0 \\ 1 & 0 & 1 & 0 \\ 1 & 0 & 0 & 1 \end{bmatrix}$$

X* is formed from **X** by deleting duplicate rows.

Since all **L**s must be linear functions of the rows of **X***, an **L** for a single-degree-of-freedom estimate may be represented symbolically as:

L1*(1 1 0 0) + L2*(1 0 1 0) + L3*(1 0 0 1)

or

L = (L1+L2+L3, L1, L2, L3) .

For this example, $\mathbf{L}\beta$ is estimable if and only if the first element of **L** is equal to the sum of the other elements of **L**, or

(L1+L2+L3)*u + L1*A1 + L2*A2 + L3*A3

is estimable for any values of L1, L2, and L3.

If other generating sets for **L** are represented symbolically, the symbolic notation will look different. However, the inherent nature of the rules will be the same. For example, if row operations are performed on **X*** to produce an identity matrix in the first 3×3,

$$\mathbf{X}^{**} = \begin{bmatrix} 1 & 0 & 0 & 1 \\ 0 & 1 & 0 & -1 \\ 0 & 0 & 1 & -1 \end{bmatrix}$$

then **X**** is also a generating set for **L**. An **L** generated from **X**** may be represented symbolically as:

L = (L1, L2, L3, L1−L2−L3)

although, again, the first element of **L** is equal to the sum of the other elements.

With the thousands of generating sets available, the question arises as to which one is the best to represent **L** symbolically. Clearly, a generating set containing a minimum of rows (of full row rank), and a maximum of zero elements is desirable. Since the GLM procedure computes a g2 inverse of **X′ X**, such that **(X′ X)⁻X′ X** usually contains numerous zeros and such that the nonzero rows are linearly independent, GLM uses the nonzero rows of **(X′ X)⁻X′ X** to represent **L** symbolically.

If the generating set represented symbolically is of full row rank, the number of symbols (L1, L2, ...) represents the maximum rank of any testable hypothesis (in other words, the maximum number of linearly independent rows for any **L** matrix which can be constructed). By letting each symbol in turn take on the value of 1 while the others are set to 0, the original generating set can be reconstructed.

A One-Way Classification Model

For the model

$$Y = \mu + A_i + \varepsilon \ , \quad i = 1,2,3 \ ,$$

the general form of estimable functions **Lb** is (from the previous example):

$$\mathbf{L}\beta = L1^*\mu + L2^*A_1 + L3^*A_2 + (L1-L2-L3)^*A_3 \ .$$

Thus **L** = (L1, L2, L3, L1 – L2 – L3).

Tests involving only the parameters A1, A2, and A3 must have an **L** of the form

$$\mathbf{L} = (0, L2, L3, -L2-L3) \ .$$

Since the above **L** involves only two symbols, at most a two-degrees-of-freedom hypothesis may be constructed. For example, let L2=1 and L3=0; then let L2=0 and L3=1:

$$\mathbf{L} = \begin{bmatrix} 0 & 1 & 0 & -1 \\ 0 & 0 & 1 & -1 \end{bmatrix}$$

The above **L** may be used to test the hypothesis that A1=A2=A3. For this example, any **L** with two linearly independent rows with column 1 equal to zero will produce the same SS. For example, a pooled linear quadratic

$$\mathbf{L} = \begin{bmatrix} 0 & 1 & 0 & -1 \\ 0 & 1 & -2 & 1 \end{bmatrix}$$

gives the same SS. In fact, for any **L** of full row rank, and any non-singular matrix **K** of conformable dimensions:

$$SS(H0{:}\ \mathbf{L}\beta=0) = SS(H0{:}\ \mathbf{KL}\beta=0) \ .$$

A Three-Factor Main-Effects Model

Consider a three-factor main-effects model involving the CLASS variables A, B, and C in the following manner:

Obs	A	B	C
1	1	2	1
2	1	1	2
3	2	1	3
4	2	2	2
5	2	2	2

The general form of an estimable function is:

Parameter	*Coefficient*
μ	L1
A1	L2
A2	L1 − L2
B1	L4
B2	L1 − L4
C1	L6
C2	L1 + L2 − L4 − 2*L6
C3	-L2 + L4 + L6

Since only four symbols (L1, L2, L4, and L6) are involved, the maximum rank hypothesis possible will have four degrees-of-freedom. If an **L** matrix with four linearly independent rows is formed, with each row being generated using the above rules, then

$$SS(H0{:}\ \mathbf{L}\beta = 0) = R(\mu,\ A,\ B,\ C) \quad .$$

In a main-effects model, the usual hypothesis desired for a main effect is the equality of all the parameters. In this example, it is not possible to test such a hypothesis, because of confounding caused by inadequate design points. The best that can be done is to construct a maximum rank hypothesis (MRH) involving only the parameters of the main effect in question. This can be done using the general form of estimable functions. For example:

- To get an MRH involving only the parameters of A, the coefficients of **L** associated with μ, B1, B2, C1, C2, and C3 must be equated to zero. Starting at the top of the general form, let L1 = 0, then L4 = 0, then L6 = 0. If C2 and C3 are not to be involved, then L2 must also be zero. Thus A1–A2 is not estimable; that is, the MRH involving only the A parameters has zero rank and $R(B|\mu,B,C) = 0$.
- To obtain the MRH involving only the B parameters, let L1 = L2 = L6 = 0. But then to remove C2 and C3 from the comparison, L4 must also be set to 0. Thus B1–B2 is not estimable and $R(B|\mu,A,C,) = 0$.
- To obtain the MRH involving only the C parameters, let L1 = L2 = L4 = 0. Thus the MRH involving only C parameters is

  $$C1 - 2*C2 + C3 = K \quad \text{(for any K)}$$

  or any multiple of the left-hand side equal to K. Furthermore,

  $$SS(H0{:}\ C1 = 2*C2 - C3 = 0) = R(C|\mu,A,B) \ .$$

A Multiple Regression Model

Let

$$E(Y) = \beta 0 + \beta 1*X1 + \beta 2*X2 + \beta 3*X3 \quad .$$

If the **X′ X** matrix is of full rank, the general form of estimable functions is then:

Parameter	*Coefficient*
$\beta 0$	L1
$\beta 1$	L2
$\beta 2$	L3
$\beta 3$	L4

To test, for example, the hypothesis that $\beta 2=0$, let L1 = L2 = L4 = 0 and let L3 = 1. Then SS($\mathbf{L}\beta=0$) = R($\beta 2|\beta 0,\beta 1,\beta 3$). In the full-rank case, all parameters, as well as any linear combination of parameters, are estimable.

Suppose, however, that X3 = 2*X1 + 3*X2. The general form of estimable functions is then:

Parameter	*Coefficient*
$\beta 0$	L1
$\beta 1$	L2
$\beta 2$	L3
$\beta 3$	2*L2 + 3*L3

For this example it is possible to test H0: $\beta 0=0$. However, neither $\beta 1$, $\beta 2$, nor $\beta 3$ is estimable, that is,

$$R(\beta 1|\beta 0,\beta 2,\beta 3) = 0$$
$$R(\beta 2|\beta 0,\beta 1,\beta 3) = 0$$
$$R(\beta 3|\beta 0,\beta 1,\beta 2) = 0$$

Note on Symbolic Notation

The preceding examples demonstrate the ability to manipulate the symbolic representation of a generating set. It should be noted that any operations performed on the symbolic notation have corresponding row operations which are performed on the generating set itself.

ESTIMABLE FUNCTIONS

Type I SS and Estimable Functions

The Type I SS and the associated hypotheses they test are by-products of the modified sweep operator used to compute a g2 inverse of **X′ X** and a solution to the normal equations. For the model E(Y) = X1*B1 + X2*B2 + X3*B3 where B1, B2 and B3 are vectors, the Type I SS for each effect correspond to:

Effect	*Type I SS*
B1	R(B1)
B2	R(B2|B1)
B3	R(B3|B1,B2)

The Type I SS are model-order-dependent; each effect is adjusted only for the preceding effects in the model.

There are numerous ways to obtain a Type I hypothesis matrix **L** for each effect.

One way is to form the **X′X** matrix and then do a Forward-Doolittle on **X′X**, skipping over any rows with a zero diagonal. The nonzero rows of the resulting matrix associated with X1 provide an **L** such that

$$SS(H0\colon \mathbf{L}\beta = 0) = R(B1) \quad .$$

The nonzero rows of the Doolittle matrix associated with X2 provide an **L** such that SS(H0: $\mathbf{L}\beta = 0$) = R(B1|B2).The last set of nonzero rows (associated with X3) provides an **L** such that

$$SS(H0\colon \mathbf{L}\beta = 0) = R(B3|B1,B2).$$

Another more formalized representation of Type I generating sets for B1, B2, and B3 respectively is:

$$\mathbf{G1} = (\mathbf{X1'\,X1} | \mathbf{X1'\,X2} | \mathbf{X1'\,X3})$$

$$\mathbf{G2} = (0 | \mathbf{X2'\,M1\,X2} | \mathbf{X2'\,M1\,X3})$$

$$\mathbf{G3} = (0 | 0 | \mathbf{X3'\,M2\,X3})$$

where

$$\mathbf{M1} = \mathbf{I} - \mathbf{X1(X1'\,X1)^{-}X1'} \quad \text{and}$$

$$\mathbf{M2} = \mathbf{M1} - \mathbf{M1\,X2(X2'\,M1\,X2)^{-}X2'\,M1} \quad .$$

Using the Type I generating set **G2** (for example),if an **L** is formed from linear combinations of the rows of **G2** such that **L** is of full row rank and of the same row rank as **G2**, then SS(H0: $\mathbf{L}\beta = 0$) = R(B2|B1).

In the GLM procedure, the Type I estimable functions printed symbolically when the E1 option is requested are:

$$\mathbf{G1^*} = \mathbf{(X1'\,X1)^{-}G1}$$

$$\mathbf{G2^*} = \mathbf{(X2'\,M1\,X2)^{-}G2}$$

$$\mathbf{G3^*} = \mathbf{(X3'\,M2\,X3)^{-}G3} \quad .$$

As can be seen from the nature of the generating sets **G1**, **G2**, and **G3**, only the Type I estimable functions for B3 are guaranteed not to involve the B1 and B2 parameters. The Type I hypothesis for B2 may (and usually does for unbalanced data) involve B3 parameters. The Type I hypothesis for B1 usually involves B2 and B3 parameters.

There are, however, a number of models for which the Type I hypotheses are considered appropriate. These are:

1. Balanced ANOVA models specified in proper sequence (that is, interactions don't precede main effects in the MODEL statement, etc.)
2. Purely nested models (specified in the proper sequence)
3. Polynomial regression models (in proper sequence).

Type II SS and Estimable Functions

For main-effects models and regression models, the general form of estimable functions may be manipulated to provide tests of hypotheses involving only the parameters of the effect in question. The same result can also be obtained by entering each effect in turn as the last effect in the model and obtaining the Type I SS for that effect. Using a modified reversible sweep operator, it is possible to obtain the same results without actually re-running the model.

Thus the Type II SS correspond to the R notation in which each effect is adjusted for all other effects "possible." For a regression model such as

E(Y) = X1*B1 + X2*B2 + X3*B3 ,

the Type II SS correspond to:

Effect	*SS*
B1	R(B1|B2,B3)
B2	R(B2|B1,B3)
B3	R(B3|B1,B2)

For a main-effects model (A, B, and C as classification variables), the Type II SS correspond to:

Effect	*SS*
A	R(A|B,C)
B	R(B|A,C)
C	R(C|A,B)

From our earlier discussion, the Type II SS provide (for regression and main-effects models) an MRH for each effect that does not involve the parameters of the other effects.

For models involving interactions and nested effects, it is not possible to obtain a test of a hypothesis for a main effect free of parameters of higher-level effects with which the main effect is involved (unless a *priori* parametric restrictions are assumed).

It is reasonable to assume, then, that any test of a hypothesis concerning an effect should involve the parameters of that effect and only those other parameters with which that effect is involved.

Definition: Given an effect E1 and another effect E2, E1 is contained in E2 provided that:

1. both effects involve the same continuous variables, if any.
2. E2 has more CLASS variables than does E1 and if E1 has CLASS variables, they all appear in E2.

   Note: the effect μ is contained in all pure CLASS effects, but is not contained in any effect involving a continuous variable. No effect is contained by μ.

Type II, Type III, and Type IV estimable functions rely on this definition, and all have one thing in common: the estimable functions involving an effect E1 also involve the parameters of all effects which contain E1, and do not involve the parameters of effects which do not contain E1 (other than E1).

Definition: The Type II estimable functions for an effect E1 have an **L** (before reduction to full row rank) of the following form:

1. All columns of **L** associated with effects not containing E1 (except E1) should be zero.
2. The submatrix of **L** associated with effect E1 should be $(\mathbf{X1}'\,\mathbf{M}\,\mathbf{X1})^{-}(\mathbf{X1}'\,\mathbf{M}\,\mathbf{X1})$.

3. Each of the remaining submatrices of **L** associated with an effect E2 which contains E1 should be $(\mathbf{X1'MX1})^{-}(\mathbf{X1'MX2})$, where

   X0 = the columns of **X** whose associated effects do not contain E1
   X1 = the columns of **X** associated with E1
   X2 = the columns of **X** associated with an effect E2 which contains E1
   $\mathbf{M} = \mathbf{I} - \mathbf{X0}(\mathbf{X0'X0})^{-}\mathbf{X0'}$.

For the model Y=A B A*B, the Type II SS correspond to

$R(A|\mu,B)$, $R(B|\mu,A)$, $R(A*B|\mu,A,B)$.

For the model Y=A B(A) C(A B), the Type II SS correspond to

$R(A|,\mu)$, $R(B(A)|\mu,A)$, $R(C(A\ B)|\mu,A,B(A))$.

For the model Y=X X*X, the Type II SS correspond to:

$R(X|\mu,X*X)$ and $R(X*X|\mu,X)$.

Example of Type II Estimable Functions

For a 2×2 factorial with *w* observations per cell, the general form of estimable functions is:

Effect	*Coefficient*
μ	L1
A1	L2
A2	L1 – L2
B1	L4
B2	L1 – L4
AB11	L6
AB12	L2 – L6
AB21	L4 – L6
AB22	L1 – L2 – L4 + L6

The Type II estimable functions are:

	Coefficients for effect		
Effect	A	B	A*B
μ	0	0	0
A1	L2	0	0
A2	–L2	0	0
B1	0	L4	0
B2	0	–L4	0
AB11	.5*L2	.5*L4	L6
AB12	.5*L2	–.5*L4	–L6
AB21	–.5*L2	.5*L4	–L6
AB22	–.5*L2	–.5*L4	L6

Any nonzero values for L2, L4, and L6 can be used to construct **L** vectors for computing the Type II SS for A, B, and A*B respectively.

For an unbalanced 2×2 factorial (with 2 observations in every cell except the AB22 cell, which contains only 1 observation), the general form of estimable functions is the same as if it were balanced, since the same effects are still estimable. However, the Type II estimable functions for A and B are not the same as they were for the balanced design.

The Type II estimable functions for this unbalanced 2×2 are:

	Coefficients for effect		
Effect	A	B	A*B
μ	0	0	0
A1	L2	0	0
A2	−L2	0	0
B1	0	L4	0
B2	0	−L4	0
AB11	.6*L2	.6*L4	L6
AB12	.4*L2	−.6*L4	−L6
AB21	−.6*L2	.4*L4	−L6
AB22	−.4*L2	−.4*L4	L6

By comparing the hypothesis being tested in the balanced case with the hypothesis being tested in the unbalanced case for effects A and B, it can be noted that the Type II hypotheses for A and B are dependent on the cell frequencies in the design. For unbalanced designs in which the cell frequencies are not proportional to the background population, the Type II hypotheses for effects which are contained in other effects are of questionable merit.

However, if an effect is not contained in any other effect, the Type II hypothesis for that effect is an MRH which does not involve any parameters except those associated with the effect in question.

Thus Type II SS are appropriate for:

1. any balanced model
2. any main-effects model
3. any pure regression model
4. an effect not contained in any other effect (regardless of the model).

In addition to the above, the Type II SS is generally accepted by most statisticians for purely nested models.

Type III SS and Estimable Functions

It has been demonstrated that when an effect is contained in another effect, the Type II hypotheses for that effect are dependent on the cell frequencies. The philosophy behind both the Type III and Type IV hypotheses is that the tests of hypotheses made for any given effect should be the same for all designs with the same general form of estimable functions.

To demonstrate this concept, recall the hypothesis being tested by the Type II SS in the balanced 2×2 factorial shown earlier. Those hypotheses are precisely the ones which the Type III and Type IV employ for all 2×2 factorials that have at least one observation per cell. The Type II and Type IV hypotheses for a design without missing cells usually differ from the hypothesis employed for the same design with missing cells, since the general form of estimable functions usually differ.

Construction of Type III Hypotheses

Type III hypotheses are constructed by working directly with the general form of estimable functions. The following steps are used to construct a hypothesis for an effect E1.

1. For every effect in the model except E1 and those effects that contain E1, equate the coefficients in the general form of estimable functions to zero.

   Note: If E1 is not contained in any other effect, this step defines the Type III hypothesis (as well as the Type II and Type IV hypotheses). If E1 is contained in other effects, go on to step 2.
2. If necessary, equate new symbols to compound expressions in the E1 block in order to obtain the simplest form for the E1 coefficients.
3. Equate all symbolic coefficients outside of the E1 block to a linear function of the symbols in the E1 block in order to make the E1 hypothesis orthogonal to hypotheses associated with effects that contain E1.

By once again observing the Type II hypotheses being tested in the balanced 2×2 factorial, it is possible to verify that the A and A*B hypotheses are orthogonal and also that the B and A*B hypotheses are orthogonal. This principle of orthogonality between an effect and any effect that contains it holds for all balanced designs. Thus construction of Type III hypotheses for any design is a logical extension of a process that is used for balanced designs.

The Type III hypotheses are precisely the hypotheses being tested by programs that reparameterize using the "usual assumptions." When no missing cells exist in a factorial model, Type III SS coincide with Yates' weighted squares-of-means technique. When cells are missing in factorial models, the Type III SS coincide with those produced by Walter Harvey's fixed-effects linear models program. See the HARVEY procedure in the *SAS Supplemental Library User's Guide*.

The following steps illustrate the construction of Type III estimable functions for a 2×2 factorial with no missing cells.

To obtain the A*B interaction hypothesis, start with the general form and equate the coefficients for effects μ, A, and B to zero as shown in the next column.

Effect	*General form*	L1=L2=L4=0
μ	L1	0
A1	L2	0
A2	L1−L2	0
B1	L4	0
B2	L1−L4	0
AB11	L6	L6
AB12	L2−L6	−L6
AB21	L4−L6	−L6
AB22	L1−L2−L4+L6	L6

The last column above represents the form of the MRH for A*B.

To obtain the Type III hypothesis for A, first start with the general form and equate the coefficients for effects μ and B to zero (let L1=L4=0). Next let L6=K*L2, and find the value of K which makes the A hypothesis orthogonal to the A*B hypothesis. In this case, K=.5. Each of these steps is shown below:

Effect	*General form*	L1=L4=0	L6=K*L2	K=.5
μ	L1	0	0	0
A1	L2	L2	L2	L2
A2	L1 – L2	–L2	–L2	–L2
B1	L4	0	0	0
B2	L1 – L4	0	0	0
AB11	L6	L6	K*L2	.5*L2
AB12	L2 – L6	L2 – L6	(1 – K)*L2	.5*L2
AB21	L4 – L6	–L6	–K*L2	–.5*L2
AB22	L1 – L2 – L4 + L6	–L2 + L6	(K – 1)*L2	–.5*L2

In the above table, the fourth column (under L6=K*L2) represents the form of all estimable functions not involving μ, B1, or B2. The prime difference between the Type II and Type III hypotheses for A is the way K is determined. Type II chooses K as a function of the cell frequencies, whereas Type III chooses K such that the estimable functions for A are orthogonal to the estimable functions for A*B.

An example of Type III estimable functions in a 3×3 factorial with unequal cell frequencies and missing diagonals is given below (N1–N6 represent the nonzero cell frequencies):

		B 1	B 2	B 3
A	1		N1	N2
	2	N3		N4
	3	N5	N6	

For any nonzero values of N1–N6, the Type III estimable functions for each effect are:

Effect	A	B	A*B
μ	0	0	0
A1	L2	0	0
A2	L3	0	0
A3	–L2 – L3	0	0
B1	0	L5	0
B2	0	L6	0
B3	0	–L5 – L6	0
AB12	.667*L2 + .333*L3	.333*L5 + .667*L6	L8
AB13	.333*L2 – .333*L3	–.333*L5 – .667*L6	–L8
AB21	.333*L2 + .667*L3	.667*L5 + .333*L6	–L8
AB23	–.333*L2 + .333*L3	–.667*L5 – .333*L6	L8
AB31	–.333*L2 – .667*L3	.333*L5 – .333*L6	L8
AB32	–.667*L2 – .333*L3	–.333*L5 + .333*L6	–L8

Type IV Estimable Functions

By once again looking at the Type II hypotheses being tested in the balanced 2×2 factorial, yet another characteristic of the hypotheses employed for balanced designs may be seen: the coefficients of lower-order effects are averaged across each higher-level effect involving the same subscripts. For example, in the A hypothesis, the coefficients of AB11 and AB12 are equal to one-half the coefficient of A1; and the coefficients of AB21 and AB22 are equal to one-half the coefficient of A2. With this in mind, then the basic concept used to construct Type IV hypotheses is that the coefficients of any effect, say E1, are distributed equitably across higher-level effects which "contain" E1. When missing cells occur, this same general philosophy is adhered to, but care must be taken in the way the distributive concept is applied.

Construction of Type IV hypotheses begins as does the construction of the Type III hypotheses. That is, for an effect E1, equate to zero all coefficients in the general form that do not belong to E1 or to any other effect containing E1. If E1 is not contained in any other effect, then the Type IV hypothesis (and Type II and Type III) has been found. If E1 is contained in other effects, then simplify, if necessary, the coefficients associated with E1, such that they are all "free" coefficients or functions of other free coefficients in the E1 block.

To illustrate the method of resolving the free coefficients outside of the E1 block, suppose that we are interested in the estimable functions for an effect A and that A is contained in AB, AC, AD, and ABC.

With missing cells, the coefficients of intermediate effects (here they are AB and AC) do not always have an equal "distribution" of the lower-order coefficients, so the coefficients of the highest-order effects are determined first (here they are ABC and AD). Once the highest-order coefficients are determined, the coefficients of intermediate effects are automatically determined.

The following process is performed for each free coefficient of A in turn. The resulting symbolic vectors are then added together to give the Type IV estimable functions for A.

1. Select a free coefficient of A, and set all other free coefficients of A to zero.
2. If any of the levels of A have zero as a coefficient, equate all of the coefficients of higher-level effects involving that level of A to zero. This step alone usually resolves most of the free coefficients remaining.
3. Check to see if any higher-level coefficients are now zero, when the coefficient of the associated level of A is not zero. If this situation occurs, the Type IV estimable functions for A are not unique.
4. For each level of A in turn, if the A coefficient for that level is nonzero, count the number of times that level occurs in the higher-level effect. Then equate each of the higher-level coefficients to the coefficient of that level of A divided by the count.

An example of a 3×3 factorial with four missing cells (N1–N5 represent positive cell frequencies):

		B 1	2	3
	1	N1	N2	
A	2	N3	N4	
	3			N6

The Type IV estimable functions are shown below:

Effect	A	B	A*B
μ	0	0	0
A1	−L3	0	0
A2	L3	0	0
A3	0	0	0
B1	0	L5	0
B2	0	L6	0
B3	0	−L5 − L6	0
AB11	−.5*L3	.5*L5	L8
AB12	−.5*L3	−.5*L5	L8
AB21	.5*L3	.5*L5	L8
AB22	.5*L3	−.5*L5	L8
AB33	0	0	0

A Comparison of Type III and Type IV Hypotheses

For the vast majority of designs, Type III and Type IV hypotheses for a given effect are the same. Specifically, they are the same for any effect E1 that is not contained in other effects for any design (with or without missing cells), the Type III and Type IV hypotheses coincide for all effects. When there are missing cells, the hypotheses may differ. By using the GLM procedure, it is possible to study the differences in the hypotheses. Each user must decide on the appropriateness of the hypotheses for a particular model.

The Type III hypotheses for three-factor and higher completely nested designs with unequal Ns in the bottommost level differ from the Type II hypotheses; however, the Type IV hypotheses do correspond to the Type II hypotheses in this case.

When missing cells occur in a design, the Type IV hypotheses may not be unique. If this occurs in GLM, you are notified, and it may be advisable for you to consider defining your own specific comparisons.

REFERENCES

Freund, Rudolf J. and Littell, Ramon C. (1981), *SAS for Linear Models.* Cary, N.C.: SAS Institute Inc.

Goodnight, J.H. (1978), "Tests of Hypotheses in Fixed Effects Linear Models," *SAS Technical Report R-101*, Cary, N.C.: SAS Institute Inc.

CATEGORICAL DATA ANALYSIS

Introduction

FUNCAT

PROBIT

Introduction to SAS Categorical Procedures

A *categorical variable* is a variable with a small number of discrete levels where the levels are treated as names rather than as representations of some ordered scale. Categorical data are also called *nominal*, *qualitative*, *discrete*, *classification*, or *count* data.

Three procedures in SAS that model categorical responses are:

FREQ is a descriptive frequency-counting procedure that also produces chi-square (χ^2) tests and measures of association for two-way tables with measures of association. PROC FREQ is documented in *SAS User's Guide: Basics, 1982 Edition* with SAS descriptive procedures.

FUNCAT models categorical responses with a linear model like analysis-of-variance procedures.

PROBIT models binary responses by the inverse cumulative normal distribution function (probit function) applied to a linear function of a predictor (dose) variable. This procedure is specialized for dose-response problems in bioassays.

The Multinomial Distribution

Suppose you stand on a corner and ask people that come by what their favorite color is: red, blue, or green? After you have interviewed 1000 people, you tabulate the results:

	Favorite Color			
	Red	Blue	Green	Total
Count	520	310	170	1000
Prob	.52	.31	.17	1.00

In the population you are sampling, there is an unknown proportion that prefers each of the three choices. The random variability in this case is attributed to random sampling of individuals with definite preferences. The variability could also be modeled as a random choice made by each individual in a homogeneous population. The underlying probabilities can be estimated by dividing each count by the total count:

$$p_j = n_j / n$$

where n_j is the count on the jth response and n is the total count.

The distribution of the counts is called the *multinomial distribution*. The probability that the first cell has $k1$ counts, the second cell has $k2$, and the third has $k3=n\text{-}k1\text{-}k2$ is

$$P(k1,k2,k3) = n!/(k1!k2!k3!)\pi_1^{k1}\pi_2^{k2}\pi_3^{k3}$$

where π_i is the true probability of choosing the ith response level. This can be generalized to any number of response levels. The special case of two response levels is called the *binomial distribution*.

Multiple Populations

Suppose that you collect data as above but keep separate tallies according to the sex of the respondent. You are now sampling two different populations that may have different response probabilities. Suppose your data are:

Favorite Color

	Sex	Red	Blue	Green	Total
Count	Male	200	50	150	400
	Female	200	100	300	600
Prob	Male	.500	.125	.375	1.00
	Female	.333	.167	.500	1.00

With two populations, it is natural to test the hypothesis that the true response probabilities are the same across populations, that is, males and females have the same response probabilities for the three colors.

This hypothesis (called *marginal homogeneity*) can be tested in PROC FREQ using the CHISQ option. The statements to read the data and invoke PROC FREQ are:

```
DATA A;
  INPUT SEX $ COLOR $ COUNT @@;
  CARDS;
MALE RED 200 MALE BLUE 50 MALE GREEN 150
FEMALE RED 200 FEMALE BLUE 100 FEMALE GREEN 300
;
PROC FREQ;
  WEIGHT COUNT;
  TABLES SEX*COLOR / CHISQ NOCOL NOPERCENT;
```

The printed output is:

```
                                                                    1
                   TABLE OF SEX BY COLOR

SEX                COLOR

FREQUENCY|
 ROW PCT |BLUE     |GREEN    |RED      |  TOTAL
---------+---------+---------+---------+
FEMALE   |     100 |     300 |     200 |    600
         |   16.67 |   50.00 |   33.33 |
---------+---------+---------+---------+
MALE     |      50 |     150 |     200 |    400
         |   12.50 |   37.50 |   50.00 |
---------+---------+---------+---------+
TOTAL          150       450       400     1000
```

(continued on next page)

(continued from previous page)

```
                     STATISTICS FOR 2-WAY TABLES

CHI-SQUARE                      27.778    DF=    2   PROB=0.0001
PHI                              0.167
CONTINGENCY COEFFICIENT          0.164
CRAMER'S V                       0.167
LIKELIHOOD RATIO CHISQUARE      27.689    DF=    2   PROB=0.0001
```

The FREQ procedure produces a χ^2 value of 27.778, which indicates a highly significant difference in response probabilities across the two populations.

The FUNCAT procedure can test the same hypothesis in a different way with these statements:

```
PROC FUNCAT;
  WEIGHT COUNT;
  MODEL COLOR=SEX;
```

The printed output is:

```
                          FUNCAT PROCEDURE                                  1

RESPONSE: COLOR                               RESPONSE LEVELS (R)=     3
WEIGHT VARIABLE: COUNT                        POPULATIONS     (S)=     2
DATA SET: A                                   TOTAL COUNT     (N)=  1000
                                              OBSERVATIONS  (OBS)=     6
```

```
                                                                            2
             SOURCE                    DF   CHI-SQUARE       PROB

             INTERCEPT                  2       131.34     0.0001
             SEX                        2        27.45     0.0001

             RESIDUAL                   0         0.00     1.0000

EFFECT                 PARAMETER DF      ESTIMATE     CHI-SQ       PROB         STD

INTERCEPT                      1  1    -1.03972       108.10     0.0001         0.1
                               2  1    0.0588915        0.69     0.4049   0.0707107
SEX                            3  1    0.346574        12.01     0.0005         0.1
                               4  1    0.346574        24.02     0.0001   0.0707107
```

Like analysis-of-variance and regression procedures, FUNCAT uses a MODEL statement to specify the response variable and the design effects. FUNCAT reports a χ^2 value of 27.45, which is very close to the value reported by PROC FREQ.

Independence

In the previous section, the chi-square test is used to test the hypothesis that the true response probabilities are the same for two populations. You can also use the chi-square test for a two-dimensional contingency table to test for the independence of two different responses.

Suppose that you ask 1000 people their favorite color and their favorite dessert. The counts are summarized below:

		Favorite Color			
	Counts	Red	Blue	Green	Total
Favorite Dessert	Ice cream	205	400	40	645
	Pie	100	50	25	175
	Cake	50	105	25	180
	Total	355	555	90	1000

		Favorite Color			
	Prob	Red	Blue	Green	Total
Favorite	Ice cream	.205	.400	.040	.645
Dessert	Pie	.100	.050	.025	.175
	Cake	.050	.105	.025	.180
	Total	.355	.555	.090	1.000

The hypothesis of independence is that the true cell probabilities are the product of the true marginals. In this example there are four degrees of freedom to test relations of the form:

P(ice,red) = P(ice)*P(red)

where

P(ice) = P(ice,red) + P(ice,blue) + P(ice,green), and
P(red) = P(ice,red) + P(pie,red) + P(cake,red).

This is a one-population problem. Both variables are responses. To test the hypothesis of independence, set up a χ^2 test in PROC FREQ similar to the previous example.

```
LENGTH DESSERT $ 4 COLOR $ 5;
DATA B;
  DO DESSERT='ICE ','PIE','CAKE';
    DO COLOR='RED ','BLUE','GREEN';
        INPUT COUNT @;
        OUTPUT;
        END;
    END;
  CARDS;
205  400   40
100   50   25
 50  105   25
;
PROC PRINT;
PROC FREQ;
  WEIGHT COUNT;
  TABLES DESSERT*COLOR / CHISQ;
```

The printed output:

```
                                                         1

          OBS    DESSERT    COLOR    COUNT

           1     ICE        RED       205
           2     ICE        BLUE      400
           3     ICE        GREEN      40
           4     PIE        RED       100
           5     PIE        BLUE       50
           6     PIE        GREEN      25
           7     CAKE       RED        50
           8     CAKE       BLUE      105
           9     CAKE       GREEN      25
```

```
                     TABLE OF DESSERT BY COLOR                                   2

           DESSERT        COLOR

           FREQUENCY|
            PERCENT |
           ROW PCT  |
           COL PCT  |BLUE     |GREEN    |RED      |  TOTAL
           ---------+---------+---------+---------+
           CAKE     |    105  |     25  |     50  |    180
                    |  10.50  |   2.50  |   5.00  |  18.00
                    |  58.33  |  13.89  |  27.78  |
                    |  18.92  |  27.78  |  14.08  |
           ---------+---------+---------+---------+
           ICE      |    400  |     40  |    205  |    645
                    |  40.00  |   4.00  |  20.50  |  64.50
                    |  62.02  |   6.20  |  31.78  |
                    |  72.07  |  44.44  |  57.75  |
           ---------+---------+---------+---------+
           PIE      |     50  |     25  |    100  |    175
                    |   5.00  |   2.50  |  10.00  |  17.50
                    |  28.57  |  14.29  |  57.14  |
                    |   9.01  |  27.78  |  28.17  |
           ---------+---------+---------+---------+
           TOTAL         555        90       355      1000
                       55.50      9.00     35.50    100.00

                       STATISTICS FOR 2-WAY TABLES

  CHI-SQUARE                          72.509    DF=    4   PROB=0.0001
  PHI                                  0.269
  CONTINGENCY COEFFICIENT              0.260
  CRAMER'S V                           0.190
  LIKELIHOOD RATIO CHISQUARE          73.188    DF=    4   PROB=0.0001
```

PROC FREQ reports a χ^2 value of 72.51, which indicates a highly significant departure from independence.

The chi-square test for the independence of two responses is mathematically identical to the chi-square test of the marginal homogeneity of a response with a factor. However, the distinction between response and factor variables is not always apparent in more complex problems. For example, Mantel (1979) discusses an example in the literature where tests were constructed without regard to the character of the variables. Some of the tests were thus a by-product of the sample design and not relevant to the response. PROC FUNCAT requires you to distinguish responses from factors.

FUNCAT specializes in homogeneity tests for single-response multiple-population problems. The procedure can test forms of independence for multiple responses but you must construct special response functions. For the example above, you need a bulky specification of this form:

```
PROC FUNCAT;
  WEIGHT COUNT;
  MODEL DESSERT*COLOR=;
  RESPONSE
            1 0 0 0  -1 0 -1 0 /                  /* RED ICE INDEP */
            0 1 0 0   0 -1 -1 0 /                 /* BLUE ICE INDEP */
            0 0 1 0  -1 0 0 -1 /                  /* RED PIE INDEP */
            0 0 0 1   0 -1 0 -1                   /* BLUE PIE INDEP */
              L O G   1 0 0   0 0 0   0 0 0/      /* RED ICE CELL */
                      0 1 0   0 0 0   0 0 0/      /* BLUE ICE CELL */
                      0 0 0   1 0 0   0 0 0/      /* RED PIE CELL */
                      0 0 0   0 1 0   0 0 0/      /* BLUE PIE CELL */
                      1 0 0   1 0 0   1 0 0/      /* RED MARGINAL */
                      0 1 0   0 1 0   0 1 0/      /* BLUE MARGINAL */
                      1 1 1   0 0 0   0 0 0/      /* ICE MARGINAL */
                      0 0 0   1 1 1   0 0 0;      /* ICE MARGINAL */
```

The printed output is:

```
                              FUNCAT PROCEDURE                                  1

       RESPONSE: DESSERT*COLOR                   RESPONSE LEVELS (R)=      9
       WEIGHT VARIABLE: COUNT                    POPULATIONS     (S)=      1
       DATA SET: B                               TOTAL COUNT     (N)=   1000
                                                 OBSERVATIONS  (OBS)=      9


         SOURCE                        DF    CHI-SQUARE       PROB

         INTERCEPT                      4         69.24     0.0001

         RESIDUAL                       0          0.00     1.0000


 EFFECT             PARAMETER DF     ESTIMATE      CHI-SQ       PROB        STD

 INTERCEPT                  1  1   0.0497907        0.76     0.3838  0.0571713
                            2  1    0.433865        7.41     0.0065   0.159436
                            3  1    0.111001       32.20     0.0001  0.0195603
                            4  1   -0.372425       10.91     0.0010   0.112768
```

FUNCAT reports a chi-square value of 69.24, which is close to the FREQ chi-square value, although the hypothesis is being tested in a different way.

Since tests of independence for multiple responses are unwieldy with FUNCAT, in the rest of this chapter we show only examples of multiple-population problems.

Multiple Classifications of Populations

Suppose the respondents give their age group, their sex, and their favorite color. You then can tabulate the counts in six populations corresponding to three age groups by two sex levels.

			Favorite Color			
	Age	Sex	Red	Blue	Green	Total
Populations	Young	Female	50	40	100	190
	Young	Male	55	10	30	95
	Middle	Female	100	30	100	230
	Middle	Male	95	20	100	215
	Old	Female	50	30	100	180
	Old	Male	50	20	20	90

You now have six populations and three response levels. You still ask the same initial question: are the response probabilities the same in each population? However, if response probabilites are not the same, there are several alternatives. The differences may be due to:

- SEX alone
- AGE alone
- SEX and AGE (additively)
- SEX and AGE with interaction.

Using PROC FUNCAT, the MODEL statements for the models above are:

- MODEL COLOR = SEX
- MODEL COLOR = AGE
- MODEL COLOR = SEX AGE
- MODEL COLOR = SEX AGE SEX*AGE

You can use these statements to analyze the last model shown above:

```
DATA A;
  INPUT AGE $ SEX $ RED BLUE GREEN;
  CARDS;
YOUNG    FEMALE    50  40  100
YOUNG    MALE      55  10   30
MIDDLE   FEMALE   100  30  100
MIDDLE   MALE      95  20  100
OLD      FEMALE    50  30  100
OLD      MALE      50  20   20
;
DATA B;
  SET A;
  COLOR='RED  ';  COUNT=RED;     OUTPUT;
  COLOR='BLUE ';  COUNT=BLUE;    OUTPUT;
  COLOR='GREEN';  COUNT=GREEN;   OUTPUT;
  KEEP COLOR AGE SEX COUNT;
PROC FUNCAT DATA=B;
  WEIGHT COUNT;
  MODEL COLOR = SEX AGE SEX*AGE;
```

The printed output is:

```
                              FUNCAT PROCEDURE                                    1

   RESPONSE: COLOR                              RESPONSE LEVELS (R)=     3
   WEIGHT VARIABLE: COUNT                       POPULATIONS     (S)=     6
   DATA SET: B                                  TOTAL COUNT     (N)=  1000
                                                OBSERVATIONS  (OBS)=    18

        SOURCE                      DF   CHI-SQUARE      PROB

        INTERCEPT                    2       106.70    0.0001
        SEX                          2        39.49    0.0001
        AGE                          4        11.22    0.0242
        SEX*AGE                      4        32.05    0.0001

        RESIDUAL                     0         0.00    1.0000

EFFECT             PARAMETER DF    ESTIMATE     CHI-SQ      PROB        STD

INTERCEPT                  1  1    -1.01952      95.85    0.0001   0.104133
                           2  1  -0.0141398       0.03    0.8571  0.0785355
SEX                        3  1     0.37354      12.87    0.0003   0.104133
                           4  1    0.476238      36.77    0.0001  0.0785355
AGE                        5  1   -0.361538       6.70    0.0096   0.139638
                           6  1   0.0397865       0.17    0.6838  0.0976986
                           7  1    0.305963       4.43    0.0353   0.145332
                           8  1   -0.097432       0.65    0.4185   0.120421
SEX*AGE                    9  1   -0.196454       1.98    0.1595   0.139638
                          10  1   -0.501885      26.39    0.0001  0.0976986
                          11  1   -0.170808       1.38    0.2399   0.145332
                          12  1    0.328481       7.44    0.0064   0.120421
```

The original data set A is transformed into a new data set B so that counts from several variables into one variable spread across several observations. The results show that all the effects are significant.

FUNCAT Response Model

FUNCAT models the response by calculating a function of the response probabilities to fit to a linear model. The standard response function is called a *logit function*, which models the logs of ratios of multinomial probabilities. However, you can specify your own response functions if you do not want this multivariate logit.

The design effects lead to the creation of a design matrix to estimate parameters that describe the differences in the response across populations.

If the response has only two levels, then the logit function yields a single value, and a least-squares procedure like GLM can be used to calculate a test. To demonstrate, combine the blue and green responses into one level, compute the logit and appropriate weight in a data set, and use GLM to compute sums of squares:

```
DATA C;
  SET A;
  TOTAL=RED+BLUE+GREEN;
  P1=(BLUE+GREEN)/TOTAL;
  P2=RED/TOTAL;
  LOGIT=LOG(P1/P2);
  WEIGHT=TOTAL*P1*P2;
PROC PRINT DATA=C;
PROC GLM DATA=C;
  CLASS AGE SEX;
  MODEL LOGIT=AGE SEX;
  WEIGHT W;
```

The printout of data set C is:

```
                                                                                              1
OBS   AGE      SEX      RED   BLUE   GREEN   TOTAL      P1          P2          LOGIT      WEIGHT

 1    YOUNG    FEMALE    50    40     100     190    0.736842    0.263158     1.02962    36.8421
 2    YOUNG    MALE      55    10      30      95    0.421053    0.578947    -0.31845    23.1579
 3    MIDDLE   FEMALE   100    30     100     230    0.565217    0.434783     0.26236    56.5217
 4    MIDDLE   MALE      95    20     100     215    0.558140    0.441860     0.23361    53.0233
 5    OLD      FEMALE    50    30     100     180    0.722222    0.277778     0.95551    36.1111
 6    OLD      MALE      50    20      20      90    0.444444    0.555556    -0.22314    22.2222
```

The printout from GLM is:

```
                              GENERAL LINEAR MODELS PROCEDURE                                       2

                                  CLASS LEVEL INFORMATION

                             CLASS      LEVELS     VALUES

                             AGE             3     MIDDLE OLD YOUNG

                             SEX             2     FEMALE MALE

                          NUMBER OF OBSERVATIONS IN DATA SET = 6

                              GENERAL LINEAR MODELS PROCEDURE

DEPENDENT VARIABLE: LOGIT
WEIGHT:             WEIGHT

SOURCE               DF    SUM OF SQUARES      MEAN SQUARE     F VALUE      PR > F     R-SQUARE          C.V.

MODEL                 3       27.47315976       9.15771992        0.86      0.5777     0.562912     1010.3278

ERROR                 2       21.33228425      10.66614213                  ROOT MSE                 LOGIT MEAN

CORRECTED TOTAL       5       48.80544401                                 3.26590602                 0.32325212

SOURCE               DF       TYPE I SS    F VALUE     PR > F       DF       TYPE III SS    F VALUE     PR > F

AGE                   2      3.83017709       0.18     0.8478        2        2.11476684       0.10     0.9098
SEX                   1     23.64298267       2.22     0.2750        1       23.64298267       2.22     0.2750
```

The Type III sums of squares for AGE and SEX are now distributed as chi-square (χ^2) and can be referred to a χ^2 table to obtain a significance probability. In the next example they are identical to the χ^2 values reported by FUNCAT.

To run the same model with FUNCAT, make a new data set D that combines BLUE and GREEN.

```
DATA D;
  SET B;
  IF COLOR='RED' THEN COLORG=COLOR;
    ELSE COLORG='BG';
PROC PRINT DATA=D;
PROC FUNCAT DATA=D;
  WEIGHT COUNT;
  MODEL COLORG=AGE SEX / FREQ PROB X;
```

The printout of data set D is:

```
                                                          1
  OBS    AGE        SEX       COLOR    COUNT    COLORG

    1    YOUNG      FEMALE    RED        50     RED
    2    YOUNG      FEMALE    BLU        40     BG
    3    YOUNG      FEMALE    GRE       100     BG
    4    YOUNG      MALE      RED        55     RED
    5    YOUNG      MALE      BLU        10     BG
    6    YOUNG      MALE      GRE        30     BG
    7    MIDDLE     FEMALE    RED       100     RED
    8    MIDDLE     FEMALE    BLU        30     BG
    9    MIDDLE     FEMALE    GRE       100     BG
   10    MIDDLE     MALE      RED        95     RED
   11    MIDDLE     MALE      BLU        20     BG
   12    MIDDLE     MALE      GRE       100     BG
   13    OLD        FEMALE    RED        50     RED
   14    OLD        FEMALE    BLU        30     BG
   15    OLD        FEMALE    GRE       100     BG
   16    OLD        MALE      RED        50     RED
   17    OLD        MALE      BLU        20     BG
   18    OLD        MALE      GRE        20     BG
```

The printed output is:

```
                                                                  2
                           FUNCAT PROCEDURE

RESPONSE: COLORG                          RESPONSE LEVELS (R)=     2
WEIGHT VARIABLE: COUNT                    POPULATIONS     (S)=     6
DATA SET: D                               TOTAL COUNT     (N)=  1000
                                          OBSERVATIONS  (OBS)=    18

                                        RESPONSE
               DESIGN                   FREQUENCIES          TOTAL

         SAMPLE AGE        SEX             1       2

           1    MIDDLE     FEMALE        130     100         230.0
           2    MIDDLE     MALE          120      95         215.0
           3    OLD        FEMALE        130      50         180.0
           4    OLD        MALE           40      50          90.0
           5    YOUNG      FEMALE        140      50         190.0
           6    YOUNG      MALE           40      55          95.0

                                        RESPONSE
               DESIGN                   PROBABILITIES        TOTAL

         SAMPLE AGE        SEX             1       2

           1    MIDDLE     FEMALE       0.5652 0.4348        230.0
           2    MIDDLE     MALE         0.5581 0.4419        215.0
           3    OLD        FEMALE       0.7222 0.2778        180.0
           4    OLD        MALE         0.4444 0.5556         90.0
           5    YOUNG      FEMALE       0.7368 0.2632        190.0
           6    YOUNG      MALE         0.4211 0.5789         95.0
```

(continued on next page)

(continued from previous page)

```
                 RESPONSE     DESIGN MATRIX
        SAMPLE   FUNCTION       1    2    3    4
          1      0.262364       1    1    0    1
          2      0.233615       1    1    0   -1
          3      0.955511       1    0    1    1
          4     -0.223144       1    0    1   -1
          5       1.02962       1   -1   -1    1
          6     -0.318454       1   -1   -1   -1

    SOURCE                      DF    CHI-SQUARE        PROB

    INTERCEPT                    1         27.54      0.0001
    AGE                          2          2.11      0.3474
    SEX                          1         23.64      0.0001

    RESIDUAL                     2         21.33      0.0001

EFFECT           PARAMETER DF     ESTIMATE     CHI-SQ        PROB          STD

INTERCEPT                1  1     0.367155      27.54      0.0001    0.0699676
AGE                      2  1    -0.129144       2.11      0.1461    0.0888541
                         3  1    0.0615285       0.36      0.5484     0.102512
SEX                      4  1      0.32683      23.64      0.0001    0.0672158
```

Sampling Issues

It is easy to distinguish between design factors and responses in continuous response models. However, in categorical models it is more difficult, because a response with respect to a sample scheme may not be a response with respect to the individual responding units.

Prepare a questionnaire with three items for the three variables smoking, age, and sex. The goal is to answer the question: is the likelihood that a person smokes tobacco a function of his or her age and sex?

The null hypothesis is that the probability of smoking is the same in each age*sex group.

Consider five sampling schemes:

(a) Stand on a corner for two hours and ask people their age, sex, and whether they smoke. Each person falls into one cell of the three-way response. The total count (and all subclass counts) is random.

(b) Stand on a corner and ask each person the three questions until you have 1000 people. Each person falls in one cell of the three-way response. The total number is fixed, but all subclass counts are random.

(c) Find 1000 smokers and 1000 non-smokers and ask their age and sex. The smoking counts are fixed. The age*sex counts are random.

(d) Find 100 people in each age*sex group and ask if they smoke. The age*sex counts are fixed. The smoking counts are random.

(e) Find 200 people in each smoking-by-sex group and ask their age. The age counts are random. Smoking and sex are fixed.

The question is what constitutes a response. Is it the way you collect the sample? Or is it the way the responding units behave? Both may be important, although we are usually more directly concerned with the behavior of the individual units. Run the model with PROC FUNCAT:

```
MODEL SMOKING = SEX AGE;
```

This model can be applied to data collected in cases (a), (b), and (d). It can be valid under (c) if certain assumptions are met (Manski and McFadden, 1981). However, the model cannot be applied to case (e), since the counts of interest are fixed by the sampling design.

The responses of interest are not the responses to the age, sex, and smoking questions: you are only interested in whether the individual smokes. The person cannot control his age or sex, but does choose whether or not to smoke.

Another issue is random sampling. The experiments assume that units are chosen randomly from a background population about which you want to generalize. The background population is also assumed to be infinite, so sampling with or without replacement is not relevant. Thus you are dealing with multinomial rather than hypergeometric distributions. (See Manski and McFadden (1981) for more information.)

Multiple Dependencies

Suppose you include another variable: whether or not the person has lung cancer. Smoking can then become a factor variable for cancer as well as a response variable for age and sex. The complete table of the four variables leads to two models and a path question:

- Is smoking a function of age and sex?
- Is cancer a function of smoking and age and sex?
- Is cancer a function of age and sex only through the path that makes smoking a function of age and sex?

Smoking becomes a response, a pure design factor, or a partial factor depending on the model.

REFERENCES

Bishop, Y., Fienberg, S.E., and Holland, P.W. (1975), *Discrete Multivariate Analysis: Theory and Practice*, Cambridge: The MIT Press, Chapter 3.

Manski, Charles F. and McFadden, Daniel (Eds.), (1981), "Alternative Estimators and Sample Designs for Discrete Choice Analysis," in *Structural Analysis of Discrete Data with Econometric Applications*, Cambridge: The MIT Press.

Mantel, Nathan (1979), "Letter to the Editor: Multidimensional Contingency Table Analysis," *The American Statistician*, 33, 93.

The FUNCAT Procedure

ABSTRACT

FUNCAT models **FUN**ctions of **CAT**egorical responses as a linear model including linear and log-linear categorical models and logistic regression. FUNCAT uses weighted least squares to produce minimum chi-square estimates according to the methods proposed by Grizzle, Starmer, and Koch (GSK) (1969). The procedure can also produce maximum-likelihood estimates for a standard response function.

INTRODUCTION

FUNCAT is specified like an analysis-of-variance procedure except that the response is categorical rather than continuous.

The model can be described by the response and the design effects. The response is categorical since what is measured are the frequency counts for each level. These counts are assumed to follow a multinomial distribution. The design effects group the experimental units into populations or samples. The observations for a population have the same values for all the design variables. Each population has a different multinomial distribution for the response counts as shown below:

sample	*response1*	. . .	*responser*	*sample size*
1	n_{11}	. . .	n_{1r}	n_1
2	n_{21}	. . .	n_{2r}	n_2
⋮				
S	n_{s1}	. . .	n_{sr}	n_s

For each sample i, the probability of the jth response (π_{ij}) is estimated by $p_{ij}= n_{ij}/n_i$. These estimates are used to construct values for some function defined on the response probabilities. This function of the true probabilities, f( ), is assumed to follow a linear model in terms of the design structure of the samples. FUNCAT applies the same function to each sample population. Thus, the model can be written:

$$f(\pi_i) = \mathbf{X}_i\beta + \varepsilon_i \quad \text{for } i=1\ldots s$$

where

$$\pi_i = (\pi_{i1}, \pi_{i2}, \ldots, \pi_{ir})$$

The function f can be any composition of log, exponential, and linear transformations. The design matrix **X** and the form of the parameters are determined by design effects, which can be main effects, crossed effects, nested effects, and special nested effects that specify values.

FUNCAT provides two estimation methods:

1. The weighted least-squares method minimizes the weighted sum of squares of the error in the model. The weights are constructed for each population and are the asymptotic inverse variance estimates of the function with respect to the raw probability estimates. This approach was developed by Grizzle, Starmer, and Koch (GSK)(1969) and others. Weighted least squares is available for any response function.
2. The maximum-likelihood method adjusts the parameters of the linear model to maximize the value of the joint multinomial likelihood function of the responses. Maximum likelihood is only available for the standard default log-linear response function.

Acknowledgments

FUNCAT is similar in capabilities to the GENCAT program by Landis, Stanish, Freeman, and Koch (1976), but is controlled by specifications similar to ANOVA and GLM in SAS. The author of FUNCAT benefited from comments by G. Koch of the University of North Carolina, D. Freeman of Yale University School of Medicine, C. Proctor of N.C. State University, J. Dunn from University of Arkansas, and others.

Cautions

Since the method depends on asymptotic approximations, you should use it only when you have a lot of data and you expect each cell to accumulate counts.

For sparse problems, avoid the log response function. If you use weighted least squares and use a LOG response function, the sample should not contain any cells with zero count. If a logarithm of a zero proportion is taken, FUNCAT prints a warning and then proceeds to take the log of a small value ($.5/n_i$ for the probability) in order to continue. FUNCAT may produce invalid results if the cells contain very few observations. The ADDCELL option on the MODEL statement is available for adding small amounts to each cell, but this may seriously bias the variance estimates for problems where the number of cells is large but the total frequency is small.

It is not clear how sparse your data can be before the statistics become invalid. A recent simulation study (Stedl, 1981) of a design with 2 response levels and 2 by 10 design levels (20 populations) obtained rejection rates for small samples when the null hypothesis was true. For the trials done by Stedl and extended at SAS, the tests based on both GSK and ML estimators did not come close to the desired rates until the average number of observations per cell was 15 or more. The ML tests did better, but tended to reject too often; the GSK tests did not reject often enough.

If the function on the responses does not have an inverse, it may not be possible to compute the fitted probability estimates resulting from the parameter estimates. If the function is invertible, the resulting probability estimates may not be in the range 0 to 1.

GSK Methods Contrasted with Contingency Table Methods

The FUNCAT procedure (using the GSK method) is most often used in experimental situations where there are clearly defined response and design effects. The emphasis is on estimation and hypothesis testing of parameters. With the GSK method, it is easy to make tests of marginal homogeneity, but difficult to test independence. The GSK approach is very natural to ANOVA users.

In contrast, the contingency table literature (for example, Bishop et al., 1975) does not make a strong distinction between response effects and design effects, although model specifications that account for the distinction can be made. The emphasis is on model building and goodness-of-fit tests with respect to a hierarchy of dependencies that are undertaken in terms of a log-linear framework. The underlying parameters are usually ignored, and the test statistics are computed by adjusting the tables to fit various marginals. With these methods, it is easy to test for independence, but may be awkward to specify tests for general hypotheses concerning design parameters.

Both methods can be used in many situations involving log-linear models, and the results are usually very similar.

SPECIFICATIONS

The following statements can be used with PROC FUNCAT:

PROC FUNCAT *option;*
DIRECT *variables;*
MODEL *response = effects / options;*
RESPONSE *function;*
WEIGHT *variable;*
CONTRAST *specification;*
BY *variables;*

The PROC and MODEL statements are required. The PROC statement invokes the procedure. The MODEL statement specifies the response and design effects of a model. The DIRECT statement, which must precede the MODEL statement if used, specifies variables containing design matrix values. A RESPONSE statement specifies custom-programmed response functions. Any number of RESPONSE statements may be used. Use the WEIGHT statement to specify a variable containing frequency counts. The CONTRAST statement specifies a hypothesis to test. The BY statement can be used to process groups of data. The RESPONSE, WEIGHT, and BY statements can be specified anywhere after the PROC statement.

PROC FUNCAT Statement

PROC FUNCAT *option;*

This statement invokes the procedure. One option is available:

DATA=*SASdataset* names the SAS data set containing the data to be analyzed. If DATA= is not given, FUNCAT uses the most recently created SAS data set.

DIRECT Statement

DIRECT *variables;*

You can optionally use the DIRECT statement to declare numeric variables to be treated in a **direct** quantitative rather than qualitative way. Variables mentioned in a DIRECT statement are inserted **directly** as columns of the design matrix **X**, rather than inducing a classification.

The DIRECT statement is useful for logistic regression, described below. If used, the DIRECT statement must precede the MODEL statement.

MODEL Statement

MODEL *response* = *effects / options;*

The MODEL statement **must** be used with PROC FUNCAT. Only one MODEL statement may appear. The response effect determines the response categories. The design effects determine the samples (populations) and the parametric structure of the samples.

The response effect is either a single variable or a crossed effect having several variables joined by asterisks.

Design effects can use four types of terms, plus combinations of these four types:

1. Crossed effects are effects joined by asterisks: A*B*C .
2. Nested effects are effects subscripted by other effects in parentheses: B(A) .
3. Direct effects are effects specified in a DIRECT statement: X .
4. Special nested-by-value effects are nested effects subscripted by a variable equal to a value: B(A=1) .

Examples of MODEL statements using these four types are listed below:

MODEL R = A B;	main effects only
MODEL R = A B A*B;	main effects with interaction
MODEL R = A B(A);	nested effects
MODEL R = A B(A=1) B(A=2);	special nested-by-value effects
DIRECT X1 X2; MODEL R = X1 X2;	quantitative effects
MODEL R = A B(A) C(B A);	multiple nesting effects

All four types of specifications can be combined into one effect as in X*A*B(C D=1). Whenever an effect is specified, all other effects contained by the effect should also be specified. For example, whenever A*B is in the model, then A and B should also be in the model as main effects. The details of how effects are generated are described later.

The options below may be specified in the MODEL statement after a slash (/).

FREQ prints the table frequencies for the design by the response classification.

PROB prints the table of probability estimates for the design by the response classifications. These estimates add to 1 across the response categories for each sample.

ONEWAY produces a one-way table of counts for each variable used in the analysis. This table is useful in determining the order of the levels in each classification.

X prints **X**, the design matrix in the linear model.

XPX prints $\mathbf{X'S^{-1}X}$, the crossproducts matrix for the normal equations. $\mathbf{X'S^{-1}F}$ appears as the last row.

COV prints $\mathbf{S}_i$ and $\mathbf{S}^i$, the covariance and inverse covariance matrices of the function of the probabilities for each population. If both COV and X are specified, $\mathbf{S}_i$ and $\mathbf{S}^i$ are printed between the rows of **X**.

COVB	prints the estimated covariance matrix of the parameter estimates.
CORRB	prints the estimated correlation matrix of the parameter estimates.
PREDICT P	prints the predicted values of the function generated by the linear model.
NOINT	suppresses the intercept term in the model.
ML	requests maximum-likelihood estimates. This option is available when the standard response function is used. Thus ML cannot be used if a RESPONSE statement is used.
NOGLS	suppresses the generalized least-squares estimates when only the maximum-likelihood estimates are needed. The parameter estimates start out at zero. This option is useful if you want to use FUNCAT for logistic regression.
ADDCELL=*number*	requests *n* to be added to the frequency count in each cell. *N* can be any positive number. Some practitioners routinely recommend ADDCELL=.5. You can also use a small ADDCELL value to avoid trouble with logs of empty cells.

RESPONSE Statement

RESPONSE *function*;

You can use the RESPONSE statement to define a response function on the response probabilities to be modeled. FUNCAT then uses this response function instead of the standard (default) response function.

The response function can be any combination of these four operations:

operation	**specification**
1. linear combination	matrix literal
2. logarithm	LOG
3. exponential	EXP
4. adding constant	+ matrix literal

These operations should be listed in the reverse order from the way they are to be applied. A matrix literal is a series of numbers with each row of numbers separated from the next row by a slash (/). For example:

```
RESPONSE 1 0 -1 / 0 1 -1;
```

specifies a simple linear response function. The matrix literal specifies a 2×3 matrix:

$$\begin{bmatrix} 1 & 0 & -1 \\ 0 & 1 & -1 \end{bmatrix}.$$

The example assumes that there are three response categories from which two function values result for each population. In this case the first function value compares the first and third categories 1 0 –1, and the second function value compares the second and third response categories 0 1 –1.

If two matrix literals appear next to each other, they should be separated by an asterisk (*).

The response function

```
RESPONSE 1 -1 LOG;
```

first takes the logs of the two response probabilities, then subtracts the second from the first. As a programming assignment, this would have appeared as:

function = LOG(P1) – LOG(P2);

which is:

function = LOG(P1/(1 – P1));

known as the *logit function*.

An example of a compound response function is:

```
RESPONSE 1 -1 EXP 0 1 1 0 / 1 0 0 1 LOG;
```

which is equivalent to the matrix expression

```
F= A*EXP(B*LOG(P));
```

where

$$A = [1 \quad -1] \quad \text{and} \quad B = \begin{bmatrix} 0 & 1 & 1 & 0 \\ 1 & 0 & 0 & 1 \end{bmatrix}.$$

This is equivalent to

```
F= P2*P3– P1*P4;
```

These functions are always evaluated from right to left. Further examples are described later in this chapter.

More than one RESPONSE statement can be specified. Each RESPONSE statement produces a separate analysis.

If no RESPONSE statement is specified, a default response function is used. The default response function compares the logs of each response probability with the log of the probability for the last response category. If there are four response categories, the default response function is equivalent to the specification:

```
RESPONSE 1 0 0 -1/
         0 1 0 -1/
         0 0 1 -1 LOG;
```

If there are *r* response categories, the default response function evaluates *r*–1 values.

Examples of RESPONSE Statements

Linear Example response functions of the form f(π) = **A***(π):

```
RESPONSE 1 -1;
```

A = [1 –1]

RESPONSE 1 0 -1/ 0 1 -1;

$$\mathbf{A} = \begin{bmatrix} 1 & 0 & -1 \\ 0 & 1 & -1 \end{bmatrix}$$

RESPONSE 1 2 3;

$$\mathbf{A} = [1 \ 2 \ 3]$$

Loglinear Example response functions of the form $f(\pi) = \mathbf{K}\log(\mathbf{A}\pi)$:

RESPONSE 1 -1 LOG;

$$\mathbf{K} = [1 \ -1]$$

$$\mathbf{A} = \begin{bmatrix} 1 & 0 \\ 0 & 1 \end{bmatrix}$$

which is equivalent to $\log(p/(1-p))$, the logit function.

RESPONSE 1 2 3 LOG 1 1 0 0 0 / 0 0 1 1 0 / 0 0 0 0 1;

$$\mathbf{K} = [1 \ 2 \ 3]$$

$$\mathbf{A} = \begin{bmatrix} 1 & 1 & 0 & 0 & 0 \\ 0 & 0 & 1 & 1 & 0 \\ 0 & 0 & 0 & 0 & 1 \end{bmatrix}$$

which is equivalent to $f(p) = \log(p_1 + p_2) + 2^*\log(p_3 + p_4) + 3^*\log(p_5)$.

RESPONSE 1 0 -1 / 0 1 -1 LOG;

$$\mathbf{K} = \begin{bmatrix} 1 & 0 & -1 \\ 0 & 1 & -1 \end{bmatrix}$$

$$\mathbf{A} = \begin{bmatrix} 1 & 0 & 0 \\ 0 & 1 & 0 \\ 0 & 0 & 1 \end{bmatrix}$$

Compounded Example response functions of the form $f(\pi) = \mathbf{A}^*\exp(\mathbf{B}^*\log(\pi))$:

RESPONSE 1 -1 EXP 0 1 1 0 / 1 0 0 1 LOG;

$$\mathbf{A} = [1 \ -1]$$

$$\mathbf{B} = \begin{bmatrix} 0 & 1 & 1 & 0 \\ 1 & 0 & 0 & 1 \end{bmatrix}$$

(equivalent to $p_{21}p_{12} - p_{11}p_{22}$ for a two-way response, which can be used to test forms of conditional independence).

WEIGHT Statement

WEIGHT *variable*;

A WEIGHT statement can be used to refer to a variable containing counts to weight the data. The counts need not be integers. The WEIGHT statement lets you use summary data sets containing a count variable.

CONTRAST Statement

CONTRAST *'label' @n effect values...@n effect values..., ...;*

The CONTRAST statement allows you to construct and test linear functions of the parameters. To use the CONTRAST statement properly, you must be familiar with how FUNCAT parameterizes models.

These terms are specified in the CONTRAST statement:

'label' up to 20 characters of identifying information printed with the test.

@n a notation for pointing to parameters associated with the nth function value, for multiple-valued response functions. If @n is not given, @1 is assumed.

effect one of the effects in the model. The effect labeled INTERCEPT can be specified for the automatic intercept parameter.

values a series of numbers that form the coefficients of the parameters associated with the given effect. If there are fewer values than parameters for an effect, the remaining coefficients become zero.

Several contrast rows can be specified for a joint test. Separate each row from the next by a comma.

Suppose A has four levels. Then A has three parameters; each compares one of the first three levels with the fourth:

```
parm1 = A1-A4
parm2 = A2-A4
parm3 = A3-A4
```

Examples of CONTRAST statements are:

```
CONTRAST 'A1 VS A4' 1 0 0;
CONTRAST 'A1 VS A2' 1 -1 0;
CONTRAST 'A1&A2 VS A3' .5 .5 -1;
CONTRAST 'A1&A2 VS A4' .5 .5;
```

Since FUNCAT is parameterized differently than GLM, you must be careful not to use the same contrasts that you would with GLM. Each parameter represents not a level, but the difference between a level and the last level. As a by-product of the parameterization, main-effect parameters are directly estimable without involving interaction parameters.

BY Statement

BY *variables;*

A BY statement may be used with PROC FUNCAT to obtain separate analyses on observations in groups defined by the BY variables. When a BY statement appears, the procedure expects the input data set to be sorted in order of the BY variables. If your input data set is not sorted in ascending order, use the SORT procedure with a similar BY statement to sort the data, or, if appropriate, use the BY statement options NOTSORTED or DESCENDING. For more information, see the discussion of the BY statement in Chapter 8, "Statements Used in the PROC Step," in *SAS User's Guide: Basics, 1982 Edition*.

DETAILS

Missing Values

Observations with missing values for any variable listed in the MODEL statement are omitted from the analysis.

Input Data Set

Data to be analyzed by FUNCAT can be either a SAS data set containing data values or a SAS data set containing frequency counts. FUNCAT assumes the SAS data set contains raw data values to be counted unless a WEIGHT statement is included.

If raw data are used, FUNCAT first counts the number of observations having each combination of values for all variables used in the MODEL statement. For example, suppose that the variables A and B each take on the values 1 and 2, and their counts can be represented in the table below:

	A=1	A=2
B=1	2	1
B=2	3	4

The SAS data set containing the raw data might look like:

data set RAW:

A	B
1	1
1	1
1	2
1	2
1	2
2	1
2	2
2	2
2	2
2	2

and the statements for FUNCAT:

```
PROC FUNCAT DATA=RAW;
  MODEL A=B;
```

If the frequency counts have already been accumulated and stored in a SAS data set, the WEIGHT statement is used to specify the variable containing the counts. Suppose the data above are processed with PROC FREQ to produce the data set SUMMARY:

```
PROC FREQ DATA=RAW;
  TABLES A*B / OUT=SUMMARY;
```

data set SUMMARY:

A	B	COUNT
1	1	2
1	2	3
2	1	1
2	2	4

Then FUNCAT is specified:

```
PROC FUNCAT DATA=SUMMARY;
  WEIGHT COUNT;
  MODEL A=B;
```

Sometimes the frequencies are stored in a data set in the form of a table with the counts for each response category contained in separate variables. In this case, reshape the data set so that the counts are in one variable across several observations. For example, for the data set TABLE:

data set TABLE:

B	A1	A2
1	2	1
2	3	4

a new data set SUMMARY is formed using these SAS statements:

```
DATA SUMMARY;
  SET TABLE;
  A=1;  COUNT=A1;  OUTPUT;
  A=2;  COUNT=A2;  OUTPUT;
  DROP A1 A2;
```

The data set SUMMARY is shown below:

data set SUMMARY:

A	B	COUNT
1	1	2
2	1	1
1	2	3
2	2	4

Then these statements

```
PROC FUNCAT DATA=SUMMARY;
  WEIGHT COUNT;
  MODEL A=B;
```

produce the analysis.

Generation of the Design Matrix

Each unique combination of classification values generates a row in the design matrix (a population, or sample). The columns of the design matrix are produced from the effect specifications in the MODEL statement.

Intercept All design matrices start with a column of 1s for the intercept unless the NOINT option is specified.

Main effects Main effects are written into the MODEL statement by name. Suppose that A is a variable with three values (levels). Then the design matrix has two columns for the main effect A corresponding to the two degrees of freedom. For each level except the last, a 1 appears in the column representing that level. For the last level, a -1 appears in every column. This approach is suggested by the programming statement:

$$A(i) = (A{=}A_i) - (A{=}A_k) \quad \text{for } i = 1 \text{ to } k-1 \ .$$

Two examples:

variable A	design columns A(1)	A(2)
1	1	0
2	0	1
3	-1	-1

variable B	design columns B(1)
1	1
2	-1

Crossed effects (interactions) Crossed effects are formed by the horizontal direct products of main effects.

data		main-A		main-B	crossed	
A	B	A(1)	A(2)	B(1)	AB(1)	AB(2)
1	1	1	0	1	1	0
1	2	1	0	-1	-1	0
2	1	0	1	1	0	1
2	2	0	1	-1	0	-1
3	1	-1	-1	1	-1	-1
3	2	-1	-1	-1	1	1

The degrees of freedom for a crossed effect are equal to the product of the degrees of freedom for each separate effect.

Nested effects The effect B(A) is read "B within A" and is like specifying a B main effect for every value of A. If n_a and n_b are the number of levels in A and B, respectively, then the number of columns for B(A) is $(n_b - 1)n_a$ if every combination of levels exists in the data.

For example:

		B(A)			
A	B	B(A=1)		B(A=2)	
1	1	1	0	0	0
2	1	0	0	1	0
1	2	0	1	0	0
2	2	0	0	0	1
1	3	-1	-1	0	0
2	3	0	0	-1	-1

FUNCAT actually allocates a column for all possible combinations of values even though some combinations may not be present in the data.

Nested effects with values Instead of nesting an effect within all values of the main effect, you can nest an effect within specified values of the nested variable. The B(A) effect shown above can be specified by the two separate nested effects with values:

B(A=1) and B(A=2),

each with n_b-1 degrees of freedom assuming a complete combination.

Direct effects To request that the actual values of a variable be inserted into the design matrix, declare the variable in a DIRECT statement. For example, you can specify the design matrix directly rather than letting SAS generate it:

X1	X2
1	0
0	1
-1	-1

where X1 and X2 are declared in a DIRECT statement:

```
DIRECT X1 X2;
```

These variables still help induce the classification for the sample populations to be defined. The variables cannot be continuous in the sense that every count has a different value for the continuous variable, since this would create a separate population for each individual.

Ordering of parameters within an effect The variables for crossed and nested effects remain in the order in which they are first encountered. For example, in the model:

```
MODEL R=B A A*B C(A B);
```

the effect A*B is actually reported as B*A and the effect C(A B) is actually reported as C(B A), since B was mentioned before A in the statement.

The ordering of parameters in the model is described by which classification "moves fastest." For example, in the classification: (A1B1 A1B2 A2B1 A2B2) the B effect is said to move fastest since its subscript changes with each element.

When a completely general effect is crossed with a multivalued response function, the order of the parameter values from fastest to slowest is:

1. response function values
2. crossed effects (interactions)
3. nested effects.

Direct effects and nested values do not affect the order of the parameters. Within each type of effect variable, the left-most variables "move fastest."

For example, consider the effect A*B(C). The order of the parameters for a two-valued response function (labeled F1, F2) is:

```
A1,   B1,   C1,   F1
A1,   B1,   C1,   F2
A2,   B1,   C1,   F1
A2,   B1,   C1,   F2
A1,   B2,   C1,   F1
A1,   B2,   C1,   F2
A2,   B2,   C1,   F1
A2,   B2,   C1,   F2
A1,   B1,   C2,   F1
A1,   B1,   C2,   F2
 .     .     .     .
 .     .     .     .
 .     .     .     .
```

Logistic Regression

If you have continuous effect variables, then each observation represents a separate sample. At this extreme of sparseness, the generalized least-squares method is unworkable; you cannot avoid logs or division by zero. Instead, use the maximum-likelihood method. FUNCAT was not designed for logistic regression and is less efficient than more specialized procedures.

FUNCAT works for responses with two or more levels, whereas most other logistic regression programs work only with dichotomous (binary) responses.

To use FUNCAT for logistic regression, specify the continuous predictors in a DIRECT statement as well as the MODEL statement. Then use the ML and NOGLS options. For example:

```
PROC FUNCAT;
  DIRECT X1 X2 X3;
  MODEL RESPONSE=X1 X2 X3 / ML NOGLS;
```

The parameter estimates from FUNCAT are the same as from a logistic regression program such as LOGIST except that the signs are reversed. This is because FUNCAT normalizes on the lowest response value of 0 while LOGIST normalizes on the 1 response. (The LOGIST procedure, contributed by Frank Harrell, is described in the *SAS Supplemental Library User's Guide*.)

Computational Method

The notation used in FUNCAT differs slightly from that used in other literature. Since the method is not generally described in commonly used textbooks, a complete description of the method follows.

Summary of basic dimensions

r	number of categories in the response
s	number of populations or samples
q	number of values from response function
d	number of design parameters on populations
$q \times d$	total number of parameters after stacking design with function

Notation A column vector of 1s is denoted by $\mathbf{j}$. A square matrix of 1s is denoted by $\mathbf{J}$. $\Sigma_{k=1}^{n}$ is the sum as k goes from 1 to n. $\mathrm{Diag}_n(\mathbf{p})$ is the diagonal matrix formed from the first n elements of the vector $\mathbf{p}$. The inverse is $\mathrm{diag}_n^{-1}(\mathbf{p})$.

The ⊗ symbol denotes the Kronecker product (also known as the *direct product*) and in this case inserts the right operand into every element of the left.
For example:

$$\mathbf{A}\otimes\mathbf{B} = \begin{bmatrix} a_{11}\mathbf{B} & \dots & a_{1n}\mathbf{B} \\ \vdots & & \\ a_{m1}\mathbf{B} & \dots & a_{mn}\mathbf{B} \end{bmatrix}$$

Input data involve counts in the form:

$$\text{s samples (populations)} \quad \overset{\text{r responses}}{\begin{bmatrix} n_{11} & n_{12} & \dots & n_{1r} \\ n_{21} & n_{22} & \dots & n_{2r} \\ \vdots & & & \vdots \\ n_{s1} & n_{s2} & \dots & n_{sr} \end{bmatrix}}$$

Let n_i denote the row sum $\Sigma_j n_{ij}$.

The following calculations are shown for each population, then for all populations composed together.

Probability estimates **(dimension)**

j^{th} response $\quad p_{ij} = n_{ij}/n_i$

i^{th} population $\quad \mathbf{p}_i = \begin{bmatrix} p_{i1} \\ p_{i2} \\ \vdots \\ p_{ir} \end{bmatrix}$ $\quad$ dim $r \times 1$

all populations $\quad \mathbf{p} = \begin{bmatrix} \mathbf{p}_1 \\ \mathbf{p}_2 \\ \vdots \\ \mathbf{p}_s \end{bmatrix}$ $\quad$ dim $sr \times 1$

Variance of probability estimates

i^{th} population $\quad \mathbf{V}_i = (\text{diag}(\mathbf{p}_i) - \mathbf{p}_i\mathbf{p}_i')/n_i$ $\quad$ dim $r \times r$

all populations $\quad \mathbf{V} = \text{diag}(\mathbf{V}_1, \mathbf{V}_2, \dots, \mathbf{V}_s)$ $\quad$ dim $sr \times sr$

Response function

i^{th} population $\quad \mathbf{F}_i = \text{F}(\mathbf{p}_i)$ $\quad$ dim $q \times 1$

all populations $\quad \mathbf{F} = \begin{bmatrix} \mathbf{F}_1 \\ \mathbf{F}_2 \\ \vdots \\ \mathbf{F}_s \end{bmatrix}$ $\quad$ dim $sq \times 1$

Derivative of function with respect to probability estimates

i^{th} population $\quad \mathbf{H}_i = \partial\,\text{F}(\mathbf{p}_i)/\partial\,\mathbf{p}_i$ $\quad$ dim $q \times r$

all populations $\quad \mathbf{H} = \text{diag}(\mathbf{H}_1, \mathbf{H}_2, \dots, \mathbf{H}_s)$ $\quad$ dim $sq \times sr$

Variance of function

i^{th} population	$\mathbf{S}_i = \mathbf{H}_i\mathbf{V}_i\mathbf{H}_i'$	dim $q \times q$
all populations	$\mathbf{S} = \text{diag}(\mathbf{S}_1, \mathbf{S}_2, \ldots, \mathbf{S}_s)$	dim $sq \times sq$

Inverse variance of function

ith population	$\mathbf{S}^i = (\mathbf{S}_i)^{-1}$	dim $q \times q$
all populations	$\mathbf{S}^{-1} = \text{diag}(\mathbf{S}^1, \mathbf{S}^2, \ldots, \mathbf{S}^s)$	dim $sq \times sq$

Basic design

i^{th} population	$\mathbf{Z}_i$	dim $1 \times d$
all populations	$\mathbf{Z} = \begin{bmatrix} \mathbf{Z}_1 \\ \mathbf{Z}_2 \\ \cdot \\ \cdot \\ \mathbf{Z}_s \end{bmatrix}$	dim $s \times d$

Expanded design

i^{th} population	$\mathbf{X}_i = (\mathbf{Z}_i \otimes \mathbf{I}_q)$	dim $q \times dq$
all populations	$\mathbf{X} = \mathbf{Z} \otimes \mathbf{I}_q$	dim $sq \times dq$

Crossproducts of design

i^{th} population	$\mathbf{C}_i = \mathbf{X}_i'\mathbf{S}^i\mathbf{X}_i$	
	$= (\mathbf{Z}_i' \otimes \mathbf{I}_q)\mathbf{S}^i(\mathbf{Z}_i \otimes \mathbf{I}_q)$	
	$= (\mathbf{Z}_i'\mathbf{Z}_i) \otimes \mathbf{S}^i$	dim $dq \times dq$
all populations	$\mathbf{C} = \mathbf{X}'\mathbf{S}^{-1}\mathbf{X} = \Sigma_{i=1}^{s} \mathbf{X}_i'\mathbf{S}^i\mathbf{X}_i = \Sigma_{i=1}^{s} \mathbf{C}_i$	dim $dq \times dq$

Crossproduct of design with function

all populations	$\mathbf{R} = \mathbf{X}'\mathbf{S}^{-1}\mathbf{F} = \Sigma_{i=1}^{s} \mathbf{X}_i'\mathbf{S}^i\mathbf{F}_i$	dim $dq \times 1$

Parameter estimates

$\mathbf{b} = \mathbf{C}^{-1}\mathbf{R} = (\mathbf{X}'\mathbf{S}^{-1}\mathbf{X})^{-1}(\mathbf{X}'\mathbf{S}^{-1}\mathbf{F})$ dim $dq \times 1$

Estimate of covariance of parameter estimates

$\text{Cov}(\mathbf{b}) = \mathbf{C}^{-1}$ dim $dq \times dq$

Residual chi-square

$$\begin{aligned} \text{RSS} &= \mathbf{F}'\mathbf{S}^{-1}\mathbf{F} - \mathbf{F}'\mathbf{S}^{-1}\mathbf{X}(\mathbf{X}'\mathbf{S}^{-1}\mathbf{X})^{-1}\mathbf{X}'\mathbf{S}^{-1}\mathbf{F} \\ &= \mathbf{F}'\mathbf{S}^{-1}\mathbf{F} - \mathbf{b}'(\mathbf{X}'\mathbf{S}^{-1}\mathbf{X})\mathbf{b} \end{aligned}$$ dim 1×1

Chi-square for H0: $\mathbf{L}\beta = 0$

$\text{SS}(\mathbf{L}\beta = 0) = (\mathbf{Lb})'(\mathbf{L}(\mathbf{X}'\mathbf{S}^{-1}\mathbf{X})^{-1}\mathbf{L}')^{-1}(\mathbf{Lb})$ dim 1×1

Derivative table for compounded functions: y = f(g(p))

function	f(g(p))	derivative
multiply matrix	Y = AG	$\partial \mathbf{Y}/\partial \mathbf{p} = \mathbf{A}\partial \mathbf{G}/\partial \mathbf{p}$
logarithm	Y = LOG(G)	$\partial \mathbf{Y}/\partial \mathbf{p} = \mathrm{diag}^{-1}(\mathbf{G})\partial \mathbf{G}/\partial \mathbf{p}$
exponential	Y = EXP(G)	$\partial \mathbf{Y}/\partial \mathbf{p} = \mathrm{diag}(\mathbf{Y})\partial \mathbf{G}/\partial \mathbf{p}$
add constant	Y = G + A	$\partial \mathbf{Y}/\partial \mathbf{p} = \partial \mathbf{G}/\partial \mathbf{p}$

Default response function: every response vs. last response

$f_j = \log(p_j/p_r)$ for $j = 1, \ldots, (r-1)$
for each population $i = 1, \ldots, s$

In the following discussion, subscripts i for the population are suppressed.

Inverse of response function

$$p_j = \exp(f_j)/(1+\Sigma_{k=1}^{r-1}\exp(f_k)) \qquad \text{for } j = 1, \ldots, r-1$$
$$p_r = 1/(1+\Sigma_{k=1}^{r-1}\exp(f_k))$$

Form of f and derivative for one of the populations

$$\mathbf{f} = \mathbf{K}\log(\mathbf{p}) = (\mathbf{I}_{r-1} \mid -\mathbf{j})\log(\mathbf{p})$$
$$\mathbf{H} = \partial \mathbf{f}/\partial \mathbf{p} = (\mathrm{diag}_{r-1}^{-1}(\mathbf{p}) \mid (-1/p_r)\mathbf{j})$$

Covariance results

$$\mathbf{S} = (1/n)\mathbf{H}\mathbf{V}(\mathbf{p})\mathbf{H}' = (\mathrm{diag}_{r-1}^{-1}(\mathbf{p}) + (1/p_r)\mathbf{J}_{r-1})/n$$
$$\mathbf{S}^{-1} = n(\mathrm{diag}_{r-1}(\mathbf{p}) - \mathbf{pp}') \qquad \dim r-1$$

$$\mathbf{S}^{-1}\mathbf{f} = n\begin{bmatrix} p_1 f_1 \\ \vdots \\ p_{r-1} f_{r-1} \end{bmatrix} - \begin{bmatrix} p_1 \\ \vdots \\ p_{r-1} \end{bmatrix} n_i \Sigma_{j=1}^{r-1} p_j f_j$$

$$\mathbf{f}'\mathbf{S}^{-1}\mathbf{f} = n\Sigma_{i=1}^{r-1}\mathbf{p}_i\mathbf{f}_i^2 - n(\Sigma_{i=1}^{r-1}\mathbf{p}_i\mathbf{f}_i)^2$$

Maximum-Likelihood Method

Using the Newton-Raphson method, let **C** be the Hessian matrix, **G** be the gradient (both functions of π and the parameters β). Let λ be a step-shortening factor that is usually 1. Then, starting with the least squares estimates $\mathbf{b}_0$ of β, construct the probabilities $\pi(\mathbf{b})$ and evaluate a sequence of iterative calculations of **b** until it converges. The method:

$$\mathbf{b}_{k+1} = \mathbf{b}_k - \lambda \mathbf{C}^{-1}\mathbf{G}$$

where

$$\mathbf{C} = \mathbf{X}'\mathbf{S}(\pi)^{-1}\mathbf{X}$$

$$\mathbf{N} = \begin{bmatrix} n_1(p_1 - \pi_1) \\ \vdots \\ n_s(p_s - \pi_s) \end{bmatrix}$$

$$\mathbf{G} = \mathbf{X}'\mathbf{N} \quad .$$

Printed Output

PROC FUNCAT always prints these problem details:

1. the RESPONSE variable
2. the WEIGHT variable, if given
3. the DATA SET name
4. the number of RESPONSE LEVELS (R)
5. the number of samples or POPULATIONS (S)
6. the TOTAL COUNT (N)
7. the number of OBSERVATIONS (OBS) from the data set.

The ONEWAY option prints:

8. ONE-WAY FREQUENCIES, which are useful to check the classification levels and their order.

The FREQ option (Example 2) prints for each population:

9. the SAMPLE or population number
10. the DESIGN, or the levels of the factors
11. the RESPONSE FREQUENCIES across response levels
12. the TOTAL count for the population.

The PROB option prints for each population:

13. the SAMPLE or population number
14. the DESIGN, or the levels of the factors
15. the raw RESPONSE PROBABILITIES, which are the counts divided by the population total count
16. the TOTAL count for the population.

The X option (Example 2) prints the design matrix, including:

17. the SAMPLE or population number
18. the RESPONSE FUNCTION, typically a logit function of the response probabilities
19. the DESIGN MATRIX, with a column for each parameter including the intercept.

You can use the XPX and COV options to print additional calculations at this point (not shown).

A chi-square analysis table is always printed, including:

20. the SOURCE, the effect in the model
21. the degrees of freedom DF for the effect
22. the CHI-SQUARE value
23. the significance PROBability for the chi-square test
24. a term for the RESIDUAL, which measures the remaining variation in the response across populations after fitting the other effects. If the model is saturated (in other words, all degrees of freedom are used) this term is zero.

A table of parameter estimates is always printed, including:

25. the EFFECT in the model for which parameters are formed
26. the PARAMETER number
27. the degree of freedom DF. (This is always 1 unless the model is not full rank, in which case the parameters marked DF 0 and other parameters before it are not estimable.)
28. the parameter ESTIMATE

29. a CHI-SQuare test that the parameter is zero
30. a significance PROBability for the chi-square test
31. STD, an estimate of the standard deviation of the parameter estimate.

You can use the COVB and CORRB options to print the covariances and correlations of the estimates.

You can use the PREDICT option to print a table of predicted and actual response function values. For the standard response function, the predicted response probabilities are also reported.

If the ML option is specified, then these items are printed:

32. the MAXIMUM LIKELIHOOD ITERATIONS, with the iteration number (ITER), the number of step-halving subiterations (SUBIT), the LOGLIKELIhood criteria to maximize, the convergence CRITERION, and the PARAMETER ESTIMATES.
33. a chi-square table similar to the one produced for the least-squares results. The CHI-SQUARE tests for each effect are Wald tests constructed by measuring the linear hypothesis with respect to the information (Hessian) matrix from the likelihood calculations. The LIKELIHOOD RATIO chi-square tests the model against a simple model with a different set of responses for each population. This is interpreted in the same way as the RESIDUAL chi-square from the least-squares method.
34. the table of estimates includes the maximum-likelihood estimates, with standard deviations and chi-square tests based on the information matrix.

EXAMPLES

Detergent Preference Study: Example 1

The choice of detergent brand is related to three other categorical variables. First, run a completely saturated model. Then run just the main effects, using maximum likelihood as well as generalized least squares.

```
TITLE DETERGENT PREFERENCE STUDY — CATEGORICAL MODELS;

* -------------------------DETERGENT PREFERENCE STUDY------------------------- *
|   CONSUMER BLIND TRIAL FOR COMPARING TWO LAUNDRY DETERGENTS                  |
|   FROM RIES AND SMITH, "THE USE OF THE CHI-SQUARE FOR PREFERENCE             |
|   TESTING IN MULTIDIMENSIONAL PROBLEMS," CHEMICAL ENGINEERING                |
|   PROGRESS, 59, 39-43, (1963). SEE ALSO COX(1970 P. 38).                     |
|   VARIABLES: SOFTNESS  SOFTNESS OF LAUNDRY WATER                             |
|              PREV      PREVIOUS USER OF BRAND M?                             |
|              TEMP      TEMPERATURE OF LAUNDRY WATER                          |
|              BRAND     BRAND PREFERRED: M OR X                               |
* ---------------------------------------------------------------------------- *;
DATA  DETERG;
  INPUT SOFTNESS :$4. BRAND :$1. PREV :$3. TEMP :$4. COUNT;
  CARDS;
SOFT  X  YES  HIGH  19
SOFT  X  YES  LOW  57
SOFT  X  NO  HIGH  29
SOFT  X  NO  LOW  63
SOFT  M  YES  HIGH  29
SOFT  M  YES  LOW  49
```

```
SOFT M NO HIGH 27
SOFT M NO LOW 53
MED X YES HIGH 23
MED X YES LOW 47
MED X NO HIGH 33
MED X NO LOW 66
MED M YES HIGH 47
MED M YES LOW 55
MED M NO HIGH 23
MED M NO LOW 50
HARD X YES HIGH 24
HARD X YES LOW 37
HARD X NO HIGH 42
HARD X NO LOW 68
HARD M YES HIGH 43
HARD M YES LOW 52
HARD M NO HIGH 30
HARD M NO LOW 42
;
PROC FUNCAT DATA=DETERG;
   WEIGHT COUNT;
   MODEL BRAND=SOFTNESS PREV TEMP
     SOFTNESS*PREV SOFTNESS*TEMP PREV*TEMP
     SOFTNESS*PREV*TEMP  /PROB ONEWAY;
   TITLE2 COMPLETELY SATURATED FUNCAT MODEL;

PROC FUNCAT DATA=DETERG;
   WEIGHT COUNT;
   MODEL BRAND=SOFTNESS PREV TEMP / ML;
   TITLE2 MAIN EFFECTS MODEL WITH MAXIMUM LIKELIHOOD;
```

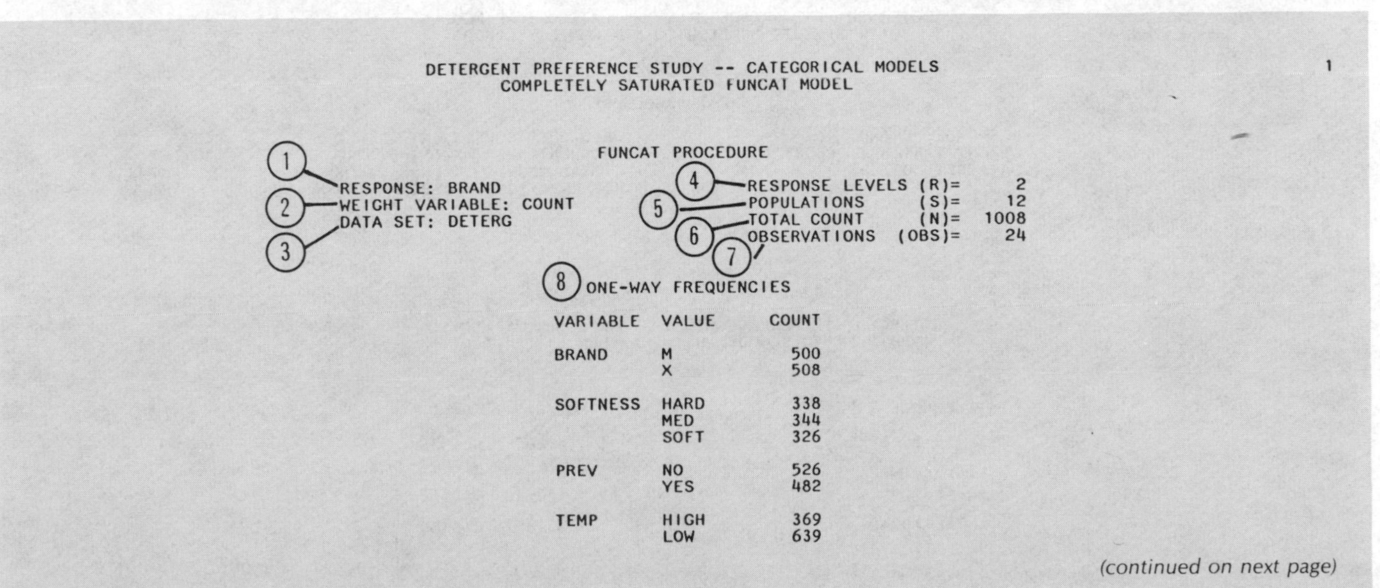

```
                 DETERGENT PREFERENCE STUDY -- CATEGORICAL MODELS                    1
                       COMPLETELY SATURATED FUNCAT MODEL

                                FUNCAT PROCEDURE
 (1)                                          (4)
   RESPONSE: BRAND                               RESPONSE LEVELS (R)=     2
(2)WEIGHT VARIABLE: COUNT                    (5) POPULATIONS     (S)=    12
   DATA SET: DETERG                              TOTAL COUNT     (N)=  1008
(3)                                          (6) OBSERVATIONS  (OBS)=    24
                                              (7)
                          (8) ONE-WAY FREQUENCIES

                          VARIABLE   VALUE      COUNT

                          BRAND      M            500
                                     X            508

                          SOFTNESS   HARD         338
                                     MED          344
                                     SOFT         326

                          PREV       NO           526
                                     YES          482

                          TEMP       HIGH         369
                                     LOW          639
```

(continued on next page)

(continued from previous page)

(13) SAMPLE	(14) DESIGN SOFTNESS	PREV	TEMP	(15) RESPONSE PROBABILITIES 1	2	(16) TOTAL
1	HARD	NO	HIGH	0.4167	0.5833	72.0
2	HARD	NO	LOW	0.3818	0.6182	110.0
3	HARD	YES	HIGH	0.6418	0.3582	67.0
4	HARD	YES	LOW	0.5843	0.4157	89.0
5	MED	NO	HIGH	0.4107	0.5893	56.0
6	MED	NO	LOW	0.4310	0.5690	116.0
7	MED	YES	HIGH	0.6714	0.3286	70.0
8	MED	YES	LOW	0.5392	0.4608	102.0
9	SOFT	NO	HIGH	0.4821	0.5179	56.0
10	SOFT	NO	LOW	0.4569	0.5431	116.0
11	SOFT	YES	HIGH	0.6042	0.3958	48.0
12	SOFT	YES	LOW	0.4623	0.5377	106.0

DETERGENT PREFERENCE STUDY -- CATEGORICAL MODELS 2
COMPLETELY SATURATED FUNCAT MODEL

(20) SOURCE	(21) DF	(22) CHI-SQUARE	(23) PROB
INTERCEPT	1	0.21	0.6505
SOFTNESS	2	0.10	0.9522
PREV	1	21.80	0.0001
TEMP	1	3.63	0.0567
SOFTNESS*PREV	2	3.80	0.1493
SOFTNESS*TEMP	2	0.20	0.9066
PREV*TEMP	1	2.25	0.1335
SOFTNESS*PREV*TEMP	2	0.74	0.6916
(24) RESIDUAL	0	0.00	1.0000

(25) EFFECT	(26) PARAMETER	(27) DF	(28) ESTIMATE	(29) CHI-SQ	(30) PROB	(31) STD
INTERCEPT	1	1	0.0304733	0.21	0.6505	0.0672532
SOFTNESS	2	1	-.00418287	0.00	0.9645	0.0939794
	3	1	0.0278252	0.09	0.7688	0.094676
PREV	4	1	-0.314016	21.80	0.0001	0.0672532
TEMP	5	1	0.128145	3.63	0.0567	0.0672532
SOFTNESS*PREV	6	1	-0.121429	1.67	0.1963	0.0939794
	7	1	-0.0636048	0.45	0.5017	0.094676
SOFTNESS*TEMP	8	1	-0.0310988	0.11	0.7407	0.0939794
	9	1	-0.0096238	0.01	0.9190	0.094676
PREV*TEMP	10	1	-0.100917	2.25	0.1335	0.0672532
SOFTNESS*PREV*TEMP	11	1	0.0765537	0.66	0.4153	0.0939794
	12	1	-0.059295	0.39	0.5311	0.094676

DETERGENT PREFERENCE STUDY -- CATEGORICAL MODELS 3
MAIN EFFECTS MODEL WITH MAXIMUM LIKELIHOOD

FUNCAT PROCEDURE

```
RESPONSE: BRAND                      RESPONSE LEVELS (R)=     2
WEIGHT VARIABLE: COUNT               POPULATIONS     (S)=    12
DATA SET: DETERG                     TOTAL COUNT     (N)=  1008
                                     OBSERVATIONS  (OBS)=    24
```

DETERGENT PREFERENCE STUDY -- CATEGORICAL MODELS 4
MAIN EFFECTS MODEL WITH MAXIMUM LIKELIHOOD

SOURCE	DF	CHI-SQUARE	PROB
INTERCEPT	1	0.20	0.6519
SOFTNESS	2	0.20	0.9033
PREV	1	19.28	0.0001
TEMP	1	3.63	0.0566
RESIDUAL	7	8.18	0.3169

(continued on next page)

(continued from previous page)

```
EFFECT              PARAMETER DF    ESTIMATE      CHI-SQ        PROB          STD

INTERCEPT                   1  1   0.0301634        0.20      0.6519    0.0668634
SOFTNESS                    2  1  -.00990954        0.01      0.9133    0.0910619
                            3  1   0.0391404        0.19      0.6643    0.0901922
PREV                        4  1   -0.281692       19.28      0.0001    0.0641546
TEMP                        5  1      0.1277        3.63      0.0566     0.066998
```

```
                  DETERGENT PREFERENCE STUDY -- CATEGORICAL MODELS                    5
                     MAIN EFFECTS MODEL WITH MAXIMUM LIKELIHOOD

(32) MAXIMUM LIKELIHOOD ITERATIONS

ITER SUBIT LOGLIKELI CRITERION ---PARAMETER ESTIMATES---
   0     0  -4.11458         1   0.0301634 -.00990954  0.0391404   -0.281692     0.1277
   1     0  -4.11399.000142607   0.0301777 -0.0094684  0.0400479   -0.283508   0.128326
   2     0  -4.11399 1.162E-11   0.0301777 -.00946827  0.0400483   -0.283508   0.128326

                      (33) SOURCE                  DF   CHI-SQUARE        PROB

                           INTERCEPT                1         0.21      0.6491
                           SOFTNESS                 2         0.22      0.8976
                           PREV                     1        19.70      0.0001
                           TEMP                     1         3.73      0.0534

                           LIKELIHOOD RATIO         7         8.23      0.3129

(34) EFFECT              PARAMETER DF    ESTIMATE      CHI-SQ        PROB          STD

     INTERCEPT                   1  1   0.0301777        0.21      0.6491    0.0663141
     SOFTNESS                    2  1  -.00946827        0.01      0.9165    0.0903011
                                 3  1   0.0400483        0.20      0.6554    0.0897255
     PREV                        4  1   -0.283508       19.70      0.0001    0.0638731
     TEMP                        5  1    0.128326        3.73      0.0534    0.0664314
```

Three-Valued Response in a Clinical Trial: Example 2

The response, a three-valued classification, is run with two 4-level main effects: operation treatment and hospital. Try three different response functions: the standard log-linear one (here a two-valued function), and scored responses both log-linear and linear in the probabilities. When you use the RESPONSE statement, be aware of the ordering of the response levels.

```
TITLE SYMPTOMS OF DUMPING SYNDROME AFTER OPERATIONS
    FOR DUODENAL ULCERS;

* -----------------------------SCORING RESPONSE FUNCTION----------------------------- *
|  FOUR DIFFERENT SURGICAL OPERATIONS FOR DUODENAL ULCERS WERE COM-                    |
|  PARED IN A CLINICAL TRIAL. THE RESPONSE IS AN UNDESIRABLE COMPLICATION              |
|  CALLED "DUMPING SYNDROME." FOUR DIFFERENT HOSPITAL LOCATIONS WERE                   |
|  ALSO CONSIDERED IN THE TRIAL. THE FOUR OPERATIONS ARE:                              |
|  OPERATN A DRAINAGE AND VAGOTOMY                                                     |
|          B 25% RESECTION + VAGOTOMY                                                  |
|          C 50% RESECTION + VAGOTOMY                                                  |
|          D 75% RESECTION                                                             |
|  DATA FROM GRIZZLE ET AL., 1969 BIOMETRICS, P489-504                                 |
|  A SCORING RESPONSE FUNCTION WAS USED TO IMPROVE THE POWER AND TO                    |
|  AVOID DISTRIBUTIONAL PROBLEMS BECAUSE THE LAST RESPONSE CATEGORY                    |
|  HAD SUCH SMALL FREQUENCY COUNTS.                                                    |
* ---------------------------------------------------------------------------------- * ;
```

```
DATA ULCER;
  INPUT HOSPITAL OPERATN $ RESPONSE $ COUNT @@;
  CARDS;
1 A NONE 23  1 A SLIGHT  7  1 A MODERATE 2
1 B NONE 23  1 B SLIGHT 10  1 B MODERATE 5
1 C NONE 20  1 C SLIGHT 13  1 C MODERATE 5
1 D NONE 24  1 D SLIGHT 10  1 D MODERATE 6
2 A NONE 18  2 A SLIGHT  6  2 A MODERATE 1
2 B NONE 18  2 B SLIGHT  6  2 B MODERATE 2
2 C NONE 13  2 C SLIGHT 13  2 C MODERATE 2
2 D NONE  9  2 D SLIGHT 15  2 D MODERATE 2
3 A NONE  8  3 A SLIGHT  6  3 A MODERATE 3
3 B NONE 12  3 B SLIGHT  4  3 B MODERATE 4
3 C NONE 11  3 C SLIGHT  6  3 C MODERATE 2
3 D NONE  7  3 D SLIGHT  7  3 D MODERATE 4
4 A NONE 12  4 A SLIGHT  9  4 A MODERATE 1
4 B NONE 15  4 B SLIGHT  3  4 B MODERATE 2
4 C NONE 14  4 C SLIGHT  8  4 C MODERATE 3
4 D NONE 13  4 D SLIGHT  6  4 D MODERATE 4
;
PROC FUNCAT;
  WEIGHT COUNT;
  MODEL RESPONSE=OPERATN HOSPITAL / ONEWAY FREQ PROB X;
  TITLE2 STANDARD LOGISTIC RESPONSE FUNCTION;
PROC FUNCAT;
  WEIGHT COUNT;
  MODEL RESPONSE=OPERATN HOSPITAL;
  *—RESPONSES ORDERED: MODERATE NONE SLIGHT—;
  RESPONSE 1 0 .5 LOG;
  TITLE2 SCORING LOGLINEAR RESPONSE FUNCTION;
PROC FUNCAT;
  WEIGHT COUNT;
  MODEL RESPONSE=OPERATN HOSPITAL ;
  *—RESPONSES ORDERED: MODERATE NONE SLIGHT—;
  RESPONSE 1 0 .5;
  TITLE2 SCORING LINEAR RESPONSE FUNCTION;
```

```
SYMPTOMS OF DUMPING SYNDROME AFTER OPERATIONS FOR DUODENAL ULCERS                1
              STANDARD LOGISTIC RESPONSE FUNCTION


                        FUNCAT PROCEDURE

RESPONSE: RESPONSE                        RESPONSE LEVELS (R)=     3
WEIGHT VARIABLE: COUNT                    POPULATIONS     (S)=    16
DATA SET: ULCER                           TOTAL COUNT     (N)=   417
                                          OBSERVATIONS  (OBS)=    48

                    ONE-WAY FREQUENCIES

                  VARIABLE  VALUE      COUNT

                  RESPONSE  MODERATE      48
                            NONE         240
                            SLIGHT       129

                  OPERATN   A             96
                            B            104
                            C            110
                            D            107

                  HOSPITAL          1    148
                                    2    105
                                    3     74
                                    4     90
```

(continued on next page)

(continued from previous page)

(9) SAMPLE	DESIGN (10) OPERATN	HOSPITAL	(11) RESPONSE FREQUENCIES 1	2	3	(12) TOTAL
1	A	1	2	23	7	32.0
2	A	2	1	18	6	25.0
3	A	3	3	8	6	17.0
4	A	4	1	12	9	22.0
5	B	1	5	23	10	38.0
6	B	2	2	18	6	26.0
7	B	3	4	12	4	20.0
8	B	4	2	15	3	20.0
9	C	1	5	20	13	38.0
10	C	2	2	13	13	28.0
11	C	3	2	11	6	19.0
12	C	4	3	14	8	25.0
13	D	1	6	24	10	40.0
14	D	2	2	9	15	26.0
15	D	3	4	7	7	18.0
16	D	4	4	13	6	23.0

SAMPLE	DESIGN OPERATN	HOSPITAL	RESPONSE PROBABILITIES 1	2	3	TOTAL

SYMPTOMS OF DUMPING SYNDROME AFTER OPERATIONS FOR DUODENAL ULCERS 2
STANDARD LOGISTIC RESPONSE FUNCTION

SAMPLE	DESIGN OPERATN	HOSPITAL	RESPONSE PROBABILITIES 1	2	3	TOTAL
1	A	1	0.0625	0.7188	0.2188	32.0
2	A	2	0.0400	0.7200	0.2400	25.0
3	A	3	0.1765	0.4706	0.3529	17.0
4	A	4	0.0455	0.5455	0.4091	22.0
5	B	1	0.1316	0.6053	0.2632	38.0
6	B	2	0.0769	0.6923	0.2308	26.0
7	B	3	0.2000	0.6000	0.2000	20.0
8	B	4	0.1000	0.7500	0.1500	20.0
9	C	1	0.1316	0.5263	0.3421	38.0
10	C	2	0.0714	0.4643	0.4643	28.0
11	C	3	0.1053	0.5789	0.3158	19.0
12	C	4	0.1200	0.5600	0.3200	25.0
13	D	1	0.1500	0.6000	0.2500	40.0
14	D	2	0.0769	0.3462	0.5769	26.0
15	D	3	0.2222	0.3889	0.3889	18.0
16	D	4	0.1739	0.5652	0.2609	23.0

SYMPTOMS OF DUMPING SYNDROME AFTER OPERATIONS FOR DUODENAL ULCERS 3
STANDARD LOGISTIC RESPONSE FUNCTION

(17) SAMPLE	(18) RESPONSE FUNCTION		(19) DESIGN MATRIX 1	2	3	4	5	6	7
1	-1.25276	1.18958	1	1	0	0	1	0	0
2	-1.79176	1.09861	1	1	0	0	0	1	0
3	-0.693147	0.287682	1	1	0	0	0	0	1
4	-2.19722	0.287682	1	1	0	0	-1	-1	-1
5	-0.693147	0.832909	1	0	1	0	1	0	0
6	-1.09861	1.09861	1	0	1	0	0	1	0
7	0	1.09861	1	0	1	0	0	0	1
8	-0.405465	1.60944	1	0	1	0	-1	-1	-1
9	-0.955511	0.430783	1	0	0	1	1	0	0
10	-1.8718	0	1	0	0	1	0	1	0
11	-1.09861	0.606136	1	0	0	1	0	0	1
12	-0.980829	0.559616	1	0	0	1	-1	-1	-1
13	-0.510826	0.875469	1	-1	-1	-1	1	0	0
14	-2.0149	-0.510826	1	-1	-1	-1	0	1	0
15	-0.559616	0	1	-1	-1	-1	0	0	1
16	-0.405465	0.77319	1	-1	-1	-1	-1	-1	-1

SOURCE	DF	CHI-SQUARE	PROB
INTERCEPT	2	98.14	0.0001
OPERATN	6	9.34	0.1553
HOSPITAL	6	8.13	0.2290
RESIDUAL	18	12.30	0.8313

(continued on next page)

(continued from previous page)

EFFECT	PARAMETER	DF	ESTIMATE	CHI-SQ	PROB	STD
INTERCEPT	1	1	-1.03566	31.13	0.0001	0.185636
	2	1	0.607483	27.06	0.0001	0.116788
OPERATN	3	1	-0.364608	1.06	0.3042	0.354879
	4	1	0.126135	0.40	0.5296	0.200663
	5	1	0.40374	1.75	0.1863	0.305521
	6	1	0.435055	4.44	0.0351	0.206464
	7	1	-0.1654	0.31	0.5774	0.296834
	8	1	-0.252665	1.87	0.1718	0.184925
HOSPITAL	9	1	0.195659	0.52	0.4697	0.270653
	10	1	0.210429	1.38	0.2394	0.178874
	11	1	-0.746127	4.75	0.0292	0.342231
	12	1	-0.236887	1.52	0.2179	0.192263
	13	1	0.478576	2.40	0.1213	0.308891
	14	1	-0.0997073	0.20	0.6538	0.222291

SYMPTOMS OF DUMPING SYNDROME AFTER OPERATIONS FOR DUODENAL ULCERS 4
SCORING LOGLINEAR RESPONSE FUNCTION

FUNCAT PROCEDURE

```
RESPONSE: RESPONSE               RESPONSE LEVELS (R)=    3
WEIGHT VARIABLE: COUNT           POPULATIONS     (S)=   16
DATA SET: ULCER                  TOTAL COUNT     (N)=  417
                                 OBSERVATIONS  (OBS)=   48
```

SYMPTOMS OF DUMPING SYNDROME AFTER OPERATIONS FOR DUODENAL ULCERS 5
SCORING LOGLINEAR RESPONSE FUNCTION

SOURCE	DF	CHI-SQUARE	PROB
INTERCEPT	1	362.04	0.0001
OPERATN	3	2.92	0.4037
HOSPITAL	3	5.23	0.1556
RESIDUAL	9	2.87	0.9692

EFFECT	PARAMETER	DF	ESTIMATE	CHI-SQ	PROB	STD
INTERCEPT	1	1	-2.79622	362.04	0.0001	0.146958
OPERATN	2	1	-0.341289	1.44	0.2301	0.284412
	3	1	-0.0634856	0.08	0.7800	0.227255
	4	1	0.0870819	0.14	0.7111	0.235086
HOSPITAL	5	1	0.0100706	0.00	0.9612	0.206898
	6	1	-0.427102	2.18	0.1395	0.289044
	7	1	0.491823	4.84	0.0278	0.223612

SYMPTOMS OF DUMPING SYNDROME AFTER OPERATIONS FOR DUODENAL ULCERS 6
SCORING LINEAR RESPONSE FUNCTION

FUNCAT PROCEDURE

```
RESPONSE: RESPONSE               RESPONSE LEVELS (R)=    3
WEIGHT VARIABLE: COUNT           POPULATIONS     (S)=   16
DATA SET: ULCER                  TOTAL COUNT     (N)=  417
                                 OBSERVATIONS  (OBS)=   48
```

SYMPTOMS OF DUMPING SYNDROME AFTER OPERATIONS FOR DUODENAL ULCERS 7
SCORING LINEAR RESPONSE FUNCTION

SOURCE	DF	CHI-SQUARE	PROB
INTERCEPT	1	248.77	0.0001
OPERATN	3	8.90	0.0307
HOSPITAL	3	2.33	0.5065
RESIDUAL	9	6.33	0.7069

(continued on next page)

(continued from previous page)

EFFECT	PARAMETER	DF	ESTIMATE	CHI-SQ	PROB	STD
INTERCEPT	1	1	0.272431	248.77	0.0001	0.0172726
OPERATN	2	1	-0.0552356	4.17	0.0411	0.0270395
	3	1	-0.0364925	1.59	0.2073	0.0289366
	4	1	0.0248148	0.78	0.3757	0.0280114
HOSPITAL	5	1	-0.020408	0.60	0.4388	0.0263603
	6	1	-0.0178223	0.44	0.5055	0.0267672
	7	1	0.0530541	2.28	0.1312	0.0351529

Logistic Regression: Example 3

For logistic regression, declare your effects in a DIRECT statement and use NOGLS and ML options.

```
* ------------------MAXIMUM LIKELIHOOD LOGISTIC REGRESSION------------------ *
|  INGOTS ARE TESTED FOR READINESS TO ROLL AFTER DIFFERENT                  |
|  TREATMENTS OF HEATING TIME AND SOAKING TIME.                             |
|  FROM COX(1970) 67-68.                                                    |
* ------------------------------------------------------------------------- * ;
DATA INGOTS;
  INPUT HEAT SOAK NREADY NTOTAL @@;
  COUNT=NREADY;
  Y=1;
  OUTPUT;
  COUNT=NTOTAL-NREADY;
  Y=0;
  OUTPUT;
  DROP NREADY NTOTAL;
  CARDS;
7 1.0 0 10   14 1.0 0 31   27 1.0 1 56   51 1.0 3 13
7 1.7 0 17   14 1.7 0 43   27 1.7 4 44   51 1.7 0  1
7 2.2 0  7   14 2.2 2 33   27 2.2 0 21   51 2.2 0  1
7 2.8 0 12   14 2.8 0 31   27 2.8 1 22
7 4.0 0  9   14 4.0 0 19   27 4.0 1 16   51 4.0 0  1
;
PROC FUNCAT;
   WEIGHT COUNT;
   DIRECT HEAT SOAK;
   MODEL Y=HEAT SOAK / ML NOGLS;
   TITLE LOGISTIC REGRESSION USING PROC FUNCAT;
```

```
          LOGISTIC REGRESSION USING PROC FUNCAT                                1

                         FUNCAT PROCEDURE

RESPONSE: Y                              RESPONSE LEVELS (R)=      2
WEIGHT VARIABLE: COUNT                   POPULATIONS     (S)=     19
DATA SET: INGOTS                         TOTAL COUNT     (N)=    387
                                         OBSERVATIONS  (OBS)=     25
```

```
                                   LOGISTIC REGRESSION USING PROC FUNCAT                                2

MAXIMUM LIKELIHOOD ITERATIONS

ITER SUBIT LOGLIKELI  CRITERION ---PARAMETER ESTIMATES---
  0    0   -227.451           1          0          0          0
  1    0   -35.4983     0.84393    2.15941 -0.0138784 -0.0037327
  2    0   -12.5838    0.645509    3.53344 -0.0363154 -0.0119734
  3    0   -7.54959    0.400056     4.7489 -0.0640013 -0.0299201
  4    0   -6.89542 0.0866501     5.41382 -0.0790272   -0.04982
  5    0   -6.876340.00276749     5.55393 -0.0819276 -0.0564395
  6    0   -6.87631 3.348E-06     5.55916 -0.0820307 -0.0567708
  7    0   -6.87631 5.357E-12     5.55917 -0.0820308 -0.0567713

                                   SOURCE                    DF    CHI-SQUARE       PROB

                                   INTERCEPT                  1         24.65     0.0001
                                   HEAT                       1         11.95     0.0005
                                   SOAK                       1          0.03     0.8639

                                   LIKELIHOOD RATIO          16         13.75     0.6171

                         EFFECT             PARAMETER DF    ESTIMATE     CHI-SQ       PROB         STD

                         INTERCEPT                  1  1     5.55917      24.65     0.0001     1.11969
                         HEAT                       2  1 -0.0820308      11.95     0.0005   0.0237345
                         SOAK                       3  1 -0.0567713       0.03     0.8639    0.331213
```

One-Population Response Function Testing Vision Symmetry: Example 4

This example runs one population with a response function that tests for symmetry across a 4 by 4 response design.

```
TITLE VISION SYMMETRY;
* ---------------------------TESTING VISION: RIGHT EYE VS LEFT--------------------------- *
|   7477 WOMEN AGED 30-39 WERE TESTED FOR VISION ON BOTH                                  |
|   RIGHT AND LEFT EYES. THE RESPONSE FUNCTION TESTS THE                                  |
|   THREE DEGREES OF FREEDOM FOR SYMMETRY BETWEEN THE RIGHT                               |
|   AND LEFT EYES. FROM GRIZZLE, STARMER AND KOCH,                                        |
|   BIOMETRICS, 1969, P493. AND BHAPKAR, JASA 1966, P228                                  |
* --------------------------------------------------------------------------------------- * ;

DATA VISION;
  INPUT RIGHT LEFT COUNT @@;
  CARDS;
1 1 1520 1 2  266 1 3  124 1 4  66
2 1  234 2 2 1512 2 3  432 2 4  78
3 1  117 3 2  362 3 3 1772 3 4 205
4 1   36 4 2   82 4 3  179 4 4 492
PROC FUNCAT;
  WEIGHT COUNT;
  MODEL RIGHT*LEFT= / ONEWAY;
  RESPONSE
    0  1  1 1    -1 0  0 0    -1  0 0 0    -1  0  0 0/
    0 -1  0 0     1 0  1 1     0 -1 0 0     0 -1  0 0/
    0  0 -1 0     0 0 -1 0     1  1 0 1     0  0 -1 0;
```

```
                    VISION SYMMETRY                                              1

                    FUNCAT PROCEDURE

RESPONSE: RIGHT*LEFT                     RESPONSE LEVELS (R)=     16
WEIGHT VARIABLE: COUNT                   POPULATIONS      (S)=      1
DATA SET: VISION                         TOTAL COUNT      (N)=   7477
                                         OBSERVATIONS  (OBS)=     16

                  ONE-WAY FREQUENCIES

               VARIABLE  VALUE      COUNT

               RIGHT           1    1976
                               2    2256
                               3    2456
                               4     789

               LEFT            1    1907
                               2    2222
                               3    2507
                               4     841
```

```
                    VISION SYMMETRY                                              2

        SOURCE                  DF   CHI-SQUARE       PROB

        INTERCEPT                3        11.98     0.0075

        RESIDUAL                 0         0.00     1.0000

EFFECT           PARAMETER DF    ESTIMATE      CHI-SQ       PROB         STD

INTERCEPT                1  1   0.0092283        5.65     0.0174   0.0038817
                         2  1  0.00454728        0.80     0.3726  0.00509955
                         3  1  -.00682092        1.83     0.1757  0.00503745
```

One-Population Repeated Measures: Example 5

Subjects are given different treatments at different times. The individual responses are combined into a multivariate response profile, as shown. The different responses are not independent and should not be modeled as coming from different populations.

```
*------------------------------------------------------------------------*
|  WHEN REPEATED MEASURES ARE TAKEN, THE RESPONSE PROFILE                |
|  BECOMES THE RESPONSE THAT IS ANALYZED. EACH PERSON IN THIS            |
|  STUDY RESPONDS TO EACH OF THREE DRUGS: A,B, AND C. THE                |
|  RESPONSES HAVE TWO LEVELS EACH (F=FAVORABLE U=UNFAVOR-                |
|  ABLE), WHICH LEADS TO 8 POSSIBLE RESPONSE PROFILES.                   |
|                                                                        |
|  THERE IS ONLY ONE POPULATION. TESTS COMPARE THE                       |
|  PROBABILITIES OF OBTAINING EACH OF THE 8 PROFILES.                    |
|                                                                        |
|  FROM: KOCH, LANDIS, FREEMAN, FREEMAN, AND LEHNEN,                     |
|  "A GENERAL METHODOLOGY FOR THE ANALYSIS OF EXPERIMENTS WITH           |
|  REPEATED MEASUREMENT OF CATEGORICAL DATA,"                            |
|  BIOMETRICS V 33 MAR 1977.                                             |
*------------------------------------------------------------------------*
```

```
DATA DRUGS;
  INPUT DRUGA $ DRUGB $ DRUGC $ COUNT @@;
  CARDS;
F F F 6    F F U 16
F U F 2    F U U  4
U F F 2    U F U  4
U U F 6    U U U  6
;
PROC FUNCAT;
  WEIGHT COUNT;
  MODEL DRUGA*DRUGB*DRUGC= ;

* -------------------------------------------------------------------- *
|  TEST FOR TOTAL SYMMETRY, I.E., THAT THE PROBABILITY OF             |
|  ANY PARTICULAR RESPONSE PROFILE IS THE SAME WITHIN                 |
|  THE NUMBER OF F''S AND U''S                                        |
|                                                                     |
|  TEST PI(FFU)=PI(FUF)=PI(UFF) 2 F''S AND 1 U                        |
|  AND PI(FUU)=PI(UFU)=PI(UUF) 1 F AND 2 U''S                         |
|                                                                     |
|        A:   F  F  F  F  U  U  U  U                                  |
|        B:   F  F  U  U  F  F  U  U                                  |
|        C:   F  U  F  U  F  U  F  U                                  |
* -------------------------------------------------------------------- * ;
  RESPONSE   1 -1  0  0  0  0  0  0/
             0  1  0  0 -1  0  0  0/
             0  0  0  1  0 -1  0  0/
             0  0  0  1  0  0 -1  0;

* -------------------------------------------------------------------- *
|  TEST THAT THE TOTAL PROBABILITY OF AN F IS THE SAME FOR            |
|  EACH OF THE THREE DRUGS.                                           |
|                                                                     |
|  CALLED FIRST ORDER MARGINAL SYMMETRY                               |
|      AF=PI(FFF)+PI(FFU)+PI(FUF)+PI(FUU)                             |
|      BF=PI(FFF)+PI(FFU)+PI(UFF)+PI(UFU)                             |
|      CF=PI(FFF)+PI(FUF)+PI(UFF)+PI(UUF)                             |
|  TEST: AF=BF=CF                                                     |
|                          A:  F  F  F  F  U  U  U  U                 |
|                          B:  F  F  U  U  F  F  U  U                 |
|              A  B  C     C:  F  U  F  U  F  U  F  U                 |
* -------------------------------------------------------------------- * ;
  RESPONSE 1 -1  0/
           1  0 -1   *   1  1  1  1  0  0  0  0/
                         1  1  0  0  1  1  0  0/
                         1  0  1  0  1  0  1  0;
```

```
                              FUNCAT PROCEDURE                                  1

RESPONSE: DRUGA*DRUGB*DRUGC            RESPONSE LEVELS (R)=     8
WEIGHT VARIABLE: COUNT                 POPULATIONS     (S)=     1
DATA SET: DRUGS                        TOTAL COUNT     (N)=    46
                                       OBSERVATIONS  (OBS)=     8
```

```
            SOURCE                     DF    CHI-SQUARE        PROB                          2

            INTERCEPT                   4         16.29      0.0027

            RESIDUAL                    0          0.00      1.0000

EFFECT               PARAMETER DF      ESTIMATE      CHI-SQ        PROB         STD

INTERCEPT                    1  1     0.304348       14.27      0.0002    0.080579
                             2  1     0.304348       14.27      0.0002    0.080579
                             3  1  -5.204E-18         0.00      1.0000   0.0614875
                             4  1  -0.0434783         0.40      0.5253   0.0684456
```

```
            SOURCE                     DF    CHI-SQUARE        PROB                          3

            INTERCEPT                   2          6.58      0.0372

            RESIDUAL                    0          0.00      1.0000

EFFECT               PARAMETER DF      ESTIMATE      CHI-SQ        PROB         STD

INTERCEPT                    1  1             0       0.00      1.0000   0.0753066
                             2  1       0.26087       5.79      0.0161    0.108412
```

REFERENCES

Bock, R.D. (1975), *Multivariate Statistical Methods in Behavioral Research*, Chapter 8, New York: McGraw-Hill.

Cox, D.R. (1970), *The Analysis of Binary Data*, New York: Halsted Press.

Forthofer, R.N. and Koch, G.G. (1973), "An Analysis of Compounded Functions of Categorical Data," *Biometrics*, 29, 143–157.

Forthofer, R.N. and Lehnen R.G. (1981), *Public Program Analysis: A New Categorical Data Approach*, Belmont, CA: Wadsworth.

Grizzle, J.E., Starmer, C.F., and Koch, G.G. (1969), "Analysis of Categorical Data by Linear Models," *Biometrics*, 25, 489–504.

Grizzle, J.E. and Williams, O.D. (1972), "Log Linear Models and Tests of Independence for Contingency Tables," *Biometrics*, 28, 137–156.

Kritzer, Herbert (1979), "Approaches to the Analysis of Complex Contingency Tables: A Guide for the Perplexed," *Sociological Methods and Research*, 7, 305–329.

Landis, J.R., Stanish, W.M., Freeman, J.L. and Koch, G.G. (1976), "A Computer Program for the Generalized Chi-Square Analysis of Categorical Data Using Weighted Least Squares, (GENCAT)," *Computer Programs in Biomedicine*, 6, 196–231.

Stedl, J.L. (1981), "The Effect of Sample Size on the True α Level with FUNCAT—A Simulation Study," *Proceedings of the Sixth Annual SAS Users Group*, Cary, NC: SAS Institute Inc.

The PROBIT Procedure

ABSTRACT

The PROBIT procedure calculates maximum-likelihood estimates of the intercept, slope, and natural (threshold) response rate for biological assay data.

INTRODUCTION

The maximum-likelihood estimates are calculated for the parameters A and B>0 in the probit equation

$$\Phi^{-1}(y) + 5 = (A + Bx)$$

where

Φ is the cumulative distribution function of the standard normal distribution,

x is the level of the dose, and

y is the probability of a response.

A modified Gauss-Newton algorithm is used to compute the estimates. You may request that the natural (threshold) response rate C be estimated. If you have an initial estimate of C from a control group, it may be specified.

The data set to be used by PROBIT must include

- a variable specifying the dose (level of the stimulus)
- a variable giving the number of subjects tested at that dose
- a variable giving the number of subjects responding to the dose.

For small chi-square values ($p > .10$), the fiducial limits are computed using a t value of 1.96. In the case of large chi-squares, variances and covariances are multiplied by the heterogeneity factor, and the usual t value is used to compute the fiducial limits.

SPECIFICATIONS

The statements used to control PROC PROBIT are:

PROC PROBIT *options;*
VAR *dose subjects response;*
BY *variables;*

PROC PROBIT Statement

PROC PROBIT *options;*

The options that may appear in the PROC PROBIT statement are:

DATA=*SASdataset* names the SAS data set to be used by PROBIT. If DATA= is omitted, the most recently created SAS data set is used.

OPTC
C=*rate* control how the threshold response is handled. Specify OPTC to request that the estimation of the natural (threshold) response rate C be optimized. Specify C=*rate* for the threshold rate or initial estimate of threshold rate. The threshold rate value must be a number between 0 and 1.

If neither OPTC nor C= is specified, a threshold rate of 0 is assumed.

If you specify both OPTC and C=, the threshold rate should be the ratio of the number of responses to the number of subjects in a control group; this value is taken as the initial estimate of the threshold rate.

If C= is specified but not OPTC, then C specifies a constant threshold rate.

If OPTC is specified but not C=, PROBIT's action depends on the response variable. When all the responses are greater than 0, the initial estimate of threshold rate is the ratio of the smallest response value to the total number of subjects in the experiment. When one or more of the responses is zero, the initial estimate of threshold rate is the reciprocal of the total number of subjects in the experiment. (The total number of subjects is the sum of the values of the subjects variable.)

HPROB=*p* specifies a probability level other than .10 to indicate a good fit. Then for chi-square values with probability less than the HPROB value, the fiducial limits will be computed using $t=1.96$.

LOG
LN requests PROC PROBIT to analyze the data using LN(DOSE). In addition to the usual output, the estimated DOSE values and 95% fiducial limits for DOSE are also printed. If OPTC is specified, any observations with a dose less than or equal to zero are used in the estimation as a control group.

LOG10 specifies an analysis like LN above except that the LOG10(DOSE) is used rather than LN(DOSE).

VAR Statement

VAR *dose subjects response;*

A VAR statement must accompany the PROC PROBIT statement. The three variables representing the level of stimulus (*dose*), the number of subjects (*subjects*), and the number of subjects responding to that dose (*response*) must appear in that order in the VAR statement.

BY Statement

BY *variables;*

A BY statement may be used with PROC PROBIT to obtain separate analyses on observations in groups defined by the BY variables. When a BY statement appears, the procedure expects the input data set to be sorted in order of the BY variables. If your input data set is not sorted in ascending order, use the SORT procedure with a similar BY statement to sort the data, or, if appropriate, use the BY statement options NOTSORTED or DESCENDING. For more information, see the discussion of the BY statement in Chapter 8, ''Statements Used in the PROC Step,'' in *SAS User's Guide: Basics, 1982 Edition.*

DETAILS

Missing Values

PROBIT does not include any observations having missing values for either the dose or the number of subjects. If the number of responses is missing, the response variable is assumed to be 0.

Printed Output

For each iteration, PROBIT prints:

1. the current estimate of the parameter A, called INTERCEPT
2. the current estimate of the parameter B, called SLOPE
3. for a threshold model, the current estimate of the parameter C (not shown in this example)
4. the mean MU of the stimulus tolerance
5. the standard deviation SIGMA of the stimulus tolerance.

PROBIT also prints:

6. the estimated dose along with the 95% fiducial limits for probability levels .01-.10, .15, .20, .25, .85, and .90-.99.
7. the covariance matrix for INTERCEPT and SLOPE
8. the covariance matrix for MU and SIGMA

PROBIT also produces these two plots:

9. a plot of the empirical probit at each level of stimulus (*dosevariable*) superimposed on the probit line
10. a plot of points computed as the raw probability at each level of stimulus superimposed on the normal probability sigmoid curve.

EXAMPLE

DOSE in this example is the variable representing the level of the stimulus, N represents the number of subjects tested at each level of the stimulus, and RESPONSE is the number of subjects responding to that level of the stimulus.

```
DATA A;
  INPUT DOSE N RESPONSE;
  CARDS;
```

```
1 10 1
2 12 2
3 10 4
4 10 5
5 12 8
6 10 8
7 10 10
;
PROC PROBIT LOG10;
   VAR DOSE N RESPONSE;
   TITLE OUTPUT FROM PROBIT PROCEDURE;
```

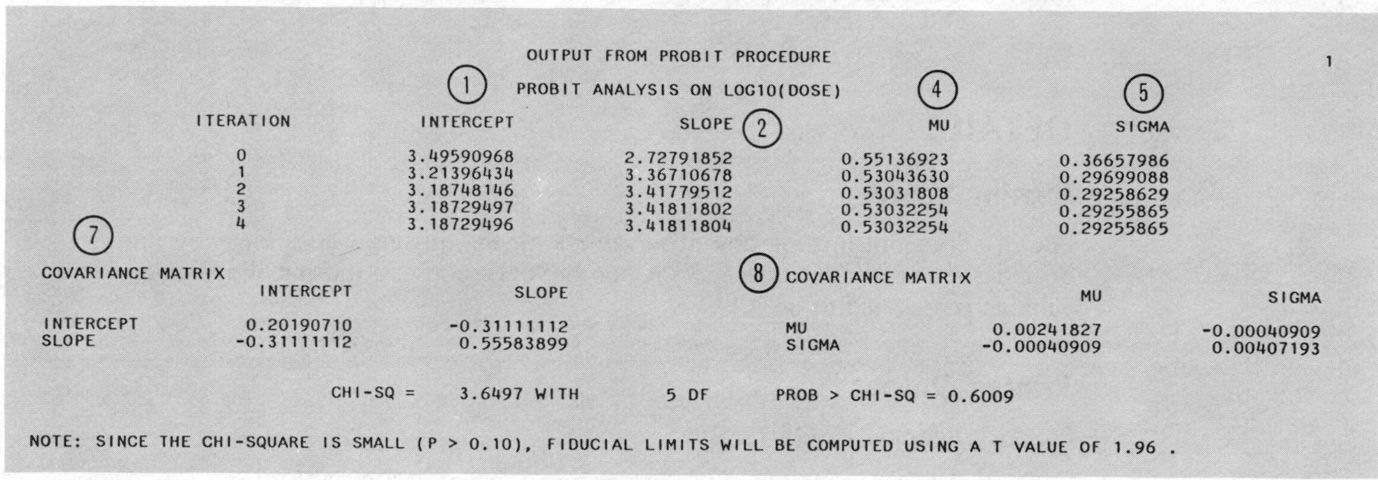

```
                              OUTPUT FROM PROBIT PROCEDURE                                        1
                          (1) PROBIT ANALYSIS ON LOG10(DOSE)         (4)                (5)
          ITERATION           INTERCEPT           SLOPE (2)            MU                SIGMA

              0              3.49590968          2.72791852        0.55136923         0.36657986
              1              3.21396434          3.36710678        0.53043630         0.29699088
              2              3.18748146          3.41779512        0.53031808         0.29258629
              3              3.18729497          3.41811802        0.53032254         0.29255865
    (7)       4              3.18729496          3.41811804        0.53032254         0.29255865

COVARIANCE MATRIX                                     (8) COVARIANCE MATRIX
                  INTERCEPT            SLOPE                                    MU               SIGMA

INTERCEPT         0.20190710        -0.31111112           MU              0.00241827       -0.00040909
SLOPE            -0.31111112         0.55583899           SIGMA          -0.00040909        0.00407193

                  CHI-SQ =      3.6497 WITH        5 DF        PROB > CHI-SQ = 0.6009

NOTE: SINCE THE CHI-SQUARE IS SMALL (P > 0.10), FIDUCIAL LIMITS WILL BE COMPUTED USING A T VALUE OF 1.96 .
```

```
                              OUTPUT FROM PROBIT PROCEDURE                                        2
PROBIT                    (9) PROBIT ANALYSIS ON LOG10(DOSE)
  10 +                                                                   X
     I
     I
   9 +
     I
     I
   8 +                                                                                          ...
     I                                                                                     .....
   7 +                                                                                .....
     I                                                                           .....
     I                                                                      .....
   6 +                                                                 .....
     I                                                          .....
     I                                                    ...X.....
   5 +                                               ...X.
     I                                          .....X
     I                                    .....
   4 +                              ...X.
     I                         ....
     I                   ....X
   3 +             X  ....
     I            ....
     I        ....
   2 +...  ....
     I ....
     I
   1 +
     I
     I
   0 +
     --------------+-----------------+----------+----------+----------+----------+-----------------+--------
                 LD01               LD10       LD25       LD50       LD75       LD90              LD99
                -0.150              0.155      0.333      0.530      0.728      0.905             1.211  LOG10(DOSE)
```

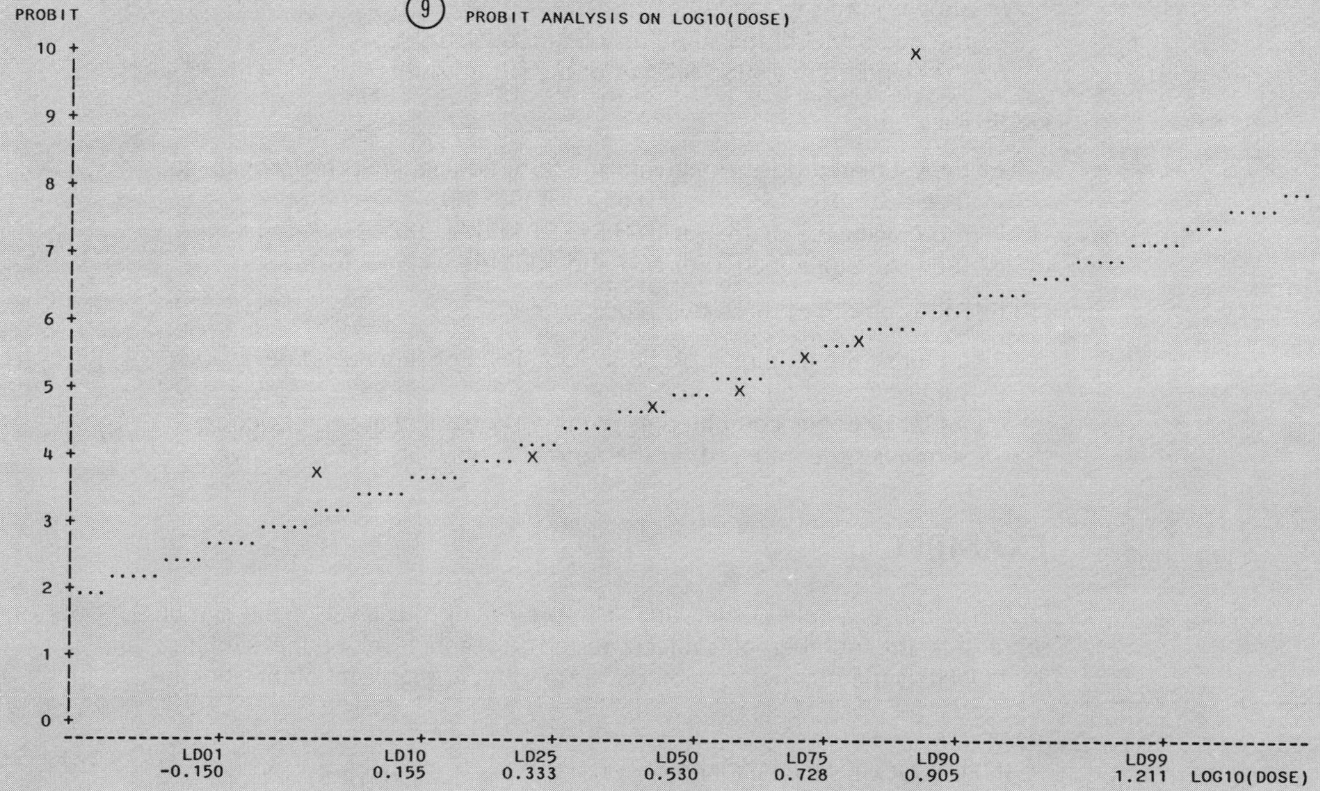

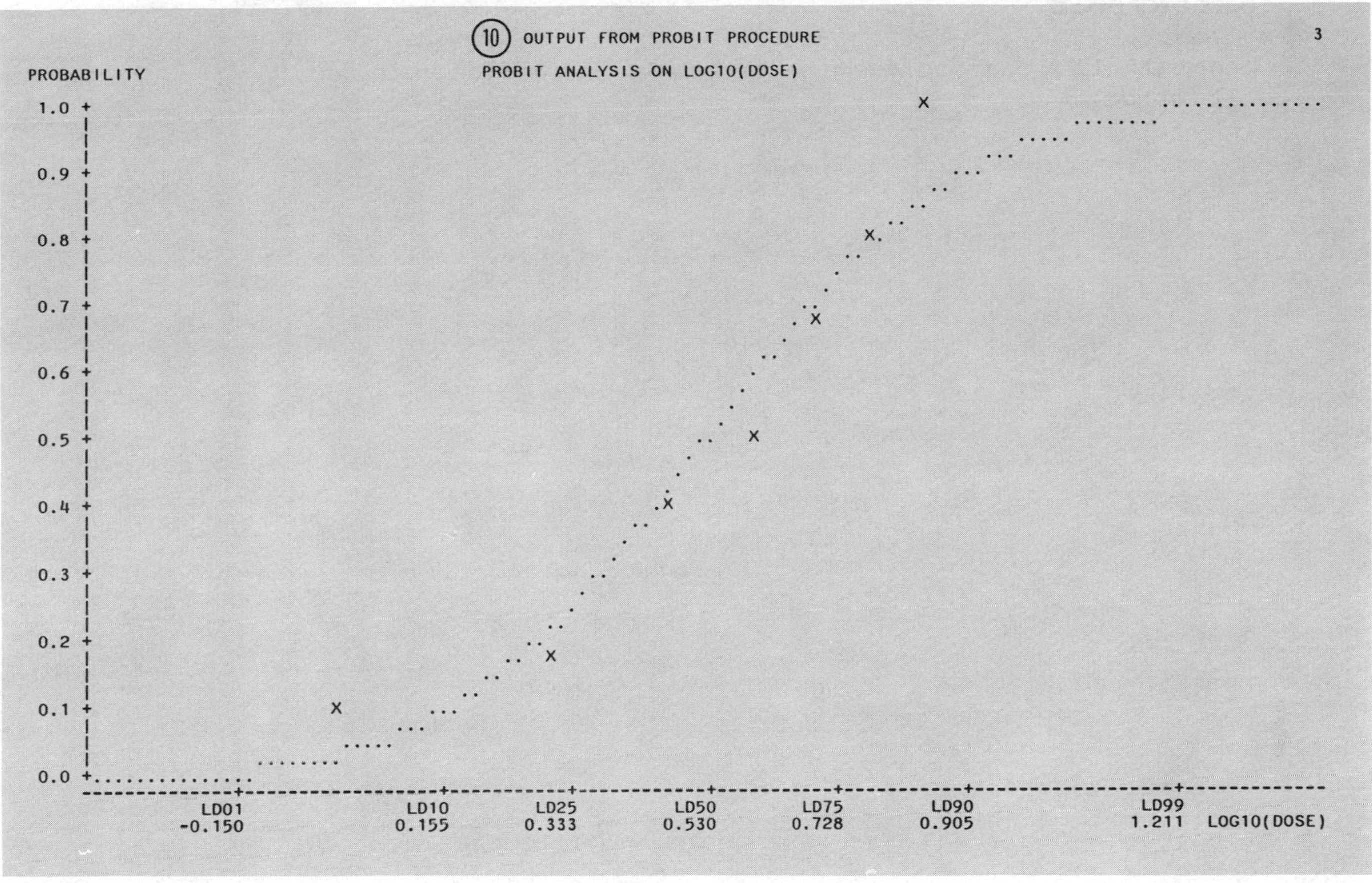

OUTPUT FROM PROBIT PROCEDURE 4

(6) PROBIT ANALYSIS ON LOG10(DOSE)

PROBABILITY	LOG10(DOSE)	95 PERCENT FIDUCIAL LIMITS LOWER	UPPER	DOSE	95 PERCENT FIDUCIAL LIMITS LOWER	UPPER
0.01	-0.15027065	-0.69520058	0.07710499	0.70750473	0.20174344	1.19427677
0.02	-0.07051947	-0.55767745	0.13475446	0.85012058	0.27689974	1.36381184
0.03	-0.01991990	-0.47065849	0.17156625	0.95516874	0.33833078	1.48445229
0.04	0.01814419	-0.40535106	0.19941188	1.04266354	0.39323208	1.58274837
0.05	0.04910638	-0.35234570	0.22217934	1.11971213	0.44427748	1.66793582
0.06	0.07546008	-0.30732670	0.24165490	1.18976195	0.49280296	1.74443545
0.07	0.09856711	-0.26793800	0.25881533	1.25477862	0.53958765	1.81474383
0.08	0.11925670	-0.23274573	0.27425605	1.31600246	0.58513257	1.88042516
0.09	0.13807306	-0.20080938	0.28836839	1.37427314	0.62978255	1.94253292
0.10	0.15539354	-0.17147720	0.30142410	1.43018936	0.67378727	2.00181575
0.15	0.22710499	-0.05086498	0.35630907	1.68696078	0.88947761	2.27148078
0.20	0.28409897	0.04368359	0.40124010	1.92353002	1.10581783	2.51906920
0.25	0.33299473	0.12342032	0.44116445	2.15275561	1.32867975	2.76162336
0.30	0.37690463	0.19347555	0.47856856	2.38179639	1.56126115	3.01001428
0.35	0.41759370	0.25657599	0.51504522	2.61573476	1.80541060	3.27374781
0.40	0.45620365	0.31428484	0.55182530	2.85893086	2.06198184	3.56307777
0.45	0.49355923	0.36753513	0.58999414	3.11572576	2.33096167	3.89039893
0.50	0.53032254	0.41692823	0.63057076	3.39095900	2.61172972	4.27140503
0.55	0.56708585	0.46295509	0.67451361	3.69050547	2.90372237	4.72621649
0.60	0.60444143	0.50617659	0.72271124	4.02199407	3.20757332	5.28094010
0.65	0.64305137	0.54734298	0.77603379	4.39593614	3.52649261	5.97081734
0.70	0.68374045	0.58744572	0.83550814	4.82770190	3.86763716	6.84712312
0.75	0.72765035	0.62775880	0.90265440	5.34134156	4.24383806	7.99198028
0.80	0.77654611	0.66998827	0.98008602	5.97786511	4.67722507	9.55181749
0.85	0.83354009	0.71675292	1.07280097	6.81616496	5.20898272	11.82499505
0.90	0.90525154	0.77312953	1.19192154	8.03991644	5.93102197	15.55684546
0.91	0.92257202	0.78645406	1.22098490	8.36704340	6.11581110	16.63354803
0.92	0.94138838	0.80082817	1.25265947	8.73752393	6.32161687	17.89202400
0.93	0.96207797	0.81652514	1.28759550	9.16384997	6.55428229	19.39078975
0.94	0.98518500	0.83393814	1.32673162	9.66462490	6.82241516	21.21932771
0.95	1.01153870	0.85366509	1.37149925	10.26924927	7.13945541	23.52335439
0.96	1.04250089	0.87668630	1.42425086	11.02810501	7.52811590	26.56139396
0.97	1.08056498	0.90479395	1.48929626	12.03829491	8.03144982	30.85291929
0.98	1.13116455	0.94188765	1.57603332	13.52584941	8.74757447	37.67327018
0.99	1.21091573	0.99987058	1.71322298	16.25233377	9.99702053	51.66815832

REFERENCE

Finney, D.J. (1971), *Statistical Methods in Biological Assay*, Second Edition, London: Griffin Press.

MULTIVARIATE ANALYSIS

Introduction

CANCORR

FACTOR

PRINCOMP

Introduction to Multivariate Procedures

The procedures in this chapter investigate relationships among variables without designating some as independent and others as dependent. Principal component and common factor analysis examine relationships within a single set of variables, while canonical correlation looks at the relationship between two sets of variables. The three procedures are:

PRINCOMP	performs a principal component analysis and computes principal component scores.
FACTOR	performs principal component and common factor analyses with rotations. Factor scores can be obtained in conjunction with the SCORE procedure.
CANCORR	performs a canonical correlation analysis and computes canonical variable scores.

The purpose of principal component analysis is to derive a small number of linear combinations (principal components) of a set of variables that retain as much of the information in the original variables as possible. Often a small number of principal components can be used in place of the original variables for plotting, regression, clustering, and so on. Principal component analysis can also be viewed as an attempt to uncover approximate linear dependencies among variables.

The purpose of common factor analysis is to explain the correlations or covariances among a set of variables in terms of a limited number of unobservable, latent variables. The latent variables are not generally computable as linear combinations of the original variables. In common factor analysis it is assumed that the variables would be linearly related were it not for uncorrelated random error or unique variation in each variable; both the linear relations and the amount of unique variation can be estimated. Principal component analysis should not be used if a common factor solution is desired (Dziuban and Harris, 1973; Lee and Comrey, 1979).

The purpose of canonical correlation analysis is to explain or summarize the relationship between two sets of variables by finding a small number of linear combinations from each set of variables that have the highest possible between-set correlations. Plots of the canonical variables may be useful in examining multivariate dependencies. With appropriate input, the CANCORR procedure can be used for maximum redundancy analysis (van den Wollenberg, 1977) or principal components of instrumental variables (Rao, 1964); contingency table analysis and optimal scaling (Mardia, Kent, and Bibby, 1979, pp. 290-295; Kshirsagar, 1972;

Nishisato, 1980); orthogonal Procrustes rotation (Mulaik, 1972; Hanson and Norris, 1981); and finding a polynomial transformation of the dependent variable to minimize interaction in an analysis of variance.

Comparison of PROC PRINCOMP and PROC FACTOR

Although PROC FACTOR performs common factor analysis, the default method of FACTOR is principal component analysis. FACTOR produces the same results as PRINCOMP except that the scoring coefficients from FACTOR are normalized to give principal component scores with unit variance, while PRINCOMP produces scores with variance equal to the corresponding eigenvalue.

PRINCOMP has the following advantages over FACTOR:

- PRINCOMP produces scores directly, while FACTOR must be followed by the SCORE procedure to compute principal component scores.
- PRINCOMP handles BY-groups more easily.
- PRINCOMP is faster if a small number of components are requested.

FACTOR has the following advantages over PRINCOMP for principal component analysis:

- FACTOR has more output than PRINCOMP, including the scree plot, pattern matrix, and residual correlations.
- FACTOR has options for printing matrices in more easily interpretable forms than PRINCOMP.
- FACTOR does rotations. *Rotation* is the application of a non-singular linear transformation to common factors to aid interpretation. An orthogonal rotation leaves originally uncorrelated factors uncorrelated. An oblique rotation causes originally uncorrelated factors to become correlated. The FACTOR procedure can rotate principal components from PRINCOMP and canonical variables from CANCORR or CANDISC.

REFERENCES

Dziuban, C.D. and Harris, C.W. (1973), "On the Extraction of Components and the Applicability of the Factor Model," *American Educational Research Journal*, 10, 93-99.

Hanson, R.J. and Norris, M.J. (1981), "Analysis of Measurements Based on the Singular Value Decomposition," *SIAM Journal on Scientific and Statistical Computing*, 2, 363-373.

Kshirsagar, A.M. (1972), *Multivariate Analysis*, New York: Marcel Dekker.

Lee, B.L. and Comrey, A.L. (1979), "Distortions in a Commonly Used Factor Analytic Procedure," *Multivariate Behavioral Research*, 14, 301-321.

Mardia, K.V., Kent, J.T., and Bibby, J.M. (1979), *Multivariate Analysis*, London: Academic Press.

Mulaik, S.A. (1972), *The Foundations of Factor Analysis*. New York: McGraw-Hill Book Co.

Nishisato, S. (1980), *Analysis of Categorical Data: Dual Scaling and Its Applications*. Toronto: University of Toronto Press.

Rao, C.R. (1964), "The Use and Interpretation of Principal Component Analysis in Applied Research," *Sankhya A*, 26, 329-358.

van den Wollenberg, A.L. (1977), "Redundancy Analysis—An Alternative to Canonical Correlation Analysis," *Psychometrika*, 42, 207-219.

The CANCORR Procedure

ABSTRACT

The CANCORR procedure performs canonical correlation, partial canonical correlation, and canonical redundancy analysis. CANCORR can create output data sets containing canonical coefficients and scores on canonical variables.

INTRODUCTION

Canonical correlation is a technique for analyzing the relationship between two sets of variables. Each set may contain several variables. Simple and multiple correlation are special cases of canonical correlation in which one or both sets contain a single variable.

CANCORR tests a series of hypotheses that each canonical correlation and all smaller canonical correlations are 0 in the population. CANCORR uses an *F* approximation (Rao, 1973; Kshirsagar, 1972) that gives better small sample results than the usual χ^2 approximation. At least one of the two sets of variables should have an approximate multivariate normal distribution for the probability levels to be valid.

Both standardized and unstandardized canonical coefficients are produced, as well as all correlations between canonical variables and the original variables. A canonical redundancy analysis (Stewart and Love, 1968; Cooley and Lohnes, 1971) can also be performed.

CANCORR can produce a data set containing the scores on each canonical variable. You can use the PRINT procedure to list these values. A plot of each canonical variable against its counterpart in the other group is often useful, and PROC PLOT can be used with the output data set to produce these plots.

A second output data set contains the canonical coefficients, which can be rotated by the FACTOR procedure.

Background

Canonical correlation was developed by Hotelling (1936). The application of canonical correlation is discussed by Cooley and Lohnes (1971), Tatsuoka (1971), and Mardia, Kent and Bibby (1979). One of the best theoretical treatments is given by Kshirsagar (1972).

Given two sets of variables, CANCORR finds a linear combination from each set, called a canonical variable, such that the correlation between the two canonical variables is maximized. This correlation between the two canonical variables is the first canonical correlation. The coefficients of the linear combinations are canonical coefficients or canonical weights. It is customary to normalize the canonical coefficients so that each canonical variable has a variance of one.

CANCORR continues by finding a second set of canonical variables, uncorrelated with the first pair, that produces the second highest correlation coefficient. The process of constructing canonical variables continues until the number of pairs of canonical variables equals the number of variables in the smaller group.

Each canonical variable is uncorrelated with all the other canonical variables of either set except for the one corresponding canonical variable in the opposite set. The canonical coefficients are not generally orthogonal, however, so the canonical variables do not represent jointly perpendicular directions through the space of the original variables.

The first canonical correlation is at least as large as the multiple correlation between any variable and the opposite set of variables. It is possible for the first canonical correlation to be very high even if all the multiple correlations are low. It is also possible for the first canonical correlation to be very large while all the multiple correlations for predicting one of the original variables from the opposite set of canonical variables are small. Canonical redundancy analysis (Stewart and Love, 1968; Cooley and Lohnes, 1971; van den Wollenberg, 1977), available with the CANCORR procedure, examines how well the original variables can be predicted from the canonical variables.

CANCORR can also perform partial canonical correlation, a multivariate generalization of ordinary partial correlation (Cooley and Lohnes, 1971; Timm, 1975). Most commonly used parametric statistical methods, ranging from *t* tests to multivariate analysis of covariance, are special cases of partial canonical correlation.

SPECIFICATIONS

The CANCORR procedure is invoked by the following statements:

PROC CANCORR *options;*
  VAR *variables;*
  WITH *variables;*
  PARTIAL *variables;*
  FREQ *variable;*
  WEIGHT *variable;*
  BY *variables;*

Usually only the VAR and WITH statements are needed in addition to the PROC CANCORR statement. The WITH statement is required.

PROC CANCORR Statement

PROC CANCORR *options;*

The options that may appear in the PROC CANCORR statement are listed below:

DATA=*SASdataset* names the SAS data set to be analyzed by CANCORR. It may be an ordinary SAS data set or a TYPE=CORR or TYPE=COV data set. If DATA= is omitted, the most recently created SAS data set is used.

OUT=*SASdataset* names an output SAS data set to contain all the original data plus scores on the canonical variables. If you want to create a permanent SAS data set, you must specify a two-level name. See Chapter 12, "SAS

Data Sets'', in the *SAS User's Guide: Basics, 1982 Edition*, for more information on permanent SAS data sets.

OUTSTAT= *SASdataset* — produces a SAS data set containing various statistics including the canonical correlations and coefficients. If you want to create a permanent SAS data set, you must specify a two-level name. See Chapter 12, ''SAS Data Sets'', in the *SAS User's Guide: Basics, 1982 Edition*, for more information on permanent SAS data sets.

SIMPLE
S — prints means and standard deviations.

CORR
C — prints correlations among the original variables.

REDUNDANCY
RED — prints redundancy statistics.

ALL — prints all optional output.

NCAN=*n* — specifies the number of canonical variables for which full output is desired.

EDF=*errordf* — specifies the error degrees of freedom from the regression analysis if the input observations are residuals from a regression. The effective number of observations is the EDF= value plus one. If you have 100 observations, then EDF=99 has the same effect as omitting EDF=.

RDF=*regressiondf* — specifies the regression degrees of freedom if the input observations are residuals from a regression analysis. The effective number of observations is the actual number minus the RDF= value. The degree of freedom for the intercept should not be included in RDF=.

NOINT — indicates that the model should not contain the intercept.

TOLERANCE=*p*
TOL=*p* — specifies tolerance for determining singularity where $0<p<1$. The default value is 1E-8.

VPREFIX=*name*
VP=*name* — specifies a prefix for naming canonical variables from the VAR statement. By default, these canonical variables are given the names V1, V2, and so on. If VPREFIX=ABC is specified, then the names are ABC1, ABC2, and so forth.

WPREFIX=*name*
WP=*name* — specifies a prefix for naming canonical variables from the WITH statement. By default, these canonical variables are given the names W1, W2, and so on. If WPREFIX=XYZ is specified, then the names are XYZ1, XYZ2, and so forth.

VNAME=*'label'*
VN=*'label'* — specifies a character constant up to 40 characters long to refer to variables from the VAR statement on the printout. The constant should be enclosed in single quotes. If VNAME= is omitted, these variables are referred to as the 'VAR' VARIABLES.

WNAME=*'label'*
WN=*'label'* — specifies a character constant up to 40 characters long to refer to variables from the WITH statement on the

printout. The constant should be enclosed in single quotes. If WNAME= is omitted, these variables are referred to as the 'WITH' VARIABLES.

VAR Statement

VAR *variables;*

The VAR statement lists the variables in the first of the two sets of variables to be analyzed. The variables must be numeric. If the VAR statement is omitted, all numeric variables not mentioned in other statements make up the first set of variables.

WITH Statement

WITH *variables;*

The WITH statement lists the variables in the second set of variables to be analyzed. The variables must be numeric. The WITH statement must be present.

PARTIAL Statement

PARTIAL *variables;*

The PARTIAL statement may be used to base the canonical analysis on partial correlations. The variables in the PARTIAL statement are partialled out of the VAR and WITH variables.

WEIGHT Statement

WEIGHT *variable;*

If you want to compute weighted product-moment correlation coefficients, give the name of the weighting variable in a WEIGHT statement.

If the WEIGHT statement is specified, the divisor used to compute variances is the sum of the weights, rather than the number of observations minus one.

FREQ Statement

FREQ *variable;*

If one variable in your input data set represents the frequency of occurrence for other values in the observation, specify the variable's name in a FREQ statement. CANCORR then treats the data set as if each observation appeared *n* times, where *n* is the value of the FREQ variable for the observation. The total number of observations is considered equal to the sum of the FREQ variable when CANCORR calculates significance probabilities.

WEIGHT and FREQ statements have a similar effect except in calculating degrees of freedom.

BY Statement

BY *variables;*

A BY statement may be used with PROC CANCORR to obtain separate analyses on observations in groups defined by the BY variables. When a BY statement appears, the procedure expects the input data set to be sorted in order of the BY variables. If your input data set is not sorted in ascending order, use the SORT procedure with a similar BY statement to sort the data, or, if appropriate, use the BY statement options NOTSORTED or DESCENDING. For more information, see the discussion of the BY statement in Chapter 8, "Statements Used in the PROC Step," in the *SAS User's Guide: Basics, 1982 Edition*.

DETAILS

Missing Values

If an observation has a missing value for any of the variables in the analysis, that observation is omitted from the analysis.

Output Data Sets

OUT= Data Set The OUT= data set contains all the variables in the original data set plus new variables containing the canonical variable scores. The number of new variables is twice that specified by the NCAN= option. The names of the new variables are formed by concatenating the values given by the VPREFIX= and WPREFIX= options (the defaults are V and W) with the numbers 1, 2, 3, and so on. The new variables have mean 0 and variance equal to 1. An OUT= data set cannot be created if the DATA= data set is TYPE=CORR or TYPE=COV or if a PARTIAL statement is used.

OUTSTAT= Data Set The OUTSTAT= data set is similar to the TYPE=CORR data set produced by the CORR procedure, but contains several results in addition to those produced by CORR.

The new data set contains the following variables:

- the BY variables, if any
- two new character variables, _TYPE_ and _NAME_
- the variables analyzed (those in the VAR statement and WITH statement).

Each observation in the new data set contains some type of statistic as indicated by the _TYPE_ variable. The values of the _TYPE_ variable are as follows:

TYPE	**Contents**
MEAN	means.
STD	standard deviations.
N	number of observations.
CORR	correlations.
CANCORR	canonical correlations.
SCORE	standardized canonical coefficients.
RAWSCORE	raw canonical coefficients.
STRUCTUR	canonical structure.

Computational Resources

Let:

n = number of observations

v = number of VAR variables

w = number of WITH variables
p = max(v,w)
q = min(v,w).

The time required to compute the correlation matrix is roughly proportional to $n(p+q)^2$.

The time required for the canonical analysis is roughly proportional to:

$$p^3/6 + p^2q + 3pq^2/2 + 5q^3$$

but the coefficient for q^3 varies depending on the number of QR iterations in the singular value decomposition.

Printed Output

If specifically requested, CANCORR prints:

1. SIMPLE STATISTICS of the input variables (MEAN, ST. DEV., SKEWNESS, KURTOSIS)
2. CORRELATIONS AMONG THE input variables.

CANCORR automatically prints the following:

3. CANONICAL CORRELATIONS
4. ADJUSTED CANCORR, adjusted canonical correlations (Lawley, 1959), which are asymptotically less biased than the raw correlations, and may be negative. The adjusted canonical correlations may not be computable and are printed as missing values if two canonical correlations are nearly equal or if some are close to zero. A missing value is also printed if an adjusted canonical correlation is larger than a previous adjusted canonical correlation.
5. APPROX STD ERROR, the approximate standard error of the canonical correlations
6. VARIANCE RATIO, the ratio of the regression variance to the error variance for each pair of canonical variables, which is equal to the reciprocal of 1 minus the squared canonical correlation.
7. CANONICAL R-SQUARED, the squared canonical correlations
8. LIKELIHOOD RATIO for the hypothesis that the current canonical correlation and all smaller ones are 0 in the population. The likelihood ratio for all canonical correlations equals Wilks' lambda.
9. F STATISTIC based on Rao's approximation (Rao, 1973, p. 556; Kshirsagar, 1972, p. 326)
10. NUM DF and DEN DF (numerator and denominator degrees of freedom) and PROB>F (probability level) associated with the F statistic
11. WILKS' LAMBDA, PILLAI'S TRACE, HOTELLING-LAWLEY TRACE, and ROY'S GREATEST ROOT with F approximations and the associated NUM DF, DEN DF (degrees of freedom), and PROB>F (probability level)
12. both STANDARDIZED and RAW (unstandardized) CANONICAL COEFFICIENTS normalized to give canonical variables with unit variance
13. all four CANONICAL STRUCTURE matrices, giving CORRELATIONS BETWEEN THE canonical variables and the original variables

If requested, CANCORR also prints:

14. the CANONICAL REDUNDANCY ANALYSIS (Stewart and Love, 1968; Cooley and Lohnes, 1971), including RAW (unstandardized) and STANDARDIZED PROPORTION and CUMULATIVE PROPORTION of the

VARIANCE OF each set of variables EXPLAINED BY THEIR OWN CANONICAL VARIABLES and the OPPOSITE CANONICAL VARIABLES

15. the SQUARED MULTIPLE CORRELATIONS of each variable with the first *m* canonical variables of the opposite set, where *m* varies from 1 to the number of canonical correlations.

EXAMPLE

Three physiological and three exercise variables were measured on twenty middle-aged men in a fitness club. CANCORR can be used to determine if the physiological variables are related in any way to the exercise variables.

```
DATA FIT;
  INPUT WEIGHT WAIST PULSE CHINS SITUPS JUMPS;
  CARDS;
  191 36 50  5 162  60
  189 37 52  2 110  60
  193 38 58 12 101 101
  162 35 62 12 105  37
  189 35 46 13 155  58
  182 36 56  4 101  42
  211 38 56  8 101  38
  167 34 60  6 125  40
  176 31 74 15 200  40
  154 33 56 17 251 250
  169 34 50 17 120  38
  166 33 52 13 210 115
  154 34 64 14 215 105
  247 46 50  1  50  50
  193 36 46  6  70  31
  202 37 62 12 210 120
  176 37 54  4  60  25
  157 32 52 11 230  80
  156 33 54 15 225  73
  138 33 68  2 110  43
  ;
PROC CANCORR DATA=FIT ALL
  VPREFIX=PHYS VNAME='PHYSIOLOGICAL MEASUREMENTS'
  WPREFIX=EXER WNAME='EXERCISES';
  VAR WEIGHT WAIST PULSE;
  WITH CHINS SITUPS JUMPS;
  TITLE MIDDLE-AGE MEN IN A HEALTH FITNESS CLUB;
  TITLE2 DATA COURTESY OF DR. A. C. LINNERUD, NC STATE UNIV;
```

```
                         MIDDLE-AGE MEN IN A HEALTH FITNESS CLUB                          1
                   DATA COURTESY OF DR. A. C. LINNERUD, NC STATE UNIV

                             CANONICAL CORRELATION ANALYSIS

20 OBSERVATIONS
 3 PHYSIOLOGICAL MEASUREMENTS
 3 EXERCISES
```

(continued on next page)

(continued from previous page)

(1)

SIMPLE STATISTICS

VARIABLE	MEAN	ST. DEV.	SKEWNESS	KURTOSIS
WEIGHT	178.60000000	24.690505313	0.9698740166	1.802346254
WAIST	35.40000000	3.201973076	1.8721345109	5.662099021
PULSE	56.10000000	7.210372645	0.8460998408	0.606913027
CHINS	9.45000000	5.286278165	-.1930287524	-1.413520979
SITUPS	145.55000000	62.566575068	0.2236427744	-1.329139344
JUMPS	70.30000000	51.277470173	2.4799104223	7.623492371

(2) CORRELATIONS AMONG THE PHYSIOLOGICAL MEASUREMENTS

	WEIGHT	WAIST	PULSE
WEIGHT	1.0000	0.8702	-.3658
WAIST	0.8702	1.0000	-.3529
PULSE	-.3658	-.3529	1.0000

CORRELATIONS AMONG THE EXERCISES

	CHINS	SITUPS	JUMPS
CHINS	1.0000	0.6957	0.4958
SITUPS	0.6957	1.0000	0.6692
JUMPS	0.4958	0.6692	1.0000

CORRELATIONS BETWEEN THE PHYSIOLOGICAL MEASUREMENTS AND THE EXERCISES

	CHINS	SITUPS	JUMPS
WEIGHT	-.3897	-.4931	-.2263
WAIST	-.5522	-.6456	-.1915
PULSE	0.1506	0.2250	0.0349

MIDDLE-AGE MEN IN A HEALTH FITNESS CLUB
DATA COURTESY OF DR. A. C. LINNERUD, NC STATE UNIV 2

CANONICAL CORRELATION ANALYSIS

(3) (4) (5) (6) (7) (8) (9) (10)

CANONICAL CORRELATIONS AND TESTS OF H0: THE CANONICAL CORRELATION IN THE CURRENT ROW AND ALL THAT FOLLOW ARE ZERO

	CANONICAL CORRELATION	ADJUSTED CAN CORR	APPROX STD ERROR	VARIANCE RATIO	CANONICAL R-SQUARED	LIKELIHOOD RATIO	F STATISTIC	NUM DF	DEN DF	PROB>F
1	0.795608154	0.705498736	0.084197333	1.7247	0.632992335	0.350390533	2.0482	9	34.223	0.0635
2	0.200556041	-0.580144410	0.220188008	0.0419	0.040222726	0.954722659	0.1758	4	30	0.9491
3	0.072570286	-1.261851594	0.228207528	0.0053	0.005266446	0.994733554	0.0847	1	16	0.7748

MULTIVARIATE TEST STATISTICS AND F APPROXIMATIONS

(11)

STATISTIC	VALUE	F	NUM DF	DEN DF	PROB>F
WILKS' LAMBDA	0.3503905	2.048234	9	34.22293	0.06353094
PILLAI'S TRACE	0.6784815	1.558707	9	48	0.1551082
HOTELLING-LAWLEY TRACE	1.771941	2.493844	9	38	0.02384017
ROY'S GREATEST ROOT	1.724739	9.198607	3	16	0.0009016772

NOTE: F STATISTIC FOR ROY'S GREATEST ROOT IS AN UPPER BOUND

RAW CANONICAL COEFFICIENTS FOR THE PHYSIOLOGICAL MEASUREMENTS

	PHYS1	PHYS2	PHYS3
WEIGHT	-.0314046879	-.0763195063	-.0077350467
WAIST	0.4932416756	0.3687229894	0.1580336471
PULSE	-.0081993154	-.0320519942	0.1457322421

RAW CANONICAL COEFFICIENTS FOR THE EXERCISES

(12)

	EXER1	EXER2	EXER3
CHINS	-.0661139864	-.0710412111	-.2452753473
SITUPS	-.0168462308	0.0019737454	0.0197676373
JUMPS	0.0139715689	0.0207141063	-.0081674724

STANDARDIZED CANONICAL COEFFICIENTS FOR THE PHYSIOLOGICAL MEASUREMENTS

	PHYS1	PHYS2	PHYS3
WEIGHT	-0.7754	-1.8844	-0.1910
WAIST	1.5793	1.1806	0.5060
PULSE	-0.0591	-0.2311	1.0508

MIDDLE-AGE MEN IN A HEALTH FITNESS CLUB
DATA COURTESY OF DR. A. C. LINNERUD, NC STATE UNIV 3

CANONICAL CORRELATION ANALYSIS

(12) STANDARDIZED CANONICAL COEFFICIENTS FOR THE EXERCISES

	EXER1	EXER2	EXER3
CHINS	-0.3495	-0.3755	-1.2966
SITUPS	-1.0540	0.1235	1.2368
JUMPS	0.7164	1.0622	-0.4188

MIDDLE-AGE MEN IN A HEALTH FITNESS CLUB
DATA COURTESY OF DR. A. C. LINNERUD, NC STATE UNIV 4

(13) CANONICAL STRUCTURE

CORRELATIONS BETWEEN THE PHYSIOLOGICAL MEASUREMENTS AND THEIR CANONICAL VARIABLES

	PHYS1	PHYS2	PHYS3
WEIGHT	0.6206	-0.7724	-0.1350
WAIST	0.9254	-0.3777	-0.0310
PULSE	-0.3328	0.0415	0.9421

CORRELATIONS BETWEEN THE EXERCISES AND THEIR CANONICAL VARIABLES

	EXER1	EXER2	EXER3
CHINS	-0.7276	0.2370	-0.6438
SITUPS	-0.8177	0.5730	0.0544
JUMPS	-0.1622	0.9586	-0.2339

CORRELATIONS BETWEEN THE PHYSIOLOGICAL MEASUREMENTS AND THE CANONICAL VARIABLES OF THE EXERCISES

	EXER1	EXER2	EXER3
WEIGHT	0.4938	-0.1549	-0.0098
WAIST	0.7363	-0.0757	-0.0022
PULSE	-0.2648	0.0083	0.0684

CORRELATIONS BETWEEN THE EXERCISES AND THE CANONICAL VARIABLES OF THE PHYSIOLOGICAL MEASUREMENTS

	PHYS1	PHYS2	PHYS3
CHINS	-0.5789	0.0475	-0.0467
SITUPS	-0.6506	0.1149	0.0040
JUMPS	-0.1290	0.1923	-0.0170

MIDDLE-AGE MEN IN A HEALTH FITNESS CLUB
DATA COURTESY OF DR. A. C. LINNERUD, NC STATE UNIV 5

(14) CANONICAL REDUNDANCY ANALYSIS

RAW VARIANCE OF THE PHYSIOLOGICAL MEASUREMENTS EXPLAINED BY

	THEIR OWN CANONICAL VARIABLES			THE OPPOSITE CANONICAL VARIABLES	
	PROPORTION	CUMULATIVE PROPORTION	CANONICAL R-SQUARED	PROPORTION	CUMULATIVE PROPORTION
1	0.3712	0.3712	0.6330	0.2349	0.2349
2	0.5436	0.9148	0.0402	0.0219	0.2568
3	0.0852	1.0000	0.0053	0.0004	0.2573

RAW VARIANCE OF THE EXERCISES EXPLAINED BY

	THEIR OWN CANONICAL VARIABLES			THE OPPOSITE CANONICAL VARIABLES	
	PROPORTION	CUMULATIVE PROPORTION	CANONICAL R-SQUARED	PROPORTION	CUMULATIVE PROPORTION
1	0.4111	0.4111	0.6330	0.2602	0.2602
2	0.5635	0.9746	0.0402	0.0227	0.2829
3	0.0254	1.0000	0.0053	0.0001	0.2830

(continued on next page)

(continued from previous page)

STANDARDIZED VARIANCE OF THE PHYSIOLOGICAL MEASUREMENTS EXPLAINED BY

	THEIR OWN CANONICAL VARIABLES			THE OPPOSITE CANONICAL VARIABLES	
	PROPORTION	CUMULATIVE PROPORTION	CANONICAL R-SQUARED	PROPORTION	CUMULATIVE PROPORTION
1	0.4508	0.4508	0.6330	0.2854	0.2854
2	0.2470	0.6978	0.0402	0.0099	0.2953
3	0.3022	1.0000	0.0053	0.0016	0.2969

STANDARDIZED VARIANCE OF THE EXERCISES EXPLAINED BY

	THEIR OWN CANONICAL VARIABLES			THE OPPOSITE CANONICAL VARIABLES	
	PROPORTION	CUMULATIVE PROPORTION	CANONICAL R-SQUARED	PROPORTION	CUMULATIVE PROPORTION
1	0.4081	0.4081	0.6330	0.2584	0.2584
2	0.4345	0.8426	0.0402	0.0175	0.2758
3	0.1574	1.0000	0.0053	0.0008	0.2767

MIDDLE-AGE MEN IN A HEALTH FITNESS CLUB 6
DATA COURTESY OF DR. A. C. LINNERUD, NC STATE UNIV

CANONICAL REDUNDANCY ANALYSIS

SQUARED MULTIPLE CORRELATIONS BETWEEN THE PHYSIOLOGICAL MEASUREMENTS AND THE FIRST 'M' CANONICAL VARIABLES OF THE EXERCISES

	M	1	2	3
15	WEIGHT	0.2438	0.2678	0.2679
	WAIST	0.5421	0.5478	0.5478
	PULSE	0.0701	0.0702	0.0749

SQUARED MULTIPLE CORRELATIONS BETWEEN THE EXERCISES AND THE FIRST 'M' CANONICAL VARIABLES OF THE PHYSIOLOGICAL MEASUREMENTS

M	1	2	3
CHINS	0.3351	0.3374	0.3396
SITUPS	0.4233	0.4365	0.4365
JUMPS	0.0167	0.0536	0.0539

The simple statistics show that WAIST and JUMPS exhibit high kurtosis (5.662 and 7.623). Since this is a small sample, the significance tests should be interpreted cautiously.

The correlations between the physiological and exercise variables are moderate, the largest being -.6456 between WAIST and SITUPS. There are larger within-set correlations: 0.8702 between WEIGHT and WAIST, 0.6957 between CHINS and SITUPS, and 0.6692 between SITUPS and JUMPS.

The first canonical correlation is 0.7956, which would appear to be substantially larger than any of the between-set correlations. The probability level for the null hypothesis that all the canonical correlations are 0 in the population is only .0635, so no firm conclusions can be drawn. The remaining canonical correlations are not worthy of consideration, as can be seen from the probability levels and especially from the negative adjusted canonical correlations.

Because the variables are not measured in the same units, the standardized coefficients rather than the raw coefficients should be interpreted. The correlations given in the canonical structure matrices should also be examined.

The first canonical variable for the physiological variables is a weighted difference of WAIST (1.5793) and WEIGHT (-.07754), with more emphasis on WAIST. The coefficient for PULSE is near 0. The correlations between WAIST and WEIGHT and the first canonical variable are both positive, 0.9254 for WAIST and 0.6206 for

WEIGHT. WEIGHT is therefore a suppressor variable, meaning that its coefficient and its correlation have opposite signs.

The first canonical variable for the exercise variables also shows a mixture of signs, subtracting SITUPS (-1.0540) and CHINS (-0.3495) from JUMPS (0.7164), with the most weight on SITUPS. All the correlations are negative, indicating that JUMPS is also a suppressor variable.

It may seem contradictory that a variable should have a coefficient of opposite sign from that of its correlation with the canonical variable. In order to explain how this can happen, consider a simplified situation: predicting SITUPS from WAIST and WEIGHT by multiple regression. In informal terms, it seems plausible that fat people should do fewer situps than skinny people. Assume that the men in the sample do not vary much in height, so there is a strong correlation between WAIST and WEIGHT (0.8702). Examine the relationships between fatness and the independent variables:

- People with large WAISTs tend to be fatter than people with small WAISTs. Hence the correlation between WAIST and SITUPS should be negative.
- People with high WEIGHTs tend to be fatter than people with low WEIGHTs. Therefore WEIGHT should correlate negatively with SITUPS.
- For a fixed value of WEIGHT, people with large WAISTs tend to be shorter and fatter. Thus the multiple regression coefficient for WAIST should be negative.
- For a fixed value of WAIST, people with higher WEIGHTS tend to be taller and skinnier. The multiple regression coefficient for WEIGHT should therefore be positive, of opposite sign from the correlation between WEIGHT and SITUPS.

The general interpretation of the first canonical correlation is therefore that WEIGHT and JUMPS act as suppressor variables to enhance the correlation between WAIST and SITUPS. This canonical correlation may be strong enough to be of practical interest, but the sample size is not large enough to draw definite conclusions.

The canonical redundancy analysis shows that neither of the first pair of canonical variables is a good overall predictor of the opposite set of variables, the proportions of variance explained being 0.2854 and 0.2584. The second and third canonical variables add virtually nothing, with cumulative proportions for all three canonical variables being 0.2969 and 0.2767. The squared multiple correlations indicate that the first canonical variable of the physiological measurements has some predictive power for CHINS (0.3351) and SITUPS (0.4233) but almost none for JUMPS (0.0167). The first canonical variable of the exercises is a fairly good predictor of WAIST (0.5421), a poorer predictor of WEIGHT (0.2438), and nearly useless for predicting PULSE (0.0701).

REFERENCES

Cooley, W.W. and Lohnes, P.R. (1971), *Multivariate Data Analysis*, New York: John Wiley & Sons.

Hotelling, H. (1936), "Relations Between Two Sets of Variables," *Biometrika*, 28, 321-377.

Kshirsagar, A.M. (1972), *Multivariate Analysis*, New York: Marcel Dekker.

Lawley, D.N. (1959), "Tests of Significance in Canonical Analysis," *Biometrika*, 46, 59-66.

Mardia, K.V., Kent, J.T., and Bibby, J.M. (1979), *Multivariate Analysis*, London: Academic Press.

Nishisato, S. (1980), *Analysis of Categorical Data: Dual Scaling and Its Applications*, Toronto: University of Toronto Press.

Rao, C.R. (1973), *Linear Statistical Inference*, New York: John Wiley & Sons.

Stewart, D.K. and Love, W.A. (1968), "A General Canonical Correlation Index," *Psychological Bulletin*, 70, 160-163.

Tatsuoka, M.M. (1971), *Multivariate Analysis*, New York: John Wiley & Sons.

Timm, N.H. (1975), *Multivariate Analysis*, Monterey, California: Brooks/Cole Publishing Co.

van den Wollenberg, A.L. (1977), "Redundancy Analysis—An Alternative to Canonical Correlation Analysis," *Psychometrika*, 42, 207-219.

The FACTOR Procedure

ABSTRACT

The FACTOR procedure performs several types of common factor and component analyses. Both orthogonal and oblique rotations are available. Scoring coefficients can be computed and written to an output data set for use by the SCORE procedure to compute factor scores.

INTRODUCTION

The FACTOR procedure performs a variety of common factor and component analyses and rotations. Input can be data, a correlation matrix, a covariance matrix, a factor pattern, or a matrix of scoring coefficients. Either the correlation or covariance matrix can be factored. Most results can be saved in an output data set.

FACTOR can process output from other procedures. For example, the canonical coefficients from the CANDISC procedure can be rotated with FACTOR.

The methods for factor extraction are principal component analysis, principal factor analysis, iterated principal factor analysis, unweighted least squares factor analysis, maximum likelihood (canonical) factor analysis, alpha factor analysis, image component analysis, and Harris component analysis. A variety of methods for prior communality estimation are also available.

The methods for rotation are varimax, quartimax, equamax, orthomax with user-specified gamma, promax with user-specified exponent, Harris-Kaiser case II with user-specified exponent, and oblique Procrustean with a user-specified target pattern.

Output includes means, standard deviations, correlations, Kaiser's measure of sampling adequacy, eigenvalues, scree plot, eigenvectors, prior and final communality estimates, unrotated factor pattern, residual and partial correlations, rotated primary factor pattern, primary factor structure, inter-factor correlations, reference structure, reference axis correlations, variance explained by each factor both ignoring and eliminating other factors, plots of both rotated and unrotated factors, squared multiple correlation of each factor with the variables, and scoring coefficients.

Any topics that are not given explicit references are discussed in Mulaik (1972) or Harman (1976).

Background

See the PRINCOMP procedure for a discussion of principal component analysis.

Common factor analysis was invented by Spearman (1904). Gould (1981) gives an interesting non-technical history of factor analysis. Kim and Mueller (1978) provide a very elementary discussion of the common factor model. Gorsuch (1974) contains a broad survey of factor analysis, and Gorsuch (1974) and Cattell (1978) are useful as a guide to practical research methodology. Harman (1976) gives a lucid discussion of many of the more technical aspects of factor analysis, especially oblique rotation. Morrison (1976) and Mardia, Kent, and Bibby (1979) provide excellent statistical treatments of common factor analysis. Mulaik (1972) is the most thorough and authoritative general reference on factor analysis and is highly recommended to anyone comfortable with matrix algebra.

A frequent source of confusion in the field of factor analysis is the term *factor*. It sometimes refers to a hypothetical, unobservable variable, as in the phrase *common factor*. In this sense, *factor analysis* must be distinguished from component analysis, since a component is an observable linear combination. *Factor* is also used in the sense of *matrix factor*, in that one matrix is a factor of a second matrix if the first matrix multiplied by its transpose equals the second matrix. In this sense, *factor analysis* refers to all methods of data analysis using matrix factors, including component analysis and common factor analysis.

A *common factor* is an unobservable, hypothetical variable that contributes to the variance of at least two of the observed variables. The unqualified term "factor" often refers to a common factor. A *unique factor* is an unobservable, hypothetical variable that contributes to the variance of only one of the observed variables. The model for common factor analysis posits one unique factor for each observed variable.

The equation for the common factor model is:

$$y_{ij} = x_{i1}b_{1j} + x_{i2}b_{2j} + \ldots + x_{iq}b_{qj} + e_{ij}$$

where:

- y_{ij} is the value of the ith observation on the jth variable,
- x_{ik} is the value of the ith observation on the kth common factor,
- b_{kj} is the regression coefficient of the kth common factor for predicting the jth variable,
- e_{ij} is the value of the ith observation on the jth unique factor,
- q is the number of common factors,

and it is assumed for convenience that all variables have a mean of 0. In matrix terms these equations reduce to:

$$\mathbf{Y} = \mathbf{XB} + \mathbf{E} \ .$$

X is the matrix of factor scores, and **B′** is the factor pattern.

There are two critical assumptions:

- the unique factors are uncorrelated with each other, and
- the unique factors are uncorrelated with the common factors.

In principal component analysis, the residuals are generally correlated with each other. In common factor analysis, the unique factors play the role of residuals, and are defined to be uncorrelated both with each other and with the common factors. Each common factor is assumed to contribute to at least two variables, otherwise it would be a unique factor.

When the factors are initially extracted, it is also assumed for convenience that the common factors are uncorrelated with each other and have unit variance. In this case, the common factor model implies that the covariance s_{jk} between the jth and k^{th} variables, $j \neq k$, is given by:

$$s_{jk} = b_{1j}b_{1k} + b_{2j}b_{2k} + \ldots + b_{qj}b_{qk}$$

or

$$\mathbf{S} = \mathbf{B'B} + \mathbf{U}^2$$

where **S** is the covariance matrix of the observed variables and $\mathbf{U}^2$ is the diagonal covariance matrix of the unique factors.

If the original variables were standardized to unit variance, the above formula would yield correlations instead of covariances. It is in this sense that common factors "explain" the correlations among the observed variables. The difference between the correlation predicted by the common factor model and the actual correlation is the *residual correlation*. A good way to assess the goodness-of-fit of the common factor model is to examine the residual correlations.

The common factor model implies that the partial correlations among the variables, removing the effects of the common factors, must all be 0. When the common factors are removed, only unique factors remain, which are by definition uncorrelated.

The assumptions of common factor analysis imply that the common factors are, in general, not linear combinations of the observed variables. In fact, even if the data contain measurements on the entire population of observations, we cannot compute the scores of the observations on the common factors. Although the common factor scores cannot be computed directly, they can be estimated in a variety of ways.

The problem of factor score indeterminacy has led several factor analysts to propose methods that yield components that can be considered approximations to common factors, but are computable because they are defined as linear combinations. These methods include Harris component analysis and image component analysis. The advantage of producing determinate component scores is offset by the fact that, even if the data fit the common factor model perfectly, component methods do not generally recover the correct factor solution. You should not use any type of component analysis if you really want a common factor analysis (Dziuban and Harris, 1973; Lee and Comrey, 1979).

Outline of Use

Principal components The most important type of analysis performed by the FACTOR procedure is principal component analysis. The statement:

```
PROC FACTOR;
```

results in a principal component analysis. The output includes all the eigenvalues and the pattern matrix for eigenvalues greater than one.

Most applications require additional output. You may, for example, want to compute principal component scores for use in subsequent analyses or obtain a graphical aid to help decide how many components to keep. It is recommended that you save the results of the analysis in a permanent SAS data library by using the OUTSTAT= option. Assuming your SAS data library has the DDname SAVE and the data are in a SAS data set called RAW, you could do a principal component analysis as follows:

```
PROC FACTOR DATA=RAW SCREE MINEIGEN=0 SCORE
   OUTSTAT=SAVE.FACT_ALL;
```

The SCREE option produces a plot of the eigenvalues that is helpful in deciding how many components to use. The MINEIGEN=0 option causes all components with variance greater than zero to be retained. The SCORE option requests that scoring coefficients be computed. The OUTSTAT= option saves the results in a specially structured SAS data set. The name of the data set, in this case FACT__ALL, is arbitrary. To compute principal component scores, use the SCORE procedure:

```
PROC SCORE DATA=RAW SCORE=SAVE.FACT__ALL
  OUT=SAVE.SCORES;
```

The SCORE procedure uses the data and the scoring coefficients that were saved in SAVE.FACT__ALL to compute principal component scores. The component scores are placed in variables named FACTOR1, FACTOR2, ..., and saved in the data set SAVE.SCORES. To plot the scores for the first three components use the PLOT procedure:

```
PROC PLOT;
  PLOT FACTOR2*FACTOR1 FACTOR3*FACTOR1 FACTOR3*FACTOR2;
```

Principal factor analysis The simplest and most computationally efficient method of common factor analysis is principal factor analysis, which is obtained like principal component analysis except for the use of a PRIORS statement. The usual form of the initial analysis is:

```
PROC FACTOR DATA=RAW SCREE MINEIGEN=0
    OUTSTAT=SAVE.FACT__ALL;
  PRIORS SMC;
```

The squared multiple correlations (SMC) of each variable with all the other variables are used as the prior communality estimates. If your correlation matrix is singular, you should specify PRIORS MAX instead of PRIORS SMC. The SCREE and MINEIGEN= options serve the same purpose as in the principal component analysis above. Saving the results with the OUTSTAT= option allows you to examine the eigenvalues and scree plot before deciding how many factors to rotate, and to try several different rotations without re-extracting the factors. The OUTSTAT= data set is automatically marked TYPE=FACTOR so the FACTOR procedure realizes that it contains statistics from a previous analysis instead of data.

After looking at the eigenvalues to estimate the number of factors, you can try some rotations. Two and three factors can be rotated with the statements:

```
PROC FACTOR DATA=SAVE.FACT__ALL N=2 ROTATE=PROMAX
    ROUND REORDER SCORE OUTSTAT=SAVE.FACT__2;
PROC FACTOR DATA=SAVE.FACT__ALL N=3 ROTATE=PROMAX
    ROUND REORDER SCORE OUTSTAT=SAVE.FACT__3;
```

The output data set from the previous run is used as input for these analyses. The options N=2 and N=3 specify the number of factors to be rotated. ROTATE=PROMAX requests a promax rotation, which has the advantage of providing both orthogonal and oblique rotations with only one invocation of FACTOR. The ROUND option causes the various factor matrices to be printed in an easily-interpretable format, and the REORDER option causes the variables to be reordered on the printout so that variables associated with the same factor appear next to each other.

You can now compute and plot factor scores for the two-factor promax-rotated solution as follows:

```
PROC SCORE DATA=RAW SCORE=SAVE.FACT_2 OUT=SAVE.SCORES;
PROC PLOT;
  PLOT FACTOR2*FACTOR1;
```

Maximum likelihood factor analysis While principal factor analysis is perhaps the most commonly used method of common factor analysis, most statisticians prefer maximum likelihood (ML) factor analysis (Lawley and Maxwell, 1971). ML estimation has desirable asymptotic properties (Bickel and Doksum, 1977) and gives better estimates than principal factor analysis in large samples. You can test hypotheses about the number of common factors using the ML method.

The ML solution is equivalent to Rao's (1955) canonical factor solution and Howe's solution maximizing the determinant of the partial correlation matrix (Morrison, 1976). Thus, as a descriptive method, ML factor analysis does not require a multivariate normal distribution. The validity of the χ^2 test for the number of factors does require approximate normality plus additional regularity conditions that are usually satisfied in practice (Geweke and Singleton, 1980).

The ML method is more expensive than principal factor analysis for two reasons. First, the communalities are estimated iteratively, and each iteration takes about as much computer time as principal factor analysis. The number of iterations typically ranges from about five to twenty. Second, if you want to extract different numbers of factors, as is often the case, you must run the FACTOR procedure once for each number of factors. Therefore, an ML analysis may well take 100 times as long as a principal factor analysis.

It is a good idea to use principal factor analysis to get a rough idea of the number of factors before doing an ML analysis. If you think that there are between one and three factors, you can use the following statements for the ML analysis:

```
PROC FACTOR DATA=RAW METHOD=ML N=1
    OUTSTAT=SAVE.FACT1;
PROC FACTOR DATA=RAW METHOD=ML N=2 ROTATE=PROMAX
    OUTSTAT=SAVE.FACT2;
PROC FACTOR DATA=RAW METHOD=ML N=3 ROTATE=PROMAX
    OUTSTAT=SAVE.FACT3;
```

The output data sets can be used for trying different rotations, computing scoring coefficients, or restarting the procedure in case it does not converge within the allotted number of iterations.

The ML method cannot be used with a singular correlation matrix and is especially prone to Heywood cases. If you have problems with ML, the best alternative is METHOD=ULS for unweighted least squares factor analysis.

SPECIFICATIONS

The FACTOR procedure is invoked by the following statements:

PROC FACTOR *options;*
  PRIORS *communalities;*
  VAR *variables;*
  PARTIAL *variables;*
  FREQ *variable;*
  WEIGHT *variable;*
  BY *variables;*

Usually only the PRIORS and VAR statements are needed in addition to the PROC FACTOR statement.

PROC FACTOR Statement

PROC FACTOR *options*;

Since the FACTOR procedure has an unusually large number of options, they are discussed in five sections:

1. Data set options
2. Factor extraction options
3. Rotation options
4. Output options
5. Miscellaneous options.

Data set options

DATA=*SASdataset* names the input data set, which may be an ordinary SAS data set or a specially structured SAS data set as described in **Input Data Set**. If DATA= is omitted, the most recently created SAS data set is used.

OUTSTAT=*SASdataset* names an output data set containing most of the results of the analysis. The output data set is described in detail in **Output Data Set**. If you want to create a permanent SAS data set, you must specify a two-level name, (see Chapter 12, "SAS Data Sets," in *SAS User's Guide: Basics, 1982 Edition*, for more information on permanent data sets).

TARGET=*SASdataset* names a data set containing the target pattern for Procrustes rotation. See the ROTATE= option below. The TARGET= data set must contain variables with the same names as those being factored. Each observation in the TARGET= data set becomes one column of the target factor pattern. Missing values are treated as zeros. _NAME_ and _TYPE_ variables are not required and are ignored if present.

Factor extraction options

METHOD=*name*
M=*name* specifies the method for extracting factors. The default is METHOD=PRINCIPAL unless the DATA= data set is TYPE=FACTOR, in which case the default is METHOD=PATTERN.

METHOD=PRINCIPAL
METHOD=PRIN
METHOD=P yields principal components if no PRIORS statement is used or if PRIORS ONE is specified, principal factors otherwise.

METHOD=PRINIT yields iterated principal factor analysis.

METHOD=ULS
METHOD=U produces unweighted least squares factor analysis.

METHOD=ALPHA
METHOD=A produces alpha factor analysis.

METHOD=ML
METHOD=M performs maximum likelihood factor analysis with an algorithm due, except for minor details, to Wayne A.

Fuller (personal communication). METHOD= ML requires a nonsingular correlation matrix.

METHOD= HARRIS
METHOD= H yields Harris component analysis of $\mathbf{S}^{-1}\mathbf{R}\mathbf{S}^{-1}$ (Harris, 1962), a noniterative approximation to canonical component analysis. This method is equivalent to METHOD= IMAGE in SAS79.5, and requires a nonsingular correlation matrix.

METHOD= IMAGE
METHOD= I yields principal component analysis of the image covariance matrix, not Kaiser's (1963, 1970, 1974) image analysis. A nonsingular correlation matrix is required.

METHOD= PATTERN reads a factor pattern from a TYPE= FACTOR, CORR, or COV data set. If you create a TYPE= FACTOR data set in a DATA step, only observations containing the factor pattern (__TYPE__='PATTERN') and the interfactor correlations (__TYPE__='FCORR'), if any, are required.

METHOD= SCORE reads scoring coefficients (__TYPE__='SCORE') from a TYPE= FACTOR, CORR, or COV data set. The data set must also contain either a correlation or a covariance matrix.

COVARIANCE
COV requests that the covariance matrix be factored instead of the correlation matrix. The COV option may be used only with METHOD= PRINCIPAL, PRINIT, ULS, or IMAGE.

WEIGHT requests that a weighted correlation or covariance matrix be factored. The WEIGHT option may be used only with METHOD= PRINCIPAL, PRINIT, ULS, or IMAGE. The input data set must be TYPE= CORR, COV or FACTOR, and the variable weights are obtained from an observation with __TYPE__='WEIGHT'.

MAXITER=*n* specifies the maximum number of iterations with METHOD= PRINIT, ULS, ALPHA, or ML. The default is 30.

CONVERGE=*n*
CONV=*n* specifies the convergence criterion for METHOD= PRINIT, ULS, ALPHA, or ML. Iteration stops when the maximum change in the communalities is less than the CONVERGE value. The default value is .001.

The following options jointly control the number of factors extracted. If two or more of NFACTORS=, MINEIGEN=, and PROPORTION= are specified, the number of factors retained is the minimum number satisfying any of the criteria.

NFACTORS=*n*
NFACT=*n*
N=*n* specifies the maximum number of factors to be extracted and determines the amount of core storage to be allocated for factor matrices. The default is the number of variables. Specifying a number that is small relative to the number of variables can substantially decrease the region required to run FACTOR, especially with oblique rotations. If NFACTORS=0 is specified, eigenvalues are computed but no factors are extracted. If NFACTORS=–1 is specified, neither eigenvalues nor factors are computed. This option may be used with

METHOD= PATTERN or METHOD= SCORE to specify a smaller number of factors than are present in the data set.

PROPORTION=*n*
PERCENT=*n*
P=*n*
: specifies the proportion of common variance to be accounted for by the retained factors using the prior communality estimates. If the value is greater than one, it is interpreted as a percentage and divided by 100. PROPORTION=0.75 and PERCENT=75 are equivalent. The default is 1.0 or 100%. You cannot specify PROPORTION= with METHOD= PATTERN or METHOD= SCORE.

MINEIGEN=*n*
MIN=*n*
: specifies the smallest eigenvalue for which a factor is retained. This option may not be used with METHOD= PATTERN or METHOD= SCORE. The default is 0 unless neither NFACTORS= nor PROPORTION= is specified and one of the following conditions holds:

If METHOD= ALPHA or METHOD= HARRIS, then MINEIGEN=1.

If METHOD= IMAGE, then

$$\text{MINEIGEN} = \frac{\text{total image variance}}{\text{number of variables}} \ .$$

For any other METHOD, if prior communality estimates of 1.0 are used, then

$$\text{MINEIGEN} = \frac{\text{total weighted variance}}{\text{number of variables}} \ .$$

When factoring an unweighted correlation matrix, this value is 1.

By default, METHOD= PRINIT, ULS, ALPHA, and ML stop iterating and set the number of factors to zero if an estimated communality exceeds one. The following options allow processing to continue:

HEYWOOD
HEY
: any communality greater than 1 is set to 1 and iterations proceed.

ULTRAHEYWOOD
ULTRA
: communalities are allowed to exceed 1. The ULTRAHEYWOOD option may cause convergence problems because communalities may become extremely large and ill-conditioned Hessians may occur.

Rotation options

ROTATE=*name*
R=*name*
: gives the rotation method. The default is ROTATE= NONE.

ROTATE= VARIMAX
ROTATE= V
: varimax rotation.

ROTATE=QUARTIMAX
ROTATE= Q
: quartimax rotation.

ROTATE= EQUAMAX
ROTATE= E
: equamax rotation.

ROTATE=ORTHOMAX
: general orthomax rotation with the weight specified by the GAMMA= option.

ROTATE=HK — Harris-Kaiser case II orthoblique rotation. The HKPOWER= option can be used to set the power of the square roots of the eigenvalues by which the eigenvectors are scaled.

ROTATE=PROMAX
ROTATE=P — promax rotation. The PREROTATE= and POWER= options can be used with ROTATE=PROMAX.

ROTATE=PROCRUSTES — oblique Procrustes rotation with target pattern given by the TARGET= data set. The unrestricted least squares method is used with factors scaled to unit length after rotation.

ROTATE=NONE
ROTATE=N — no rotation is performed.

GAMMA=*n* — specifies the orthomax weight. This option may be used only with ROTATE=ORTHOMAX or PREROTATE=ORTHOMAX.

HKPOWER=*n*
HKP=*n* — specifies the power of the square roots of the eigenvalues used to rescale the eigenvectors for Harris-Kaiser (ROTATE=HK) rotation. Values between 0.0 and 1.0 are reasonable. The default value is 0.0, yielding the independent cluster solution. A value of 1.0 is equivalent to a varimax rotation. The HKPOWER option may also be specified with ROTATE=QUARTIMAX, VARIMAX, EQUAMAX, or ORTHOMAX, in which case the Harris-Kaiser rotation uses the specified orthogonal rotation method.

POWER=*n* — specifies the power to be used in computing the target pattern for ROTATE=PROMAX. The default value is 3.

PREROTATE=*name*
PRE=*name* — specifies the prerotation method for ROTATE=PROMAX. Any rotation method other than PROMAX or PROCRUSTES can be used. The default is VARIMAX. If a previously rotated pattern is read using METHOD=PATTERN, then PREROTATE=NONE should be specified.

NORM=*name* — specifies the method for normalizing the rows of the factor pattern for rotation.

If NORM=KAISER is specified, Kaiser's normalization is used. If NORM=WEIGHT is used, the rows are weighted by the Cureton-Mulaik technique (Cureton and Mulaik, 1975). If NORM=NONE or NORM=RAW is specified, normalization is not performed. The default is NORM=KAISER.

Output Options

SIMPLE
S — prints means and standard deviations.

CORR
C — prints the correlation matrix.

MSA — prints the partial correlations between each pair of variables controlling for all other variables (the negative anti-image correlations), and Kaiser's measure of sampling adequacy (Kaiser, 1970; Kaiser and Rice, 1974; Cerny and Kaiser, 1977).

SCREE prints a scree plot of the eigenvalues (Cattell, 1966; Cattell, 1978; Cattell and Vogelman, 1977; Horn and Engstrom, 1979).

EIGENVECTORS
EV prints the eigenvectors.

PRINT prints input factor pattern or scoring coefficients and related statistics. In oblique cases, the reference and factor structures are computed and printed. The PRINT option may be used only with METHOD= PATTERN or METHOD= SCORE.

RESIDUALS
RES prints the residual correlation matrix and the associated partial correlation matrix.

PREPLOT plots the factor pattern before rotation.

PLOT plots the factor pattern after rotation.

NPLOT=*n* specifies the number of factors to be plotted. The default is all the factors. The smallest allowable value is 2. If NPLOT=*n* is specified, then all pairs of the first *n* factors are plotted, giving a total of $n(n-1)/2$ plots.

SCORE prints the factor scoring coefficients. The squared multiple correlation of each factor with the variables is also printed except in the case of unrotated principal components.

ALL prints all optional output except plots and, when the input data set is TYPE= CORR, COV, or FACTOR, simple statistics, correlations, and MSA are not printed.

REORDER
RE causes the rows (variables) of various factor matrices to be reordered on the printout. Variables with their highest absolute loading (reference structure loading for oblique rotations) on the first factor are printed first, from largest to smallest loading, followed by variables with their highest absolute loading on the second factor, and so on. The order of the variables in the output data set is not affected. The factors are not reordered.

ROUND prints correlation and loading matrices with entries multiplied by 100 and rounded to the nearest integer. The exact values can be obtained from the output data set. (See also the FLAG= option.)

FLAG=*n* is used with the ROUND option to cause absolute values larger than *n* to be flagged by an asterisk. The default value is the root mean square of all the values in the matrix being printed.

FUZZ=*n* causes correlations and factor loadings with absolute values less than the specified number to print as missing values. For partial correlations the FUZZ value is divided by 2. For residual correlations the FUZZ value is divided by 4. The exact values in any matrix may be obtained from the output data set.

Miscellaneous options

NOINT requests that no intercept be used; covariances or correlations are not corrected for the mean.

NOCORR	can be used when METHOD= PATTERN or SCORE and OUTSTAT= are specified to prevent the correlation matrix from being transferred to the OUTSTAT= data set. NOCORR greatly reduces core requirements when there are many variables but few factors.
TOLERANCE=*n* TOL=*n*	specifies the tolerance for determining singularity of the correlation matrix. The default value is 1E-7.

PRIORS Statement

PRIORS *communalities*;

The PRIORS statement specifies numeric values between 0.0 and 1.0 for the prior communality estimates for each variable. The first numeric value corresponds to the first variable in the VAR statement, the second value to the second variable, and so on. The number of numeric values must equal the number of variables, for example:

```
PROC FACTOR;
  VAR X Y Z;
  PRIORS .7 .8 .9;
```

One of the following options can be specified instead of the list of constants:

ONE	all prior communalities are set to 1.0.
MAX	the prior communality estimate for each variable is its maximum absolute correlation with any other variable.
SMC	the prior communality estimate for each variable is its squared multiple correlation with all other variables.
ASMC	the prior communality estimates are proportional to the squared multiple correlations but are adjusted so that their sum is equal to that of the maximum absolute correlations (Cureton, 1968).
INPUT	the prior communality estimates are read from the DATA= data set (which must be TYPE= FACTOR), from the first observation with either _TYPE_='PRIORS' or _TYPE_= 'COMMUNAL'.
RANDOM	the prior communality estimates are set to random numbers uniformly distributed between 0 and 1.

The default prior communality estimates are as follows:

METHOD	**PRIORS**
PRINCIPAL	ONE
PRINIT	ONE
ALPHA	SMC
ULS	SMC
ML	SMC
HARRIS	(not applicable)
IMAGE	(not applicable)
PATTERN	(not applicable)
SCORE	(not applicable)

VAR Statement

VAR *variables;*

The VAR statement lists the numeric variables to be analyzed. If it is omitted, all numeric variables not given in other statements are analyzed.

PARTIAL Statement

PARTIAL *variables;*

If you want the analysis to be based on a partial correlation or covariance matrix, use the PARTIAL statement to list the variables to be partialled out.

FREQ Statement

FREQ *variable;*

If a variable in your data set represents the frequency of occurrence for the other values in the observation, include the variable's name in a FREQ statement. The procedure then treats the data set as if each observation appears *n* times, where *n* is the value of the FREQ variable for the observation. The total number of observations is considered equal to the sum of the FREQ variable when the procedure computes significance probabilities.

The WEIGHT and FREQ statements have a similar effect, except in determining the number of observations.

WEIGHT Statement

WEIGHT *variable;*

If you want to use relative weights for each observation in the input data set, specify a variable containing weights in a WEIGHT statement. This is often done when the variance associated with each observation is different and the values of the weight variable are proportional to the reciprocals of the variances.

If the WEIGHT statement is specified, the divisor used to compute variances is the sum of the weights, rather than the number of observations minus one.

BY Statement

BY *variables;*

A BY statement may be used with PROC FACTOR to obtain separate analyses on observations in groups defined by the BY variables. When a BY statement appears, the procedure expects the input data set to be sorted in order of the BY variables. If your input data set is not sorted in ascending order, use the SORT procedure with a similar BY statement to sort the data, or, if appropriate, use the BY statement options NOTSORTED or DESCENDING. For more information, see the discussion of the BY statement in Chapter 8, "Statements Used in the PROC Step," in *SAS User's Guide: Basics, 1982 Edition*.

DETAILS

Input Data Set

The FACTOR procedure can read an ordinary SAS data set containing raw data or a special TYPE=CORR, TYPE=COV, or TYPE=FACTOR data set containing previously computed statistics. A TYPE=CORR data set can be created by the CORR procedure or various other procedures such as PRINCOMP. It contains means, standard deviations, the sample size, the correlation matrix, and possibly other statistics if created by some procedure other than CORR. A TYPE=COV data set is similar to a TYPE=CORR data set but contains a covariance matrix. A TYPE=FACTOR data set can be created by the FACTOR procedure and is described in **Output Data Set**.

If your data set has many observations and you plan to run FACTOR several times, you can save computer time by first creating a TYPE=CORR data set and using it as input to FACTOR:

```
PROC CORR DATA=RAW OUTP=CORREL;
  * create TYPE=CORR data set;
PROC FACTOR DATA=CORREL METHOD=ML;
  * maximum likelihood;
PROC FACTOR DATA=CORREL;
  * principal components;
```

The data set created by the CORR procedure is automatically given the TYPE=CORR attribute, so you do not have to specify TYPE=CORR. However, if you use a DATA step with a SET statement to modify the correlation data set, you must use the TYPE=CORR attribute on the new data set. You can use a VAR statement with FACTOR when reading a TYPE=CORR data set to select a subset of the variables or change the order of the variables.

Problems can arise from using the CORR procedure when there are missing data. By default, CORR computes each correlation from all observations that have values present for the pair of variables involved (pairwise deletion). The resulting correlation matrix may have negative eigenvalues. If the NOMISS option is used with CORR, observations with any missing values are completely omitted from the calculations (listwise deletion), and there is no possibility of negative eigenvalues.

You can also have FACTOR create a TYPE=FACTOR data set, which includes all the information in a TYPE=CORR data set, and use it for repeated analyses. For a TYPE=FACTOR data set, the default value of METHOD is PATTERN. The following statements give the same FACTOR results as the previous example:

```
PROC FACTOR DATA=RAW METHOD=ML OUTSTAT=FACT;
  * maximum likelihood;
PROC FACTOR DATA=FACT METHOD=PRIN;
  * principal components;
```

A TYPE=FACTOR data set can be used to try several different rotation methods on the same data without repeatedly extracting the factors. In the following example, the second and third PROC FACTOR statements use the data set FACT created by the first PROC FACTOR statement:

```
PROC FACTOR DATA=RAW OUTSTAT=FACT;
  * principal components;
PROC FACTOR ROTATE=VARIMAX;
  * varimax rotation;
PROC FACTOR ROTATE=QUARTIMAX;
  * quartimax rotation;
```

You can create a TYPE=CORR or TYPE=FACTOR data set in a DATA step. Be sure to specify the TYPE= option in parentheses after the data set name in the

DATA statement, and include the _TYPE_ and _NAME_ variables. In a TYPE=CORR data set only the correlation matrix (_TYPE_='CORR') is necessary. It may contain missing values as long as every pair of variables has at least one nonmissing value:

```
DATA CORREL(TYPE=CORR);
  _TYPE_='CORR';
  INPUT _NAME_ $ X Y Z;
  CARDS;
X 1.0  .   .
Y  .7 1.0  .
Z  .5  .4 1.0
PROC FACTOR;
```

You can create a TYPE=FACTOR data set containing only a factor pattern (_TYPE_='PATTERN') and use the FACTOR procedure to rotate it:

```
DATA PAT(TYPE=FACTOR);
  _TYPE_='PATTERN';
  INPUT _NAME_ $ X Y Z;
  CARDS;
FACTOR1 .5 .7 .3
FACTOR2 .8 .2 .8
PROC FACTOR ROTATE=PROMAX PREROTATE=NONE;
```

If the input factors are oblique, you must also include the inter-factor correlation matrix with _TYPE_='FCORR':

```
DATA PAT(TYPE=FACTOR);
  INPUT _TYPE_ $ _NAME_ $ X Y Z;
  CARDS;
PATTERN FACTOR1  .5  .7 .3
PATTERN FACTOR2  .8  .2 .8
FCORR   FACTOR1 1.0  .2 .
FCORR   FACTOR2  .2 1.0 .
PROC FACTOR ROTATE=PROMAX PREROTATE=NONE;
```

Some procedures, such as PRINCOMP and CANDISC, produce TYPE=CORR data sets containing scoring coefficients (_TYPE_='SCORE'). These coefficients may be input to FACTOR and rotated by using the METHOD=SCORE option. The input data set **must** contain the correlation matrix as well as the scoring coefficients:

```
PROC PRINCOMP DATA=RAW N=2 OUTSTAT=PRIN;
PROC FACTOR DATA=PRIN METHOD=SCORE ROTATE=VARIMAX;
```

Missing Values

If the DATA= data set contains data (rather than a matrix or factor pattern), then observations with missing values for any variables in the analysis are omitted from the computations. If a correlation or covariance matrix is read, it may contain missing values as long as every pair of variables has at least one nonmissing entry. Missing values in a pattern or scoring coefficient matrix are treated as zeros.

Cautions

- The amount of time that FACTOR takes is roughly proportional to the cube of the number of variables. Factoring 100 variables may therefore take 1000 times as long as factoring ten variables. Iterative methods (PRINIT, ALPHA, ULS, ML) may also take 100 times as long as non-iterative methods (PRINCIPAL, IMAGE, HARRIS).
- No computer program is capable of reliably determining the optimal number of factors, since the decision is ultimately subjective. Do not blindly accept the number of factors obtained by default. Use your own judgment to make an intelligent decision.
- Singular correlation matrices cause problems with PRIORS SMC and METHOD=ML. Singularities may result from using a variable that is the sum of other variables, coding too many dummy variables from a classification variable, or having more variables than observations.
- If the CORR procedure is used to compute the correlation matrix, and there are missing data and the NOMISS option is not specified, then the correlation matrix may have negative eigenvalues.
- If a TYPE=CORR or TYPE=FACTOR data set is copied or modified using a DATA step, the new data set does not automatically have the same TYPE as the old data set. You must specify the TYPE data set option on the DATA statement. If you try to analyze a data set that has lost its TYPE=CORR attribute, FACTOR prints a warning message saying that the data set contains _NAME_ and _TYPE_ variables, but analyzes the data set as an ordinary SAS data set.
- For a TYPE=FACTOR data set, the default is METHOD=PATTERN, not METHOD=PRIN.
- In SAS79, the factor pattern in the OUT= data set was given _TYPE_='FACTOR'. In SAS82 it is _TYPE_='PATTERN'.

Output Data Set

The OUTSTAT= data set is similar to the TYPE=CORR data set produced by the CORR procedure, but is TYPE=FACTOR and contains many results in addition to those produced by CORR.

The output data set contains the following variables:

- the BY variables, if any
- two new character variables, _TYPE_ and _NAME_
- the variables analyzed, that is, those in the VAR statement, or, if there is no VAR statement, all numeric variables not listed in any other statement.

Each observation in the output data set contains some type of statistic as indicated by the _TYPE_ variable. The _NAME_ variable is blank except where otherwise indicated. The values of the _TYPE_ variable are as follows:

TYPE	**Contents**
MEAN	means.
STD	standard deviations.
N	sample size.
CORR	correlations. The _NAME_ variable contains the name of the variable corresponding to each row of the correlation matrix.

IMAGE	image coefficients. The _NAME_ variable contains the name of the variable corresponding to each row of the image coefficient matrix.
IMAGECOV	image covariance matrix. The _NAME_ variable contains the name of the variable corresponding to each row of the image covariance matrix.
COMMUNAL	final communality estimates.
PRIOR	prior communality estimates, or estimates from the last iteration for iterative methods.
WEIGHT	variable weights.
EIGENVAL	eigenvalues.
UNROTATE	unrotated factor pattern. The _NAME_ variable contains the name of the factor.
RESIDUAL	residual correlations. The _NAME_ variable contains the name of the variable corresponding to each row of the residual correlation matrix.
TRANSFOR	transformation matrix from rotation. The _NAME_ variable contains the name of the factor.
FCORR	inter-factor correlations. The _NAME_ variable contains the name of the factor.
PATTERN	factor pattern. The _NAME_ variable contains the name of the factor.
RCORR	reference axis correlations. The _NAME_ variable contains the name of the factor.
REFERENC	reference structure. The _NAME_ variable contains the name of the factor.
STRUCTUR	factor structure. The _NAME_ variable contains the name of the factor.
SCORE	scoring coefficients. The _NAME_ variable contains the name of the factor.

Factor Scores

To compute factor scores for each observation:

- Use the SCORE option on the PROC FACTOR statement.
- Create a TYPE=FACTOR output data set with the OUTSTAT= option.
- Use the SCORE procedure with both the raw data and the TYPE=FACTOR data set.
- Do **not** use the TYPE= option on the PROC SCORE statement.

For example:

```
PROC FACTOR DATA=RAW SCORE OUTSTAT=FACT;
PROC SCORE DATA=RAW SCORE=FACT OUT=SCORES;
```

or:

```
PROC CORR DATA=RAW OUTP=CORREL;
PROC FACTOR DATA=CORREL SCORE OUTSTAT=FACT;
PROC SCORE DATA=RAW SCORE=FACT OUT=SCORES;
```

A component analysis (principal, image, or Harris), produces scores with mean zero and variance one. If you have done a common factor analysis, then the true factor scores have mean zero and variance one, but the computed factor scores are only estimates of the true factor scores. These estimates have mean zero, but variance equal to the squared multiple correlation of the factor with the variables. If the true factors are uncorrelated, the estimated factor scores may have small non-zero correlations.

Variable Weights and Variance Explained

A principal component analysis of a correlation matrix treats all variables as equally important. A principal component analysis of a covariance matrix gives more weight to variables with larger variances. A principal component analysis of a covariance matrix is equivalent to an analysis of a weighted correlation matrix, where the weight of each variable is equal to its variance. Variables with large weights tend to have larger loadings on the first component and smaller residual correlations than variables with small weights.

You may want to give variables weights other than their variances. Mulaik (1972) explains how to obtain a maximally reliable component by means of a weighted principal component analysis. With the FACTOR procedure, you can indirectly give arbitrary weights to the variables by using the COV option and rescaling the variables to have variance equal to the desired weight, or directly by using the WEIGHT option and including the weights in a TYPE=CORR data set.

Arbitrary variable weights can be used with METHOD=PRINCIPAL, PRINIT, ULS, or IMAGE. Alpha and ML factor analyses compute variable weights based on the communalities (Harman, 1976, pp. 217–218). For alpha factor analysis, the weight of a variable is the reciprocal of its communality. In ML factor analysis, the weight is the reciprocal of the uniqueness. Harris component analysis uses weights equal to the reciprocal of one minus the squared multiple correlation of each variable with the other variables.

For uncorrelated factors, the variance explained by a factor can be computed with or without taking the weights into account. The usual method for computing variance accounted for by a factor is to take the sum of squares of the corresponding column of the factor pattern, yielding an unweighted result. If the square of each loading is multiplied by the weight of the variable before the sum is taken, the result is the weighted variance explained, which is equal to the corresponding eigenvalue, except in image analysis. Whether the weighted or unweighted result is more important depends on the purpose of the analysis.

In the case of correlated factors, the variance explained by a factor can be computed with or without taking the other factors into account. If you want to ignore the other factors, the variance explained is given by the weighted or unweighted sum of squares of the appropriate column of the factor structure, since the factor structure contains simple correlations. If you want to subtract the variance explained by the other factors from the amount explained by the factor in question (the "Type II" variance explained), you can take the weighted or unweighted sum of squares of the appropriate column of the reference structure, since the reference structure contains semi-partial correlations. There are other ways of measuring the variance explained. For example, given a prior ordering of the factors, you could compute a "Type I" variance explained by eliminating from each factor the variance explained by previous factors. Another method, based on direct and joint contributions, is given by Harman (1976, pp. 268–270).

Heywood Cases and Other Anomalies

Since communalities are squared correlations, one would expect them always to lie between 0 and 1. It is a mathematical peculiarity of the common factor model, however, that final communality estimates may exceed 1. If a communality equals 1, the situation is referred to as a Heywood case, and if a communality exceeds 1, it is an ultra-Heywood case. An ultra-Heywood case implies that some unique factor has negative variance, a clear indication that something is wrong. Possible causes include:

- bad prior communality estimates
- too many common factors
- too few common factors
- not enough data to provide stable estimates
- the common factor model is not an appropriate model for the data.

Whatever the cause, an ultra-Heywood case renders a factor solution invalid. Factor analysts disagree about whether or not a factor solution with a Heywood case can be considered legitimate.

Theoretically, the communality of a variable should not exceed its reliability. Violation of this condition is called a quasi-Heywood case and should be regarded with the same suspicion as an ultra-Heywood case.

Elements of the factor structure and reference structure matrices may exceed 1 only in the presence of an ultra-Heywood case. On the other hand, an element of the factor pattern may exceed 1 in an oblique rotation.

The maximum likelihood method is especially susceptible to (quasi- or ultra-) Heywood cases. During the iteration process, a variable with high communality is given a high weight, which tends to increase its communality, which increases its weight, and so on.

It is often stated that the squared multiple correlation of a variable with the other variables is a lower bound to its communality. This is true if the common factor model fits the data perfectly, but is not generally the case with real data. A final communality estimate that is less than the squared multiple correlation may therefore indicate poor fit, possibly due to not enough factors. It is by no means as serious a problem as an ultra-Heywood case. Factor methods using the Newton-Raphson method may actually produce communalities less than 0, a result even more disastrous than an ultra-Heywood case.

The squared multiple correlation of a factor with the variables may exceed 1, even in the absence of ultra-Heywood cases. This situation is also cause for alarm. Alpha factor analysis seems to be especially prone to this problem, but it does not occur with maximum likelihood. If a squared multiple correlation is negative, too many factors have been retained.

With data that do not fit the common factor model perfectly, you can expect some of the eigenvalues to be negative. If an iterative factor method converges properly, the sum of the eigenvalues corresponding to rejected factors should be 0; hence, some eigenvalues are positive and some negative. If a principal factor analysis fails to yield any negative eigenvalues, the prior communality estimates are probably too large. Negative eigenvalues cause the cumulative proportion of variance explained to exceed 1 for a sufficiently large number of factors. The cumulative proportion of variance explained by the retained factors should be approximately 1 for principal factor analysis and should converge to 1 for iterative methods. Occasionally a single factor may explain more than 100% of the common variance in a principal factor analysis, indicating that the prior communality estimates are too low.

If a squared canonical correlation or a coefficient alpha is negative, too many factors have been retained.

Principal component analysis, unlike common factor analysis, has none of the above problems if the covariance or correlation matrix is computed correctly from a data set with no missing values. Various methods for missing value correlation may produce negative eigenvalues in principal components, as may severe rounding of the correlations.

Computer Resources

Let:

n = number of observations
v = number of variables
f = number of factors
i = number of iterations during factor extraction
r = number of iterations during factor rotation.

The overall time for a factor analysis is very roughly proportional to iv^3.

The time required to compute the correlation matrix is roughly proportional to nv^2.

The time required for PRIORS SMC or ASMC is roughly proportional to v^3. The time required for PRIORS MAX is roughly proportional to v^2.

The time required to compute eigenvalues is roughly proportional to v^3.

The time required to compute final eigenvectors is roughly proportional to fv^2.

Each iteration in METHOD= PRINIT or METHOD= ALPHA requires computation of eigenvalues and f eigenvectors.

Each iteration in METHOD= ML or METHOD= ULS requires computation of eigenvalues and $v - f$ eigenvectors.

The time required for ROTATE= VARIMAX, QUARTIMAX, EQUAMAX, ORTHOMAX, PROMAX, or HK is roughly proportional to rvf^2.

ROTATE= PROCRUSTES takes time roughly proportional to vf^2.

Printed Output

FACTOR's output includes:

1. MEAN and STD DEV (standard deviation) of each variable and the number of OBSERVATIONS if SIMPLE is specified
2. CORRELATIONS if CORR is specified
3. INVERSE CORRELATION MATRIX if ALL is specified
4. PARTIAL CORRELATIONS CONTROLLING ALL OTHER VARIABLES (negative anti-image correlations) if MSA is specified. If the data are appropriate for the common factor model, the partial correlations should be small.
5. KAISER'S MEASURE OF SAMPLING ADEQUACY (Kaiser, 1970; Kaiser and Rice, 1974; Cerny and Kaiser, 1977) if MSA is specified, both OVERALL and for each variable. The MSA is a summary of how small the partial correlations are relative to the ordinary correlations. Values greater than .8 can be considered good. Values less than .5 require remedial action, either by deleting the offending variables or including other variables related to the offenders.
6. PRIOR COMMUNALITY ESTIMATES, unless 1.0s are used or METHOD= IMAGE, HARRIS, PATTERN, or SCORE
7. SQUARED MULTIPLE CORRELATIONS of each variable with all the other variables if METHOD= IMAGE or HARRIS
8. IMAGE COEFFICIENTS if METHOD= IMAGE

9. IMAGE COVARIANCE MATRIX if METHOD= IMAGE
10. PRELIMINARY EIGENVALUES based on the prior communalities if METHOD= PRINIT, ALPHA, ML, or ULS, including the TOTAL and the AVERAGE of the eigenvalues, the DIFFERENCE between successive eigenvalues, the PROPORTION of variation represented, and the CUMULATIVE proportion of variation
11. the number of FACTORS that WILL BE RETAINED unless METHOD= PATTERN or SCORE
12. A SCREE PLOT OF EIGENVALUES if SCREE is specified. The preliminary eigenvalues are used if METHOD= PRINIT, ALPHA, ML, or ULS.
13. the iteration history if METHOD= PRINIT, ALPHA, ML, or ULS, containing the iteration number (ITER); the CRITERION being optimized (Joreskog, 1977) and the RIDGE value for the iteration if METHOD= ML or ULS; the maximum CHANGE in any communality estimate; and the COMMUNALITIES
14. SIGNIFICANCE TESTS if METHOD= ML, including CHI-SQUARED, DF, and PROB>CHI**2 for H0: NO COMMON FACTORS and H0: the factors retained ARE SUFFICIENT to explain the correlations. The variables should have an approximate multivariate normal distribution for the probability levels to be valid. Lawley and Maxwell (1921) suggest that the number of observations should exceed the number of variables by 50 or more, although Geweke and Singleton (1980) claim that as few as ten observations are adequate with five variables and one common factor. Certain regularity conditions must also be satisfied for the χ^2 test to be valid (Geweke and Singleton, 1980), but in practice these conditions usually are satisfied. The notation PROB>CHI**2 means "the probability under the null hypothesis of obtaining a greater χ^2 statistic than that observed."
15. AKAIKE'S INFORMATION CRITERION if METHOD= ML. Akaike's information criterion (AIC) (Akaike, 1973; Akaike, 1974) is a general criterion for estimating the best number of parameters to include in a model when maximum likelihood estimation is used. The number of factors that yields the smallest value of AIC is considered best. AIC, like the chi-square test, tends to include factors that are statistically significant but inconsequential for practical purposes.
16. SCHWARZ'S BAYESIAN CRITERION if METHOD= ML. Schwarz's Bayesian criterion (SBC) (Schwarz, 1978) is another criterion, similar to AIC, for determining the best number of parameters. The number of factors that yields the smallest value of SBC is considered best. SBC seems to be less inclined to include trivial factors than either AIC or the chi-square test.
17. SQUARED CANONICAL CORRELATIONS if METHOD= ML, equal to the squared multiple correlation for predicting each factor from the variables
18. COEFFICIENT ALPHA FOR EACH FACTOR if METHOD= ALPHA
19. EIGENVECTORS if EIGENVECTORS or ALL is specified, unless METHOD= PATTERN or SCORE
20. EIGENVALUES OF THE (WEIGHTED) (REDUCED) (IMAGE) CORRELATION or COVARIANCE MATRIX, unless METHOD= PATTERN or SCORE. Included are the TOTAL and the AVERAGE of the eigenvalues, the DIFFERENCE between successive eigenvalues, the PROPORTION of variation represented, and the CUMULATIVE proportion of variation.

21. the FACTOR PATTERN, which is equal to both the matrix of standardized regression coefficients for predicting variables from common factors and the matrix of correlations between variables and common factors, since the extracted factors are uncorrelated
22. VARIANCE EXPLAINED BY EACH FACTOR, both WEIGHTED and UNWEIGHTED if variable weights are used.
23. FINAL COMMUNALITY ESTIMATES, including the TOTAL communality; or FINAL COMMUNALITY ESTIMATES AND VARIABLE WEIGHTS, including the TOTAL communality, both WEIGHTED and UNWEIGHTED, if variable weights are used. Final communality estimates are the squared multiple correlations for predicting the variables from the estimated factors, and can be obtained by taking the sum of squares of each row of the factor pattern, or a weighted sum of squares if variable weights are used.
24. RESIDUAL CORRELATIONS WITH UNIQUENESS ON THE DIAGONAL if RESIDUAL or ALL is SPECIFIED
25. ROOT-MEAN-SQUARE OFF-DIAGONAL RESIDUALS, both OVER-ALL and for each variable, if RESIDUAL or ALL is SPECIFIED
26. PARTIAL CORRELATIONS CONTROLLING FACTORS if RESIDUAL or ALL is SPECIFIED
27. ROOT-MEAN-SQUARE OFF-DIAGONAL PARTIALS, both OVER-ALL and for each variable, if RESIDUAL or ALL is SPECIFIED
28. PLOT OF FACTOR PATTERN for unrotated factors if PREPLOT is specified, the number of plots determined by NPLOT=
29. VARIABLE WEIGHTS FOR ROTATION if NORM=WEIGHT is specified
30. FACTOR WEIGHTS FOR ROTATION if HKPOWER= is specified
31. ORTHOGONAL TRANSFORMATION MATRIX if an orthogonal rotation is requested
32. ROTATED FACTOR PATTERN if an orthogonal rotation is requested
33. VARIANCE EXPLAINED BY EACH FACTOR after rotation. If an orthogonal rotation is requested both weighted and unweighted if variable weights are used
34. TARGET MATRIX FOR PROCRUSTEAN TRANSFORMATION if ROTATE=PROCRUSTES or PROMAX
35. PROCRUSTEAN TRANSFORMATION MATRIX if ROTATE=PROCRUSTES or PROMAX
36. OBLIQUE TRANSFORMATION MATRIX if an oblique rotation is requested, which for ROTATE=PROMAX is the product of the prerotation and the Procrustean rotation
37. INTER-FACTOR CORRELATIONS if an oblique rotation is requested
38. ROTATED FACTOR PATTERN (STD REG COEFS) if an oblique rotation is requested, giving standardized regression coefficients for predicting the variables from the factors
39. REFERENCE AXIS CORRELATIONS if an oblique rotation is requested, which are the partial correlations between the primary factors when all factors other than the two being correlated are partialled out.
40. REFERENCE STRUCTURE (SEMIPARTIAL CORRELATIONS) if an oblique rotation is requested. The reference structure is the matrix of semipartial correlations (Kerlinger and Pedhazur, 1973) between variables and common factors, removing from each common factor the effects of other common factors. If the common factors are uncorrelated, the reference structure is equal to the factor pattern.

41. VARIANCE EXPLAINED BY EACH FACTOR ELIMINATING the effects of all OTHER FACTORS if an oblique rotation is requested, both WEIGHTED and UNWEIGHTED if variable weights are used, equal to the (weighted) sum of the squared elements of the reference structure corresponding to that factor.
42. FACTOR STRUCTURE (CORRELATIONS) if an oblique rotation is requested. The (primary) factor structure is the matrix of correlations between variables and common factors. If the common factors are uncorrelated, the factor structure is equal to the factor pattern.
43. VARIANCE EXPLAINED BY EACH FACTOR IGNORING the effects of all OTHER FACTORS if an oblique rotation is requested, both WEIGHTED and UNWEIGHTED if variable weights are used, equal to the (weighted) sum of the squared elements of the factor structure corresponding to that factor
44. FINAL COMMUNALITY ESTIMATES for the rotated factors if ROTATE= is specified, which should equal the unrotated communalities
45. SQUARED MULTIPLE CORRELATIONS OF THE VARIABLES WITH EACH FACTOR if SCORE or ALL is specified, except for unrotated principal components
46. STANDARDIZED SCORING COEFFICIENTS if SCORE or ALL is specified
47. PLOT OF FACTOR PATTERN for rotated factors if PLOT is specified and an orthogonal rotation is requested, the number of plots determined by NPLOT=
48. PLOT OF REFERENCE STRUCTURE for rotated factors if PLOT is specified and an oblique rotation is requested, the number of plots determined by NPLOT= . Included are the REFERENCE AXIS CORRELATION and the ANGLE between the reference axes for each pair of factors plotted.

If ROTATE= PROMAX is used, the output includes results for both the prerotation and the Procrustean rotation.

EXAMPLES

Principal Components: Example 1

The data in the example below are five socio-economic variables for twelve census tracts in the Los Angeles Standard Metropolitan Statistical Area as given by Harman (1976).

The first analysis is a principal component analysis. Simple descriptive statistics and correlations are also printed.

```
DATA SOCECON;
  TITLE FIVE SOCIO-ECONOMIC VARIABLES;
  TITLE2 SEE PAGE 14 OF HARMAN: MODERN FACTOR ANALYSIS, 3RD ED;
  INPUT POP SCHOOL EMPLOY SERVICES HOUSE;
  CARDS;
5700 12.8 2500 270 25000
1000 10.9  600  10 10000
3400  8.8 1000  10  9000
3800 13.6 1700 140 25000
4000 12.8 1600 140 25000
8200  8.3 2600  60 12000
1200 11.4  400  10 16000
9100 11.5 3300  60 14000
```

```
9900 12.5 3400 180 18000
9600 13.7 3600 390 25000
9600  9.6 3300  80 12000
9400 11.4 4000 100 13000
;
PROC FACTOR DATA=SOCECON SIMPLE CORR;
  TITLE3 PRINCIPAL COMPONENTS ANALYSIS;
```

There are two large eigenvalues, 2.873314 and 1.796660, which together account for 93.4% of the standardized variance. Thus the first two principal components provide an adequate summary of the data for most purposes. Three components, explaining 97.7% of the variation should satisfy anybody. FACTOR retains two components on the basis of the eigenvalues-greater-than-one rule, since the third eigenvalue is only 0.214837.

The first component has large positive loadings for all five variables, with an especially high correlation with SERVICES (0.93239). The second component is a contrast of POP (0.80642) and EMPLOY (0.72605) against SCHOOL (−0.54476) and HOUSE (−0.55818), with a very small loading on SERVICES (−0.10431).

The final communality estimates show that all the variables are well accounted for by two components, with final communality estimates ranging from 0.880236 for SERVICES to 0.987826 for POP.

```
                              FIVE SOCIO-ECONOMIC VARIABLES                                1
                    SEE PAGE 14 OF HARMAN: MODERN FACTOR ANALYSIS, 3RD ED
                              PRINCIPAL COMPONENTS ANALYSIS

 (1) MEANS AND STANDARD DEVIATIONS FROM              12 OBSERVATIONS

                  POP        SCHOOL       EMPLOY     SERVICES        HOUSE
MEAN          6241.67       11.4417      2333.33      120.833        17000
STD DEV       3439.99       1.78654      1241.21      114.928      6367.53

                      (2) CORRELATIONS

                   POP      SCHOOL      EMPLOY    SERVICES       HOUSE

POP            1.00000     0.00975     0.97245     0.43887     0.02241
SCHOOL         0.00975     1.00000     0.15428     0.69141     0.86307
EMPLOY         0.97245     0.15428     1.00000     0.51472     0.12193
SERVICES       0.43887     0.69141     0.51472     1.00000     0.77765
HOUSE          0.02241     0.86307     0.12193     0.77765     1.00000
```

```
                              FIVE SOCIO-ECONOMIC VARIABLES                                2
                    SEE PAGE 14 OF HARMAN: MODERN FACTOR ANALYSIS, 3RD ED
                              PRINCIPAL COMPONENTS ANALYSIS

INITIAL FACTOR METHOD: PRINCIPAL COMPONENTS

                    (6) PRIOR COMMUNALITY ESTIMATES:   ONE

(20) EIGENVALUES OF THE CORRELATION MATRIX:   TOTAL =   5.000000    AVERAGE =   1.000000

                      1           2           3           4           5
EIGENVALUE     2.873314    1.796660    0.214837    0.099934    0.015255
DIFFERENCE     1.076654    1.581823    0.114903    0.084679
PROPORTION       0.5747      0.3593      0.0430      0.0200      0.0031
CUMULATIVE       0.5747      0.9340      0.9770      0.9969      1.0000

     (11) 2 FACTORS WILL BE RETAINED BY THE MINEIGEN CRITERION

                      (21) FACTOR PATTERN

                                FACTOR1     FACTOR2

                  POP           0.58096     0.80642
                  SCHOOL        0.76704    -0.54476
                  EMPLOY        0.67243     0.72605
                  SERVICES      0.93239    -0.10431
                  HOUSE         0.79116    -0.55818

              (22) VARIANCE EXPLAINED BY EACH FACTOR

                        FACTOR1     FACTOR2
                       2.873314    1.796660

       (23) FINAL COMMUNALITY ESTIMATES: TOTAL =      4.669974

               POP      SCHOOL      EMPLOY    SERVICES       HOUSE
          0.987826    0.885106    0.979306    0.880236    0.937500
```

Principal Factor Analysis: Example 2

The next example is a principal factor analysis using squared multiple correlations for the prior communality estimates (PRIORS SMC). Kaiser's measure of sampling adequacy (MSA) is requested. A SCREE plot of the eigenvalues is printed. The RESIDUAL correlations and partial correlations are computed. The PREPLOT option plots the unrotated factor pattern.

ROTATE=PROMAX produces an orthogonal varimax prerotation followed by an oblique rotation. The variables are REORDERed according to their largest factor loadings. SCORE requests scoring coefficients. The reference structure is PLOTted.

An output data set is created by FACTOR and printed.

```
PROC FACTOR DATA=SOCECON MSA SCREE RESIDUAL PREPLOT
                 ROTATE=PROMAX REORDER PLOT
                 OUT=FACT__ALL;
  PRIORS SMC;
  TITLE3 PRINCIPAL FACTOR ANALYSIS WITH PROMAX ROTATION;
PROC PRINT;
  TITLE3 FACTOR OUTPUT DATA SET;
```

If the data are appropriate for the common factor model, the partial correlations controlling the other variables should be small compared to the original correlations. The partial correlation between SCHOOL and HOUSE, for example, is .64, slightly less than the original correlation of .86. The partial correlation between POP and SCHOOL is -.54, which is much larger than the original correlation, an indication of trouble. Kaiser's MSA is a summary, for each variable and for all variables together, of how much smaller the partial correlations are than the original correlations. Values of .8 or .9 are considered good, while MSAs below .5 are unacceptable. POP, SCHOOL, and EMPLOY have very poor MSAs. Only SERVICES has a good MSA. The overall MSA of .57 is sufficiently bad that additional variables should be included in the analysis to better define the common factors. A commonly used rule-of-thumb is that there should be at least three variables per factor. It will be shown below that there seem to be two common factors in these data, so the problem is probably that there are not enough variables.

The SMCs are all fairly large; hence, the factor loadings do not differ greatly from the principal component analysis.

The eigenvalues show clearly that two common factors are present. There are two large positive eigenvalues that together account for 101.31% of the common variance, as close to 100% as one is ever likely to get without iterating. The scree plot shows a sharp bend at the third eigenvalue, reinforcing the above conclusion.

The principal factor pattern is similar to the principal component pattern. For example, SERVICES has the largest loading on the first factor, POP the smallest. POP and EMPLOY have large positive loadings on the second factor; HOUSE and SCHOOL have large negative loadings.

The final communality estimates are all fairly close to the priors, only HOUSE having increased appreciably from 0.847019 to 0.884950. The fact that the communality estimates appear to be good is reflected by the amount of common variance accounted for being very close to 100%.

The residual correlations are low, the largest being .03. The partial correlations are not quite as impressive, since the uniquenesses are also rather small.

The plot of the unrotated factor pattern shows two tight clusters of variables, HOUSE and SCHOOL, at the negative end of FACTOR2 and EMPLOY and POP at the positive end. SERVICES is in between but closer to HOUSE and SCHOOL. A good rotation would put the reference axes through the two clusters.

The varimax rotation puts one axis through HOUSE and SCHOOL, but misses POP and EMPLOY slightly. The promax rotation places an axis through POP and EMPLOY, but misses HOUSE and SCHOOL. Since an independent-cluster solution would be possible if it were not for SERVICES, a Harris-Kaiser rotation weighted by the Cureton-Mulaik technique should be used.

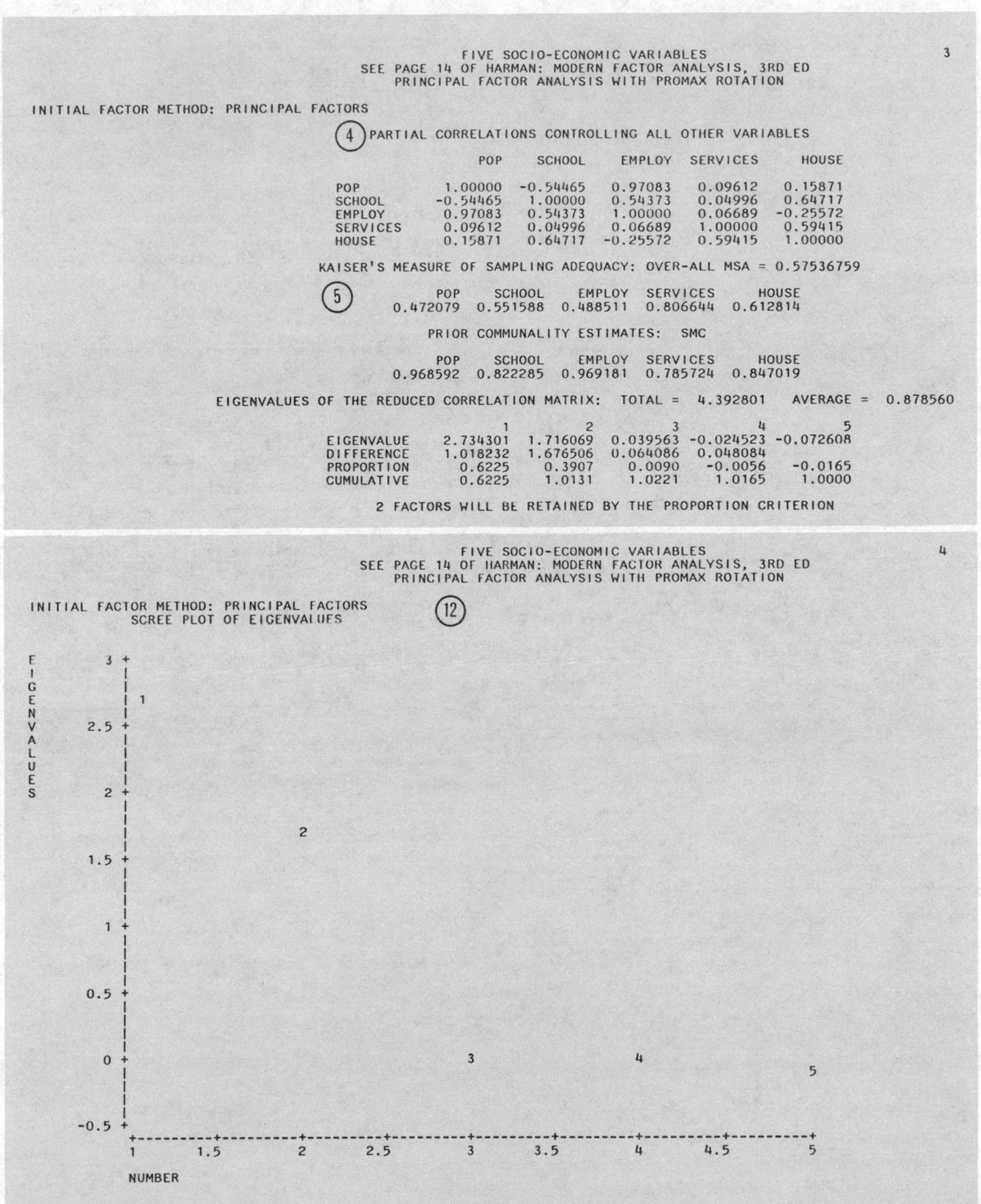

```
                              FIVE SOCIO-ECONOMIC VARIABLES                                 3
                    SEE PAGE 14 OF HARMAN: MODERN FACTOR ANALYSIS, 3RD ED
                         PRINCIPAL FACTOR ANALYSIS WITH PROMAX ROTATION

INITIAL FACTOR METHOD: PRINCIPAL FACTORS

                (4) PARTIAL CORRELATIONS CONTROLLING ALL OTHER VARIABLES

                               POP      SCHOOL     EMPLOY   SERVICES      HOUSE

               POP          1.00000   -0.54465    0.97083    0.09612    0.15871
               SCHOOL      -0.54465    1.00000    0.54373    0.04996    0.64717
               EMPLOY       0.97083    0.54373    1.00000    0.06689   -0.25572
               SERVICES     0.09612    0.04996    0.06689    1.00000    0.59415
               HOUSE        0.15871    0.64717   -0.25572    0.59415    1.00000

             KAISER'S MEASURE OF SAMPLING ADEQUACY: OVER-ALL MSA = 0.57536759

                (5)          POP     SCHOOL     EMPLOY   SERVICES      HOUSE
                        0.472079   0.551588   0.488511   0.806644   0.612814

                             PRIOR COMMUNALITY ESTIMATES:   SMC

                             POP     SCHOOL     EMPLOY   SERVICES      HOUSE
                        0.968592   0.822285   0.969181   0.785724   0.847019

     EIGENVALUES OF THE REDUCED CORRELATION MATRIX:   TOTAL =   4.392801    AVERAGE =   0.878560

                                    1          2          3          4          5
               EIGENVALUE    2.734301   1.716069   0.039563  -0.024523  -0.072608
               DIFFERENCE    1.018232   1.676506   0.064086   0.048084
               PROPORTION      0.6225     0.3907     0.0090    -0.0056    -0.0165
               CUMULATIVE      0.6225     1.0131     1.0221     1.0165     1.0000

                     2 FACTORS WILL BE RETAINED BY THE PROPORTION CRITERION

                              FIVE SOCIO-ECONOMIC VARIABLES                                 4
                    SEE PAGE 14 OF HARMAN: MODERN FACTOR ANALYSIS, 3RD ED
                         PRINCIPAL FACTOR ANALYSIS WITH PROMAX ROTATION

INITIAL FACTOR METHOD: PRINCIPAL FACTORS          (12)
            SCREE PLOT OF EIGENVALUES

E       3 +
I         I
G         I
E         I 1
N         I
V     2.5 +
A         I
L         I
U         I
E         I
S       2 +
          I
          I
          I                  2
          I
      1.5 +
          I
          I
          I
          I
        1 +
          I
          I
          I
          I
      0.5 +
          I
          I
          I
          I
        0 +                                      3                   4
          I                                                                            5
          I
          I
          I
     -0.5 +
          +---------+---------+---------+---------+---------+---------+---------+---------+
          1        1.5        2        2.5        3        3.5        4        4.5        5

          NUMBER
```

FIVE SOCIO-ECONOMIC VARIABLES 5
SEE PAGE 14 OF HARMAN: MODERN FACTOR ANALYSIS, 3RD ED
PRINCIPAL FACTOR ANALYSIS WITH PROMAX ROTATION

INITIAL FACTOR METHOD: PRINCIPAL FACTORS

FACTOR PATTERN

	FACTOR1	FACTOR2
SERVICES	0.87899	-0.15847
HOUSE	0.74215	-0.57806
EMPLOY	0.71447	0.67936
SCHOOL	0.71370	-0.55515
POP	0.62533	0.76621

VARIANCE EXPLAINED BY EACH FACTOR

FACTOR1	FACTOR2
2.734301	1.716069

FINAL COMMUNALITY ESTIMATES: TOTAL = 4.450370

POP	SCHOOL	EMPLOY	SERVICES	HOUSE
0.978113	0.817564	0.971999	0.797743	0.884950

(24) RESIDUAL CORRELATIONS WITH UNIQUENESS ON THE DIAGONAL

	POP	SCHOOL	EMPLOY	SERVICES	HOUSE
POP	0.02189	-0.01118	0.00514	0.01063	0.00124
SCHOOL	-0.01118	0.18244	0.02151	-0.02390	0.01248
EMPLOY	0.00514	0.02151	0.02800	-0.00565	-0.01561
SERVICES	0.01063	-0.02390	-0.00565	0.20226	0.03370
HOUSE	0.00124	0.01248	-0.01561	0.03370	0.11505

(25) ROOT MEAN SQUARE OFF-DIAGONAL RESIDUALS: OVER-ALL = 0.01693282

POP	SCHOOL	EMPLOY	SERVICES	HOUSE
0.008153	0.018130	0.013828	0.021517	0.019602

(26) PARTIAL CORRELATIONS CONTROLLING FACTORS

	POP	SCHOOL	EMPLOY	SERVICES	HOUSE
POP	1.00000	-0.17693	0.20752	0.15975	0.02471
SCHOOL	-0.17693	1.00000	0.30097	-0.12443	0.08614
EMPLOY	0.20752	0.30097	1.00000	-0.07504	-0.27509
SERVICES	0.15975	-0.12443	-0.07504	1.00000	0.22093
HOUSE	0.02471	0.08614	-0.27509	0.22093	1.00000

ROOT MEAN SQUARE OFF-DIAGONAL PARTIALS: OVER-ALL = 0.18550132

(27)

POP	SCHOOL	EMPLOY	SERVICES	HOUSE
0.158508	0.190259	0.231818	0.154470	0.182015

```
                                   FIVE SOCIO-ECONOMIC VARIABLES                         6
                         SEE PAGE 14 OF HARMAN: MODERN FACTOR ANALYSIS, 3RD ED
                            PRINCIPAL FACTOR ANALYSIS WITH PROMAX ROTATION

INITIAL FACTOR METHOD: PRINCIPAL FACTORS

                            PLOT OF FACTOR PATTERN FOR FACTOR1  AND FACTOR2

                                              FACTOR1
                                                 1

                                          D     .9

                                                .8
                      E
                       B                        .7                          C
                                                .6                            A

                                                .5

                                                .4

                                                .3

                                                .2
                                                                                        F
                                                .1                                      A
                                                                                        C
             -1.-.9-.8-.7-.6-.5-.4-.3-.2-.1  0 .1 .2 .3 .4 .5 .6 .7 .8 .9 1.0T
                                                                                        O
                                               -.1                                      R
                                                                                        2
                                               -.2

                                               -.3

                                               -.4

                                               -.5

                                               -.6

                                               -.7

                                               -.8

                                               -.9

                                               -1

             POP     =A   SCHOOL  =B   EMPLOY  =C   SERVICES=D   HOUSE   =E
```

(28)

```
                              FIVE SOCIO-ECONOMIC VARIABLES                              7
                    SEE PAGE 14 OF HARMAN: MODERN FACTOR ANALYSIS, 3RD ED
                        PRINCIPAL FACTOR ANALYSIS WITH PROMAX ROTATION

PREROTATION METHOD: VARIMAX

                       (31) ORTHOGONAL TRANSFORMATION MATRIX

                                          1          2

                          1         0.78895    0.61446
                          2        -0.61446    0.78895

                         ROTATED FACTOR PATTERN
                    (32)
                                    FACTOR1    FACTOR2

                        HOUSE       0.94072   -0.00004
                        SCHOOL      0.90419    0.00055
                        SERVICES    0.79085    0.41509
                        POP         0.02255    0.98874
                        EMPLOY      0.14625    0.97499

                      VARIANCE EXPLAINED BY EACH FACTOR

                     (33)       FACTOR1    FACTOR2
                               2.349857   2.100513

                  FINAL COMMUNALITY ESTIMATES: TOTAL =    4.450370

                          POP     SCHOOL     EMPLOY   SERVICES      HOUSE
                     0.978113   0.817564   0.971999   0.797743   0.884950
```

```
                              FIVE SOCIO-ECONOMIC VARIABLES                              8
                    SEE PAGE 14 OF HARMAN: MODERN FACTOR ANALYSIS, 3RD ED
                        PRINCIPAL FACTOR ANALYSIS WITH PROMAX ROTATION

PREROTATION METHOD: VARIMAX

                     PLOT OF FACTOR PATTERN FOR FACTOR1  AND FACTOR2

                                       FACTOR1
                                          1
                                         E
                                         .B

                                         .8         D

                                         .7

                                         .6

                                         .5

                                         .4

                                         .3

                                         .2
                                                                              C  F
                                         .1                                      A
                                                                                 C
           -1.-.9-.8-.7-.6-.5-.4-.3-.2-.1  0  .1  .2  .3  .4  .5  .6  .7  .8  .9 A.OT
                                                                                 O
                                        -.1                                      R
                                                                                 2
                                        -.2

                                        -.3

                                        -.4

                                        -.5

                                        -.6

                                        -.7

                                        -.8

                                        -.9

                                        -1

           POP     =A   SCHOOL  =B   EMPLOY  =C   SERVICES=D   HOUSE   =E
```

```
                    FIVE SOCIO-ECONOMIC VARIABLES                      9
          SEE PAGE 14 OF HARMAN: MODERN FACTOR ANALYSIS, 3RD ED
              PRINCIPAL FACTOR ANALYSIS WITH PROMAX ROTATION

ROTATION METHOD: PROMAX

(34) TARGET MATRIX FOR PROCRUSTEAN TRANSFORMATION

                        FACTOR1    FACTOR2

       HOUSE            1.00000   -0.00000
       SCHOOL           1.00000    0.00000
       SERVICES         0.69421    0.10045
       POP              0.00001    1.00000
       EMPLOY           0.00326    0.96793

     PROCRUSTEAN TRANSFORMATION MATRIX

(35)                          1          2

          1             1.04117   -0.09865
          2            -0.10572    0.96303

       OBLIQUE TRANSFORMATION MATRIX

(36)                          1          2

          1             0.73803    0.54202
          2            -0.70555    0.86528

         INTER-FACTOR CORRELATIONS

(37)                    FACTOR1    FACTOR2

       FACTOR1          1.00000    0.20188
       FACTOR2          0.20188    1.00000

(38) ROTATED FACTOR PATTERN (STD REG COEFS)

                        FACTOR1    FACTOR2

       HOUSE            0.95558   -0.09792
       SCHOOL           0.91842   -0.09352
       SERVICES         0.76053    0.33932
       POP             -0.07908    1.00192
       EMPLOY           0.04799    0.97509

        REFERENCE AXIS CORRELATIONS

(39)                    FACTOR1    FACTOR2

       FACTOR1          1.00000   -0.20188
       FACTOR2         -0.20188    1.00000
```

```
                    FIVE SOCIO-ECONOMIC VARIABLES                     10
          SEE PAGE 14 OF HARMAN: MODERN FACTOR ANALYSIS, 3RD ED
              PRINCIPAL FACTOR ANALYSIS WITH PROMAX ROTATION

ROTATION METHOD: PROMAX

     REFERENCE STRUCTURE (SEMIPARTIAL CORRELATIONS)

(40)                    FACTOR1    FACTOR2

       HOUSE            0.93591   -0.09590
       SCHOOL           0.89951   -0.09160
       SERVICES         0.74487    0.33233
       POP             -0.07745    0.98129
       EMPLOY           0.04700    0.95501

 VARIANCE EXPLAINED BY EACH FACTOR ELIMINATING OTHER FACTORS
(41)
                    FACTOR1   FACTOR2
                    2.248089  2.003020

(42) FACTOR STRUCTURE (CORRELATIONS)

                        FACTOR1    FACTOR2

       HOUSE            0.93582    0.09500
       SCHOOL           0.89954    0.09189
       SERVICES         0.82903    0.49286
       POP              0.12319    0.98596
       EMPLOY           0.24484    0.98478

  VARIANCE EXPLAINED BY EACH FACTOR IGNORING OTHER FACTORS

(43)                FACTOR1   FACTOR2
                    2.447349  2.202280

     FINAL COMMUNALITY ESTIMATES: TOTAL =    4.450370

(44)      POP     SCHOOL     EMPLOY   SERVICES      HOUSE
     0.978113   0.817564   0.971999   0.797743   0.884950
```

```
                              FIVE SOCIO-ECONOMIC VARIABLES                              11
                 SEE PAGE 14 OF HARMAN: MODERN FACTOR ANALYSIS, 3RD ED
                   PRINCIPAL FACTOR ANALYSIS WITH PROMAX ROTATION

ROTATION METHOD: PROMAX

           (48) PLOT OF REFERENCE STRUCTURE FOR FACTOR1  AND FACTOR2
                REFERENCE AXIS CORRELATION = -0.2019  ANGLE = 101.65

                                   FACTOR1
                                       1
                                   E
                                   B .9

                                     .8
                                                 D
                                     .7

                                     .6

                                     .5

                                     .4

                                     .3

                                     .2
                                                                          F
                                     .1                                   A
                                                                       C  C
 -1.-.9-.8-.7-.6-.5-.4-.3-.2-.1  0  .1 .2 .3 .4 .5 .6 .7 .8 .9 1.0T
                                                                          O
                                    -.1                                A  R
                                                                          2
                                    -.2

                                    -.3

                                    -.4

                                    -.5

                                    -.6

                                    -.7

                                    -.8

                                    -.9

                                    -1

 POP     =A   SCHOOL  =B   EMPLOY  =C   SERVICES=D   HOUSE   =E
```

FIVE SOCIO-ECONOMIC VARIABLES 12

SEE PAGE 14 OF HARMAN: MODERN FACTOR ANALYSIS, 3RD ED

FACTOR OUTPUT DATA SET

OBS	_TYPE_	_NAME_	POP	SCHOOL	EMPLOY	SERVICES	HOUSE
1	MEAN		6241.67	11.4417	2333.33	120.833	17000.0
2	STD		3439.99	1.7865	1241.21	114.928	6367.5
3	N		12.00	12.0000	12.00	12.000	12.0
4	CORR	POP	1.00	0.0098	0.97	0.439	0.0
5	CORR	SCHOOL	0.01	1.0000	0.15	0.691	0.9
6	CORR	EMPLOY	0.97	0.1543	1.00	0.515	0.1
7	CORR	SERVICES	0.44	0.6914	0.51	1.000	0.8
8	CORR	HOUSE	0.02	0.8631	0.12	0.778	1.0
9	COMMUNAL		0.98	0.8176	0.97	0.798	0.9
10	PRIORS		0.97	0.8223	0.97	0.786	0.8
11	EIGENVAL		2.73	1.7161	0.04	-0.025	-0.1
12	UNROTATE	FACTOR1	0.63	0.7137	0.71	0.879	0.7
13	UNROTATE	FACTOR2	0.77	-0.5552	0.68	-0.158	-0.6
14	RESIDUAL	POP	0.02	-0.0112	0.01	0.011	0.0
15	RESIDUAL	SCHOOL	-0.01	0.1824	0.02	-0.024	0.0
16	RESIDUAL	EMPLOY	0.01	0.0215	0.03	-0.006	-0.0
17	RESIDUAL	SERVICES	0.01	-0.0239	-0.01	0.202	0.0
18	RESIDUAL	HOUSE	0.00	0.0125	-0.02	0.034	0.1
19	PRETRANS	FACTOR1	0.79	-0.6145	.	.	.
20	PRETRANS	FACTOR2	0.61	0.7889	.	.	.
21	PREROTAT	FACTOR1	0.02	0.9042	0.15	0.791	0.9
22	PREROTAT	FACTOR2	0.99	0.0006	0.97	0.415	-0.0
23	TRANSFOR	FACTOR1	0.74	-0.7055	.	.	.
24	TRANSFOR	FACTOR2	0.54	0.8653	.	.	.
25	FCORR	FACTOR1	1.00	0.2019	.	.	.
26	FCORR	FACTOR2	0.20	1.0000	.	.	.
27	PATTERN	FACTOR1	-0.08	0.9184	0.05	0.761	1.0
28	PATTERN	FACTOR2	1.00	-0.0935	0.98	0.339	-0.1
29	RCORR	FACTOR1	1.00	-0.2019	.	.	.
30	RCORR	FACTOR2	-0.20	1.0000	.	.	.
31	REFERENC	FACTOR1	-0.08	0.8995	0.05	0.745	0.9
32	REFERENC	FACTOR2	0.98	-0.0916	0.96	0.332	-0.1
33	STRUCTUR	FACTOR1	0.12	0.8995	0.24	0.829	0.9
34	STRUCTUR	FACTOR2	0.99	0.0919	0.98	0.493	0.1

The output data set can be used for Harris-Kaiser rotation by deleting observations with __TYPE__='PATTERN' and __TYPE__='FCORR', which are for the promax-rotated factors, and changing __TYPE__='UNROTATE' to 'PATTERN':

```
DATA FACT2(TYPE=FACTOR);
  SET;
  IF __TYPE__='PATTERN'|__TYPE__='FCORR' THEN DELETE;
  IF __TYPE__='UNROTATE' THEN __TYPE__='PATTERN';
PROC FACTOR ROTATE=HK NORM=WEIGHT REORDER PLOT;
  TITLE3 HARRIS-KAISER ROTATION WITH CURETON-MULAIK WEIGHTS;
```

SERVICES is given a small weight, and the axes are placed as desired.

```
                                   FIVE SOCIO-ECONOMIC VARIABLES                          13
                        SEE PAGE 14 OF HARMAN: MODERN FACTOR ANALYSIS, 3RD ED
                         HARRIS-KAISER ROTATION WITH CURETON-MULAIK WEIGHTS

ROTATION METHOD: HARRIS-KAISER

                          (29) VARIABLE WEIGHTS FOR ROTATION

                          POP     SCHOOL     EMPLOY  SERVICES      HOUSE
                     0.959827  0.939454  0.997464  0.121948  0.940073

                             OBLIQUE TRANSFORMATION MATRIX

                                               1           2

                                1        0.73537     0.61899
                                2       -0.68283     0.78987

                               INTER-FACTOR CORRELATIONS

                                         FACTOR1     FACTOR2

                             FACTOR1     1.00000     0.08358
                             FACTOR2     0.08358     1.00000

                        ROTATED FACTOR PATTERN (STD REG COEFS)

                                         FACTOR1     FACTOR2

                             HOUSE       0.94048     0.00279
                             SCHOOL      0.90391     0.00327
                             SERVICES    0.75459     0.41892
                             POP        -0.06335     0.99227
                             EMPLOY      0.06152     0.97885

                               REFERENCE AXIS CORRELATIONS

                                         FACTOR1     FACTOR2

                             FACTOR1     1.00000    -0.08358
                             FACTOR2    -0.08358     1.00000

                      REFERENCE STRUCTURE (SEMIPARTIAL CORRELATIONS)

                                         FACTOR1     FACTOR2

                             HOUSE       0.93719     0.00278
                             SCHOOL      0.90075     0.00326
                             SERVICES    0.75195     0.41745
                             POP        -0.06312     0.98880
                             EMPLOY      0.06130     0.97543

               VARIANCE EXPLAINED BY EACH FACTOR ELIMINATING OTHER FACTORS

                                      FACTOR1     FACTOR2
                                     2.262854    2.103473
```

```
                                   FIVE SOCIO-ECONOMIC VARIABLES                          14
                        SEE PAGE 14 OF HARMAN: MODERN FACTOR ANALYSIS, 3RD ED
                         HARRIS-KAISER ROTATION WITH CURETON-MULAIK WEIGHTS

ROTATION METHOD: HARRIS-KAISER

                              FACTOR STRUCTURE (CORRELATIONS)

                                         FACTOR1     FACTOR2

                             HOUSE       0.94071     0.08139
                             SCHOOL      0.90419     0.07882
                             SERVICES    0.78960     0.48198
                             POP         0.01958     0.98698
                             EMPLOY      0.14332     0.98399
```

(continued on next page)

(continued from previous page)

```
                    VARIANCE EXPLAINED BY EACH FACTOR IGNORING OTHER FACTORS

                                     FACTOR1      FACTOR2
                                     2.346896     2.187516

                    FINAL COMMUNALITY ESTIMATES: TOTAL =     4.450370

                         POP      SCHOOL      EMPLOY  SERVICES        HOUSE
                    0.978113  0.817564    0.971999  0.797743    0.884950
```

```
                                 FIVE SOCIO-ECONOMIC VARIABLES                                  15
                      SEE PAGE 14 OF HARMAN: MODERN FACTOR ANALYSIS, 3RD ED
                        HARRIS-KAISER ROTATION WITH CURETON-MULAIK WEIGHTS

ROTATION METHOD: HARRIS-KAISER

                      PLOT OF REFERENCE STRUCTURE FOR FACTOR1  AND FACTOR2
                      REFERENCE AXIS CORRELATION = -0.0836   ANGLE =   94.79

                                          FACTOR1
                                             1
                                             E
                                            .B

                                            .8
                                                                D
                                            .7

                                            .6

                                            .5

                                            .4

                                            .3

                                            .2
                                                                                    F
                                            .1                                      A
                                                                                 C  C
                  -1.-.9-.8-.7-.6-.5-.4-.3-.2-.1   0 .1 .2 .3 .4 .5 .6 .7 .8 .9 1.0T
                                                                                 A  O
                                           -.1                                      R
                                                                                    2
                                           -.2

                                           -.3

                                           -.4

                                           -.5

                                           -.6

                                           -.7

                                           -.8

                                           -.9

                                            -1

                  POP     =A   SCHOOL  =B   EMPLOY  =C   SERVICES=D   HOUSE   =E
```

Maximum Likelihood Factor Analysis: Example 3

This example uses maximum likelihood factor analyses for one, two, and three factors. It is already apparent from the principal factor analysis that the best number of common factors is almost certainly two. The one and three factor ML solutions reinforce this conclusion and illustrate some of the numerical problems that can occur.

```
PROC FACTOR DATA=SOCECON METHOD=ML HEYWOOD N=1;
  TITLE3 MAXIMUM LIKELIHOOD FACTOR ANALYSIS WITH ONE FACTOR;

PROC FACTOR DATA=SOCECON METHOD=ML HEYWOOD N=2;
  TITLE3 MAXIMUM LIKELIHOOD FACTOR ANALYSIS WITH TWO FACTORS;

PROC FACTOR DATA=SOCECON METHOD=ML HEYWOOD N=3;
  TITLE3 MAXIMUM LIKELIHOOD FACTOR ANALYSIS WITH THREE FACTORS;
```

With one factor, the solution on the second iteration is so close to the optimum that FACTOR cannot find a better solution even though the convergence criterion has not been met, hence the message, "UNABLE TO IMPROVE CRITERION." When this message appears you should try re-running FACTOR with some different prior communality estimates to make sure that the solution is correct. In this case, other prior estimates lead to the same solution or possibly to worse local optima, as indicated by CRITERION or CHI-SQUARE.

EMPLOY has a communality of 1.0, and therefore an infinite weight that is printed below the final communality estimate as a missing value (.). The first eigenvalue is also infinite. Infinite values are ignored in computing the total of the eigenvalues and the total final communality.

The two-factor analysis converges without incident. This time, however, POP is a Heywood case.

The three-factor analysis generates the message, "WARNING: TOO MANY FACTORS FOR A UNIQUE SOLUTION." The number of parameters in the model exceeds the number of elements in the correlation matrix from which they can be estimated, so an infinite number of different perfect solutions can be obtained. The CRITERION approaches zero (8.799E-08) at an improper optimum, as indicated by "CONVERGED, BUT NOT TO A PROPER OPTIMUM." The degrees of freedom for the chi-square test are –2, so a probability level cannot be computed for three factors. Note also that EMPLOY is a Heywood case again.

The probability levels for the chi-square test are .0001 for the hypothesis of no common factors, .0002 for one common factor, and .1382 for two common factors. We are therefore inclined to accept the two-factor model as an adequate representation. Akaike's information criterion and Schwarz's Bayesian criterion attain their minimum values at two common factors, so there is little doubt that two factors are appropriate for these data.

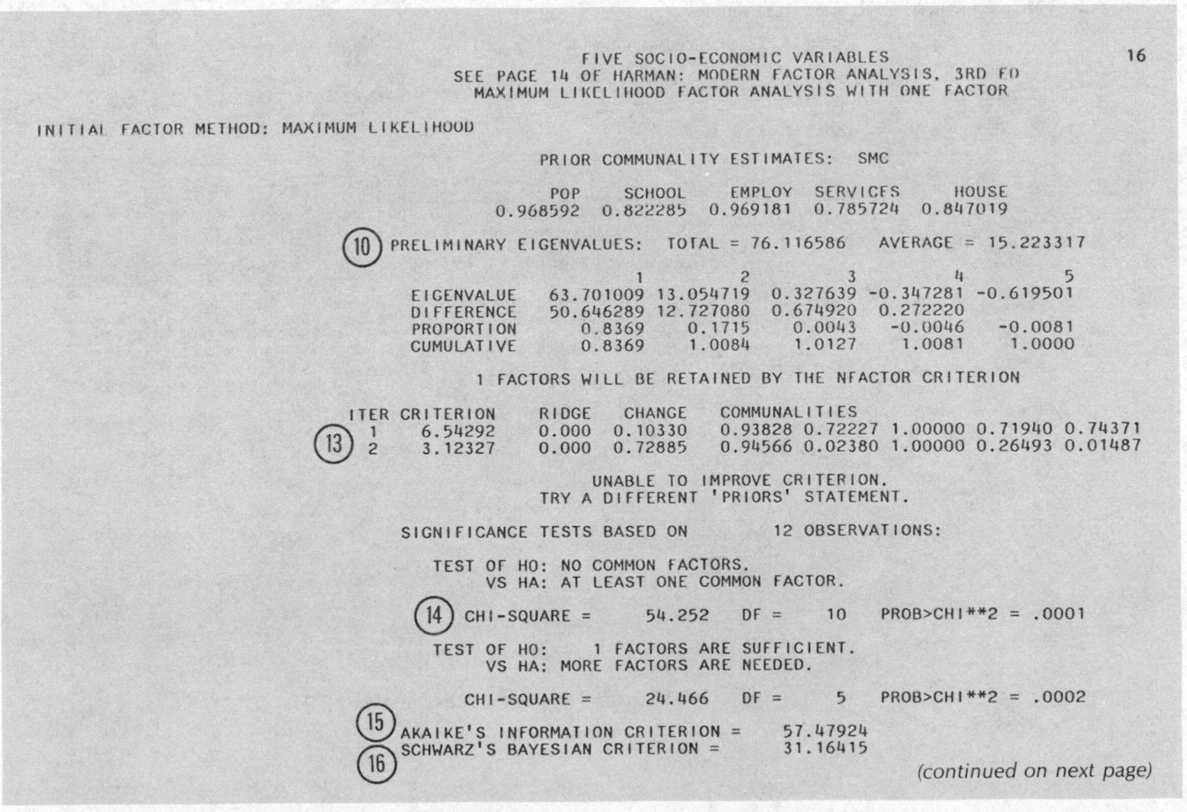

```
                              FIVE SOCIO-ECONOMIC VARIABLES                                  16
                     SEE PAGE 14 OF HARMAN: MODERN FACTOR ANALYSIS, 3RD ED
                       MAXIMUM LIKELIHOOD FACTOR ANALYSIS WITH ONE FACTOR

INITIAL FACTOR METHOD: MAXIMUM LIKELIHOOD

                          PRIOR COMMUNALITY ESTIMATES:   SMC

                         POP     SCHOOL      EMPLOY  SERVICES      HOUSE
                    0.968592  0.822285  0.969181  0.785724  0.847019

      (10) PRELIMINARY EIGENVALUES:   TOTAL = 76.116586    AVERAGE = 15.223317

                             1          2          3          4          5
          EIGENVALUE   63.701009 13.054719  0.327639 -0.347281 -0.619501
          DIFFERENCE   50.646289 12.727080  0.674920  0.272220
          PROPORTION      0.8369    0.1715    0.0043    -0.0046    -0.0081
          CUMULATIVE      0.8369    1.0084    1.0127     1.0081     1.0000

                  1 FACTORS WILL BE RETAINED BY THE NFACTOR CRITERION

      ITER CRITERION      RIDGE    CHANGE     COMMUNALITIES
  (13)  1    6.54292      0.000   0.10330     0.93828 0.72227 1.00000 0.71940 0.74371
        2    3.12327      0.000   0.72885     0.94566 0.02380 1.00000 0.26493 0.01487

                           UNABLE TO IMPROVE CRITERION.
                        TRY A DIFFERENT 'PRIORS' STATEMENT.

          SIGNIFICANCE TESTS BASED ON          12 OBSERVATIONS:

             TEST OF HO: NO COMMON FACTORS.
                  VS HA: AT LEAST ONE COMMON FACTOR.

        (14) CHI-SQUARE =        54.252    DF =    10    PROB>CHI**2 = .0001

             TEST OF HO:     1 FACTORS ARE SUFFICIENT.
                  VS HA: MORE FACTORS ARE NEEDED.

             CHI-SQUARE =        24.466    DF =     5    PROB>CHI**2 = .0002

      (15) AKAIKE'S INFORMATION CRITERION =       57.47924
      (16) SCHWARZ'S BAYESIAN CRITERION =         31.16415
```

(continued on next page)

(continued from previous page)

```
(17) SQUARED CANONICAL CORRELATIONS

                          FACTOR1
                          1.000000

EIGENVALUES OF THE WEIGHTED REDUCED CORRELATION MATRIX:  TOTAL = -0.000000    AVERAGE = -0.000000

                          1          2          3          4          5
           EIGENVALUE     .          1.927160  -0.228313  -0.792956  -0.905891
           DIFFERENCE    -1.927160   2.155473   0.564643   0.112935
```

```
                       FIVE SOCIO-ECONOMIC VARIABLES                                   17
              SEE PAGE 14 OF HARMAN: MODERN FACTOR ANALYSIS, 3RD ED
               MAXIMUM LIKELIHOOD FACTOR ANALYSIS WITH ONE FACTOR

INITIAL FACTOR METHOD: MAXIMUM LIKELIHOOD

                                FACTOR PATTERN

                                       FACTOR1

                         POP           0.97245
                         SCHOOL        0.15428
                         EMPLOY        1.00000
                         SERVICES      0.51472
                         HOUSE         0.12193

                    VARIANCE EXPLAINED BY EACH FACTOR

                                       FACTOR1
                         WEIGHTED    17.801063
                         UNWEIGHTED   2.249260

              FINAL COMMUNALITY ESTIMATES AND VARIABLE WEIGHTS
         TOTAL COMMUNALITY: WEIGHTED = 17.801063    UNWEIGHTED =   2.249260

                            POP     SCHOOL     EMPLOY   SERVICES      HOUSE
          COMMUNALITY  0.945656   0.023803   1.000000   0.264935   0.014866
          WEIGHT      18.401165   1.024384    .         1.360424   1.015090
```

```
                       FIVE SOCIO-ECONOMIC VARIABLES                                   18
              SEE PAGE 14 OF HARMAN: MODERN FACTOR ANALYSIS, 3RD ED
              MAXIMUM LIKELIHOOD FACTOR ANALYSIS WITH TWO FACTORS

INITIAL FACTOR METHOD: MAXIMUM LIKELIHOOD

                      PRIOR COMMUNALITY ESTIMATES:  SMC

                      POP     SCHOOL     EMPLOY   SERVICES      HOUSE
                 0.968592   0.822285   0.969181   0.785724   0.847019

         PRELIMINARY EIGENVALUES:  TOTAL = 76.116586    AVERAGE = 15.223317

                          1          2          3          4          5
          EIGENVALUE  63.701009  13.054719   0.327639  -0.347281  -0.619501
          DIFFERENCE  50.646289  12.727080   0.674920   0.272220
          PROPORTION     0.8369     0.1715     0.0043    -0.0046    -0.0081
          CUMULATIVE     0.8369     1.0084     1.0127     1.0081     1.0000

              2 FACTORS WILL BE RETAINED BY THE NFACTOR CRITERION

     ITER CRITERION    RIDGE  CHANGE     COMMUNALITIES
        1  0.343122    0.000  0.04710    1.00000 0.80672 0.95058 0.79348 0.89412
        2  0.307218    0.000  0.03068    1.00000 0.80821 0.96023 0.81048 0.92480
        3  0.306786    0.000  0.00629    1.00000 0.81149 0.95948 0.81677 0.92023
        4  0.306737    0.000  0.00218    1.00000 0.80985 0.95963 0.81498 0.92241
        5  0.306732    0.000  0.00071    1.00000 0.81019 0.95955 0.81569 0.92187

                        CONVERGENCE CRITERION SATISFIED.

         SIGNIFICANCE TESTS BASED ON          12 OBSERVATIONS:

             TEST OF HO: NO COMMON FACTORS.
                  VS HA: AT LEAST ONE COMMON FACTOR.

                 CHI-SQUARE =      54.252    DF =     10    PROB>CHI**2 = .0001

             TEST OF HO:     2 FACTORS ARE SUFFICIENT.
                  VS HA: MORE FACTORS ARE NEEDED.

                 CHI-SQUARE =       2.198    DF =      1    PROB>CHI**2 = .1382

         AKAIKE'S INFORMATION CRITERION =     31.68079
         SCHWARZ'S BAYESIAN CRITERION =       19.23474

                        SQUARED CANONICAL CORRELATIONS

                              FACTOR1   FACTOR2
                              1.000000  0.951889
```

(continued on next page)

(continued from previous page)

```
EIGENVALUES OF THE WEIGHTED REDUCED CORRELATION MATRIX:  TOTAL = 19.785316    AVERAGE =   3.957063

                        1          2          3          4          5
      EIGENVALUE        .      19.785314  0.543185  -0.039771  -0.503412
      DIFFERENCE  -19.785314  19.242129  0.582956   0.463641
      PROPORTION      0.0000     1.0000     0.0275    -0.0020    -0.0254
      CUMULATIVE      0.0000     1.0000     1.0275     1.0254     1.0000
```

```
                              FIVE SOCIO-ECONOMIC VARIABLES                          19
                     SEE PAGE 14 OF HARMAN: MODERN FACTOR ANALYSIS, 3RD ED
                      MAXIMUM LIKELIHOOD FACTOR ANALYSIS WITH TWO FACTORS

INITIAL FACTOR METHOD: MAXIMUM LIKELIHOOD

                                          FACTOR PATTERN

                                              FACTOR1    FACTOR2

                                    POP       1.00000    0.00000
                                    SCHOOL    0.00975    0.90003
                                    EMPLOY    0.97245    0.11797
                                    SERVICES  0.43887    0.78930
                                    HOUSE     0.02241    0.95989

                                   VARIANCE EXPLAINED BY EACH FACTOR

                                              FACTOR1    FACTOR2
                                    WEIGHTED  24.432971 19.785314
                                  UNWEIGHTED  2.138861   2.368353

                          FINAL COMMUNALITY ESTIMATES AND VARIABLE WEIGHTS
                 TOTAL COMMUNALITY: WEIGHTED = 44.218285    UNWEIGHTED =   4.507214

                                  POP     SCHOOL     EMPLOY   SERVICES      HOUSE
                COMMUNALITY   1.000000  0.810145   0.959571   0.815603   0.921894
                WEIGHT            .     5.268294  24.724667   5.425646  12.799679
```

```
                    FIVE SOCIO-ECONOMIC VARIABLES                                    20
           SEE PAGE 14 OF HARMAN: MODERN FACTOR ANALYSIS, 3RD ED
           MAXIMUM LIKELIHOOD FACTOR ANALYSIS WITH THREE FACTORS

INITIAL FACTOR METHOD: MAXIMUM LIKELIHOOD

                    PRIOR COMMUNALITY ESTIMATES:  SMC

                    POP      SCHOOL     EMPLOY   SERVICES     HOUSE
               0.968592  0.822285  0.969181  0.785724  0.847019

        PRELIMINARY EIGENVALUES:  TOTAL = 76.116586    AVERAGE = 15.223317

                         1          2          3          4          5
      EIGENVALUE   63.701009  13.054719  0.327639  -0.347281  -0.619501
      DIFFERENCE   50.646289  12.727080  0.674920   0.272220
      PROPORTION      0.8369     0.1715     0.0043    -0.0046    -0.0081
      CUMULATIVE      0.8369     1.0084     1.0127     1.0081     1.0000

               3 FACTORS WILL BE RETAINED BY THE NFACTOR CRITERION
                 WARNING: TOO MANY FACTORS FOR A UNIQUE SOLUTION.

     ITER CRITERION      RIDGE   CHANGE    COMMUNALITIES
      1    0.160126      0.031   0.05102   0.96382 0.84123 1.00000 0.80346 0.89804
      2 0.00340681       0.031   0.05878   0.98216 0.87692 1.00000 0.80295 0.95682
      3 .000041434       0.031   0.01013   0.98334 0.88004 1.00000 0.80507 0.96695
      4   1.472E-06      0.031   0.00155   0.98316 0.88054 1.00000 0.80480 0.96850
      5   8.799E-08      0.031   0.00029   0.98311 0.88065 1.00000 0.80462 0.96879

                  CONVERGED, BUT NOT TO A PROPER OPTIMUM.
                  TRY A DIFFERENT 'PRIORS' STATEMENT.

       SIGNIFICANCE TESTS BASED ON        12 OBSERVATIONS:

          TEST OF HO: NO COMMON FACTORS.
               VS HA: AT LEAST ONE COMMON FACTOR.

            CHI-SQUARE =      54.252   DF =     10   PROB>CHI**2 = .0001

          TEST OF HO:     3 FACTORS ARE SUFFICIENT.
               VS HA: MORE FACTORS ARE NEEDED.

            CHI-SQUARE =       0.000   DF =     -2   PROB>CHI**2 = .0000

       AKAIKE'S INFORMATION CRITERION =           34
       SCHWARZ'S BAYESIAN CRITERION =       21.12171

                      SQUARED CANONICAL CORRELATIONS

                        FACTOR1    FACTOR2    FACTOR3
                        1.000000   0.975846   0.699066

EIGENVALUES OF THE WEIGHTED REDUCED CORRELATION MATRIX:  TOTAL = 42.724206    AVERAGE =   8.544841

                         1          2          3          4          5
      EIGENVALUE         .      40.401173  2.322986  0.000319  -0.000273
      DIFFERENCE   -40.401173  38.078186  2.322667  0.000591
      PROPORTION      0.0000     0.9456     0.0544     0.0000    -0.0000
      CUMULATIVE      0.0000     0.9456     1.0000     1.0000     1.0000
```

```
                                    FIVE SOCIO-ECONOMIC VARIABLES                     21
                           SEE PAGE 14 OF HARMAN: MODERN FACTOR ANALYSIS, 3RD ED
                           MAXIMUM LIKELIHOOD FACTOR ANALYSIS WITH THREE FACTORS
INITIAL FACTOR METHOD: MAXIMUM LIKELIHOOD

                                         FACTOR PATTERN

                                        FACTOR1   FACTOR2   FACTOR3

                              POP       0.97245  -0.11188  -0.15792
                              SCHOOL    0.15428   0.88881   0.25859
                              EMPLOY    1.00000  -0.00000  -0.00000
                              SERVICES  0.51472   0.72429  -0.12272
                              HOUSE     0.12193   0.97335  -0.08071

                                 VARIANCE EXPLAINED BY EACH FACTOR

                                       FACTOR1    FACTOR2   FACTOR3
                              WEIGHTED   58.032408 40.401173  2.322986
                              UNWEIGHTED  2.249260  2.274514  0.113382

                            FINAL COMMUNALITY ESTIMATES AND VARIABLE WEIGHTS
                      TOTAL COMMUNALITY: WEIGHTED =    100.757    UNWEIGHTED =  4.637157

                                     POP     SCHOOL     EMPLOY  SERVICES      HOUSE
                      COMMUNALITY   0.983113  0.880658  1.000000  0.804594  0.968791
                      WEIGHT       59.218853  8.378825   .        5.118261 32.040674
```

REFERENCES

Akaike, H. (1973), "Information Theory and the Extension of the Maximum Likelihood Principle," in *2nd International Symposium on Information Theory*, ed. V.N. Petrov and F. Csaki, Budapest: Akailseoniai-Kiudo, 267–281.

Akaike, H. (1974), "A New Look at the Statistical Identification Model," *IEEE Transactions on Automatic Control*, 19, 716–723.

Bickel, P.J., and Doksum, K.A. (1977), *Mathematical Statistics*, San Francisco: Holden-Day.

Cattell, R.B. (1966), "The Scree Test for the Number of Factors," *Multivariate Behavioral Research*, 1, 245–276.

Cattell, R.B. (1978), *The Scientific Use of Factor Analysis*, New York: Plenum.

Cattell, R.B. and Vogelman, S. (1977), "A Comprehensive Trial of the Scree and KG Criteria for Determining the Number of Factors," *Multivariate Behavioral Research*, 12, 289–325.

Cerny, B.A. and Kaiser, H.F. (1977), "A Study of a Measure of Sampling Adequacy for Factor-Analytic Correlation Matrices," *Multivariate Behavioral Research*, 12, 43–47.

Cureton, E.E. (1968), *A Factor Analysis of Project TALENT Tests and Four Other Test Batteries*, (Interim report 4 to the U. S. Office of Education, Cooperative Research Project No. 3051.) Palo Alto: Project TALENT Office, American Institutes for Research and University of Pittsburgh.

Cureton, E.E. and Mulaik, S.A. (1975), "The Weighted Varimax Rotation and the Promax Rotation," *Psychometrika*, 40, 183–195.

Dziuban, C.D. and Harris, C.W. (1973), "On the Extraction of Components and the Applicability of the Factor Model," *American Educational Research Journal*, 10, 93–99.

Geweke, J.F. and Singleton, K.J. (1980), "Interpreting the Likelihood Ratio Statistic in Factor Models When Sample Size is Small," *Journal of the American Statistical Association*, 75, 133–137.

Gorsuch, R.L. (1974), *Factor Analysis*, Philadelphia: W. B. Saunders Co.

Gould, S.J. (1981), *The Mismeasure of Man*, New York: W. W. Norton & Co.

Harman, H.H. (1976), *Modern Factor Analysis, 3rd Edition*, Chicago: University of Chicago Press.

Harris, C.W. (1962), "Some Rao-Guttman Relationships," *Psychometrika*, 27, 247–263.

Horn, J.L. and Engstrom, R. (1979), "Cattell's Scree Test in Relation to Bartlett's Chi-Square Test and Other Observations on the Number of Factors Problem," *Multivariate Behavioral Research*, 14, 283–300.

Joreskog, K.G. (1962), "On the Statistical Treatment of Residuals in Factor Analysis," *Psychometrika*, 27, 335–354.

Joreskog, K.G. (1977), "Factor Analysis by Least-Squares and Maximum Likelihood Methods," in *Statistical Methods for Digital Computers*, ed. K. Enslein, A. Ralston, and H.S. Wilf, New York: John Wiley & Sons.

Kaiser, H.F. (1963), "Image Analysis," in *Problems in Measuring Change*, ed. C.W. Harris, Madison: University of Wisconsin Press.

Kaiser, H.F. (1970), "A Second Generation Little Jiffy," *Psychometrika*, 35, 401–415.

Kaiser, H.F. and Cerny, B.A. (1979), "Factor Analysis of the Image Correlation Matrix," *Educational and Psychological Measurement*, 39, 711–714.

Kaiser, H.F. and Rice, J. (1974), "Little Jiffy, Mark IV," *Educational and Psychological Measurement*, 34, 111–117.

Kerlinger, F.N. and Pedhazur, E.J. (1973), *Multiple Regression in Behavioral Research*, New York: Holt, Rinehart & Winston.

Kim, J.O. and Mueller, C.W. (1978), *Introduction to Factor Analysis: What It Is and How To Do It*, Sage University Paper series on Quantitative Applications in the Social Sciences, series no. 07–013, Beverly Hills and London: Sage Publications.

Lawley, D.N. and Maxwell, A.E. (1971), *Factor Analysis as a Statistical Method*, New York: Macmillan.

Lee, B.L. and Comrey, A.L. (1979), "Distortions in a Commonly Used Factor Analytic Procedure," *Multivariate Behavioral Research*, 14, 301–321.

Mardia, K.V., Kent, J.T., and Bibby, J.M. (1979), *Multivariate Analysis*, London: Academic Press.

McDonald, R.P. (1975), "A Note on Rippe's Test of Significance in Common Factor Analysis," *Psychometrika*, 40, 117–119.

Morrison, D.F. (1976), *Multivariate Statistical Methods, 2nd edition*, New York: McGraw-Hill.

Mulaik, S.A. (1972), *The Foundations of Factor Analysis*, New York: McGraw-Hill.

Rao, C.R. (1955), "Estimation and Tests of Significance in Factor Analysis," *Psychometrika*, 20, 93–111.

Schwarz, G. (1978), "Estimating the Dimension of a Model," *Annals of Statistics*, 6, 461–464.

Spearman, C. (1904), "General Intelligence Objectively Determined and Measured," *American Journal of Psychology*, 15, 201–293.

The PRINCOMP Procedure

ABSTRACT

The PRINCOMP procedure performs principal component analysis. Input may be raw data, a correlation matrix, or a covariance matrix; either the correlation matrix or the covariance matrix can be analyzed. Output data sets containing eigenvalues, eigenvectors, and standardized or unstandardized principal component scores may be created.

INTRODUCTION

Principal component analysis is a multivariate technique for examining relationships among several quantitative variables. It is used for summarizing data and detecting linear relationships. Plots of principal components are especially valuable tools in exploratory data analysis. Principal components can be used to reduce the number of variables in regression, clustering, and so on.

Background

Principal component analysis was originated by Pearson (1901) and later developed by Hotelling (1933). The application of principal components is discussed by Rao (1964), Cooley and Lohnes (1971), and Gnanadesikan (1977). Excellent statistical treatments of principal components are found in Kshirsagar (1972), Morrison (1976), and Mardia, Kent, and Bibby (1979).

Given a data set with p numeric variables, p principal components may be computed. Each principal component is a linear combination of the original variables, with coefficients equal to the eigenvectors of the correlation or covariance matrix. The eigenvectors are customarily taken with unit-norm. The principal components are sorted by descending order of the eigenvalues, which are equal to the variances of the components.

Principal components have a variety of useful properties (Rao, 1964; Kshirsagar, 1972):

- The eigenvectors are orthogonal, so the principal components represent jointly perpendicular directions through the space of the original variables.
- The principal component scores are jointly uncorrelated. Note that this property is quite distinct from the previous one.
- The first principal component has the largest variance of any unit-length linear combination of the observed variables. The jth principal compo-

nent has the largest variance of any unit-length linear combination orthogonal to the first j-1 principal components. The last principal component has the smallest variance of any linear combination of the original variables.

- The scores on the first j principal components have the highest possible generalized variance of any set of unit-norm linear combinations of the original variables.
- The first j principal components give a least-squares solution to the model

$$\mathbf{Y}=\mathbf{XB}+\mathbf{E}$$

where **Y** is an n by p matrix of the centered observed variables; **X** is the n by j matrix of scores on the first j principal components; **B** is the j by p matrix of eigenvectors; **E** is an n by p matrix of residuals; and it is desired to minimize trace **(E′E)**, the sum of all the squared elements in **E**. In other words, the first j principal components are the best linear predictors of the original variables among all possible sets of j variables, although any non-singular linear transformation of the first j principal components would provide equally good prediction. The same result is obtained if you want to minimize the determinant or the Euclidean (Schur, Frobenious) norm of **E′E**, rather than the trace.

- In geometric terms, the j-dimensional linear subspace spanned by the first j principal components gives the best possible fit to the data points as measured by the sum of squared perpendicular distances from each data point to the subspace.

Principal components analysis can also be used for exploring polynomial relationships and for multivariate outlier detection (Gnanadesikan, 1977), and is related to factor analysis, correspondence analysis, allometry, and biased regression techniques (Mardia, Kent, and Bibby, 1979).

SPECIFICATIONS

The PRINCOMP procedure is invoked by the following statements:

PROC PRINCOMP *options;*
  VAR *variables;*
  PARTIAL *variables;*
  FREQ *variable;*
  WEIGHT *variable;*
  BY *variables;*

Usually only the VAR statement is used in addition to the PROC PRINCOMP statement.

PROC PRINCOMP Statement

PROC PRINCOMP *options;*

The following options may appear on the PROC statement:

DATA=*SASdataset* names the SAS data set to be analyzed. The data set may be an ordinary SAS data set, or a TYPE=CORR or TYPE=COV data set. If DATA= is omitted, the most recently created SAS data set is used.

OUT=*SASdataset* names an output SAS data set that contains all the original data as well as the principal component scores. If you want to create a permanent SAS data set, you must specify a two-level name, (see Chapter 12, "SAS Data Sets," in *SAS User's Guide: Basics, 1982 Edition*, for information on permanent SAS data sets).

OUTSTAT=*SASdataset* names an output SAS data set that contains means, standard deviations, number of observations, correlations or covariances, eigenvalues, and eigenvectors. If the COV option is specified, the data set is TYPE=COV and contains covariances; otherwise it is TYPE=CORR and contains correlations. If you want to create a permanent SAS data set, you must specify a two-level name, (see Chapter 12, "SAS Data Sets," in *SAS User's Guide: Basics, 1982 Edition*, for information on permanent SAS data sets).

NOINT requests that the covariance or correlation matrix not be corrected for the mean, that is, that no intercept be used in the model.

COVARIANCE
COV requests that the principal components be computed from the covariance matrix. If COV is not specified, the correlation matrix is analyzed.

N=*n* specifies the number of principal components to be computed.

STANDARD
STD requests that the principal component scores in the OUT= data set be standardized to unit variance. If STANDARD is not specified the scores have variance equal to the corresponding eigenvalue.

PREFIX=*name* specifies a prefix for naming the principal components. By default the names are PRIN1, PRIN2, ..., PRIN*n*. If PREFIX=ABC is specified, the components are named ABC1, ABC2, ABC3, and so on. The number of characters in the prefix plus the number of digits required to designate the components should not exceed eight.

NOPRINT suppresses the printout.

VAR Statement

VAR *variables;*

The VAR statement lists the numeric variables to be analyzed. If it is omitted, all numeric variables not specified in other statements are analyzed.

PARTIAL Statement

PARTIAL *variables;*

If you want to analyze a partial correlation or covariance matrix, specify the names of the numeric variables to be partialled out in the PARTIAL statement.

FREQ Statement

FREQ *variable;*

If a variable in your data set represents the frequency of occurrence for the other values in the observation, include the variable's name in a FREQ statement. The procedure then treats the data set as if each observation appears *n* times, where *n* is the value of the FREQ variable for the observation. The total number of observations is considered equal to the sum of the FREQ variable when the procedure determines degrees of freedom for significance probabilities.

The WEIGHT and FREQ statements have a similar effect, except in the calculation of degrees of freedom.

WEIGHT Statement

WEIGHT *variable;*

If you want to use relative weights for each observation in the input data set, place the weights in a variable in the data set and specify the name in a WEIGHT statement. This is often done when the variance associated with each observation is different and the values of the weight variable are proportional to the reciprocals of the variances. When the WEIGHT statement is specified, the divisor used to compute variances is the sum of the weights rather than the number of observations minus one.

BY Statement

BY *variables;*

A BY statement may be used with PROC PRINCOMP to obtain separate analyses on observations in groups defined by the BY variables. When a BY statement appears, the procedure expects the input data set to be sorted in order of the BY variables. If your input data set is not sorted in ascending order, use the SORT procedure with a similar BY statement to sort the data, or, if appropriate, use the BY statement options NOTSORTED or DESCENDING. For more information, see the discussion of the BY statement in Chapter 8, "Statements Used in the PROC Step," in *SAS User's Guide: Basics, 1982 Edition*.

DETAILS

Missing Values

Observations with missing values are omitted from the analysis and are given missing values for principal component scores in the OUT= data set.

Output Data Sets

OUT= Data Set The OUT= data set contains all the variables in the original data set plus new variables containing the principal component scores. The N= option determines the number of new variables. The names of the new variables are formed by concatenating the value given by the PREFIX= option (or PRIN if PREFIX= is omitted) and the numbers 1, 2, 3, and so on. The new variables have mean 0 and variance equal to the corresponding eigenvalue, unless the STANDARD option is specified to standardize the scores to unit variance.

An OUT= data set cannot be created if the DATA= data set is TYPE=CORR or TYPE=COV or if a PARTIAL statement is used.

OUTSTAT= Data Set The OUTSTAT= data set is similar to the TYPE=CORR data set produced by the CORR procedure. The OUTSTAT= data set is TYPE=CORR unless the COV option is specified, in which case it is TYPE=COV.

The new data set contains the following variables:

- the BY variables, if any
- two new character variables, __TYPE__ and __NAME__
- the variables analyzed, that is, those in the VAR statement, or, if there is no VAR statement, all numeric variables not listed in any other statement.

Each observation in the new data set contains some type of statistic as indicated by the __TYPE__ variable. The values of the __TYPE__ variable are as follows:

__TYPE__	Contents
MEAN	mean of each variable.
STD	standard deviations. This observation is omitted if the COV option is specified so the SCORE procedure does not standardize the variables before computing scores.
N	number of observations on which the analysis is based. This value is the same for each variable.
CORR	correlations between each variable and the variable named by the __NAME__ variable. The number of observations with __TYPE__='CORR' is equal to the number of variables being analyzed. If the COV option is specified, no __TYPE__= 'CORR' observations are produced.
COV	covariances between each variable and the variable named by the __NAME__ variable. __TYPE__='COV' observations are produced only if the COV option is specified.
EIGENVAL	eigenvalues. If the N= option requested fewer than the maximum number of principal components, only the specified number of eigenvalues are produced, with missing values filling out the observation.
SCORE	eigenvectors. The __NAME__ variable contains the name of the corresponding principal component as constructed from the PREFIX= option. The number of observations with __TYPE__='SCORE' equals the number of principal components computed.

The data set may be used with the SCORE procedure to compute principal component scores, or may be used as input to the FACTOR procedure with METHOD=SCORE to rotate the components.

Computational Resources

Let:

n = number of observations
v = number of variables
c = number of components.

The time required to compute the correlation matrix is roughly proportional to nv^2.

The time required to compute eigenvalues is roughly proportional to v^3.

The time required to compute eigenvectors is roughly proportional to cv^2.

Printed Output

The PRINCOMP procedure prints:

1. SIMPLE STATISTICS, including the MEAN and ST DEV (standard deviation) for each variable, if the DATA= data set is neither TYPE=CORR nor TYPE=COV.
2. the CORRELATIONS or COVARIANCES among the variables if the DATA= data set is neither TYPE=CORR nor TYPE=COV
3. the TOTAL VARIANCE if the COV option is used
4. EIGENVALUES of the correlation or covariance matrix, as well as the DIFFERENCE between successive eigenvalues, the PROPORTION of variance explained by each eigenvalue, and the CUMULATIVE proportion of variance explained
5. the EIGENVECTORS.

EXAMPLES

January and July Temperatures: Example 1

The first example analyzes mean daily temperatures in selected cities in January and July. Both the raw data and the principal components are plotted to illustrate how principal components are orthogonal rotations of the original variables.

Note that since the COV option is used and JANUARY has a higher standard deviation than JULY, JANUARY receives a higher loading on the first component.

```
DATA TEMPERAT;
  TITLE 'MEAN TEMPERATURE IN JANUARY AND JULY FOR SELECTED CITIES';
  INPUT CITY $1-15 JANUARY JULY;
  CARDS;
MOBILE           51.2 81.6
PHOENIX          51.2 91.2
LITTLE ROCK      39.5 81.4
SACRAMENTO       45.1 75.2
DENVER           29.9 73.0
HARTFORD         24.8 72.7
WILMINGTON       32.0 75.8
WASHINGTON, DC   35.6 78.7
JACKSONVILLE     54.6 81.0
MIAMI            67.2 82.3
ATLANTA          42.4 78.0
BOISE            29.0 74.5
CHICAGO          22.9 71.9
PEORIA           23.8 75.1
INDIANAPOLIS     27.9 75.0
DES MOINES       19.4 75.1
WICHITA          31.3 80.7
LOUISVILLE       33.3 76.9
NEW ORLEANS      52.9 81.9
PORTLAND, MAINE  21.5 68.0
BALTIMORE        33.4 76.6
BOSTON           29.2 73.3
```

```
DETROIT            25.5 73.3
SAULT STE MARIE    14.2 63.8
DULUTH              8.5 65.6
MINNEAPOLIS        12.2 71.9
JACKSON            47.1 81.7
KANSAS CITY        27.8 78.8
ST LOUIS           31.3 78.6
GREAT FALLS        20.5 69.3
OMAHA              22.6 77.2
RENO               31.9 69.3
CONCORD            20.6 69.7
ATLANTIC CITY      32.7 75.1
ALBUQUERQUE        35.2 78.7
ALBANY             21.5 72.0
BUFFALO            23.7 70.1
NEW YORK           32.2 76.6
CHARLOTTE          42.1 78.5
RALEIGH            40.5 77.5
BISMARCK            8.2 70.8
CINCINNATI         31.1 75.6
CLEVELAND          26.9 71.4
COLUMBUS           28.4 73.6
OKLAHOMA CITY      36.8 81.5
PORTLAND, OREG     38.1 67.1
PHILADELPHIA       32.3 76.8
PITTSBURGH         28.1 71.9
PROVIDENCE         28.4 72.1
COLUMBIA           45.4 81.2
SIOUX FALLS        14.2 73.3
MEMPHIS            40.5 79.6
NASHVILLE          38.3 79.6
DALLAS             44.8 84.8
EL PASO            43.6 82.3
HOUSTON            52.1 83.3
SALT LAKE CITY     28.0 76.7
BURLINGTON         16.8 69.8
NORFOLK            40.5 78.3
RICHMOND           37.5 77.9
SPOKANE            25.4 69.7
CHARLESTON, WV     34.5 75.0
MILWAUKEE          19.4 69.9
CHEYENNE           26.6 69.1
;
PROC PLOT;
  PLOT JULY*JANUARY=CITY/VPOS=36;
PROC PRINCOMP COV OUT=PRIN;
  VAR JULY JANUARY;
PROC PLOT;
  PLOT PRIN2*PRIN1=CITY/VPOS=26;
  TITLE2 PLOT OF PRINCIPAL COMPONENTS;
```

MEAN TEMPERATURE IN JANUARY AND JULY FOR SELECTED CITIES 1

PLOT OF JULY*JANUARY SYMBOL IS VALUE OF CITY

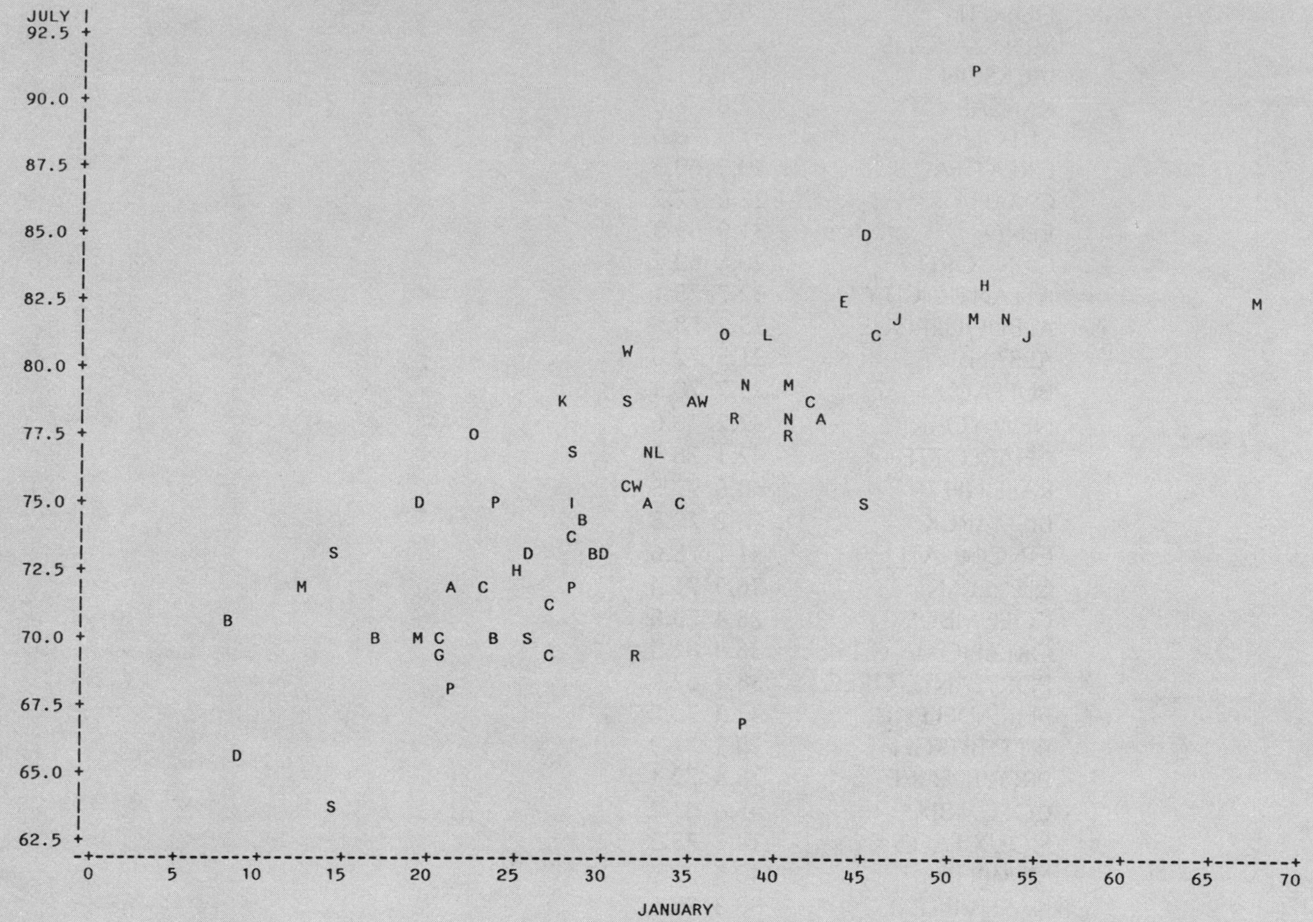

JULY: 92.5, 90.0, 87.5, 85.0, 82.5, 80.0, 77.5, 75.0, 72.5, 70.0, 67.5, 65.0, 62.5

JANUARY: 0, 5, 10, 15, 20, 25, 30, 35, 40, 45, 50, 55, 60, 65, 70

NOTE: 3 OBS HIDDEN

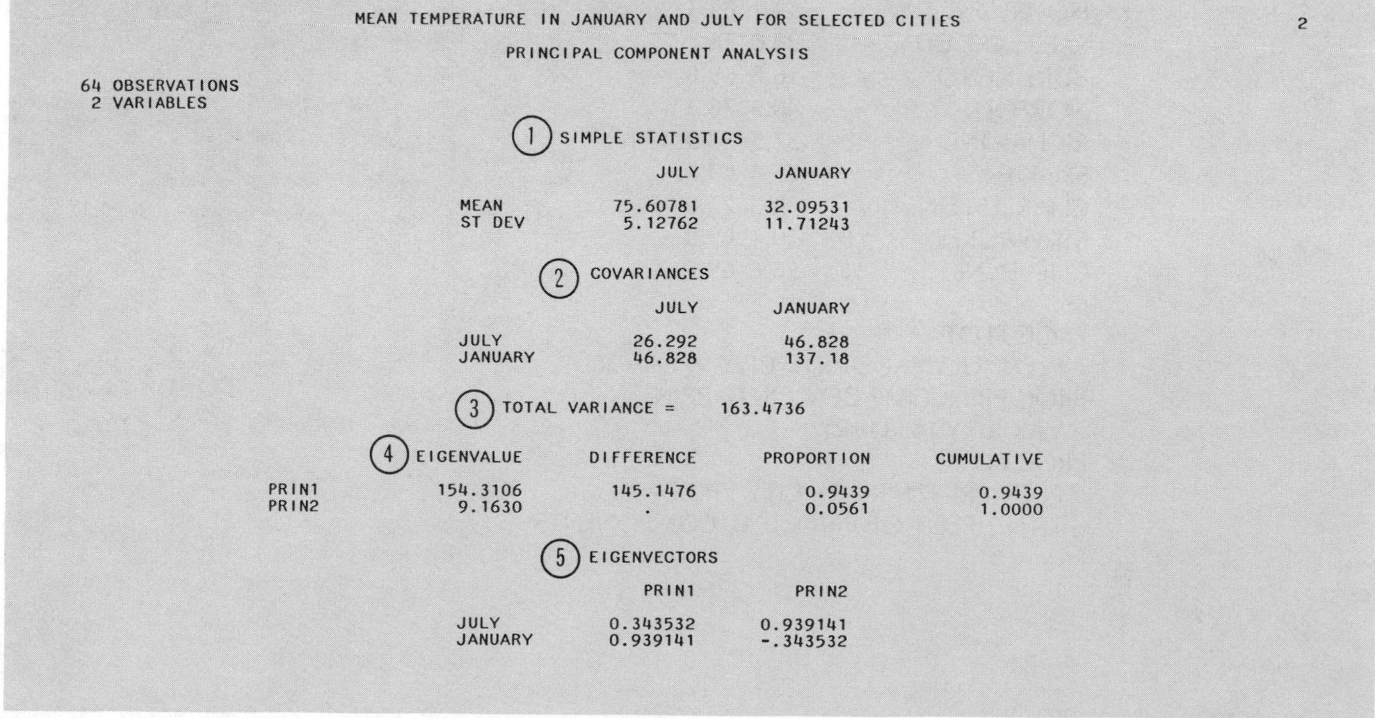

MEAN TEMPERATURE IN JANUARY AND JULY FOR SELECTED CITIES 2

PRINCIPAL COMPONENT ANALYSIS

64 OBSERVATIONS
2 VARIABLES

(1) SIMPLE STATISTICS

	JULY	JANUARY
MEAN	75.60781	32.09531
ST DEV	5.12762	11.71243

(2) COVARIANCES

	JULY	JANUARY
JULY	26.292	46.828
JANUARY	46.828	137.18

(3) TOTAL VARIANCE = 163.4736

	(4) EIGENVALUE	DIFFERENCE	PROPORTION	CUMULATIVE
PRIN1	154.3106	145.1476	0.9439	0.9439
PRIN2	9.1630	.	0.0561	1.0000

(5) EIGENVECTORS

	PRIN1	PRIN2
JULY	0.343532	0.939141
JANUARY	0.939141	-.343532

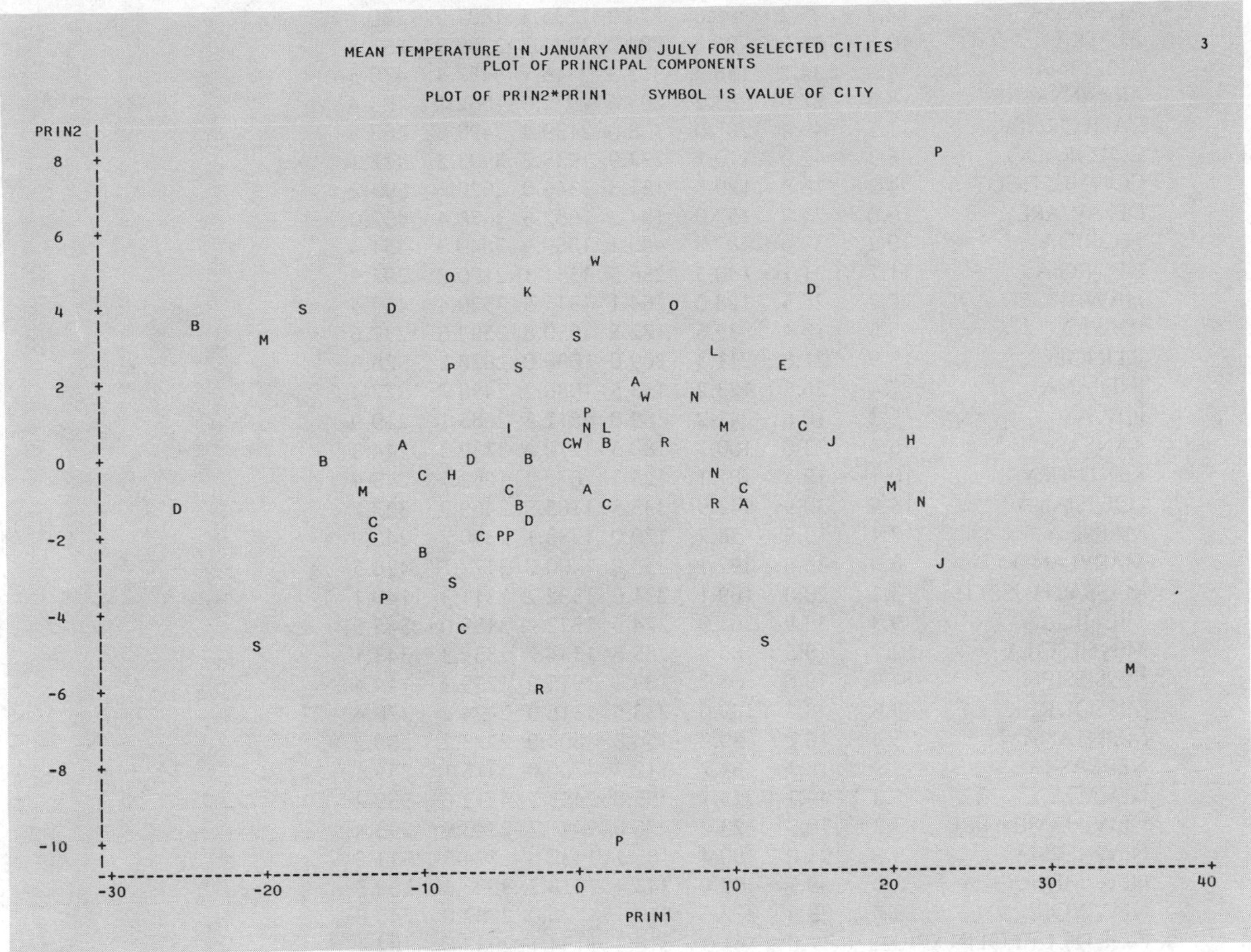

Crime Rates: Example 2

The data below give crime rates per 100,000 people in seven categories for each of the fifty states. Since there are seven variables it is impossible to plot all the variables simultaneously. Principal components can be used to summarize the data in two or three dimensions and help to visualize the data.

```
DATA CRIME;
  TITLE CRIME RATES PER 100,000 POPULATION BY STATE;
  INPUT STATE $1-15 MURDER RAPE ROBBERY ASSAULT BURGLARY LARCENY
    AUTO;
  CARDS;
```

```
ALABAMA          14.2  25.2   96.8  278.3 1135.5 1881.9  280.7
ALASKA           10.8  51.6   96.8  284.0 1331.7 3369.8  753.3
ARIZONA           9.5  34.2  138.2  312.3 2346.1 4467.4  439.5
ARKANSAS          8.8  27.6   83.2  203.4  972.6 1862.1  183.4
CALIFORNIA       11.5  49.4  287.0  358.0 2139.4 3499.8  663.5
COLORADO          6.3  42.0  170.7  292.9 1935.2 3903.2  477.1
CONNECTICUT       4.2  16.8  129.5  131.8 1346.0 2620.7  593.2
DELAWARE          6.0  24.9  157.0  194.2 1682.6 3678.4  467.0
FLORIDA          10.2  39.6  187.9  449.1 1859.9 3840.5  351.4
GEORGIA          11.7  31.1  140.5  256.5 1351.1 2170.2  297.9
HAWAII            7.2  25.5  128.0   64.1 1911.5 3920.4  489.4
IDAHO             5.5  19.4   39.6  172.5 1050.8 2599.6  237.6
ILLINOIS          9.9  21.8  211.3  209.0 1085.0 2828.5  528.6
INDIANA           7.4  26.5  123.2  153.5 1086.2 2498.7  377.4
IOWA              2.3  10.6   41.2   89.8  812.5 2685.1  219.9
KANSAS            6.6  22.0  100.7  180.5 1270.4 2739.3  244.3
KENTUCKY         10.1  19.1   81.1  123.3  872.2 1662.1  245.4
LOUISIANA        15.5  30.9  142.9  335.5 1165.5 2469.9  337.7
MAINE             2.4  13.5   38.7  170.0 1253.1 2350.7  246.9
MARYLAND          8.0  34.8  292.1  358.9 1400.0 3177.7  428.5
MASSACHUSETTS     3.1  20.8  169.1  231.6 1532.2 2311.3 1140.1
MICHIGAN          9.3  38.9  261.9  274.6 1522.7 3159.0  545.5
MINNESOTA         2.7  19.5   85.9   85.8 1134.7 2559.3  343.1
MISSISSIPPI      14.3  19.6   65.7  189.1  915.6 1239.9  144.4
MISSOURI          9.6  28.3  189.0  233.5 1318.3 2424.2  378.4
MONTANA           5.4  16.7   39.2  156.8  804.9 2773.2  309.2
NEBRASKA          3.9  18.1   64.7  112.7  760.0 2316.1  249.1
NEVADA           15.8  49.1  323.1  355.0 2453.1 4212.6  559.2
NEW HAMPSHIRE     3.2  10.7   23.2   76.0 1041.7 2343.9  293.4
NEW JERSEY        5.6  21.0  180.4  185.1 1435.8 2774.5  511.5
NEW MEXICO        8.8  39.1  109.6  343.4 1418.7 3008.6  259.5
NEW YORK         10.7  29.4  472.6  319.1 1728.0 2782.0  745.8
NORTH CAROLINA   10.6  17.0   61.3  318.3 1154.1 2037.8  192.1
NORTH DAKOTA      0.9   9.0   13.3   43.8  446.1 1843.0  144.7
OHIO              7.8  27.3  190.5  181.1 1216.0 2696.8  400.4
OKLAHOMA          8.6  29.2   73.8  205.0 1288.2 2228.1  326.8
OREGON            4.9  39.9  124.1  286.9 1636.4 3506.1  388.9
PENNSYLVANIA      5.6  19.0  130.3  128.0  877.5 1624.1  333.2
RHODE ISLAND      3.6  10.5   86.5  201.0 1489.5 2844.1  791.4
SOUTH CAROLINA   11.9  33.0  105.9  485.3 1613.6 2342.4  245.1
SOUTH DAKOTA      2.0  13.5   17.9  155.7  570.5 1704.4  147.5
TENNESSEE        10.1  29.7  145.8  203.9 1259.7 1776.5  314.0
TEXAS            13.3  33.8  152.4  208.2 1603.1 2988.7  397.6
UTAH              3.5  20.3   68.8  147.3 1171.6 3004.6  334.5
VERMONT           1.4  15.9   30.8  101.2 1348.2 2201.0  265.2
VIRGINIA          9.0  23.3   92.1  165.7  986.2 2521.2  226.7
WASHINGTON        4.3  39.6  106.2  224.8 1605.6 3386.9  360.3
WEST VIRGINIA     6.0  13.2   42.2   90.9  597.4 1341.7  163.3
WISCONSIN         2.8  12.9   52.2   63.7  846.9 2614.2  220.7
WYOMING           5.4  21.9   39.7  173.9  811.6 2772.2  282.0
;
PROC PRINCOMP OUT=CRIMCOMP;
```

The eigenvalues indicate that two or three components provide a good summary of the data, two components accounting for 76% of the standardized variance and three components explaining 87%. Subsequent components contribute less than 5% each.

The first component is a measure of overall crime rate. The first eigenvector shows approximately equal loadings on all variables. The second eigenvector has high positive loadings on AUTO and LARCENY, and high negative loadings on MURDER and ASSAULT. There is also a small positive loading on BURGLARY and a small negative loading on RAPE. This component seems to measure the preponderance of property crime over violent crime. The interpretation of the third component is not obvious.

A simple way to examine the principal components in more detail is to print the output data set sorted by each of the large components.

```
PROC SORT;
  BY PRIN1;
PROC PRINT;
  ID STATE;
  VAR PRIN1 PRIN2 MURDER RAPE ROBBERY ASSAULT BURGLARY LARCENY
    AUTO;
  TITLE2 STATES LISTED IN ORDER OF OVERALL CRIME RATE;
  TITLE3 AS DETERMINED BY THE FIRST PRINCIPAL COMPONENT;
PROC SORT;
  BY PRIN2;
PROC PRINT;
  ID STATE;
  VAR PRIN1 PRIN2 MURDER RAPE ROBBERY ASSAULT BURGLARY LARCENY
    AUTO;
  TITLE2 STATES LISTED IN ORDER OF PROPERTY VS. VIOLENT CRIME;
  TITLE3 AS DETERMINED BY THE SECOND PRINCIPAL COMPONENT;
```

Another recommended procedure is to make scatter plots of the first few components. The sorted listings help to identify observations on the plots.

```
PROC PLOT;
  PLOT PRIN2*PRIN1=STATE;
  TITLE2 PLOT OF THE FIRST TWO PRINCIPAL COMPONENTS;
PROC PLOT;
  PLOT PRIN3*PRIN1=STATE;
  TITLE2 PLOT OF THE FIRST AND THIRD PRINCIPAL COMPONENTS;
```

It is possible to identify regional trends on the plot of the first two components. Nevada and California are at the extreme right with high overall crime rates but an average ratio of property crime to violent crime. North and South Dakota are on the extreme left with low over all crime rates. Southeastern states tend to be in the bottom of the plot, with a higher than average ratio of violent to property crime. New England states tend to be in the upper part of the plot, with a greater than average ratio of property crime to violent crime.

The most striking feature of the plot of the first and third principal components is that Massachusetts and New York are outliers on the third component.

CRIME RATES PER 100,000 POPULATION BY STATE

PRINCIPAL COMPONENT ANALYSIS

1

50 OBSERVATIONS
7 VARIABLES

SIMPLE STATISTICS

	MURDER	RAPE	ROBBERY	ASSAULT	BURGLARY	LARCENY	AUTO
MEAN	7.444000	25.73400	124.0920	211.3000	1291.904	2671.288	377.5260
ST DEV	3.866769	10.75963	88.3486	100.2530	432.456	725.909	193.3944

CORRELATIONS

	MURDER	RAPE	ROBBERY	ASSAULT	BURGLARY	LARCENY	AUTO
MURDER	1.0000	0.6012	0.4837	0.6486	0.3858	0.1019	0.0688
RAPE	0.6012	1.0000	0.5919	0.7403	0.7121	0.6140	0.3489
ROBBERY	0.4837	0.5919	1.0000	0.5571	0.6372	0.4467	0.5907
ASSAULT	0.6486	0.7403	0.5571	1.0000	0.6229	0.4044	0.2758
BURGLARY	0.3858	0.7121	0.6372	0.6229	1.0000	0.7921	0.5580
LARCENY	0.1019	0.6140	0.4467	0.4044	0.7921	1.0000	0.4442
AUTO	0.0688	0.3489	0.5907	0.2758	0.5580	0.4442	1.0000

	EIGENVALUE	DIFFERENCE	PROPORTION	CUMULATIVE
PRIN1	4.114960	2.876238	0.587851	0.587851
PRIN2	1.238722	0.512905	0.176960	0.764812
PRIN3	0.725817	0.409385	0.103688	0.868500
PRIN4	0.316432	0.058458	0.045205	0.913704
PRIN5	0.257974	0.035935	0.036853	0.950558
PRIN6	0.222039	0.097983	0.031720	0.982278
PRIN7	0.124056	.	0.017722	1.000000

EIGENVECTORS

	PRIN1	PRIN2	PRIN3	PRIN4	PRIN5	PRIN6	PRIN7
MURDER	0.300279	-.629174	0.178245	-.232114	0.538123	0.259117	0.267593
RAPE	0.431759	-.169435	-.244198	0.062216	0.188471	-.773271	-.296485
ROBBERY	0.396875	0.042247	0.495861	-.557989	-.519977	-.114385	-.003903
ASSAULT	0.396652	-.343528	-.069510	0.629804	-.506651	0.172363	0.191745
BURGLARY	0.440157	0.203341	-.209895	-.057555	0.101033	0.535987	-.648117
LARCENY	0.357360	0.402319	-.539231	-.234890	0.030099	0.039406	0.601690
AUTO	0.295177	0.502421	0.568384	0.419238	0.369753	-.057298	0.147046

CRIME RATES PER 100,000 POPULATION BY STATE
STATES LISTED IN ORDER OF OVERALL CRIME RATE
AS DETERMINED BY THE FIRST PRINCIPAL COMPONENT

2

STATE	PRIN1	PRIN2	MURDER	RAPE	ROBBERY	ASSAULT	BURGLARY	LARCENY	AUTO
NORTH DAKOTA	-3.9641	0.3877	0.9	9.0	13.3	43.8	446.1	1843.0	144.7
SOUTH DAKOTA	-3.1720	-0.2545	2.0	13.5	17.9	155.7	570.5	1704.4	147.5
WEST VIRGINIA	-3.1477	-0.8143	6.0	13.2	42.2	90.9	597.4	1341.7	163.3
IOWA	-2.5816	0.8248	2.3	10.6	41.2	89.8	812.5	2685.1	219.9
WISCONSIN	-2.5030	0.7808	2.8	12.9	52.2	63.7	846.9	2614.2	220.7
NEW HAMPSHIRE	-2.4656	0.8250	3.2	10.7	23.2	76.0	1041.7	2343.9	293.4
NEBRASKA	-2.1507	0.2257	3.9	18.1	64.7	112.7	760.0	2316.1	249.1
VERMONT	-2.0643	0.9450	1.4	15.9	30.8	101.2	1348.2	2201.0	265.2
MAINE	-1.8263	0.5788	2.4	13.5	38.7	170.0	1253.1	2350.7	246.9
KENTUCKY	-1.7269	-1.1466	10.1	19.1	81.1	123.3	872.2	1662.1	245.4
PENNSYLVANIA	-1.7201	-0.1959	5.6	19.0	130.3	128.0	877.5	1624.1	333.2
MONTANA	-1.6680	0.2710	5.4	16.7	39.2	156.8	804.9	2773.2	309.2
MINNESOTA	-1.5543	1.0564	2.7	19.5	85.9	85.8	1134.7	2559.3	343.1
MISSISSIPPI	-1.5074	-2.5467	14.3	19.6	65.7	189.1	915.6	1239.9	144.4
IDAHO	-1.4325	-0.0080	5.5	19.4	39.6	172.5	1050.8	2599.6	237.6
WYOMING	-1.4246	0.0627	5.4	21.9	39.7	173.9	811.6	2772.2	282.0
ARKANSAS	-1.0544	-1.3454	8.8	27.6	83.2	203.4	972.6	1862.1	183.4
UTAH	-1.0500	0.9366	3.5	20.3	68.8	147.3	1171.6	3004.6	334.5
VIRGINIA	-0.9162	-0.6927	9.0	23.3	92.1	165.7	986.2	2521.2	226.7
NORTH CAROLINA	-0.6993	-1.6703	10.6	17.0	61.3	318.3	1154.1	2037.8	192.1
KANSAS	-0.6341	-0.0280	6.6	22.0	100.7	180.5	1270.4	2739.3	244.3

(continued on next page)

(continued from previous page)

CONNECTICUT	-0.5413	1.5012	4.2	16.8	129.5	131.8	1346.0	2620.7	593.2
INDIANA	-0.4999	0.0000	7.4	26.5	123.2	153.5	1086.2	2498.7	377.4
OKLAHOMA	-0.3214	-0.6243	8.6	29.2	73.8	205.0	1288.2	2228.1	326.8
RHODE ISLAND	-0.2016	2.1466	3.6	10.5	86.5	201.0	1489.5	2844.1	791.4
TENNESSEE	-0.1366	-1.1350	10.1	29.7	145.8	203.9	1259.7	1776.5	314.0
ALABAMA	-0.0499	-2.0961	14.2	25.2	96.8	278.3	1135.5	1881.9	280.7
NEW JERSEY	0.2179	0.9642	5.6	21.0	180.4	185.1	1435.8	2774.5	511.5
OHIO	0.2395	0.0905	7.8	27.3	190.5	181.1	1216.0	2696.8	400.4
GEORGIA	0.4904	-1.3808	11.7	31.1	140.5	256.5	1351.1	2170.2	297.9
ILLINOIS	0.5129	0.0942	9.9	21.8	211.3	209.0	1085.0	2828.5	528.6
MISSOURI	0.5564	-0.5585	9.6	28.3	189.0	233.5	1318.3	2424.2	378.4
HAWAII	0.8231	1.8239	7.2	25.5	128.0	64.1	1911.5	3920.4	489.4
WASHINGTON	0.9306	0.7378	4.3	39.6	106.2	224.8	1605.6	3386.9	360.3
DELAWARE	0.9646	1.2967	6.0	24.9	157.0	194.2	1682.6	3678.4	467.0
MASSACHUSETTS	0.9784	2.6311	3.1	20.8	169.1	231.6	1532.2	2311.3	1140.1
LOUISIANA	1.1202	-2.0833	15.5	30.9	142.9	335.5	1165.5	2469.9	337.7
NEW MEXICO	1.2142	-0.9508	8.8	39.1	109.6	343.4	1418.7	3008.6	259.5
TEXAS	1.3970	-0.6813	13.3	33.8	152.4	208.2	1603.1	2988.7	397.6
OREGON	1.4490	0.5860	4.9	39.9	124.1	286.9	1636.4	3506.1	388.9
SOUTH CAROLINA	1.6034	-2.1621	11.9	33.0	105.9	485.3	1613.6	2342.4	245.1
MARYLAND	2.1828	-0.1947	8.0	34.8	292.1	358.9	1400.0	3177.7	428.5
MICHIGAN	2.2733	0.1549	9.3	38.9	261.9	274.6	1522.7	3159.0	545.5
ALASKA	2.4215	0.1665	10.8	51.6	96.8	284.0	1331.7	3369.8	753.3
COLORADO	2.5093	0.9166	6.3	42.0	170.7	292.9	1935.2	3903.2	477.1
ARIZONA	3.0141	0.8449	9.5	34.2	138.2	312.3	2346.1	4467.4	439.5
FLORIDA	3.1118	-0.6039	10.2	39.6	187.9	449.1	1859.9	3840.5	351.4
NEW YORK	3.4525	0.4329	10.7	29.4	472.6	319.1	1728.0	2782.0	745.8
CALIFORNIA	4.2838	0.1432	11.5	49.4	287.0	358.0	2139.4	3499.8	663.5
NEVADA	5.2670	-0.2526	15.8	49.1	323.1	355.0	2453.1	4212.6	559.2

CRIME RATES PER 100,000 POPULATION BY STATE
STATES LISTED IN ORDER OF PROPERTY VS. VIOLENT CRIME
AS DETERMINED BY THE SECOND PRINCIPAL COMPONENT

3

STATE	PRIN1	PRIN2	MURDER	RAPE	ROBBERY	ASSAULT	BURGLARY	LARCENY	AUTO
MISSISSIPPI	-1.5074	-2.5467	14.3	19.6	65.7	189.1	915.6	1239.9	144.4
SOUTH CAROLINA	1.6034	-2.1621	11.9	33.0	105.9	485.3	1613.6	2342.4	245.1
ALABAMA	-0.0499	-2.0961	14.2	25.2	96.8	278.3	1135.5	1881.9	280.7
LOUISIANA	1.1202	-2.0833	15.5	30.9	142.9	335.5	1165.5	2469.9	337.7
NORTH CAROLINA	-0.6993	-1.6703	10.6	17.0	61.3	318.3	1154.1	2037.8	192.1
GEORGIA	0.4904	-1.3808	11.7	31.1	140.5	256.5	1351.1	2170.2	297.9
ARKANSAS	-1.0544	-1.3454	8.8	27.6	83.2	203.4	972.6	1862.1	183.4
KENTUCKY	-1.7269	-1.1466	10.1	19.1	81.1	123.3	872.2	1662.1	245.4
TENNESSEE	-0.1366	-1.1350	10.1	29.7	145.8	203.9	1259.7	1776.5	314.0
NEW MEXICO	1.2142	-0.9508	8.8	39.1	109.6	343.4	1418.7	3008.6	259.5
WEST VIRGINIA	-3.1477	-0.8143	6.0	13.2	42.2	90.9	597.4	1341.7	163.3
VIRGINIA	-0.9162	-0.6927	9.0	23.3	92.1	165.7	986.2	2521.2	226.7
TEXAS	1.3970	-0.6813	13.3	33.8	152.4	208.2	1603.1	2988.7	397.6
OKLAHOMA	-0.3214	-0.6243	8.6	29.2	73.8	205.0	1288.2	2228.1	326.8
FLORIDA	3.1118	-0.6039	10.2	39.6	187.9	449.1	1859.9	3840.5	351.4
MISSOURI	0.5564	-0.5585	9.6	28.3	189.0	233.5	1318.3	2424.2	378.4
SOUTH DAKOTA	-3.1720	-0.2545	2.0	13.5	17.9	155.7	570.5	1704.4	147.5
NEVADA	5.2670	-0.2526	15.8	49.1	323.1	355.0	2453.1	4212.6	559.2
PENNSYLVANIA	-1.7201	-0.1959	5.6	19.0	130.3	128.0	877.5	1624.1	333.2
MARYLAND	2.1828	-0.1947	8.0	34.8	292.1	358.9	1400.0	3177.7	428.5
KANSAS	-0.6341	-0.0280	6.6	22.0	100.7	180.5	1270.4	2739.3	244.3
IDAHO	-1.4325	-0.0080	5.5	19.4	39.6	172.5	1050.8	2599.6	237.6
INDIANA	-0.4999	0.0000	7.4	26.5	123.2	153.5	1086.2	2498.7	377.4
WYOMING	-1.4246	0.0627	5.4	21.9	39.7	173.9	811.6	2772.2	282.0
OHIO	0.2395	0.0905	7.8	27.3	190.5	181.1	1216.0	2696.8	400.4
ILLINOIS	0.5129	0.0942	9.9	21.8	211.3	209.0	1085.0	2828.5	528.6
CALIFORNIA	4.2838	0.1432	11.5	49.4	287.0	358.0	2139.4	3499.8	663.5
MICHIGAN	2.2733	0.1549	9.3	38.9	261.9	274.6	1522.7	3159.0	545.5
ALASKA	2.4215	0.1665	10.8	51.6	96.8	284.0	1331.7	3369.8	753.3
NEBRASKA	-2.1507	0.2257	3.9	18.1	64.7	112.7	760.0	2316.1	249.1
MONTANA	-1.6680	0.2710	5.4	16.7	39.2	156.8	804.9	2773.2	309.2
NORTH DAKOTA	-3.9641	0.3877	0.9	9.0	13.3	43.8	446.1	1843.0	144.7
NEW YORK	3.4525	0.4329	10.7	29.4	472.6	319.1	1728.0	2782.0	745.8
MAINE	-1.8263	0.5788	2.4	13.5	38.7	170.0	1253.1	2350.7	246.9
OREGON	1.4490	0.5860	4.9	39.9	124.1	286.9	1636.4	3506.1	388.9
WASHINGTON	0.9306	0.7378	4.3	39.6	106.2	224.8	1605.6	3386.9	360.3
WISCONSIN	-2.5030	0.7808	2.8	12.9	52.2	63.7	846.9	2614.2	220.7
IOWA	-2.5816	0.8248	2.3	10.6	41.2	89.8	812.5	2685.1	219.9
NEW HAMPSHIRE	-2.4656	0.8250	3.2	10.7	23.2	76.0	1041.7	2343.9	293.4
ARIZONA	3.0141	0.8449	9.5	34.2	138.2	312.3	2346.1	4467.4	439.5
COLORADO	2.5093	0.9166	6.3	42.0	170.7	292.9	1935.2	3903.2	477.1
UTAH	-1.0500	0.9366	3.5	20.3	68.8	147.3	1171.6	3004.6	334.5
VERMONT	-2.0643	0.9450	1.4	15.9	30.8	101.2	1348.2	2201.0	265.2
NEW JERSEY	0.2179	0.9642	5.6	21.0	180.4	185.1	1435.8	2774.5	511.5
MINNESOTA	-1.5543	1.0564	2.7	19.5	85.9	85.8	1134.7	2559.3	343.1
DELAWARE	0.9646	1.2967	6.0	24.9	157.0	194.2	1682.6	3678.4	467.0
CONNECTICUT	-0.5413	1.5012	4.2	16.8	129.5	131.8	1346.0	2620.7	593.2
HAWAII	0.8231	1.8239	7.2	25.5	128.0	64.1	1911.5	3920.4	489.4
RHODE ISLAND	-0.2016	2.1466	3.6	10.5	86.5	201.0	1489.5	2844.1	791.4
MASSACHUSETTS	0.9784	2.6311	3.1	20.8	169.1	231.6	1532.2	2311.3	1140.1

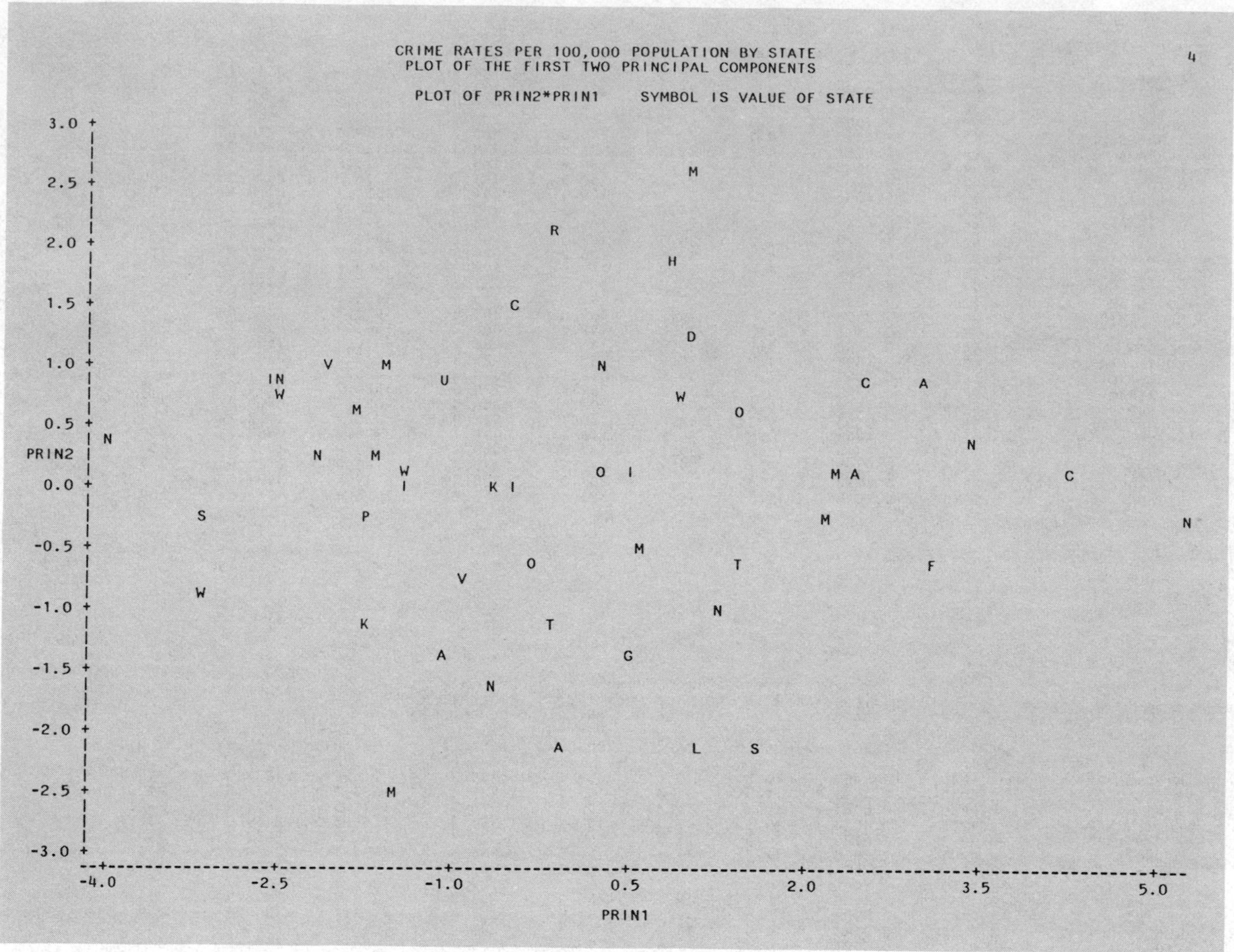
CRIME RATES PER 100,000 POPULATION BY STATE
PLOT OF THE FIRST TWO PRINCIPAL COMPONENTS
4
PLOT OF PRIN2*PRIN1 SYMBOL IS VALUE OF STATE
PRIN2
PRIN1
3.0
2.5
2.0
1.5
1.0
0.5
0.0
-0.5
-1.0
-1.5
-2.0
-2.5
-3.0
-4.0
-2.5
-1.0
0.5
2.0
3.5
5.0

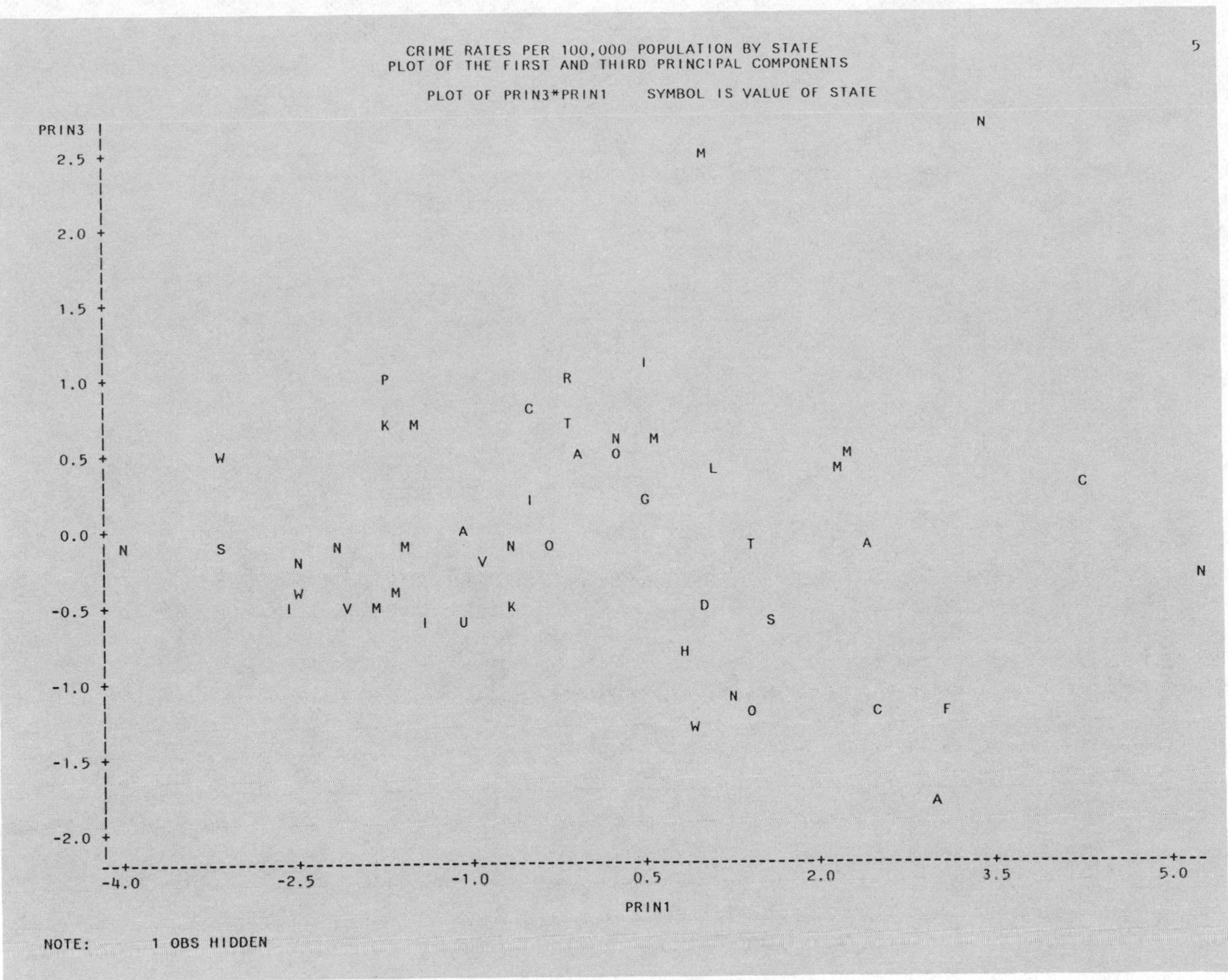

REFERENCES

Cooley, W.W. and Lohnes, P.R. (1971), *Multivariate Data Analysis*, New York: John Wiley & Sons.

Gnanadesikan, R. (1977), *Methods for Statistical Data Analysis of Multivariate Observations*, New York: John Wiley & Sons.

Hotelling, H. (1933), "Analysis of a Complex of Statistical Variables into Principal Components," *Journal of Educational Psychology*, 24, 417-441, 498-520.

Kshirsagar, A.M. (1972), *Multivariate Analysis*, New York: Marcell Dekker.

Mardia, K.V., Kent, J.T., and Bibby, J.M. (1979), *Multivariate Analysis*, London: Academic Press.

Morrison, D.F. (1976), *Multivariate Statistical Methods*, Second Edition. New York: McGraw-Hill.

Pearson, K. (1901), "On Lines and Planes of Closest Fit to Systems of Points in Space." *Philosophical Magazine*, 6 (2), 559-572.

Rao, C.R. (1964), "The Use and Interpretation of Principal Component Analysis in Applied Research." *Sankya A*, 26, 329-358.

DISCRIMINANT ANALYSIS

Introduction to SAS Discriminant Procedures

The procedures in this chapter analyze data with one classification variable and several continuous variables. The purpose may be to find a rule for placing observations into the classes from knowledge of the continuous variables, or to find a subset of variables or a set of linear combinations that best reveal the differences among the classes. The procedures are:

DISCRIM	classifies observations assuming a multivariate normal distribution within each class. The classes may or may not be assumed to have equal covariance matrices.
NEIGHBOR	classifies observations using a nonparametric nearest-neighbor method.
CANDISC	finds linear combinations of the variables that best summarize the differences among the classes and computes scores for each observation on the linear combinations.
STEPDISC	uses forward selection, backward elimination, or stepwise selection to try to find a subset of variables that best reveals differences among the classes.

The term *discriminant analysis* (Fisher, 1936; Cooley and Lohnes, 1971; Tatsuoka, 1971; Kshirsagar, 1972; Lachenbruch, 1975, 1979; Gnanadesikan, 1977; Klecka, 1980) refers to several different types of analysis. *Classificatory discriminant analysis* is used to classify observations into two or more known groups on the basis of one or more numeric variables. Classification can be done by the DISCRIM and NEIGHBOR procedures. Use NEIGHBOR when the classes have radically non-normal distributions. DISCRIM works better with approximately normal distributions. *Canonical discriminant analysis* is a dimension-reduction technique related to principal components and canonical correlation and is performed by the CANDISC procedure. *Stepwise discriminant analysis* is a variable-selection technique implemented by the STEPDISC procedure. After selecting a subset of variables with STEPDISC, you can use any of the other discriminant procedures to obtain more detailed analyses. CANDISC and STEPDISC are most appropriate for approximately normally distributed data.

If your continuous variables are not normally distributed, or if you wish to classify observations on the basis of categorical variables, you should consider using the FUNCAT procedure to fit a categorical linear model with the classification variable as the dependent variable. Press and Wilson (1978) compare logistic regression and discriminant analysis and conclude that logistic regression is preferable to discriminant methods based on normality assumptions when the variables do not have multivariate normal distributions within classes.

Another alternative to discriminant analysis is to perform a series of univariate one-way *ANOVA*s. Both CANDISC and STEPDISC provide summaries of the univariate *ANOVA*s. The advantage of the multivariate approach is that two or more groups that overlap considerably when each variable is viewed separately may be quite distinct when examined from a multivariate point of view. Consider the two groups, indicated by *H* and *O* in the following scatterplot:

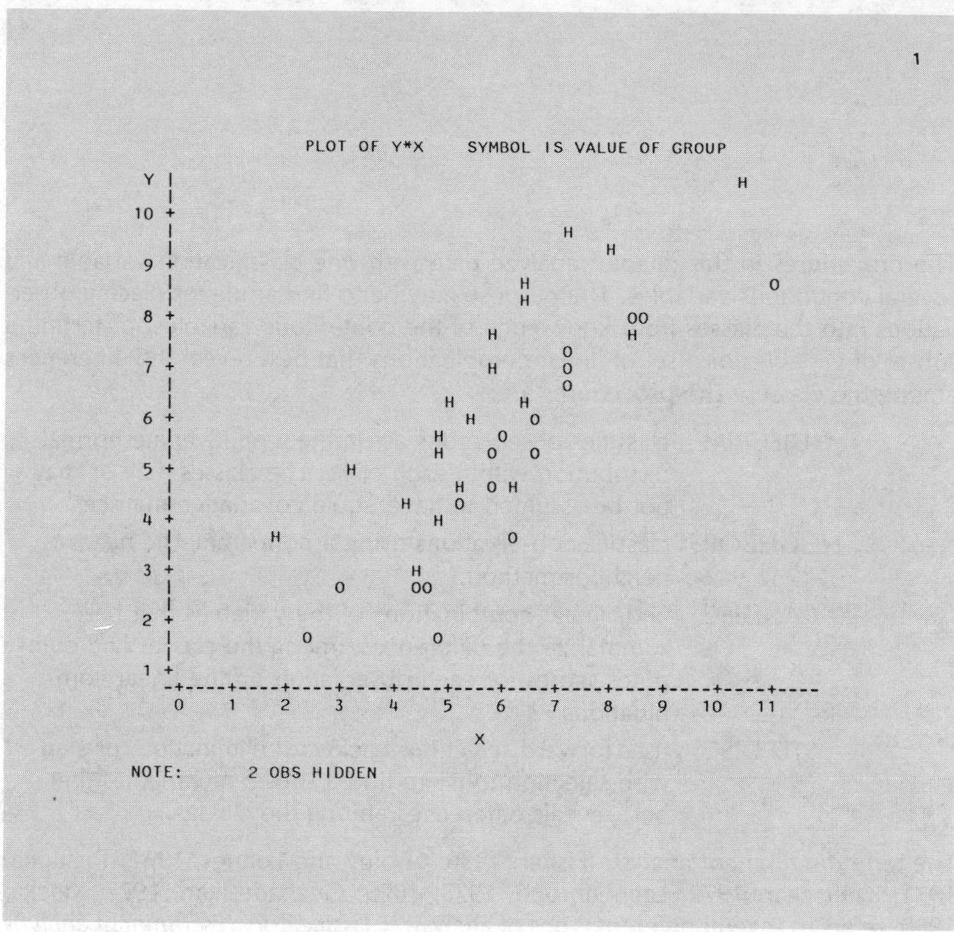

The univariate significance tests from CANDISC are as follows:

```
                                                                          2
                          UNIVARIATE STATISTICS

               R SQUARED        VAR RATIO           F         PROB GT F

X          .01591733156    .01617479107   0.6146420606   .43790513592
Y          .08754241115    .09594134809   3.6457712275   .06378348898
```

The univariate R^2s are very small, and neither variable shows a significant difference between the groups at the .05 level. Here are the most important statistics from the canonical analysis:

```
                                                                              3
          CANONICAL      LIKELIHOOD
          R-SQUARED         RATIO       F STATISTIC   NUM DF   DEN DF   PROB>F

        0.422322000   0.577678000         13.5248          2        37   0.0000
```

```
                                                                              4
                             RAW CANONICAL COEFFICIENTS

                                           CAN1                 CAN2

                   X                 -.9866887168         0.3834889396
                   Y                 0.9328326060         0.1403144570
```

The multivariate test for differences between the groups is significant beyond the .00005 level. Thus the multivariate analysis has found a highly significant difference when the univariate analyses failed to achieve even the .05 level. The canonical coefficients for the first canonical variable, CAN1, show that the groups differ most widely on the linear combination $-.9867X+.9328Y$, or approximately Y-X. The R^2 between CAN1 and the group variable is .4223 as given by the CANONICAL R-SQUARED, which is much higher than either univariate R^2.

In this example the variables are highly correlated within groups. If the within-class correlation were smaller, there would be greater agreement between the univariate and multivariate analyses.

REFERENCES

Cooley, W.W. and Lohnes, P.R. (1971), *Multivariate Data Analysis*, New York: John Wiley & Sons.

Fisher, R.A. (1936), "The Use of Multiple Measurements in Taxonomic Problems," *Annals of Eugenics*, 7, 179-188.

Gnanadesikan, R. (1977), *Methods for Statistical Data Analysis of Multivariate Observations*, New York: John Wiley & Sons.

Klecka, W.R. (1980), *Discriminant Analysis*, Sage University Paper Series on Quantitative Applications in the Social Sciences, 07-019. Beverly Hills and London: Sage Publications.

Kshirsagar, A.M. (1972), *Multivariate Analysis*, New York: Marcel Dekker.

Lachenbruch, P.A. (1975), *Discriminant Analysis*, New York: Hafner.

Lachenbruch, P.A. (1979), "Discriminant Analysis," *Biometrics*, 35, 69-85.

Press, S.J. and Wilson, S. (1978), "Choosing Between Logistic Regression and Discriminant Analysis," *Journal of the American Statistical Association*, 73, 699-705.

Tatsuoka, M.M. (1971), *Multivariate Analysis*, New York: John Wiley & Sons.

The CANDISC Procedure

ABSTRACT

The CANDISC procedure performs a canonical discriminant analysis, computes Mahalanobis distances, and does both univariate and multivariate one-way analyses of variance. Output data sets containing canonical coefficients and scores on the canonical variables can be created.

INTRODUCTION

Canonical discriminant analysis is a dimension-reduction technique related to principal component analysis and canonical correlation. Given a classification variable and several quantitative variables, CANDISC derives *canonical variables* (linear combinations of the quantitative variables) that summarize between-class variation in much the same way that principal components summarize total variation.

For each canonical correlation CANDISC tests the hypothesis that it and all smaller canonical correlations are 0 in the population. An *F* approximation (Rao, 1973; Kshirsagar, 1972) is used that gives better small sample results than the usual chi-squared approximation. The variables should have an approximate multivariate normal distribution within each class with a common covariance matrix in order for the probability levels to be valid.

Both standardized and unstandardized canonical coefficients are printed, as well as the correlations between canonical variables and the original variables, and the means of each class on the canonical variables.

CANDISC performs univariate and multivariate one-way analyses of variance, and computes and tests Mahalanobis distances for pairwise comparisons of the classes.

The procedure can produce an output data set containing the scores on each canonical variable. You can use the PRINT procedure to list these values and the PLOT procedure to plot pairs of canonical variables to aid visual interpretation of group differences. A second output data set contains canonical coefficients that can be rotated by the FACTOR procedure.

Background

Given two or more groups of observations with measurements on several quantitative variables, canonical discriminant analysis derives a linear combination of the variables that has the highest possible multiple correlation with the groups. This maximal multiple correlation is called the *first canonical correlation*. The coefficients of the linear combination are the *canonical coefficients* or *canonical weights*. The variable defined by the linear combination is the *first canonical variable* or *canonical component*. It is customary to normalize the canonical coefficients so that the pooled within-group variance of the canonical variable is one. Canonical variables are sometimes called *discriminant functions*, but this usage is ambiguous since DISCRIM produces very different functions for classification that are also called discriminant functions.

The second canonical correlation is obtained by finding the linear combination uncorrelated with the first canonical variable that has the highest possible multiple correlation with the groups. The process of extracting canonical variables can be repeated until the number of canonical variables equals the number of original variables or the number of classes minus one, whichever is smaller.

The first canonical correlation is at least as large as the multiple correlation between the groups and any of the original variables. If the original variables have low within-group correlations, then the first canonical correlation is not much greater than the largest multiple correlation. If the original variables have high within-group correlations, the first canonical correlation may be large even if all the multiple correlations are small. In other words, the first canonical variable may show substantial differences among the classes even if none of the original variables do.

Canonical discriminant analysis is equivalent to canonical correlation analysis between the quantitative variables and a set of dummy variables coded from the class variable. Canonical discriminant analysis is also equivalent to performing the following steps:

- Transform the variables so that the pooled within-class covariance matrix is an identity matrix.
- Compute class means on the transformed variables.
- Do a principal component analysis on the means, weighting each mean by the number of observations in the class. The eigenvalues are equal to the ratio of between-class variance to within-class variance in the direction of each principal component.
- Back-transform the principal components into the space of the original variables to obtain the canonical variables.

An interesting property of the canonical variables is that they are uncorrelated whether the correlation is calculated from the total sample or from the pooled within-class correlations. The canonical coefficients are not orthogonal, however, so the canonical variables do not represent perpendicular directions through the space of the original variables.

SPECIFICATIONS

The CANDISC procedure is invoked by the following statements:

PROC CANDISC *options;*
VAR *variables;*
CLASS *variable;*
PROB *variables;*
FREQ *variable;*
WEIGHT *variable;*
BY *variables;*

Usually only the VAR and CLASS statements are needed in addition to the PROC CANDISC statement.

PROC CANDISC Statement

PROC CANDISC *options;*

The following options can appear on the PROC statement:

DATA=*SASdataset* names the data set to be analyzed. The data set may be an ordinary SAS data set, or a TYPE=CORR or TYPE=COV data set produced by either the CORR procedure using a BY statement, or a previous run of CANDISC using the OUTSTAT= option. If DATA= is omitted, the most recently created SAS data set is used.

OUT=*SASdataset* names an output SAS data set that contains the original data and the canonical variable scores. If you want to create a permanent SAS data set you must specify a two-level name (see Chapter 12, "SAS Data Sets," in *SAS User's Guide: Basics, 1982 Edition*, for more information on permanent SAS data sets).

OUTSTAT=*SASdataset* names an output SAS data set that contains various statistics including class means, pooled standard deviations, correlation and covariance matrices for both the total sample and pooled within-classes, and canonical correlations, coefficients, and means. Use a two-level name if you want to create a permanent data set (see Chapter 12, "SAS Data Sets," in *SAS User's Guide: Basics, 1982 Edition*, for more information on permanent SAS data sets).

NCAN=*n* specifies the number of canonical variables to be computed. The default is the number of variables. If NCAN=0 is specified, the procedure computes the canonical correlations, but not the canonical coefficients, structures, or means. A negative value suppresses the canonical analysis entirely.

PREFIX=*name* specifies a prefix for naming the canonical variables. By default the names are CAN1, CAN2,..., CAN*n*. If PREFIX=ABC is specified, the components are named ABC1, ABC2, ABC3, and so on. The number of characters in the prefix plus the number of digits required to designate the canonical variables should not exceed eight.

UNIVARIATE
UNI prints univariate statistics including means, standard deviations, R^2s, and *F* statistics and probability levels for analyses of variance.

STDMEAN prints standardized means.

TCORR prints total sample correlations.

WCORR prints pooled within-class correlations.

BCORR prints between-class correlations.

TCOV prints total sample covariances.

WCOV prints pooled within-class covariances.

BCOV prints between-class covariances.

MAHALANOBIS
MAH prints Mahalanobis distances between classes.

ALL prints everything.

TOLERANCE=*n*
TOL=*n* specifies the tolerance for determining the singularity of the total correlation matrix. The default value is 1E-8.

EDF=*n* specifies the error degrees of freedom from the regression analysis if your input observations are residuals from a regression analysis. The effective number of observations is the EDF= value plus one. If you have 100 observations, then EDF=99 has the same effect as omitting EDF=.

RDF=*n* specifies the regression degrees of freedom if your input observations are residuals from a regression analysis. The effective number of observations is the actual number minus the RDF= value. The degree of freedom for the intercept should not be included in RDF=.

VAR Statement

VAR *variables;*

The VAR statement lists the numeric variables to be analyzed. If it is omitted, all numeric variables not specified in any other statements are used.

CLASS Statement

CLASS *variable;*

The CLASS statement specifies the name of a variable, either character or numeric, that defines the classes to be analyzed. Class levels are determined by the unformatted values of the class variable. Either a CLASS statement or a PROB statement **must** be present.

PROB Statement

PROB *variables;*

A PROB statement can be used instead of a CLASS statement. Each variable in the PROB statement defines a class. The variables must be numeric, with values summing to one for each observation. There are at least two situations in which the PROB statement can be used:

- when some observations are repeated many times. The FREQ variable (see below) gives the total number of times each observation occurs, while the PROB statement gives the proportion belonging to each class.
- when you do not know with certainty the class to which each observation belongs. The PROB variables give the probabilities of the observations coming from each class. The standard errors and significance levels printed by CANDISC are not valid in this case.

FREQ Statement

FREQ *variable;*

If a variable in your data set represents the frequency of occurrence for the other values in the observation, include the variable's name in a FREQ statement. The procedure then treats the data set as if each observation appears *n* times, where *n* is the value of the FREQ variable for the observation. The total number of observations is considered to be equal to the sum of the FREQ variable when the procedure determines degrees of freedom for significance probabilities.

The WEIGHT and FREQ statements have a similar effect, except in the calculation of degrees of freedom.

WEIGHT Statement

WEIGHT *variable*;

If you want to use relative weights for each observation in the input data set, place the weights in a variable in the data set and specify the name in a WEIGHT statement. This is often done when the variance associated with each observation is different and the values of the weight variable are proportional to the reciprocals of the variances.

If the WEIGHT statement is specified, the divisor used to compute variances is the sum of the weights, rather than the number of observations minus one.

BY Statement

BY *variables*;

A BY statement may be used with PROC CANDISC to obtain separate analyses on observations in groups defined by the BY variables. When a BY statement appears, the procedure expects the input data set to be sorted in order of the BY variables. If your input data set is not sorted in ascending order, use the SORT procedure with a similar BY statement to sort the data, or, if appropriate, use the BY statement options NOTSORTED or DESCENDING. For more information, see the discussion of the BY statement in Chapter 8, "Statements Used in the PROC Step," in *SAS User's Guide: Basics, 1982 Edition*.

DETAILS

Missing Values

If an observation has a missing value for any of the continuous variables, it is omitted from the analysis. If an observation has a missing CLASS value but is otherwise complete, it is not used in computing the canonical correlations and coefficients, but canonical variable scores are computed for the OUT= data set.

Output Data Sets

OUT= Data Set The OUT= data set contains all the variables in the original data set plus new variables containing the canonical variable scores. The NCAN= option determines the number of new variables. The names of the new variables are formed as described in the PREFIX= option. The new variables have mean 0 and pooled within-class variance equal to 1. An OUT= data set cannot be created if the DATA= data set is TYPE=CORR or TYPE=COV.

OUTSTAT= Data Set The OUTSTAT= data set is similar to the TYPE=CORR data set produced by the CORR procedure, but contains many results in addition to those produced by CORR.

The OUTSTAT= data set contains the following variables:

- the BY variables, if any
- the CLASS variable if a CLASS statement was used, or a new variable called _CLASS_ if a PROB statement was used
- two new character variables, _TYPE_ and _NAME_
- the quantitative variables, that is, those in the VAR statement, or, if there is no VAR statement, all numeric variables not listed in any other statement.

Each observation in the new data set contains some type of statistic as indicated by the __TYPE__ variable. The values of the __TYPE__ variable are as follows:

__TYPE__	Contents
COV	total sample covariance matrix.
BCOV	between-class covariance matrix.
WCOV	pooled within-class covariance matrix.
CORR	total sample correlation matrix.
BCORR	between-class correlation matrix.
WCORR	pooled within-class correlation matrix.
MEAN	means for both the total sample (CLASS variable missing) and each class (CLASS variable present).
STD	total sample standard deviations.
BSTD	between-class standard deviations.
WSTD	pooled within-class standard deviations.
RSQUARED	univariate R^2s.
N	number of observations in the total sample (CLASS variable missing) and within each class (CLASS variable present).
TSTDMEAN	total-standardized class means.
WSTDMEAN	within-standardized class means.
CANCORR	canonical correlations.
STRUCTUR	canonical structure.
SCORE	standardized canonical coefficients.
RAWSCORE	raw canonical coefficients.
CANMEAN	means of the canonical variables for each class.

Computational Resources

Let:

n = number of observations
v = number of variables
c = number of classes.

The time required to compute correlation/covariance matrices is roughly proportional to $(n+c)v^2$.

The time required to compute Mahalanobis distances is roughly proportional to cv^2.

The time required for the canonical analysis is roughly proportional to v^3.

Printed Output

If specified, the following statistics are printed:

1. MEANS for the total sample
2. TOTAL STD, total sample standard deviations
3. WITHIN STD, pooled within-class standard deviations
4. BETWEEN STD, between-class standard deviations
5. R-SQUARED, univariate R^2s
6. VAR RATIO, univariate ratios of between-class variance to within-class variance
7. univariate F values and PROB GT F, probability levels
8. CLASS MEANS
9. TOTAL-STANDARDIZED CLASS MEANS

10. WITHIN-STANDARDIZED CLASS MEANS
11. TOTAL SAMPLE COVARIANCE MATRIX
12. BETWEEN-CLASS COVARIANCE MATRIX
13. POOLED WITHIN-CLASS COVARIANCE MATRIX
14. TOTAL SAMPLE CORRELATION MATRIX
15. BETWEEN-CLASS CORRELATION MATRIX
16. POOLED WITHIN-CLASS CORRELATION MATRIX
17. MAHALANOBIS DISTANCES BETWEEN CLASSES and PROB > MAHALANOBIS DISTANCE.

By default, the printout contains these statistics:

18. CANONICAL CORRELATIONS
19. ADJUSTED CANCORR, adjusted canonical correlations (Lawley, 1959). These are asymptotically less biased than the raw correlations and can be negative. The adjusted canonical correlations may not be computable and are printed as missing values if two canonical correlations are nearly equal or if some are close to zero. A missing value is also printed if an adjusted canonical correlation is larger than a previous adjusted canonical correlation.
20. APPROX STD ERROR, approximate standard error of the canonical correlations
21. VARIANCE RATIO, the ratio of the between-class variance to the within-class variance for each canonical variable. This is equal to the squared canonical correlation divided by one minus the squared canonical correlation and is analogous to an eigenvalue in principal component analysis.
22. CANONICAL R-SQUARED, the squared canonical correlations
23. LIKELIHOOD RATIO for the hypothesis that the current canonical correlation and all smaller ones are zero in the population. The likelihood ratio for all canonical correlations equals Wilks' lambda.
24. F STATISTIC based on Rao's approximation (Rao, 1973, p. 556; Kshirsagar, 1972, p. 326)
25. NUM DF (numerator degrees of freedom), DEN DF (denominator degrees of freedom), and PROB>F, the probability level associated with the *F* statistic
26. WILKS' LAMBDA, PILLAI'S TRACE, HOTELLING-LAWLEY TRACE, and ROY'S GREATEST ROOT with *F* approximations
27. TOTAL CANONICAL STRUCTURE giving total-sample correlations between the canonical variables and the original variables
28. WITHIN CANONICAL STRUCTURE giving within-class correlations between the canonical variables and the original variables
29. STANDARDIZED CANONICAL COEFFICIENTS normalized to give canonical variables with unit within-class variance when applied to the standardized variables
30. RAW (unstandardized) CANONICAL COEFFICIENTS normalized to give canonical variables with unit within-class variance when applied to the raw variables
31. CLASS MEANS ON CANONICAL VARIABLES.

EXAMPLE

The iris data published by Fisher (1936) have been widely used for examples in discriminant analysis and cluster analysis. The sepal length, sepal width, petal length, and petal width were measured in millimeters on 50 iris specimens from each of three species, *Iris setosa, I. versicolor,* and *I. virginica*. Here we perform a canonical discriminant analysis, create an output data set containing scores on the canonical variables, and plot the canonical variables.

```
DATA IRIS;
  TITLE FISHER (1936) IRIS DATA;
  INPUT SEPALLEN SEPALWID PETALLEN PETALWID SPEC_NO@@;
  IF SPEC_NO=1 THEN SPECIES='SETOSA ';
  IF SPEC_NO=2 THEN SPECIES='VERSICOLOR';
  IF SPEC_NO=3 THEN SPECIES='VIRGINICA ';
  LABEL SEPALLEN=SEPAL LENGTH IN MM.
    SEPALWID=SEPAL WIDTH IN MM.
    PETALLEN=PETAL LENGTH IN MM.
    PETALWID=PETAL WIDTH IN MM.;
  CARDS;
50 33 14 02 1 64 28 56 22 3 65 28 46 15 2
67 31 56 24 3 63 28 51 15 3 46 34 14 03 1
69 31 51 23 3 62 22 45 15 2 59 32 48 18 2
46 36 10 02 1 61 30 46 14 2 60 27 51 16 2
65 30 52 20 3 56 25 39 11 2 65 30 55 18 3
58 27 51 19 3 68 32 59 23 3 51 33 17 05 1
57 28 45 13 2 62 34 54 23 3 77 38 67 22 3
63 33 47 16 2 67 33 57 25 3 76 30 66 21 3
49 25 45 17 3 55 35 13 02 1 67 30 52 23 3
70 32 47 14 2 64 32 45 15 2 61 28 40 13 2
48 31 16 02 1 59 30 51 18 3 55 24 38 11 2
63 25 50 19 3 64 32 53 23 3 52 34 14 02 1
49 36 14 01 1 54 30 45 15 2 79 38 64 20 3
44 32 13 02 1 67 33 57 21 3 50 35 16 06 1
58 26 40 12 2 44 30 13 02 1 77 28 67 20 3
63 27 49 18 3 47 32 16 02 1 55 26 44 12 2
50 23 33 10 2 72 32 60 18 3 48 30 14 03 1
51 38 16 02 1 61 30 49 18 3 48 34 19 02 1
50 30 16 02 1 50 32 12 02 1 61 26 56 14 3
64 28 56 21 3 43 30 11 01 1 58 40 12 02 1
51 38 19 04 1 67 31 44 14 2 62 28 48 18 3
49 30 14 02 1 51 35 14 02 1 56 30 45 15 2
58 27 41 10 2 50 34 16 04 1 46 32 14 02 1
60 29 45 15 2 57 26 35 10 2 57 44 15 04 1
50 36 14 02 1 77 30 61 23 3 63 34 56 24 3
58 27 51 19 3 57 29 42 13 2 72 30 58 16 3
54 34 15 04 1 52 41 15 01 1 71 30 59 21 3
64 31 55 18 3 60 30 48 18 3 63 29 56 18 3
49 24 33 10 2 56 27 42 13 2 57 30 42 12 2
55 42 14 02 1 49 31 15 02 1 77 26 69 23 3
60 22 50 15 3 54 39 17 04 1 66 29 46 13 2
52 27 39 14 2 60 34 45 16 2 50 34 15 02 1
44 29 14 02 1 50 20 35 10 2 55 24 37 10 2
58 27 39 12 2 47 32 13 02 1 46 31 15 02 1
69 32 57 23 3 62 29 43 13 2 74 28 61 19 3
59 30 42 15 2 51 34 15 02 1 50 35 13 03 1
56 28 49 20 3 60 22 40 10 2 73 29 63 18 3
67 25 58 18 3 49 31 15 01 1 67 31 47 15 2
```

```
63 23 44 13 2 54 37 15 02 1 56 30 41 13 2
63 25 49 15 2 61 28 47 12 2 64 29 43 13 2
51 25 30 11 2 57 28 41 13 2 65 30 58 22 3
69 31 54 21 3 54 39 13 04 1 51 35 14 03 1
72 36 61 25 3 65 32 51 20 3 61 29 47 14 2
56 29 36 13 2 69 31 49 15 2 64 27 53 19 3
68 30 55 21 3 55 25 40 13 2 48 34 16 02 1
48 30 14 01 1 45 23 13 03 1 57 25 50 20 3
57 38 17 03 1 51 38 15 03 1 55 23 40 13 2
66 30 44 14 2 68 28 48 14 2 54 34 17 02 1
51 37 15 04 1 52 35 15 02 1 58 28 51 24 3
67 30 50 17 2 63 33 60 25 3 53 37 15 02 1
;
PROC CANDISC ALL OUT=DISC;
  CLASSES SPECIES;
  VAR SEPALLEN SEPALWID PETALLEN PETALWID;
PROC PLOT;
  PLOT CAN2*CAN1=SPEC__NO;
  TITLE2 PLOT OF CANONICAL DISCRIMINANT FUNCTIONS;
```

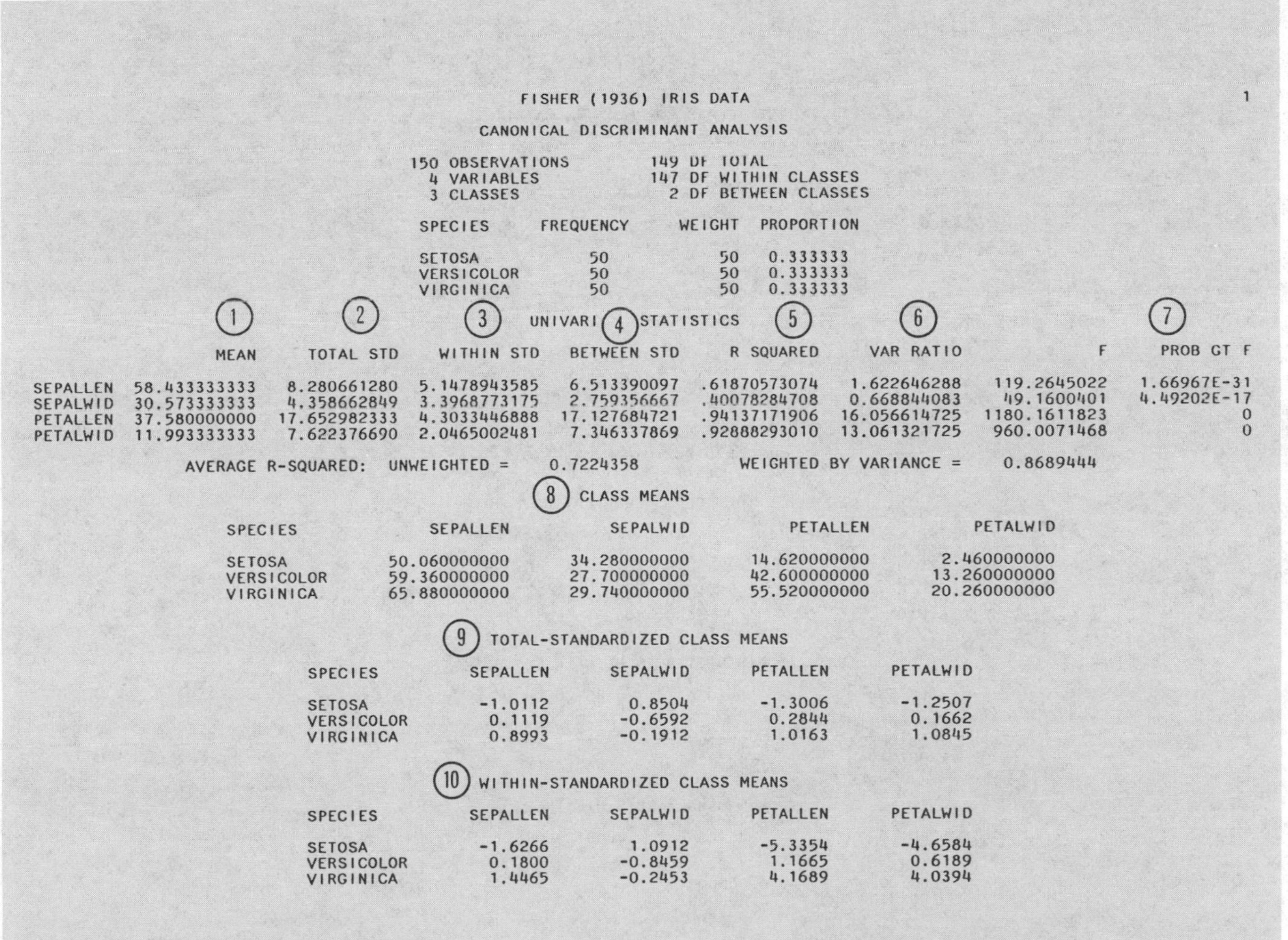

FISHER (1936) IRIS DATA 1

CANONICAL DISCRIMINANT ANALYSIS

150 OBSERVATIONS 149 DF TOTAL
4 VARIABLES 147 DF WITHIN CLASSES
3 CLASSES 2 DF BETWEEN CLASSES

SPECIES	FREQUENCY	WEIGHT	PROPORTION
SETOSA	50	50	0.333333
VERSICOLOR	50	50	0.333333
VIRGINICA	50	50	0.333333

UNIVARIATE STATISTICS

	① MEAN	② TOTAL STD	③ WITHIN STD	④ BETWEEN STD	⑤ R SQUARED	⑥ VAR RATIO	F	⑦ PROB GT F
SEPALLEN	58.433333333	8.280661280	5.1478943585	6.513390097	.61870573074	1.622646288	119.2645022	1.66967E-31
SEPALWID	30.573333333	4.358662849	3.3968773175	2.759356667	.40078284708	0.668844083	49.1600401	4.49202E-17
PETALLEN	37.580000000	17.652982333	4.3033446888	17.127684721	.94137171906	16.056614725	1180.1611823	0
PETALWID	11.993333333	7.622376690	2.0465002481	7.346337869	.92888293010	13.061321725	960.0071468	0

AVERAGE R-SQUARED: UNWEIGHTED = 0.7224358 WEIGHTED BY VARIANCE = 0.8689444

⑧ CLASS MEANS

SPECIES	SEPALLEN	SEPALWID	PETALLEN	PETALWID
SETOSA	50.060000000	34.280000000	14.620000000	2.460000000
VERSICOLOR	59.360000000	27.700000000	42.600000000	13.260000000
VIRGINICA	65.880000000	29.740000000	55.520000000	20.260000000

⑨ TOTAL-STANDARDIZED CLASS MEANS

SPECIES	SEPALLEN	SEPALWID	PETALLEN	PETALWID
SETOSA	-1.0112	0.8504	-1.3006	-1.2507
VERSICOLOR	0.1119	-0.6592	0.2844	0.1662
VIRGINICA	0.8993	-0.1912	1.0163	1.0845

⑩ WITHIN-STANDARDIZED CLASS MEANS

SPECIES	SEPALLEN	SEPALWID	PETALLEN	PETALWID
SETOSA	-1.6266	1.0912	-5.3354	-4.6584
VERSICOLOR	0.1800	-0.8459	1.1665	0.6189
VIRGINICA	1.4465	-0.2453	4.1689	4.0394

FISHER (1936) IRIS DATA 2

CANONICAL DISCRIMINANT ANALYSIS

(11) TOTAL SAMPLE COVARIANCE MATRIX

VARIABLE	SEPALLEN	SEPALWID	PETALLEN	PETALWID
SEPALLEN	68.56935123	-4.24340045	127.43154362	51.62706935
SEPALWID	-4.24340045	18.99794183	-32.96563758	-12.16393736
PETALLEN	127.43154362	-32.96563758	311.62778523	129.56093960
PETALWID	51.62706935	-12.16393736	129.56093960	58.10062640

(12) BETWEEN-CLASS COVARIANCE MATRIX

VARIABLE	SEPALLEN	SEPALWID	PETALLEN	PETALWID
SEPALLEN	42.42425056	-13.39105145	110.90496644	47.83847875
SEPALWID	-13.39105145	7.61404922	-38.41583893	-15.39105145
PETALLEN	110.90496644	-38.41583893	293.35758389	125.35167785
PETALWID	47.83847875	-15.39105145	125.35167785	53.96868009

(13) POOLED WITHIN-CLASS COVARIANCE MATRIX

VARIABLE	SEPALLEN	SEPALWID	PETALLEN	PETALWID
SEPALLEN	26.500816327	9.272108844	16.751428571	3.840136054
SEPALWID	9.272108844	11.538775510	5.524353741	3.271020408
PETALLEN	16.751428571	5.524353741	18.518775510	4.266530612
PETALWID	3.840136054	3.271020408	4.266530612	4.188163265

(14) TOTAL SAMPLE CORRELATION MATRIX

VARIABLE	SEPALLEN	SEPALWID	PETALLEN	PETALWID
SEPALLEN	1.0000	-0.1176	0.8718	0.8179
SEPALWID	-0.1176	1.0000	-0.4284	-0.3661
PETALLEN	0.8718	-0.4284	1.0000	0.9629
PETALWID	0.8179	-0.3661	0.9629	1.0000

(15) BETWEEN-CLASS CORRELATION MATRIX

VARIABLE	SEPALLEN	SEPALWID	PETALLEN	PETALWID
SEPALLEN	1.0000	-0.7451	0.9941	0.9998
SEPALWID	-0.7451	1.0000	-0.8128	-0.7593
PETALLEN	0.9941	-0.8128	1.0000	0.9962
PETALWID	0.9998	-0.7593	0.9962	1.0000

FISHER (1936) IRIS DATA 3

CANONICAL DISCRIMINANT ANALYSIS

(16) POOLED WITHIN-CLASS CORRELATION MATRIX

VARIABLE	SEPALLEN	SEPALWID	PETALLEN	PETALWID
SEPALLEN	1.0000	0.5302	0.7562	0.3645
SEPALWID	0.5302	1.0000	0.3779	0.4705
PETALLEN	0.7562	0.3779	1.0000	0.4845
PETALWID	0.3645	0.4705	0.4845	1.0000

(17)

MAHALANOBIS DISTANCES BETWEEN CLASSES

SPECIES	SETOSA	VERSICOLOR	VIRGINICA
SETOSA	.	9.4797	13.3935
VERSICOLOR	9.4797	.	4.1474
VIRGINICA	13.3935	4.1474	.

PROB > MAHALANOBIS DISTANCE

SPECIES	SETOSA	VERSICOLOR	VIRGINICA
SETOSA	.	0.000000	0.000000
VERSICOLOR	0.000000	.	0.000000
VIRGINICA	0.000000	0.000000	.

(continued on next page)

(continued from previous page)

(18) (19) (20) (21) (22) (23) (24) (25)

CANONICAL CORRELATIONS AND TESTS OF H0: THE CANONICAL CORRELATION IN THE CURRENT ROW AND ALL THAT FOLLOW ARE ZERO

	CANONICAL CORRELATION	ADJUSTED CAN CORR	APPROX STD ERROR	VARIANCE RATIO	CANONICAL R-SQUARED	LIKELIHOOD RATIO	F STATISTIC	NUM DF	DEN DF	PROB>F
1	0.984820894	0.984097357	0.002468166	32.1919	0.969872194	0.023438631	199.1453	8	288	0.0000
2	0.471197019	0.439283495	0.063734062	0.2854	0.222026631	0.777973369	13.7939	3	145	0.0000

MULTIVARIATE TEST STATISTICS AND F APPROXIMATIONS

(26)

STATISTIC	VALUE	F	NUM DF	DEN DF	PROB>F
WILKS' LAMBDA	0.02343863	199.1453	8	288	0
PILLAI'S TRACE	1.191899	53.46649	8	290	9.74216E-53
HOTELLING-LAWLEY TRACE	32.47732	580.5321	8	286	0
ROY'S GREATEST ROOT	32.19193	1166.957	4	145	0

NOTE: F STATISTIC FOR WILKS' LAMBDA IS EXACT
NOTE: F STATISTIC FOR ROY'S GREATEST ROOT IS AN UPPER BOUND

FISHER (1936) IRIS DATA 4

CANONICAL DISCRIMINANT ANALYSIS

(27) TOTAL CANONICAL STRUCTURE

	CAN1	CAN2	CAN3	CAN4
SEPALLEN	0.7919	0.2176	-0.4275	-0.3779
SEPALWID	-0.5308	0.7580	-0.3745	0.0590
PETALLEN	0.9850	0.0460	-0.1636	-0.0313
PETALWID	0.9728	0.2229	0.0428	-0.0460

(28) WITHIN CANONICAL STRUCTURE

	CAN1	CAN2	CAN3	CAN4
SEPALLEN	0.2226	0.3108	-0.6923	-0.6120
SEPALWID	-0.1190	0.8637	-0.4838	0.0763
PETALLEN	0.7061	0.1677	-0.6758	-0.1291
PETALWID	0.6332	0.7372	0.1606	-0.1725

(29) STANDARDIZED CANONICAL COEFFICIENTS

	CAN1	CAN2	CAN3	CAN4
SEPALLEN	-0.6868	0.0200	-0.2100	-2.6304
SEPALWID	-0.6688	0.9434	-0.6503	0.9715
PETALLEN	3.8858	-1.6451	-3.2263	4.2276
PETALWID	2.1422	2.1641	3.0827	-1.6091

(30) RAW CANONICAL COEFFICIENTS

	CAN1	CAN2	CAN3	CAN4
SEPALLEN	-.0829377642	0.0024102149	-.0253635563	-.3176592443
SEPALWID	-.1534473068	0.2164521235	-.1491965217	0.2228936770
PETALLEN	0.2201211656	-.0931921210	-.1827626273	0.2394822760
PETALWID	0.2810460309	0.2839187853	0.4044320622	-.2110965476

(31) CLASS MEANS ON CANONICAL VARIABLES

SPECIES	CAN1	CAN2	CAN3	CAN4
SETOSA	-7.6076	0.2151	0.0000	-0.0000
VERSICOLOR	1.8250	-0.7279	0.0000	-0.0000
VIRGINICA	5.7826	0.5128	0.0000	-0.0000

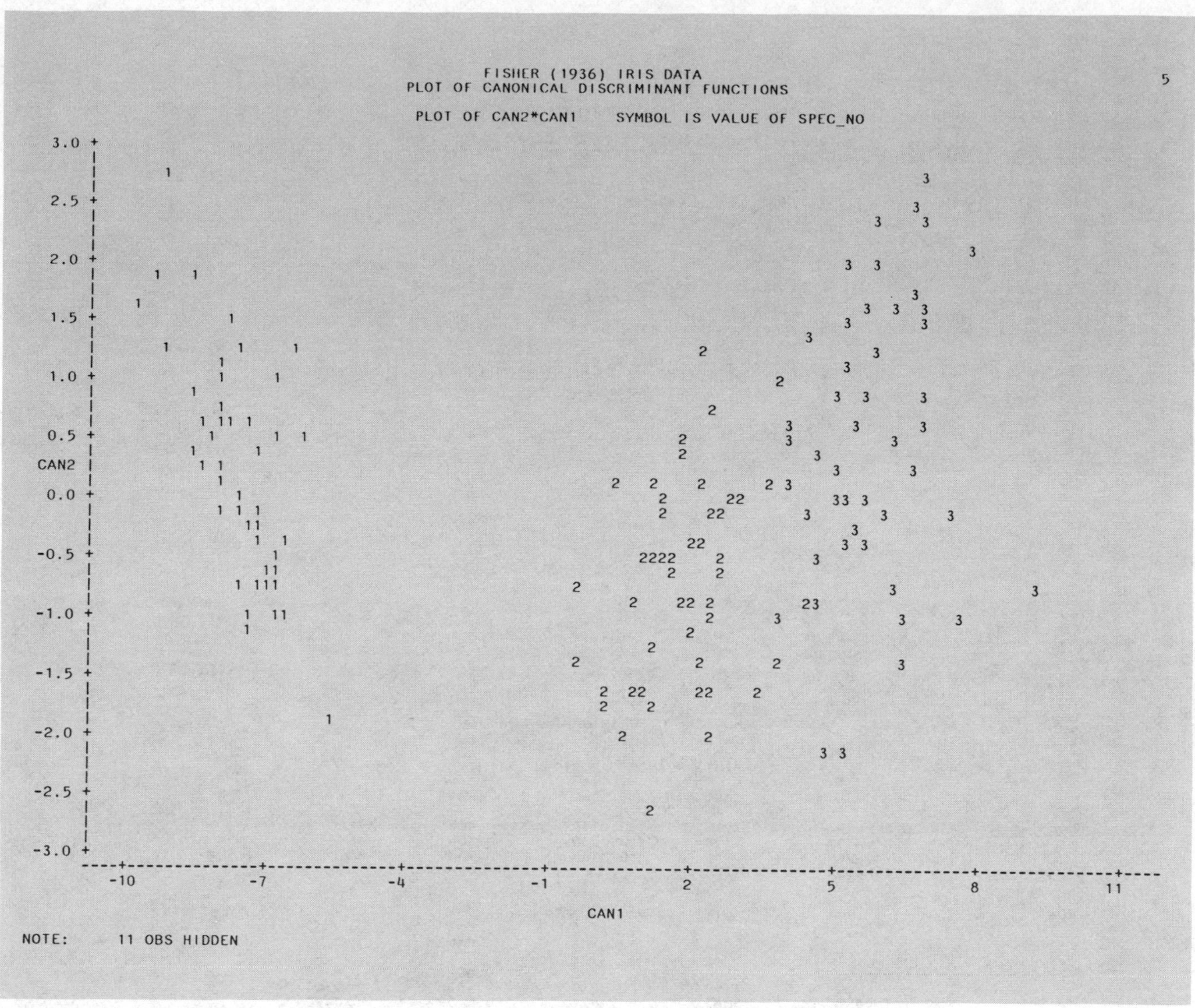

REFERENCES

Fisher, R.A. (1936), "The Use of Multiple Measurements in Taxonomic Problems," *Annals of Eugenics*, 7, 179-188.

Kshirsagar, A.M. (1972), *Multivariate Analysis*, New York: Marcel Dekker.

Lawley, D.N. (1959), "Tests of Significance in Canonical Analysis," *Biometrika*, 46, 59-66.

Rao, C.R. (1973), *Linear Statistical Inference*, New York: John Wiley & Sons.

The DISCRIM Procedure

ABSTRACT

The DISCRIM procedure computes linear or quadratic discriminant functions for classifying observations into two or more groups on the basis of one or more numeric variables. The discriminant functions can be stored in an output data set for future use.

INTRODUCTION

For a set of observations containing one or more quantitative variables and a classification variable defining groups of observations, DISCRIM develops a discriminant model to classify each observation into one of the groups. The distribution within each group should be approximately multivariate normal.

The discriminant model, also known as a classification criterion, is determined by a measure of generalized squared distance (Rao, 1973). The classification criterion can be based on either the individual within-group covariance matrices or the pooled covariance matrix; it also takes into account the prior probabilities of the groups.

Optionally, DISCRIM tests the homogeneity of the within-group covariance matrices. The results of the test determine whether the classification criterion is based on the within-group covariance matrices or the pooled covariance matrix. This test is not robust against non-normality.

The classification criterion can be applied to a second data set during the same execution of DISCRIM. DISCRIM can also store calibration information in a special SAS data set and apply it to other data sets.

Background

DISCRIM develops a discriminant model or classification criterion using a measure of generalized squared distance assuming that each class has a multivariate normal distribution. The classification criterion is based on either the individual within-group covariance matrices or the pooled covariance matrix; it also takes into account the prior probabilities of the groups. Each observation is placed in the class from which it has the smallest generalized squared distance. DISCRIM can also compute the posterior probability of an observation belonging to each class.

The notation below is used to describe the generalized squared distance:

t a subscript to distinguish the groups

$\mathbf{S}_t$ the covariance matrix within group t

$\lvert\mathbf{S}_t\rvert$	the determinant of $\mathbf{S}_t$
$\mathbf{S}$	the pooled covariance matrix
$\mathbf{x}$	a vector containing the variables of an observation
$\mathbf{m}_t$	the vector containing means of the variables in the group t
q_t	the prior probability for group t.

The generalized squared distance from $\mathbf{x}$ to group t is

$$D_t^2(\mathbf{x}) = g_1(\mathbf{x},t) + g_2(t)$$

where:

$$g_1(\mathbf{x},t) = (\mathbf{x}-\mathbf{m}_t)'\,\mathbf{S}_t^{-1}(\mathbf{x}-\mathbf{m}_t) + \log_e\lvert\mathbf{S}_t\rvert$$

if the within-group covariance matrices are used, or

$$g_1(\mathbf{x},t) = (\mathbf{x}-\mathbf{m}_t)'\,\mathbf{S}^{-1}(\mathbf{x}-\mathbf{m}_t)$$

if the pooled covariance matrix is used; and

$$g_2(t) = -2\log_e(q_t)$$

if the prior probabilities are not all equal, or

$$g_2(t) = 0$$

if the prior probabilities are all equal.

The posterior probability of an observation $\mathbf{x}$ belonging to group t is

$$p_t(\mathbf{x}) = \frac{\exp(-0.5D_t^2(\mathbf{x}))}{\Sigma_u(\exp(-0.5D_u^2(\mathbf{x})))}$$

An observation is classified in group u if setting $t=u$ produces the smallest value of $D_t^2(\mathbf{x})$ or the largest value of $p_t(\mathbf{x})$.

SPECIFICATIONS

The following statements are used with DISCRIM:

PROC DISCRIM *options;*
  CLASS *variable;*
  VAR *variables;*
  ID *variable;*
  PRIORS *probabilities;*
  TESTCLASS *variable;*
  TESTID *variable;*
  BY *variables;*

PROC DISCRIM Statement

PROC DISCRIM *options;*

The options below may appear in the PROC DISCRIM statement.

SIMPLE
S prints simple descriptive statistics for all variables.

POOL=YES
POOL=NO
POOL=TEST
determines whether the pooled or within-group covariance matrix is the basis of the measure of generalized squared distance.

When POOL=YES appears, or when the POOL= option is omitted, the measure of generalized squared distance is based on the pooled covariance matrix.

When POOL=NO is specified, the measure is based on the individual within-group covariance matrices.

When POOL=TEST is specified, a likelihood ratio test (Morrison, 1976; Kendall and Stuart, 1961; Anderson, 1958) of the homogeneity of the within-group covariance matrices is made and the result is printed. If the test statistic is significant at the level specified by the SLPOOL= option (below), the within-group matrices are used. Otherwise, the pooled covariance matrix is used.

The discriminant function coefficients are printed only when the pooled covariance matrix is used.

SLPOOL=*n* specifies the significance level for the test of homogeneity. SLPOOL= is used only when POOL=TEST is also specified.

If POOL=TEST appears but SLPOOL= is absent, .10 is used as the significance level for the test.

WCOV prints the within-group covariance matrices.

WCORR prints the within-group correlation matrices.

PCOV prints the pooled covariance matrix.

PCORR prints the partial correlation matrix based on the pooled covariance matrix.

LIST prints the classification results for each observation.

LISTERR prints only misclassified observations.

THRESHOLD=*n* specifies the minimum acceptable posterior probability for classification. If the posterior probability associated with the smallest distance is less than the THRESHOLD value, the observation is classified into group OTHER.

DATA=*SASdataset* names the data set to be used by DISCRIM. If it is omitted, DISCRIM uses the last data set created.

NOSUMMARY DISCRIM rereads the data and produces a classification summary of the discriminant model unless NOSUMMARY is specified.

OUT=*SASdataset* names the output data set. If you want to create a permanent SAS data set with PROC DISCRIM you must specify a two-level name (see Chapter 12, "SAS Data Sets," in *SAS User's Guide: Basics, 1982 Edition*, for more information on permanent SAS data sets).

TESTDATA=*SASdataset* names a second data set whose observations are to be classified. The variable names in this data set must match those in the DATA= data set.

When TESTDATA= is specified, TESTCLASS and TESTID statements can also be used (see below).

TESTLIST lists all observations in the TESTDATA= data set.

TESTLISTERR lists only misclassified observations in the TESTDATA data set.

CLASS Statement

CLASS *variable*;

The classification *variable* values define the groups for analysis. Class levels are determined by the unformatted values of the class variable. The specified variable can be numeric or character. A CLASS statement must accompany the PROC DISCRIM statement.

VAR Statement

VAR *variables*;

The VAR statement specifies the quantitative variables to be included in the analysis. If you do not use a VAR statement, the analysis includes all numeric variables not listed in other statements.

ID Statement

ID *variable*;

The ID statement is effective only when LIST or LISTERR appears in the PROC DISCRIM statement. When DISCRIM prints the classification results, the ID *variable* is printed for each observation, rather than the observation number.

PRIORS Statement

PRIORS *probabilities*;

You need a PRIORS statement whenever you do not want DISCRIM to assume that the prior probabilities are equal. If you want to set the prior probabilities proportional to the sample sizes, use:

PRIORS PROPORTIONAL;

The keyword PROPORTIONAL can be abbreviated PROP.

If you want other than equal or proportional priors, give the prior probability you want for each level of the classification variable. For example, to define prior probabilities for each level of GRADE, where GRADE's values are A, B, C, and D, you might use the statement:

PRIORS A=.1 B=.3 C=.5 D=.1;

If GRADE were numeric, with values of 1, 2, and 3, the PRIORS statement might be:

PRIORS 1=.3 2=.6 3=.1;

The prior probabilities specified should sum to 1.

TESTCLASS Statement

TESTCLASS *variable*;

The TESTCLASS statement names the *variable* in the TESTDATA= data set to use in determining whether an observation in the TESTDATA= data set is misclassified. The TESTCLASS variable should have the same type (character or numeric) and length as the variable given in the CLASS statement. DISCRIM considers an observation misclassified when the TESTCLASS variable's value does not match the group into which the TESTDATA= observation is classified.

TESTID Statement

TESTID *variable*;

When the TESTID statement appears and the TESTLIST or TESTLISTERR options also appear, DISCRIM uses the value of the TESTID variable, instead of the observation number, to identify each observation in the classification results for the TESTDATA data set. The variable given in the TESTID statement must be in the TESTDATA= data set.

BY Statement

BY *variables*;

A BY statement may be used with PROC DISCRIM to obtain separate analyses on observations in groups defined by the BY variables. When a BY statement appears, the procedure expects the input data set to be sorted in order of the BY variables. If your input data set is not sorted in ascending order, use the SORT procedure with a similar BY statement to sort the data, or, if appropriate, use the BY statement options NOTSORTED or DESCENDING. For more information, see the discussion of the BY statement in Chapter 8, "Statements Used in the PROC Step," in *SAS User's Guide: Basics, 1982 Edition*.

DETAILS

Missing Values

Observations with missing values for variables in the analysis are excluded from the development of the classification criterion. When the classification variable's values are blank or missing, the observation is excluded from the development of the classification criterion, but if no other variables in the analysis have missing values for that observation, it is classified and printed with the classification results.

Saving and Using Calibration Information

Calibration information developed by DISCRIM can be saved in a SAS data set by specifying OUT= followed by the data set name in the PROC DISCRIM statement. DISCRIM then creates a specially structured SAS data set of TYPE=DISCAL that contains the calibration information.

To use this calibration information to classify observations in another data set:

- give the calibration data set after DATA= in the PROC DISCRIM statement, and
- give the data set to be classified after TESTDATA= in the PROC DISCRIM statement.

Only the TESTLIST, TESTLISTERR, and THRESHOLD options and the TESTCLASS and TESTID statements are effective in this case.

Here is an example.

```
DATA ORIGINAL;
  INPUT POSITION X1 X2;
  CARDS;
data lines
PROC DISCRIM OUT=INFO;
  CLASS POSITION;
```

```
DATA CHECK;
  INPUT POSITION X1 X2;
  CARDS;
second set of data lines
PROC DISCRIM DATA=INFO TESTDATA=CHECK TESTLIST;
  TESTCLASS POSITION;
```

The first DATA step creates the SAS data set ORIGINAL, which DISCRIM uses to develop a classification criterion. Specifying OUT=INFO in the PROC DISCRIM statement causes DISCRIM to store the calibration information in a new data set called INFO. The next DATA step creates the data set CHECK. The second PROC DISCRIM specifies DATA=INFO and TESTDATA=CHECK, so that the classification criterion developed earlier is applied to the CHECK data set.

Printed Output

1. values of the classification variable, FREQUENCY (frequencies), and the PRIOR PROBABILITIES for each group
2. optionally, SIMPLE descriptive STATISTICS including N (the number of observations), SUM, MEAN, VARIANCE, and STANDARD DEVIATION for each group
3. optionally, the WITHIN COVARIANCE MATRICES, $\mathbf{S}_t$ for each group
4. optionally, WITHIN CORRELATION COEFFICIENTS and PROB>|R| (the within-group correlation matrix for each group)
5. optionally, the POOLED COVARIANCE MATRIX, **S**
6. optionally, PARTIAL CORRELATION COEFFICIENTS COMPUTED FROM POOLED COVARIANCE MATRIX and PROB>|R| (the partial correlation matrix based on the pooled covariance matrix)
7. WITHIN COVARIANCE MATRIX INFORMATION including COVARIANCE MATRIX RANK and NATURAL LOG OF DETERMINANT OF THE COVARIANCE MATRIX for each group (the rank of $\mathbf{S}_t$ and $\log_e|\mathbf{S}_t|$ and pooled (the rank of **S** and $\log_e|\mathbf{S}|$)
8. optionally, TEST OF HOMOGENEITY OF WITHIN COVARIANCE MATRICES (the results of a chi-square test of homogeneity of the within-group covariance matrices) (Morrison, 1976; Kendall and Stuart, 1961; Anderson, 1958)
9. the PAIRWISE SQUARED GENERALIZED DISTANCES BETWEEN GROUPS
10. if the pooled covariance matrix is used, the LINEARIZED DISCRIMINANT FUNCTION
11. optionally, the CLASSIFICATION RESULTS FOR CALIBRATION DATA including OBS, the observation number (if an ID statement is included, the values of the identification variable are printed instead of the observation number), the actual group for the observation, the group into which the developed criterion would classify it, and the POSTERIOR PROBABILITY of its MEMBERSHIP in each group
12. a CLASSIFICATION SUMMARY FOR CALIBRATION DATA, summary of the performance of the classification criterion.

EXAMPLES

Iris Data: Example 1

The iris data published by Fisher (1936) have been widely used for examples in discriminant analysis and cluster analysis. The sepal length, sepal width, petal length, and petal width were measured in millimeters on 50 iris specimens from each of three species, *Iris setosa, I. versicolor,* and *I. virginica*. DISCRIM is used to classify the irises using a quadratic classification function.

```
DATA IRIS;
  TITLE FISHER (1936) IRIS DATA;
  INPUT SEPALLEN SEPALWID PETALLEN PETALWID SPEC_NO @@;
  IF SPEC_NO=1 THEN SPECIES='SETOSA     ';
  IF SPEC_NO=2 THEN SPECIES='VERSICOLOR';
  IF SPEC_NO=3 THEN SPECIES='VIRGINICA  ';
  DROP SPEC_NO;
  LABEL SEPALLEN=SEPAL LENGTH IN MM.
        SEPALWID=SEPAL WIDTH IN MM.
        PETALLEN=PETAL LENGTH IN MM.
        PETALWID=PETAL WIDTH IN MM.;
  CARDS;
50 33 14 02 1 64 28 56 22 3 65 28 46 15 2
67 31 56 24 3 63 28 51 15 3 46 34 14 03 1
69 31 51 23 3 62 22 45 15 2 59 32 48 18 2
46 36 10 02 1 61 30 46 14 2 60 27 51 16 2
65 30 52 20 3 56 25 39 11 2 65 30 55 18 3
58 27 51 19 3 68 32 59 23 3 51 33 17 05 1
57 28 45 13 2 62 34 54 23 3 77 38 67 22 3
63 33 47 16 2 67 33 57 25 3 76 30 66 21 3
49 25 45 17 3 55 35 13 02 1 67 30 52 23 3
70 32 47 14 2 64 32 45 15 2 61 28 40 13 2
48 31 16 02 1 59 30 51 18 3 55 24 38 11 2
63 25 50 19 3 64 32 53 23 3 52 34 14 02 1
49 36 14 01 1 54 30 45 15 2 79 38 64 20 3
44 32 13 02 1 67 33 57 21 3 50 35 16 06 1
58 26 40 12 2 44 30 13 02 1 77 28 67 20 3
63 27 49 18 3 47 32 16 02 1 55 26 44 12 2
50 23 33 10 2 72 32 60 18 3 48 30 14 03 1
51 38 16 02 1 61 30 49 18 3 48 34 19 02 1
50 30 16 02 1 50 32 12 02 1 61 26 56 14 3
64 28 56 21 3 43 30 11 01 1 58 40 12 02 1
51 38 19 04 1 67 31 44 14 2 62 28 48 18 3
49 30 14 02 1 51 35 14 02 1 56 30 45 15 2
58 27 41 10 2 50 34 16 04 1 46 32 14 02 1
60 29 45 15 2 57 26 35 10 2 57 44 15 04 1
50 36 14 02 1 77 30 61 23 3 63 34 56 24 3
58 27 51 19 3 57 29 42 13 2 72 30 58 16 3
54 34 15 04 1 52 41 15 01 1 71 30 59 21 3
64 31 55 18 3 60 30 48 18 3 63 29 56 18 3
49 24 33 10 2 56 27 42 13 2 57 30 42 12 2
55 42 14 02 1 49 31 15 02 1 77 26 69 23 3
60 22 50 15 3 54 39 17 04 1 66 29 46 13 2
52 27 39 14 2 60 34 45 16 2 50 34 15 02 1
44 29 14 02 1 50 20 35 10 2 55 24 37 10 2
58 27 39 12 2 47 32 13 02 1 46 31 15 02 1
69 32 57 23 3 62 29 43 13 2 74 28 61 19 3
59 30 42 15 2 51 34 15 02 1 50 35 13 03 1
56 28 49 20 3 60 22 40 10 2 73 29 63 18 3
67 25 58 18 3 49 31 15 01 1 67 31 47 15 2
63 23 44 13 2 54 37 15 02 1 56 30 41 13 2
```

```
63 25 49 15 2 61 28 47 12 2 64 29 43 13 2
51 25 30 11 2 57 28 41 13 2 65 30 58 22 3
69 31 54 21 3 54 39 13 04 1 51 35 14 03 1
72 36 61 25 3 65 32 51 20 3 61 29 47 14 2
56 29 36 13 2 69 31 49 15 2 64 27 53 19 3
68 30 55 21 3 55 25 40 13 2 48 34 16 02 1
48 30 14 01 1 45 23 13 03 1 57 25 50 20 3
57 38 17 03 1 51 38 15 03 1 55 23 40 13 2
66 30 44 14 2 68 28 48 14 2 54 34 17 02 1
51 37 15 04 1 52 35 15 02 1 58 28 51 24 3
67 30 50 17 2 63 33 60 25 3 53 37 15 02 1
;
PROC DISCRIM SIMPLE WCOV WCORR PCOV PCORR
      LISTERR POOL=TEST;
  CLASS SPECIES;
```

In the PROC DISCRIM statement, the WCOV, WCORR, PCOV, and PCORR options ask DISCRIM to print the within-group covariance and correlation matrices, the pooled covariance matrix, and the partial correlation matrix based on the pooled covariance matrix. The LISTERR option requests the classification results for misclassified observations. POOL=TEST asks DISCRIM to test the homogeneity of the within-group covariance matrices. The test is significant at the .10 level, so the covariance matrices are not pooled and the quadratic classification criterion is used.

(1) FISHER (1936) IRIS DATA 1

DISCRIMINANT ANALYSIS

SPECIES	FREQUENCY	PRIOR PROBABILITY
SETOSA	50	0.33333333
VERSICOLOR	50	0.33333333
VIRGINICA	50	0.33333333
TOTAL	150	1.00000000

(2) FISHER (1936) IRIS DATA 2

DISCRIMINANT ANALYSIS SIMPLE STATISTICS

VARIABLE	N	SUM	MEAN	VARIANCE	STANDARD DEVIATION
SPECIES = SETOSA					
SEPALLEN	50	2503.00000000	50.06000000	12.42489796	3.52489687
SEPALWID	50	1714.00000000	34.28000000	14.36897959	3.79064369
PETALLEN	50	731.00000000	14.62000000	3.01591837	1.73663996
PETALWID	50	123.00000000	2.46000000	1.11061224	1.05385589
SPECIES = VERSICOLOR					
SEPALLEN	50	2968.00000000	59.36000000	26.64326531	5.16171147
SEPALWID	50	1385.00000000	27.70000000	9.84693878	3.13798323
PETALLEN	50	2130.00000000	42.60000000	22.08163265	4.69910977
PETALWID	50	663.00000000	13.26000000	3.91061224	1.97752680
SPECIES = VIRGINICA					
SEPALLEN	50	3294.00000000	65.88000000	40.43428571	6.35879593
SEPALWID	50	1487.00000000	29.74000000	10.40040816	3.22496638
PETALLEN	50	2776.00000000	55.52000000	30.45877551	5.51894696
PETALWID	50	1013.00000000	20.26000000	7.54326531	2.74650056

(3) FISHER (1936) IRIS DATA 3

DISCRIMINANT ANALYSIS WITHIN COVARIANCE MATRICES

SPECIES = SETOSA DF = 49

VARIABLE	SEPALLEN	SEPALWID	PETALLEN	PETALWID
SEPALLEN	12.42489796	9.92163265	1.63551020	1.03306122
SEPALWID	9.92163265	14.36897959	1.16979592	0.92979592
PETALLEN	1.63551020	1.16979592	3.01591837	0.60693878
PETALWID	1.03306122	0.92979592	0.60693878	1.11061224

SPECIES = VERSICOLOR DF = 49

VARIABLE	SEPALLEN	SEPALWID	PETALLEN	PETALWID
SEPALLEN	26.64326531	8.51836735	18.28979592	5.57795918
SEPALWID	8.51836735	9.84693878	8.26530612	4.12040816
PETALLEN	18.28979592	8.26530612	22.08163265	7.31020408
PETALWID	5.57795918	4.12040816	7.31020408	3.91061224

SPECIES = VIRGINICA DF = 49

VARIABLE	SEPALLEN	SEPALWID	PETALLEN	PETALWID
SEPALLEN	40.43428571	9.37632653	30.32897959	4.90938776
SEPALWID	9.37632653	10.40040816	7.13795918	4.76285714
PETALLEN	30.32897959	7.13795918	30.45877551	4.88244898
PETALWID	4.90938776	4.76285714	4.88244898	7.54326531

(4) FISHER (1936) IRIS DATA 4

DISCRIMINANT ANALYSIS WITHIN CORRELATION COEFFICIENTS / PROBABILITY > |R|

SPECIES = SETOSA

VARIABLE	SEPALLEN	SEPALWID	PETALLEN	PETALWID
SEPALLEN	1.000000 0.0000	0.742547 0.0001	0.267176 0.0607	0.278098 0.0505
SEPALWID	0.742547 0.0001	1.000000 0.0000	0.177700 0.2170	0.232752 0.1038
PETALLEN	0.267176 0.0607	0.177700 0.2170	1.000000 0.0000	0.331630 0.0186
PETALWID	0.278098 0.0505	0.232752 0.1038	0.331630 0.0186	1.000000 0.0000

SPECIES = VERSICOLOR

VARIABLE	SEPALLEN	SEPALWID	PETALLEN	PETALWID
SEPALLEN	1.000000 0.0000	0.525911 0.0001	0.754049 0.0001	0.546461 0.0001
SEPALWID	0.525911 0.0001	1.000000 0.0000	0.560522 0.0001	0.663999 0.0001
PETALLEN	0.754049 0.0001	0.560522 0.0001	1.000000 0.0000	0.786668 0.0001
PETALWID	0.546461 0.0001	0.663999 0.0001	0.786668 0.0001	1.000000 0.0000

SPECIES = VIRGINICA

VARIABLE	SEPALLEN	SEPALWID	PETALLEN	PETALWID
SEPALLEN	1.000000 0.0000	0.457228 0.0008	0.864225 0.0001	0.281108 0.0480
SEPALWID	0.457228 0.0008	1.000000 0.0000	0.401045 0.0039	0.537728 0.0001
PETALLEN	0.864225 0.0001	0.401045 0.0039	1.000000 0.0000	0.322108 0.0225
PETALWID	0.281108 0.0480	0.537728 0.0001	0.322108 0.0225	1.000000 0.0000

(5) FISHER (1936) IRIS DATA 5

DISCRIMINANT ANALYSIS POOLED COVARIANCE MATRIX DF = 147

VARIABLE	SEPALLEN	SEPALWID	PETALLEN	PETALWID
SEPALLEN	26.50081633	9.27210884	16.75142857	3.84013605
SEPALWID	9.27210884	11.53877551	5.52435374	3.27102041
PETALLEN	16.75142857	5.52435374	18.51877551	4.26653061
PETALWID	3.84013605	3.27102041	4.26653061	4.18816327

(6) FISHER (1936) IRIS DATA 6

DISCRIMINANT ANALYSIS PARTIAL CORRELATION COEFFICIENTS COMPUTED FROM POOLED COVARIANCE MATRIX / PROB > |R|

VARIABLE	SEPALLEN	SEPALWID	PETALLEN	PETALWID
SEPALLEN	1.000000 0.0000	0.530236 0.0001	0.756164 0.0001	0.364506 0.0001
SEPALWID	0.530236 0.0001	1.000000 0.0000	0.377916 0.0001	0.470535 0.0001
PETALLEN	0.756164 0.0001	0.377916 0.0001	1.000000 0.0000	0.484459 0.0001
PETALWID	0.364506 0.0001	0.470535 0.0001	0.484459 0.0001	1.000000 0.0000

(7) FISHER (1936) IRIS DATA 7

DISCRIMINANT ANALYSIS WITHIN COVARIANCE MATRIX INFORMATION

SPECIES	COVARIANCE MATRIX RANK	NATURAL LOG OF DETERMINANT OF THE COVARIANCE MATRIX
SETOSA	4	5.35332042
VERSICOLOR	4	7.54635570
VIRGINICA	4	9.49362227
POOLED	4	8.46214197

(8) FISHER (1936) IRIS DATA 8

DISCRIMINANT ANALYSIS TEST OF HOMOGENEITY OF WITHIN COVARIANCE MATRICES

```
NOTATION:  K     =   NUMBER OF GROUPS

           P     =   NUMBER OF VARIABLES

           N     =   TOTAL NUMBER OF OBSERVATIONS

           N(I)  =   NUMBER OF OBSERVATIONS IN THE I'TH GROUP

                                                 N(I)/2
                        ΤΤ  |WITHIN SS MATRIX(I)|
           V     =   ------------------------------------
                                            N/2
                         |POOLED SS MATRIX|

                                                           2
                           _                      _
                          |            1         1   |   2P  + 3P - 1
           RHO   =   1.0 - | SUM ------  -  ---  |  ------------
                          |_         N(I)-1      N-K _|   6(P+1)(K-1)

           DF    =   .5(K-1)P(P+1)

                                   _                    _
                                  |     PN/2             |
                                  |    N          V      |
UNDER NULL HYPOTHESIS:  -2 RHO LN |  ------------------  |  IS DISTRIBUTED APPROXIMATELY AS CHI-SQUARE(DF)
                                  |              PN(I)/2 |
                                  |_   ΤΤ  N(I)         _|
```

(continued on next page)

(continued from previous page)

```
TEST CHI-SQUARE VALUE =             143.81943870    WITH        20 DF    PROB > CHI-SQ = 0.0001

SINCE THE CHI-SQUARE VALUE IS SIGNIFICANT AT THE 0.1000 LEVEL, THE WITHIN COVARIANCE MATRICES WILL BE USED IN
THE DISCRIMINANT FUNCTION.

REFERENCE: KENDALL,M.G. AND A.STUART  THE ADVANCED THEORY OF STATISTICS VOL.3 P266 & 282.
```

(9)

```
                                  FISHER (1936) IRIS DATA                                          9

     DISCRIMINANT ANALYSIS     PAIRWISE SQUARED GENERALIZED DISTANCES BETWEEN GROUPS

                        2                      -1
                       D (I|J) = (X̄ - X̄ )' COV   (X̄ - X̄ ) + LN |COV |
                                   I   J       J    I   J          J

                          GENERALIZED SQUARED DISTANCE TO SPECIES

       FROM SPECIES                SETOSA          VERSICOLOR          VIRGINICA

           SETOSA              5.35332042        110.74017480       178.26120901
           VERSICOLOR        328.41534800          7.54635570        23.33237667
           VIRGINICA         711.43825544         25.41305990         9.49362227
```

(11)

```
                                  FISHER (1936) IRIS DATA                                          10

     DISCRIMINANT ANALYSIS     CLASSIFICATION RESULTS FOR CALIBRATION DATA: WORK.IRIS

GENERALIZED SQUARED DISTANCE FUNCTION:          POSTERIOR PROBABILITY OF MEMBERSHIP IN EACH SPECIES:

 2                 -1                                                  2                    2
D (X) = (X-X̄ )' COV  (X-X̄ ) + LN |COV |          PR(J|X) = EXP(-.5 D (X)) / SUM EXP(-.5 D (X))
 J          J       J    J            J                                J         K           K

                                                          POSTERIOR PROBABILITY OF MEMBERSHIP IN SPECIES:

               OBS   FROM          CLASSIFIED              SETOSA   VERSICOLOR     VIRGINICA
                     SPECIES       INTO SPECIES

                 5   VIRGINICA     VERSICOLOR    *         0.0000       0.6050        0.3950
                 9   VERSICOLOR    VIRGINICA     *         0.0000       0.3359        0.6641
                12   VERSICOLOR    VIRGINICA     *         0.0000       0.1543        0.8457

                                   * MISCLASSIFIED OBSERVATION
```

(12)

```
                                  FISHER (1936) IRIS DATA                                          11

     DISCRIMINANT ANALYSIS     CLASSIFICATION SUMMARY FOR CALIBRATION DATA: WORK.IRIS

GENERALIZED SQUARED DISTANCE FUNCTION:          POSTERIOR PROBABILITY OF MEMBERSHIP IN EACH SPECIES:

 2                 -1                                                  2                    2
D (X) = (X-X̄ )' COV  (X-X̄ ) + LN |COV |          PR(J|X) = EXP(-.5 D (X)) / SUM EXP(-.5 D (X))
 J          J       J    J            J                                J         K           K

                                       NUMBER OF OBSERVATIONS AND PERCENTS CLASSIFIED INTO SPECIES:
                  FROM
                  SPECIES              SETOSA  VERSICOLOR   VIRGINICA        TOTAL

                  SETOSA                   50           0           0           50
                                       100.00        0.00        0.00       100.00

                  VERSICOLOR                0          48           2           50
                                         0.00       96.00        4.00       100.00

                  VIRGINICA                 0           1          49           50
                                         0.00        2.00       98.00       100.00

                  TOTAL                    50          49          51          150
                  PERCENT               33.33       32.67       34.00       100.00

                  PRIORS               0.3333      0.3333      0.3333
```

Remote Sensing Data on Crops: Example 2

In the example below, the observations are grouped into five crops: clover, corn, cotton, soybeans, and sugar beets. Four measures called X1-X4 make up the descriptive variables. The first PROC DISCRIM statement creates a calibration data set using the OUT= option. The second DISCRIM uses the information in the calibration data set to classify a test data set. Note that the values of the identification variable, XVALUES, are obtained by rereading the X1-X4 fields in the data lines as one character variable.

```
DATA CROPS;
  TITLE REMOTE SENSING DATA ON FIVE CROPS;
  INPUT CROP $ 1-10 X1-X4 XVALUES $ 11-21;
  CARDS;
CORN        16 27 31 33
CORN        15 23 30 30
CORN        16 27 27 26
CORN        18 20 25 23
CORN        15 15 31 32
CORN        15 32 32 15
CORN        12 15 16 73
SOYBEANS    20 23 23 25
SOYBEANS    24 24 25 32
SOYBEANS    21 25 23 24
SOYBEANS    27 45 24 12
SOYBEANS    12 13 15 42
SOYBEANS    22 32 31 43
COTTON      31 32 33 34
COTTON      29 24 26 28
COTTON      34 32 28 45
COTTON      26 25 23 24
COTTON      53 48 75 26
COTTON      34 35 25 78
SUGARBEETS  22 23 25 42
SUGARBEETS  25 25 24 26
SUGARBEETS  34 25 16 52
SUGARBEETS  54 23 21 54
SUGARBEETS  25 43 32 15
SUGARBEETS  26 54  2 54
CLOVER      12 45 32 54
CLOVER      24 58 25 34
CLOVER      87 54 61 21
CLOVER      51 31 31 16
CLOVER      96 48 54 62
CLOVER      31 31 11 11
CLOVER      56 13 13 71
CLOVER      32 13 27 32
CLOVER      36 26 54 32
CLOVER      53 08 06 54
CLOVER      32 32 62 16
;
PROC DISCRIM DATA=CROPS POOL=YES LIST OUT=CROPCAL;
  CLASS CROP;
  ID XVALUES;
  VAR X1-X4;
  TITLE2 CLASSIFICATION OF CROP DATA;
```

```
DATA TEST;
  INPUT CROP $ 1-10 X1-X4 XVALUES $ 11-21;
  CARDS;
CORN       16 27 31 33
SOYBEANS   21 25 23 24
COTTON     29 24 26 28
SUGARBEETS 54 23 21 54
CLOVER     32 32 62 16
;
PROC DISCRIM DATA=CROPCAL TESTDATA=TEST TESTLIST;
  CLASS CROP;
  TESTCLASS CROP;
  TESTID XVALUES;
  VAR X1-X4;
  TITLE2 CLASSIFICATION OF TEST DATA;
```

REMOTE SENSING DATA ON FIVE CROPS
CLASSIFICATION OF CROP DATA 1

DISCRIMINANT ANALYSIS

CROP	FREQUENCY	PRIOR PROBABILITY
CLOVER	11	0.20000000
CORN	7	0.20000000
COTTON	6	0.20000000
SOYBEANS	6	0.20000000
SUGARBEETS	6	0.20000000
-----	--	----------
TOTAL	36	1.00000000

REMOTE SENSING DATA ON FIVE CROPS
CLASSIFICATION OF CROP DATA 2

DISCRIMINANT ANALYSIS   POOLED COVARIANCE MATRIX INFORMATION

COVARIANCE MATRIX RANK	NATURAL LOG OF DETERMINANT OF THE COVARIANCE MATRIX
4	21.30189392

REMOTE SENSING DATA ON FIVE CROPS
CLASSIFICATION OF CROP DATA 3

DISCRIMINANT ANALYSIS   PAIRWISE SQUARED GENERALIZED DISTANCES BETWEEN GROUPS

$$D^2(I|J) = (\bar{X}_I - \bar{X}_J)' \; COV^{-1} \; (\bar{X}_I - \bar{X}_J)$$

GENERALIZED SQUARED DISTANCE TO CROP

FROM CROP	CLOVER	CORN	COTTON	SOYBEANS	SUGARBEETS
CLOVER	0.00000000	4.25308108	0.86616669	2.58313162	1.48909745
CORN	4.25308108	0.00000000	1.88446483	0.73030740	2.89042690
COTTON	0.86616669	1.88446483	0.00000000	1.43466961	1.29555784
SOYBEANS	2.58313162	0.73030740	1.43466961	0.00000000	1.07646391
SUGARBEETS	1.48909745	2.89042690	1.29555784	1.07646391	0.00000000

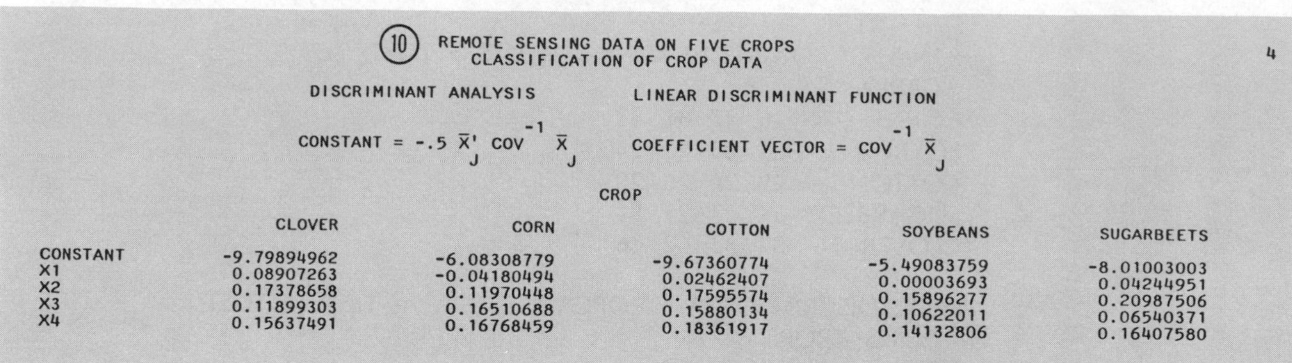

(10) REMOTE SENSING DATA ON FIVE CROPS
CLASSIFICATION OF CROP DATA 4

DISCRIMINANT ANALYSIS LINEAR DISCRIMINANT FUNCTION

$\text{CONSTANT} = -.5\ \bar{X}_J'\ \text{COV}^{-1}\ \bar{X}_J$ $\qquad$ $\text{COEFFICIENT VECTOR} = \text{COV}^{-1}\ \bar{X}_J$

	CROP				
	CLOVER	CORN	COTTON	SOYBEANS	SUGARBEETS
CONSTANT	-9.79894962	-6.08308779	-9.67360774	-5.49083759	-8.01003003
X1	0.08907263	-0.04180494	0.02462407	0.00003693	0.04244951
X2	0.17378658	0.11970448	0.17595574	0.15896277	0.20987506
X3	0.11899303	0.16510688	0.15880134	0.10622011	0.06540371
X4	0.15637491	0.16768459	0.18361917	0.14132806	0.16407580

REMOTE SENSING DATA ON FIVE CROPS
CLASSIFICATION OF CROP DATA 5

DISCRIMINANT ANALYSIS CLASSIFICATION RESULTS FOR CALIBRATION DATA: WORK.CROPS

GENERALIZED SQUARED DISTANCE FUNCTION:

$$D_J^2(X) = (X-\bar{X}_J)'\ \text{COV}^{-1}\ (X-\bar{X}_J)$$

POSTERIOR PROBABILITY OF MEMBERSHIP IN EACH CROP:

$$PR(J|X) = \text{EXP}(-.5\ D_J^2(X))\ /\ \text{SUM}_K\ \text{EXP}(-.5\ D_K^2(X))$$

XVALUES	FROM CROP	CLASSIFIED INTO CROP		CLOVER	CORN	COTTON	SOYBEANS	SUGARBEETS
				POSTERIOR PROBABILITY OF MEMBERSHIP IN CROP:				
16 27 31 33	CORN	CORN		0.0541	0.3855	0.1956	0.2653	0.0995
15 23 30 30	CORN	CORN		0.0466	0.4341	0.1579	0.2811	0.0802
16 27 27 26	CORN	SOYBEANS	*	0.0591	0.3236	0.1506	0.3390	0.1277
18 20 25 23	CORN	SOYBEANS	*	0.0637	0.3460	0.1198	0.3645	0.1060
15 15 31 32	CORN	CORN		0.0360	0.5535	0.1317	0.2342	0.0447
15 32 32 15	CORN	SOYBEANS	*	0.0583	0.3091	0.1450	0.3762	0.1112
12 15 16 73	CORN	CORN		0.0274	0.4964	0.2044	0.1521	0.1197
20 23 23 25	SOYBEANS	SOYBEANS		0.0807	0.2672	0.1307	0.3675	0.1539
24 24 25 32	SOYBEANS	SOYBEANS		0.1091	0.2407	0.1794	0.3009	0.1698
21 25 23 24	SOYBEANS	SOYBEANS		0.0900	0.2320	0.1336	0.3696	0.1748
27 45 24 12	SOYBEANS	SUGARBEETS	*	0.1452	0.0530	0.1148	0.3075	0.3795
12 13 15 42	SOYBEANS	CORN	*	0.0330	0.4487	0.1014	0.3052	0.1117
22 32 31 43	SOYBEANS	COTTON	*	0.0898	0.2494	0.2929	0.2063	0.1616
31 32 33 34	COTTON	COTTON		0.1806	0.1530	0.2795	0.2078	0.1791
29 24 26 28	COTTON	SOYBEANS	*	0.1601	0.1838	0.1780	0.2967	0.1814
34 32 28 45	COTTON	COTTON		0.2021	0.1040	0.2851	0.1609	0.2480
26 25 23 24	COTTON	SOYBEANS	*	0.1318	0.1767	0.1418	0.3469	0.2028
53 48 75 26	COTTON	COTTON		0.3407	0.0432	0.5660	0.0288	0.0214
34 35 25 78	COTTON	COTTON		0.1389	0.0768	0.4300	0.0669	0.2875
22 23 25 42	SUGARBEETS	CORN	*	0.0870	0.2947	0.2132	0.2503	0.1548
25 25 24 26	SUGARBEETS	SOYBEANS	*	0.1219	0.1994	0.1537	0.3359	0.1891
34 25 16 52	SUGARBEETS	SUGARBEETS		0.1869	0.0874	0.1949	0.1731	0.3576
54 23 21 54	SUGARBEETS	CLOVER	*	0.4743	0.0232	0.1749	0.0694	0.2582
25 43 32 15	SUGARBEETS	SOYBEANS	*	0.1398	0.1104	0.1868	0.3144	0.2486
26 54 2 54	SUGARBEETS	SUGARBEETS		0.0483	0.0072	0.0543	0.0688	0.8214
12 45 32 54	CLOVER	COTTON	*	0.0406	0.2453	0.3648	0.1570	0.1923
24 58 25 34	CLOVER	SUGARBEETS	*	0.0977	0.0351	0.1826	0.1579	0.5267
87 54 61 21	CLOVER	CLOVER		0.8835	0.0005	0.0831	0.0043	0.0287
51 31 31 16	CLOVER	CLOVER		0.5211	0.0253	0.1254	0.1380	0.1902
96 48 54 62	CLOVER	CLOVER		0.8649	0.0002	0.1039	0.0012	0.0297
31 31 11 11	CLOVER	SUGARBEETS	*	0.1566	0.0392	0.0538	0.3424	0.4080
56 13 13 71	CLOVER	CLOVER		0.4657	0.0253	0.1707	0.0567	0.2817
32 13 27 32	CLOVER	SOYBEANS	*	0.1731	0.2665	0.1797	0.2687	0.1121
36 26 54 32	CLOVER	COTTON	*	0.1717	0.2693	0.4152	0.1091	0.0347
53 08 06 54	CLOVER	CLOVER		0.4433	0.0279	0.0929	0.1073	0.3287
32 32 62 16	CLOVER	COTTON	*	0.1378	0.3183	0.3885	0.1313	0.0240

* MISCLASSIFIED OBSERVATION

REMOTE SENSING DATA ON FIVE CROPS
CLASSIFICATION OF CROP DATA 6

DISCRIMINANT ANALYSIS CLASSIFICATION SUMMARY FOR CALIBRATION DATA: WORK.CROPS

GENERALIZED SQUARED DISTANCE FUNCTION:

$$D_J^2(X) = (X-\bar{X}_J)'\ \text{COV}^{-1}\ (X-\bar{X}_J)$$

POSTERIOR PROBABILITY OF MEMBERSHIP IN EACH CROP:

$$PR(J|X) = \text{EXP}(-.5\ D_J^2(X))\ /\ \text{SUM}_K\ \text{EXP}(-.5\ D_K^2(X))$$

(continued on next page)

(continued from previous page)

NUMBER OF OBSERVATIONS AND PERCENTS CLASSIFIED INTO CROP:

FROM CROP	CLOVER	CORN	COTTON	SOYBEANS	SUGARBEETS	TOTAL
CLOVER	5 45.45	0 0.00	3 27.27	1 9.09	2 18.18	11 100.00
CORN	0 0.00	4 57.14	0 0.00	3 42.86	0 0.00	7 100.00
COTTON	0 0.00	0 0.00	4 66.67	2 33.33	0 0.00	6 100.00
SOYBEANS	0 0.00	1 16.67	1 16.67	3 50.00	1 16.67	6 100.00
SUGARBEETS	1 16.67	1 16.67	0 0.00	2 33.33	2 33.33	6 100.00
TOTAL PERCENT	6 16.67	6 16.67	8 22.22	11 30.56	5 13.89	36 100.00
PRIORS	0.2000	0.2000	0.2000	0.2000	0.2000	

REMOTE SENSING DATA ON FIVE CROPS
CLASSIFICATION OF TEST DATA 7

DISCRIMINANT ANALYSIS CLASSIFICATION RESULTS FOR TEST DATA: WORK.TEST

GENERALIZED SQUARED DISTANCE FUNCTION:

$$D_J^2(X) = (X-\bar{X}_J)' \; COV^{-1} \; (X-\bar{X}_J)$$

POSTERIOR PROBABILITY OF MEMBERSHIP IN EACH CROP:

$$PR(J|X) = EXP(-.5 \; D_J^2(X)) \; / \; SUM_K \; EXP(-.5 \; D_K^2(X))$$

POSTERIOR PROBABILITY OF MEMBERSHIP IN CROP:

XVALUES	FROM CROP	CLASSIFIED INTO CROP		CLOVER	CORN	COTTON	SOYBEANS	SUGARBEETS
16 27 31 33	CORN	CORN		0.0541	0.3855	0.1956	0.2653	0.0995
21 25 23 24	SOYBEANS	SOYBEANS		0.0900	0.2320	0.1336	0.3696	0.1748
29 24 26 28	COTTON	SOYBEANS	*	0.1601	0.1838	0.1780	0.2967	0.1814
54 23 21 54	SUGARBEETS	CLOVER	*	0.4743	0.0232	0.1749	0.0694	0.2582
32 32 62 16	CLOVER	COTTON	*	0.1378	0.3183	0.3885	0.1313	0.0240

* MISCLASSIFIED OBSERVATION

REMOTE SENSING DATA ON FIVE CROPS
CLASSIFICATION OF TEST DATA 8

DISCRIMINANT ANALYSIS CLASSIFICATION SUMMARY FOR TEST DATA: WORK.TEST

GENERALIZED SQUARED DISTANCE FUNCTION:

$$D_J^2(X) = (X-\bar{X}_J)' \; COV^{-1} \; (X-\bar{X}_J)$$

POSTERIOR PROBABILITY OF MEMBERSHIP IN EACH CROP:

$$PR(J|X) = EXP(-.5 \; D_J^2(X)) \; / \; SUM_K \; EXP(-.5 \; D_K^2(X))$$

NUMBER OF OBSERVATIONS AND PERCENTS CLASSIFIED INTO CROP:

FROM CROP	CLOVER	CORN	COTTON	SOYBEANS	SUGARBEETS	TOTAL
CLOVER	0 0.00	0 0.00	1 100.00	0 0.00	0 0.00	1 100.00
CORN	0 0.00	1 100.00	0 0.00	0 0.00	0 0.00	1 100.00
COTTON	0 0.00	0 0.00	0 0.00	1 100.00	0 0.00	1 100.00
SOYBEANS	0 0.00	0 0.00	0 0.00	1 100.00	0 0.00	1 100.00
SUGARBEETS	1 100.00	0 0.00	0 0.00	0 0.00	0 0.00	1 100.00
TOTAL PERCENT	1 20.00	1 20.00	1 20.00	2 40.00	0 0.00	5 100.00
PRIORS	0.2000	0.2000	0.2000	0.2000	0.2000	

REFERENCES

Anderson, T.W. (1958), *An Introduction to Multivariate Statistical Analysis*. New York: John Wiley & Sons.

Kendall, M.G. and Stuart, A. (1961), *The Advanced Theory of Statistics*, Vol. 3. London: Charles Griffin and Company, Ltd.

Morrison, D.F. (1976), *Multivariate Statistical Methods*. New York: McGraw-Hill.

Rao, C. Radhakrishna (1973), *Linear Statistical Inference and Its Applications*. New York: John Wiley & Sons.

The NEIGHBOR Procedure

ABSTRACT

The NEIGHBOR procedure performs a nearest neighbor discriminant analysis, classifying observations into groups according to either the nearest neighbor rule or the *k*-nearest-neighbor rule.

INTRODUCTION

The NEIGHBOR procedure can be used to classify observations when the classes do not have multivariate normal distributions. Given an observation to classify, NEIGHBOR looks at the *k* observations closest to it, that is, the *k* nearest neighbors. Either Mahalanobis distance based on the total covariance matrix or Euclidean distance may be used. The observation is placed in the class containing the highest proportion of the *k* nearest neighbors. Optionally, classification may be based on the proportion of nearest neighbors times a prior probability for each class.

Background

Nearest neighbor discriminant analysis is a nonparametric method for classifying observations into one of several classes on the basis of one or more quantitative variables.

Letting $\mathbf{x}_1$ and $\mathbf{x}_2$ represent two observation vectors, NEIGHBOR computes the Mahalanobis distance between $\mathbf{x}_1$ and $\mathbf{x}_2$ based on the total-sample covariance matrix $\mathbf{T}$:

$$d^2(\mathbf{x}_1,\mathbf{x}_2) = (\mathbf{x}_1 - \mathbf{x}_2)'\mathbf{T}^{-1}(\mathbf{x}_1 - \mathbf{x}_2)$$

or optionally the Euclidean distance:

$$d^2(\mathbf{x}_1,\mathbf{x}_2) = (\mathbf{x}_1 - \mathbf{x}_2)'(\mathbf{x}_1 - \mathbf{x}_2) \quad .$$

Using the nearest neighbor rule, $\mathbf{x}_2$ is classified into the group corresponding to the $\mathbf{x}_1$ point that yields the smallest $d^2(\mathbf{x}_1,\mathbf{x}_2)$. Using the *k*-nearest-neighbor rule, the *k* smallest distances are saved. Of these *k* distances, let n_i represent the number of distances that correspond to group *i*. The posterior probability of membership in group *i* is:

$$P_i = \frac{n_i \text{ prior}_i}{\Sigma\, n_j \text{ prior}_j}$$

Then $\mathbf{x}_2$ is assigned to the group for which P_i is a maximum, unless there is a tie for largest or unless this maximum probability is less than the threshold specified. In these cases, $\mathbf{x}_2$ is classified into group OTHER.

SPECIFICATIONS

The following statements are used with NEIGHBOR:

PROC NEIGHBOR *options;*
CLASS *variable;*
VAR *variables;*
ID *variable;*
PRIORS *probabilities;*
TESTCLASS *variable;*
TESTID *variable;*
BY *variables;*

PROC NEIGHBOR Statement

PROC NEIGHBOR *options;*

The options below may appear in the PROC NEIGHBOR statement.

K=*k*	specifies a *k* value for the *k*-nearest-neighbor rule. If K= is not specified, the default *k* value is 1, producing nearest neighbor classification.
IDENTITY	specifies use of Euclidean distances. If IDENTITY is not specified, Mahalanobis distances are used.
THRESHOLD=*p*	specifies the minimum acceptable posterior probability for classification. If the posterior probability associated with the smallest distance is less than the THRESHOLD value, the observation is classified into group OTHER. If THRESHOLD= is omitted, no observations are classified into group OTHER unless there is a tie for the largest posterior probability.
LIST	prints the classification results for each observation.
LISTERR	prints only misclassified observations.
DATA=*SASdataset*	names the data set to be analyzed. If DATA= is not specified, NEIGHBOR uses the most recently created data set.
TESTDATA= *SASdataset*	names a data set to be classified by the classification criterion developed by NEIGHBOR (using the DATA= data set). The variable names in the TESTDATA= data set must match those in the DATA= data set.
TESTLIST	lists all the observations in the TESTDATA= data set.
TESTLISTERR	lists only misclassified observations in the TESTDATA= data set.

CLASS Statement

CLASS *variable;*

The CLASS statement gives the name of a *variable*, the formatted values of which define the classes. A CLASS statement must accompany the PROC NEIGHBOR statement.

VAR Statement

VAR *variables;*

List the *variables* to be included in the analysis in the VAR statement. If the VAR statement is omitted, all numeric variables that are not specified in other statements are included in the analysis.

ID Statement

ID *variable;*

When you use an ID statement and also specify the LIST or LISTERR option, NEIGHBOR prints the value of the identification *variable* to identify observations in the classification results, rather than the observation number.

PRIORS Statement

PRIORS *probabilities;*

If you do not want NEIGHBOR to assume that the prior probabilities are equal, use a PRIORS statement. If you want NEIGHBOR to set the prior probabilities proportional to the sample sizes, use the statement:

```
PRIORS PROPORTIONAL;
```

The keyword PROPORTIONAL may be abbreviated PROP. If you want other than equal or proportional priors, give the prior probability for each level of the classification variable. For example, if the classification variable GRADE has the four values A, B, C, and D, you might write the PRIORS statement:

```
PRIORS A=.1 B=.3  C=.5 D=.1;
```

If GRADE is a numeric variable with values of 1, 2, and 3, you might write:

```
PRIORS 1=.3 2=.6  3=.1;
```

The prior probabilities specified in the PRIORS statement should sum to 1.

TESTCLASS Statement

TESTCLASS *variable;*

The TESTCLASS statement names the *variable* in the TESTDATA= data set to use in determining whether an observation in the TESTDATA= data set is misclassified. The TESTCLASS variable should have the same type (character or numeric) and length as the variable given in the CLASS statement. NEIGHBOR considers an observation misclassified when the TESTCLASS variable's value does not match the group into which the TESTDATA= observation is classified.

TESTID Statement

TESTID *variable;*

When the TESTID statement appears and the TESTLIST or TESTLISTERR options are specified, NEIGHBOR uses the value of the TESTID variable, instead of the observation number, to identify each observation in the classification results for the TESTDATA= data set. The variable given in the TESTID statement must be in the TESTDATA= data set.

BY Statement

BY *variables;*

A BY statement may be used with PROC NEIGHBOR to obtain separate analyses on observations in groups defined by the BY variables. When a BY statement appears, the procedure expects the input data set to be sorted in order of the BY variables. If your input data set is not sorted in ascending order, use the SORT procedure with a similar BY statement to sort the data, or, if appropriate, use the BY statement options NOTSORTED or DESCENDING. For more information, see the discussion of the BY statement in Chapter 8, ''Statements Used in the PROC Step,'' in *SAS User's Guide: Basics, 1982 Edition*.

DETAILS

Missing Values

When an observation has a missing value for a variable or classification value needed in the analysis, that observation is excluded from the development of the classification criterion. If the classification value is blank or missing, but if no other variables in the observation have missing values, the observation is classified according to the classification criterion.

Computer Resources

Let:

n = number of observations in the DATA= data set
t = number of observations in the TESTDATA= data set
v = number of variables.

NEIGHBOR stores the data and the covariance matrix in core, requiring:

$$120v + 4v(v + 1) + n(8v + 16)$$

bytes of array storage. The time required to classify the observations is roughly proportional to:

$$(n + t)nv$$

if IDENTITY is specified; otherwise:

$$(n + t)nv^2 \quad .$$

Printed Output

PROC NEIGHBOR produces printed output that includes:

1. values of the classification variable, and the FREQUENCY and PRIOR PROBABILITY for each group.

Optionally, the classification results for each observation are printed, including:

2. the observation number, or if an ID statement is included, the values of the ID variable are substituted for the observation number
3. the actual group for the observation
4. the group into which the developed criterion classifies it
5. the POSTERIOR PROBABILITY OF its MEMBERSHIP IN each group
6. CLASSIFICATION SUMMARY FOR CALIBRATION DATA, a summary of the performance of the classification criterion when LIST or LISTERR is specified in the PROC NEIGHBOR statement.

EXAMPLE

In the example below, the observations are grouped into five crops: clover, corn, cotton, soybeans, and sugar beets. Four measures, X1-X4, make up the descriptive variables. Specifying K=4 in the PROC NEIGHBOR statement requests a k-nearest-neighbor discriminant analysis, with a K value of 4. The LIST option prints the classification results for each observation. The TESTDATA= and TESTLIST options classify the observations in the data set TEST using the classification function based on the data set CROPS. Since the ID statement is included, values of the variable XVALUES identify the observations on the output. The classification variable is CROP. Note that the values of the identification variable, XVALUES, are obtained by rereading the X1-X4 fields in the data lines as one character variable.

```
DATA CROPS;
  TITLE REMOTE SENSING DATA ON FIVE CROPS;
  INPUT CROP $ 1-10 X1-X4 XVALUES $ 11-21;
  CARDS;
CORN       26 27 31 33
CORN       15 23 30 30
CORN       16 27 27 26
CORN       18 20 25 23
CORN       15 15 31 32
CORN       15 32 32 15
CORN       12 15 16 73
SOYBEANS   20 23 23 25
SOYBEANS   24 24 25 32
SOYBEANS   21 25 23 24
SOYBEANS   27 45 24 12
SOYBEANS   12 13 15 42
SOYBEANS   22 32 31 43
COTTON     31 32 33 34
COTTON     29 24 26 28
COTTON     34 32 28 45
COTTON     26 25 23 24
COTTON     53 48 75 26
COTTON     34 35 25 78
SUGARBEETS22 23 25 42
```

```
SUGARBEETS25 25 24 26
SUGARBEETS34 25 16 52
SUGARBEETS54 23 21 54
SUGARBEETS25 43 32 15
SUGARBEETS26 54 2 54
CLOVER    12 45 32 54
CLOVER    24 58 25 34
CLOVER    87 54 61 21
CLOVER    51 31 31 16
CLOVER    96 48 54 62
CLOVER    31 31 11 11
CLOVER    56 13 13 71
CLOVER    32 13 27 32
CLOVER    36 26 54 32
CLOVER    53 08 06 54
CLOVER    32 32 62 16
;
DATA TEST;
  INPUT CROP $ 1-10 X1-X4 XVALUES $ 11-21;
  CARDS;
CORN      16 27 31 33
SOYBEANS  21 25 23 24
COTTON    29 24 26 28
SUGARBEETS54 23 21 54
CLOVER    32 32 62 16
;
PROC NEIGHBOR DATA=CROPS TESTDATA=TEST K=4 LIST TESTLIST;
  CLASS CROP;
  ID XVALUES;
  TESTCLASS CROP;
  TESTID XVALUES;
  VAR X1-X4;
```

```
                  REMOTE SENSING DATA ON FIVE CROPS                    1
              NEAREST 4 NEIGHBORS DISCRIMINANT ANALYSIS

       CROP                    FREQUENCY     PRIOR PROBABILITY

       CLOVER                     11            0.20000000

       CORN                        7            0.20000000

       COTTON                      6            0.20000000

       SOYBEANS                    6            0.20000000

       SUGARBEETS                  6            0.20000000
       -----                      --            ----------
       TOTAL                      36            1.00000000
```

REMOTE SENSING DATA ON FIVE CROPS 2

NEAREST 4 NEIGHBORS DISCRIMINANT ANALYSIS CLASSIFICATION RESULTS FOR CALIBRATION DATA: WORK.CROPS

DISTANCE FUNCTION: $D^2(X,Y) = (X-Y)'\ COV^{-1}\ (X-Y)$

POSTERIOR PROBABILITY OF MEMBERSHIP IN CROP:

XVALUES	FROM CROP	CLASSIFIED INTO CROP		CLOVER	CORN	COTTON	SOYBEANS	SUGARBEETS
16 27 31 33	CORN	CORN		0.0000	0.5000	0.0000	0.2500	0.2500
15 23 30 30	CORN	CORN		0.0000	0.7500	0.0000	0.2500	0.0000
16 27 27 26	CORN	SOYBEANS	*	0.0000	0.2500	0.0000	0.7500	0.0000
18 20 25 23	CORN	SOYBEANS	*	0.0000	0.0000	0.0000	0.7500	0.2500
15 15 31 32	CORN	CORN		0.0000	0.7500	0.0000	0.0000	0.2500
15 32 32 15	CORN	SOYBEANS	*	0.0000	0.2500	0.0000	0.5000	0.2500
12 15 16 73	CORN	SOYBEANS	*	0.0000	0.0000	0.2500	0.5000	0.2500
20 23 23 25	SOYBEANS	OTHER	@	0.0000	0.2500	0.2500	0.2500	0.2500
24 24 25 32	SOYBEANS	SOYBEANS		0.0000	0.0000	0.2500	0.5000	0.2500
21 25 23 24	SOYBEANS	COTTON	*	0.0000	0.0000	0.5000	0.2500	0.2500
27 45 24 12	SOYBEANS	OTHER	@	0.2500	0.2500	0.2500	0.0000	0.2500
12 13 15 42	SOYBEANS	CORN	*	0.0000	0.5000	0.0000	0.2500	0.2500
22 32 31 43	SOYBEANS	COTTON	*	0.0000	0.2500	0.5000	0.0000	0.2500
31 32 33 34	COTTON	SOYBEANS	*	0.0000	0.2500	0.2500	0.5000	0.0000
29 24 26 28	COTTON	SOYBEANS	*	0.0000	0.0000	0.2500	0.5000	0.2500
34 32 28 45	COTTON	SUGARBEETS	*	0.0000	0.0000	0.2500	0.2500	0.5000
26 25 23 24	COTTON	SOYBEANS	*	0.0000	0.0000	0.2500	0.5000	0.2500
53 48 75 26	COTTON	CLOVER	*	0.5000	0.0000	0.2500	0.2500	0.0000
34 35 25 78	COTTON	OTHER	@	0.2500	0.2500	0.2500	0.2500	0.0000
22 23 25 42	SUGARBEETS	OTHER	@	0.0000	0.5000	0.0000	0.5000	0.0000
25 25 24 26	SUGARBEETS	OTHER	@	0.0000	0.0000	0.5000	0.5000	0.0000
34 25 16 52	SUGARBEETS	SUGARBEETS		0.0000	0.0000	0.2500	0.2500	0.5000
54 23 21 54	SUGARBEETS	CLOVER	*	0.5000	0.0000	0.2500	0.0000	0.2500
25 43 32 15	SUGARBEETS	OTHER	@	0.0000	0.5000	0.0000	0.5000	0.0000
26 54 2 54	SUGARBEETS	CLOVER	*	0.5000	0.0000	0.0000	0.2500	0.2500
12 45 32 54	CLOVER	COTTON	*	0.0000	0.2500	0.5000	0.2500	0.0000
24 58 25 34	CLOVER	SUGARBEETS	*	0.2500	0.0000	0.0000	0.2500	0.5000
87 54 61 21	CLOVER	OTHER	@	0.5000	0.0000	0.5000	0.0000	0.0000
51 31 31 16	CLOVER	COTTON	*	0.2500	0.0000	0.5000	0.0000	0.2500
96 48 54 62	CLOVER	CLOVER		0.5000	0.0000	0.2500	0.0000	0.2500
31 31 11 11	CLOVER	SOYBEANS	*	0.2500	0.0000	0.2500	0.5000	0.0000
56 13 13 71	CLOVER	OTHER	@	0.5000	0.0000	0.0000	0.0000	0.5000
32 13 27 32	CLOVER	CORN	*	0.0000	0.5000	0.2500	0.2500	0.0000
36 26 54 32	CLOVER	CORN	*	0.2500	0.7500	0.0000	0.0000	0.0000
53 08 06 54	CLOVER	OTHER	@	0.5000	0.0000	0.0000	0.0000	0.5000
32 32 62 16	CLOVER	CORN	*	0.2500	0.5000	0.2500	0.0000	0.0000

* MISCLASSIFIED OBSERVATION @ THRESHOLD PROBABILITY NOT MET

REMOTE SENSING DATA ON FIVE CROPS 3

NEAREST 4 NEIGHBORS DISCRIMINANT ANALYSIS CLASSIFICATION SUMMARY FOR CALIBRATION DATA: WORK.CROPS

DISTANCE FUNCTION: $D^2(X,Y) = (X-Y)'\ COV^{-1}\ (X-Y)$

NUMBER OF OBSERVATIONS AND PERCENTS CLASSIFIED INTO CROP:

FROM CROP	CLOVER	CORN	COTTON	SOYBEANS	SUGARBEETS	OTHER	TOTAL
CLOVER	1	3	2	1	1	3	11
	9.09	27.27	18.18	9.09	9.09	27.27	100.00
CORN	0	3	0	4	0	0	7
	0.00	42.86	0.00	57.14	0.00	0.00	100.00
COTTON	1	0	0	3	1	1	6
	16.67	0.00	0.00	50.00	16.67	16.67	100.00
SOYBEANS	0	1	2	1	0	2	6
	0.00	16.67	33.33	16.67	0.00	33.33	100.00
SUGARBEETS	2	0	0	0	1	3	6
	33.33	0.00	0.00	0.00	16.67	50.00	100.00
TOTAL	4	7	4	9	3	9	36
PERCENT	11.11	19.44	11.11	25.00	8.33	25.00	100.00
PRIORS	0.2000	0.2000	0.2000	0.2000	0.2000		

REMOTE SENSING DATA ON FIVE CROPS 4

NEAREST 4 NEIGHBORS DISCRIMINANT ANALYSIS CLASSIFICATION RESULTS FOR TEST DATA: WORK.TEST

DISTANCE FUNCTION: $D^2(X,Y) = (X-Y)'\ COV^{-1}\ (X-Y)$

XVALUES	FROM CROP	CLASSIFIED INTO CROP		POSTERIOR PROBABILITY OF MEMBERSHIP IN CROP: CLOVER	CORN	COTTON	SOYBEANS	SUGARBEETS
16 27 31 33	CORN	CORN		0.0000	0.7500	0.0000	0.0000	0.2500
21 25 23 24	SOYBEANS	SOYBEANS		0.0000	0.0000	0.2500	0.5000	0.2500
29 24 26 28	COTTON	COTTON		0.0000	0.0000	0.5000	0.2500	0.2500
54 23 21 54	SUGARBEETS	OTHER	@	0.5000	0.0000	0.0000	0.0000	0.5000
32 32 62 16	CLOVER	CLOVER		0.5000	0.2500	0.2500	0.0000	0.0000

* MISCLASSIFIED OBSERVATION @ THRESHOLD PROBABILITY NOT MET

REMOTE SENSING DATA ON FIVE CROPS 5

NEAREST 4 NEIGHBORS DISCRIMINANT ANALYSIS CLASSIFICATION SUMMARY FOR TEST DATA: WORK.TEST

DISTANCE FUNCTION: $D^2(X,Y) = (X-Y)'\ COV^{-1}\ (X-Y)$

NUMBER OF OBSERVATIONS AND PERCENTS CLASSIFIED INTO CROP:

FROM CROP	CLOVER	CORN	COTTON	SOYBEANS	SUGARBEETS	OTHER	TOTAL
CLOVER	1	0	0	0	0	0	1
	100.00	0.00	0.00	0.00	0.00	0.00	100.00
CORN	0	1	0	0	0	0	1
	0.00	100.00	0.00	0.00	0.00	0.00	100.00
COTTON	0	0	1	0	0	0	1
	0.00	0.00	100.00	0.00	0.00	0.00	100.00
SOYBEANS	0	0	0	1	0	0	1
	0.00	0.00	0.00	100.00	0.00	0.00	100.00
SUGARBEETS	0	0	0	0	0	1	1
	0.00	0.00	0.00	0.00	0.00	100.00	100.00
TOTAL	1	1	1	1	0	1	5
PERCENT	20.00	20.00	20.00	20.00	0.00	20.00	100.00
PRIORS	0.2000	0.2000	0.2000	0.2000	0.2000		

REFERENCES

Cover, T.M. and Hart, P.E. (1967), "Nearest Neighbor Pattern Classification," *IEEE Transactions on Information Theory*, IT-13, 21-27.

Fix, Evelyn and Hodges, J.L., Jr. (1959), "Discriminatory Analysis: Nonparametric Discrimination: Consistency Properties," *Report No. 4, Project No. 21-49-004, School of Aviation Medicine*, Randolph Air Force Base, Texas.

Hand, D.J. (1981), *Discrimination and Classification*, New York: John Wiley & Sons.

The STEPDISC Procedure

ABSTRACT

The STEPDISC procedure performs a stepwise discriminant analysis by forward selection, backward elimination, or stepwise selection.

INTRODUCTION

The STEPDISC procedure selects a subset of quantitative variables to produce a good discrimination model using forward selection, backward elimination, and stepwise selection (Klecka, 1980). The groups are assumed to be multivariate normal with a common covariance matrix.

Variables are chosen to enter or leave the model according to one of two criteria:

1. the significance level of an *F* test from an analysis of covariance, where the variables already chosen act as covariates and the variable under consideration is the dependent variable, or
2. the squared partial correlation for predicting the variable under consideration from the CLASS variable, controlling for the effects of the variables already selected for the model.

It is important to remember that when many significance tests are performed, each at a level of, say, 5%, the overall probability of rejecting at least one true null hypothesis is much larger than 5%. If you want to guard against including any variables that do not contribute to the discriminatory power of the model in the population, you should specify a very small significance level. In most applications, all variables considered have some discriminatory power, however small. If you want to choose the model that provides the best discrimination using the sample estimates, you need only guard against estimating more parameters than can be reliably estimated with the given sample size. In this case you should use a moderate significance level, perhaps in the range of 10% to 25% (Costanza and Afifi, 1979).

The significance level and the squared partial correlation criteria select variables in the same order, although they may differ in the number of variables selected. Increasing the sample size tends to increase the number of variables selected when you use significance levels, but has little effect on the number selected using squared partial correlations.

Forward selection begins with no variables in the model. At each step the variable is entered that contributes most to the discriminatory power of the model as measured by Wilks' lambda, the likelihood ratio criterion. When none of the unselected variables meet the entry criterion, the forward selection process stops.

Backward elimination begins with all variables in the model except those that are linearly dependent with previous variables in the VAR statement. At each step the variable that contributes least to the discriminatory power of the model (as measured by Wilks' lambda) is removed. When all remaining variables meet the criterion to stay in the model, the backward elimination process stops.

Stepwise selection begins with no variables in the model. At each step, if a variable already in the model fails to meet the criterion to stay, the worst such variable is removed. Otherwise, the variable that contributes most to the discriminatory power of the model (as measured by Wilks' lambda) is entered. When all variables in the model meet the criterion to stay, and none of the other variables meet the criterion to enter, the stepwise selection process stops.

The models selected by STEPDISC are not necessarily the best possible models and Wilks' lambda may not be the best measure of discriminatory power for your application. However, if STEPDISC is used carefully, in combination with your knowledge of the data and careful cross-validation, it can be a valuable aid in selecting a discrimination model.

SPECIFICATIONS

The following statements are used with STEPDISC:

PROC STEPDISC *options;*
VAR *variables;*
CLASS *variable;*
BY *variables;*

The PROC and CLASS statements are required, but the VAR and BY statements are optional.

PROC STEPDISC Statement

PROC STEPDISC *options;*

The following options can appear in the PROC statement:

DATA=*SASdataset*	names the data set to be analyzed.
STEPWISE SW	requests stepwise selection, which is the default if neither FORWARD nor BACKWARD is requested.
FORWARD FW	requests forward selection.
BACKWARD BW	requests backward elimination.
SLENTRY=*p* SLE=*p*	specifies the significance level to enter, where $0<=p<=1$. The default is .15.
SLSTAY=*p* SLS=*p*	specifies the significance level to stay, where $0<=p<=1$. The default is .15.
PR2ENTRY=*p* PR2E=*p*	specifies the partial R^2 to enter, where $p<=1$.
PR2STAY=*p* PR2S=*p*	specifies the partial R^2 to stay, where $p<=1$.
TOLERANCE=*p* TOL=*p*	specifies the tolerance for entering variables, where $0<p<1$. The default is 1E-8.
INCLUDE=*n*	requests that the first *n* variables in the VAR statement be included in every model.

MAXSTEP=*n*	specifies the maximum number of steps.
SIMPLE	prints means and standard deviations.
STDMEAN	prints within- and total-standardized class means.
TCORR	prints total sample correlations.
WCORR	prints within-group correlations.
SHORT	prints the summary table only.

VAR Statement

VAR *variables*;

The VAR statement specifies the numeric variables eligible for selection. The default is all numeric variables not mentioned in other statements.

CLASS Statement

CLASS *variable*;

The CLASS statement specifies the name of one numeric or character variable defining the groups in the discriminant analysis. Class levels are determined by the unformatted values of the class variable. This statement must be included.

BY Statement

BY *variables*;

A BY statement may be used with PROC STEPDISC to obtain separate analyses on observations in groups defined by the BY variables. When a BY statement appears, the procedure expects the input data set to be sorted in order of the BY variables. If your input data set is not sorted in ascending order, use the SORT procedure with a similar BY statement to sort the data, or, if appropriate, use the BY statement options NOTSORTED or DESCENDING. For more information, see the discussion of the BY statement in Chapter 8, "Statements Used in the PROC Step," in *SAS User's Guide: Basics, 1982 Edition*.

DETAILS

Missing Values

Observations containing missing values are omitted from the analysis.

Limitation

The number of variables plus the number of classes should not exceed 250.

Printed Output

STEPDISC prints:

1. CLASS LEVEL INFORMATION, including the values of the classification variable, the FREQUENCY of each value, and its PROPORTION in the total sample.

Optional output includes:

2. CLASS MEANS

3. STANDARD DEVIATIONS, both for the TOTAL SAMPLE and pooled WITHIN CLASS
4. WITHIN-STANDARDIZED CLASS MEANS, obtained by subtracting the grand mean from each class mean and dividing by the pooled within-class standard deviation
5. TOTAL-STANDARDIZED CLASS MEANS, obtained by subtracting the grand mean from each class mean and dividing by the total sample standard deviation
6. TOTAL SAMPLE CORRELATIONS
7. POOLED WITHIN CLASS CORRELATIONS.

At each step the following statistics are printed:

8. for each variable considered for entry or removal: (PARTIAL) R**2, the squared (partial) correlation, the F statistic, and PROB>F, the probability level, from a one-way analysis covariance
9. the TOLERANCE for each variable being considered for entry. Tolerance is one minus the squared multiple correlation of the variable with the other variables already in the model.

   A variable is entered only if its tolerance is greater than the value specified in the TOLERANCE= option. The TOLERANCE is computed using the total sample correlation matrix. It is customary to compute tolerance using the within-group correlation matrix (Jennrich, 1977), but it is possible for a variable with excellent discriminatory power to have a high total-sample tolerance and a low within-group tolerance. For example, STEPDISC will enter a variable that yields perfect discrimination (that is, produces a canonical correlation of one), but a program using within-group tolerance will not.
10. the variable LABEL, if any
11. the name of the variable chosen
12. the variable(s) already selected or removed
13. WILKS' LAMBDA and the associated F approximation with degrees of freedom and PROB>F, the associated probability level after the selected variable has been entered or removed. Wilks' lambda is close to 0 if any two groups are well separated.
14. PILLAI'S TRACE and the associated F approximation with degrees of freedom and PROB>F, the associated probability level after the selected variable has been entered or removed
15. The average squared canonical correlation (ASCC) is Pillai's trace divided by the number of groups minus 1. ASCC is close to 1 if all groups are well separated and if all or most directions in the discriminant space show good separation for at least two groups.

A summary table is printed giving statistics associated with the variable chosen at each step. The number of statistics printed depends on the LINESIZE used.

The summary table includes:

- STEP number
- VARIABLE ENTERED or REMOVED
- NUMBER of variables IN the model
- PARTIAL R**2
- F STATISTIC for entering or removing the variable
- PROB>F, the probability level for the previous F statistic
- WILKS' LAMBDA
- PROB<LAMBDA based on the F approximation to Wilks' lambda
- AVERAGE SQUARED CANONICAL CORRELATION

- PROB>ASCC based on the F approximation to Pillai's trace
- the variable LABEL, if any.

EXAMPLE

The iris data published by Fisher (1936) have been widely used for examples in discriminant analysis and cluster analysis. The sepal length, sepal width, petal length, and petal width were measured in millimeters on 50 iris specimens from each of three species, *Iris setosa, I. versicolor,* and *I. virginica*. A stepwise discriminant analysis is performed using stepwise selection.

```
DATA IRIS;
  TITLE FISHER (1936) IRIS DATA;
  INPUT SEPALLEN SEPALWID PETALLEN PETALWID SPEC_NO @@;
  IF SPEC_NO=1 THEN SPECIES='SETOSA ';
  ELSE IF SPEC_NO=2 THEN SPECIES='VERSICOLOR';
  ELSE SPECIES='VIRGINICA ';
  LABEL SEPALLEN=SEPAL LENGTH IN MM.
        SEPALWID=SEPAL WIDTH IN MM.
        PETALLEN=PETAL LENGTH IN MM.
        PETALWID=PETAL WIDTH IN MM.;
  CARDS;
50 33 14 02 1 64 28 56 22 3 65 28 46 15 2
67 31 56 24 3 63 28 51 15 3 46 34 14 03 1
69 31 51 23 3 62 22 45 15 2 59 32 48 18 2
46 36 10 02 1 61 30 46 14 2 60 27 51 16 2
65 30 52 20 3 56 25 39 11 2 65 30 55 18 3
58 27 51 19 3 68 32 59 23 3 51 33 17 05 1
57 28 45 13 2 62 34 54 23 3 77 38 67 22 3
63 33 47 16 2 67 33 57 25 3 76 30 66 21 3
49 25 45 17 3 55 35 13 02 1 67 30 52 23 3
70 32 47 14 2 64 32 45 15 2 61 28 40 13 2
48 31 16 02 1 59 30 51 18 3 55 24 38 11 2
63 25 50 19 3 64 32 53 23 3 52 34 14 02 1
49 36 14 01 1 54 30 45 15 2 79 38 64 20 3
44 32 13 02 1 67 33 57 21 3 50 35 16 06 1
58 26 40 12 2 44 30 13 02 1 77 28 67 20 3
63 27 49 18 3 47 32 16 02 1 55 26 44 12 2
50 23 33 10 2 72 32 60 18 3 48 30 14 03 1
51 38 16 02 1 61 30 49 18 3 48 34 19 02 1
50 30 16 02 1 50 32 12 02 1 61 26 56 14 3
64 28 56 21 3 43 30 11 01 1 58 40 12 02 1
51 38 19 04 1 67 31 44 14 2 62 28 48 18 3
49 30 14 02 1 51 35 14 02 1 56 30 45 15 2
58 27 41 10 2 50 34 16 04 1 46 32 14 02 1
60 29 45 15 2 57 26 35 10 2 57 44 15 04 1
50 36 14 02 1 77 30 61 23 3 63 34 56 24 3
58 27 51 19 3 57 29 42 13 2 72 30 58 16 3
54 34 15 04 1 52 41 15 01 1 71 30 59 21 3
64 31 55 18 3 60 30 48 18 3 63 29 56 18 3
49 24 33 10 2 56 27 42 13 2 57 30 42 12 2
```

```
55 42 14 02 1 49 31 15 02 1 77 26 69 23 3
60 22 50 15 3 54 39 17 04 1 66 29 46 13 2
52 27 39 14 2 60 34 45 16 2 50 34 15 02 1
44 29 14 02 1 50 20 35 10 2 55 24 37 10 2
58 27 39 12 2 47 32 13 02 1 46 31 15 02 1
69 32 57 23 3 62 29 43 13 2 74 28 61 19 3
59 30 42 15 2 51 34 15 02 1 50 35 13 03 1
56 28 49 20 3 60 22 40 10 2 73 29 63 18 3
67 25 58 18 3 49 31 15 01 1 67 31 47 15 2
63 23 44 13 2 54 37 15 02 1 56 30 41 13 2
63 25 49 15 2 61 28 47 12 2 64 29 43 13 2
51 25 30 11 2 57 28 41 13 2 65 30 58 22 3
69 31 54 21 3 54 39 13 04 1 51 35 14 03 1
72 36 61 25 3 65 32 51 20 3 61 29 47 14 2
56 29 36 13 2 69 31 49 15 2 64 27 53 19 3
68 30 55 21 3 55 25 40 13 2 48 34 16 02 1
48 30 14 01 1 45 23 13 03 1 57 25 50 20 3
57 38 17 03 1 51 38 15 03 1 55 23 40 13 2
66 30 44 14 2 68 28 48 14 2 54 34 17 02 1
51 37 15 04 1 52 35 15 02 1 58 28 51 24 3
67 30 50 17 2 63 33 60 25 3 53 37 15 02 1
;
PROC STEPDISC STEPWISE SIMPLE STDMEAN TCORR WCORR;
  CLASS SPECIES;
  VAR SEPALLEN SEPALWID PETALLEN PETALWID;
```

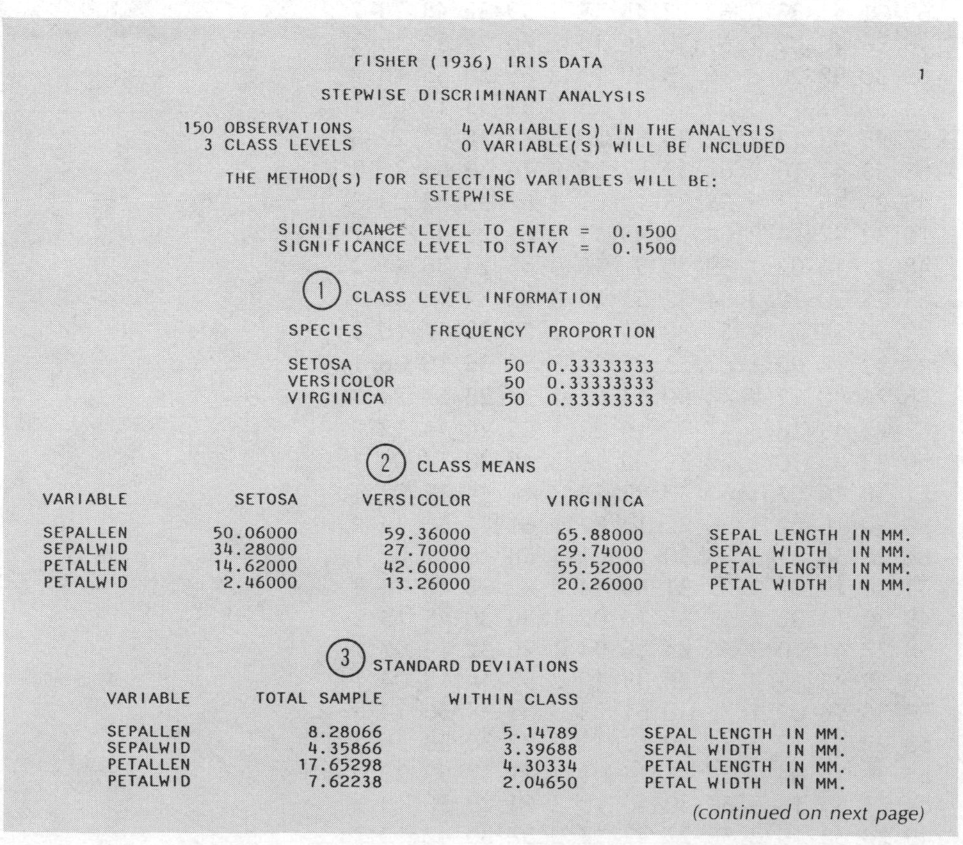

```
                         FISHER (1936) IRIS DATA                              1

                      STEPWISE DISCRIMINANT ANALYSIS

         150 OBSERVATIONS              4 VARIABLE(S) IN THE ANALYSIS
           3 CLASS LEVELS              0 VARIABLE(S) WILL BE INCLUDED

              THE METHOD(S) FOR SELECTING VARIABLES WILL BE:
                                 STEPWISE

                SIGNIFICANCE LEVEL TO ENTER =  0.1500
                SIGNIFICANCE LEVEL TO STAY  =  0.1500

                      (1) CLASS LEVEL INFORMATION

                SPECIES      FREQUENCY  PROPORTION

                SETOSA              50  0.33333333
                VERSICOLOR          50  0.33333333
                VIRGINICA           50  0.33333333


                             (2) CLASS MEANS

VARIABLE          SETOSA      VERSICOLOR       VIRGINICA

SEPALLEN        50.06000        59.36000        65.88000      SEPAL LENGTH IN MM.
SEPALWID        34.28000        27.70000        29.74000      SEPAL WIDTH  IN MM.
PETALLEN        14.62000        42.60000        55.52000      PETAL LENGTH IN MM.
PETALWID         2.46000        13.26000        20.26000      PETAL WIDTH  IN MM.


                         (3) STANDARD DEVIATIONS

      VARIABLE     TOTAL SAMPLE      WITHIN CLASS

      SEPALLEN          8.28066           5.14789     SEPAL LENGTH IN MM.
      SEPALWID          4.35866           3.39688     SEPAL WIDTH  IN MM.
      PETALLEN         17.65298           4.30334     PETAL LENGTH IN MM.
      PETALWID          7.62238           2.04650     PETAL WIDTH  IN MM.
```

(continued on next page)

(continued from previous page)

```
        (4) WITHIN-STANDARDIZED CLASS MEANS

VARIABLE        SETOSA      VERSICOLOR      VIRGINICA

SEPALLEN      -1.62656        0.18001        1.44655      SEPAL LENGTH IN MM.
SEPALWID       1.09120       -0.84587       -0.24532      SEPAL WIDTH  IN MM.
PETALLEN      -5.33538        1.16653        4.16885      PETAL LENGTH IN MM.
PETALWID      -4.65836        0.61894        4.03942      PETAL WIDTH  IN MM.
```

```
                    FISHER (1936) IRIS DATA                                   2

                 STEPWISE DISCRIMINANT ANALYSIS

              (5) TOTAL-STANDARDIZED CLASS MEANS

VARIABLE        SETOSA      VERSICOLOR      VIRGINICA

SEPALLEN      -1.01119        0.11191        0.89928      SEPAL LENGTH IN MM.
SEPALWID       0.85041       -0.65922       -0.19119      SEPAL WIDTH  IN MM.
PETALLEN      -1.30063        0.28437        1.01626      PETAL LENGTH IN MM.
PETALWID      -1.25070        0.16618        1.08453      PETAL WIDTH  IN MM.


              (6) TOTAL SAMPLE CORRELATIONS

                SEPALLEN      SEPALWID      PETALLEN      PETALWID

SEPALLEN         1.0000        -.1176        0.8718        0.8179

SEPALWID         -.1176        1.0000        -.4284        -.3661

PETALLEN         0.8718        -.4284        1.0000        0.9629

PETALWID         0.8179        -.3661        0.9629        1.0000

              (7) POOLED WITHIN CLASS CORRELATIONS

                SEPALLEN      SEPALWID      PETALLEN      PETALWID

SEPALLEN         1.0000        0.5302        0.7562        0.3645

SEPALWID         0.5302        1.0000        0.3779        0.4705

PETALLEN         0.7562        0.3779        1.0000        0.4845

PETALWID         0.3645        0.4705        0.4845        1.0000
```

```
                    FISHER (1936) IRIS DATA                                   3

                  STEPWISE SELECTION:  STEP  1

            STATISTICS FOR ENTRY, DF (9) 2,147  (10)

VARIABLE     R**2       (8) F     PROB > F   TOLERANCE   LABEL

SEPALLEN    0.6187     119.265     0.0001     1.0000     SEPAL LENGTH IN MM.
SEPALWID    0.4008      49.160     0.0001     1.0000     SEPAL WIDTH  IN MM.
PETALLEN    0.9414    1180.161     0.0001     1.0000     PETAL LENGTH IN MM.
PETALWID    0.9289     960.007     0.0001     1.0000     PETAL WIDTH  IN MM.

           (11) VARIABLE PETALLEN WILL BE ENTERED

        THE FOLLOWING VARIABLE(S) HAVE BEEN ENTERED:
                 (12) PETALLEN

                    MULTIVARIATE STATISTICS

(13) WILKS' LAMBDA  = 0.05862828    F(2,147) =       1180.161     PROB > F = 0.0000
     PILLAI'S TRACE =   0.941372    F(2,147) =       1180.161     PROB > F = 0.0000
(14)
       (15) AVERAGE SQUARED CANONICAL CORRELATION = 0.47068586
-------------------------------------------------------------------------------------

                  STEPWISE SELECTION:  STEP  2

               STATISTICS FOR REMOVAL,  DF = 2,147

VARIABLE     R**2         F      PROB > F   LABEL

PETALLEN    0.9414    1180.161    0.0001   PETAL LENGTH IN MM.
```

(continued on next page)

(continued from previous page)

```
                    NO VARIABLES CAN BE REMOVED

                    ---------------------------

                STATISTICS FOR ENTRY, DF = 2,146

                PARTIAL
VARIABLE         R**2        F      PROB > F  TOLERANCE  LABEL

SEPALLEN        0.3198     34.323    0.0001     0.2400   SEPAL LENGTH IN MM.
SEPALWID        0.3709     43.035    0.0001     0.8164   SEPAL WIDTH  IN MM.
PETALWID        0.2533     24.766    0.0001     0.0729   PETAL WIDTH  IN MM.

                VARIABLE SEPALWID WILL BE ENTERED

          THE FOLLOWING VARIABLE(S) HAVE BEEN ENTERED:
                       SEPALWID  PETALLEN
```

```
                       FISHER (1936) IRIS DATA                                        4

                      STEPWISE SELECTION:  STEP  2


                       MULTIVARIATE STATISTICS

WILKS' LAMBDA  = 0.03688411      F(4,292) =         307.105      PROB > F = 0.0000
PILLAI'S TRACE =    1.119908     F(4,294) =          93.528      PROB > F = 0.0000

            AVERAGE SQUARED CANONICAL CORRELATION = 0.55995394
------------------------------------------------------------------------------------------
                      STEPWISE SELECTION:  STEP  3

                 STATISTICS FOR REMOVAL,  DF = 2,146

               PARTIAL
      VARIABLE   R**2          F      PROB > F  LABEL

      SEPALWID   0.3709     43.035    0.0001  SEPAL WIDTH  IN MM.
      PETALLEN   0.9384   1112.954    0.0001  PETAL LENGTH IN MM.

                      NO VARIABLES CAN BE REMOVED

                      ---------------------------

                   STATISTICS FOR ENTRY, DF = 2,145

            PARTIAL
  VARIABLE    R**2        F      PROB > F  TOLERANCE  LABEL

  SEPALLEN   0.1447     12.268    0.0001     0.1323   SEPAL LENGTH IN MM.
  PETALWID   0.3229     34.569    0.0001     0.0662   PETAL WIDTH  IN MM.

                  VARIABLE PETALWID WILL BE ENTERED

             THE FOLLOWING VARIABLE(S) HAVE BEEN ENTERED:
                   SEPALWID  PETALLEN  PETALWID

                       MULTIVARIATE STATISTICS

WILKS' LAMBDA  = 0.02497554     F(6,290) =         257.503      PROB > F = 0.0000
PILLAI'S TRACE =    1.189914    F(6,292) =          71.485      PROB > F = 0.0000

            AVERAGE SQUARED CANONICAL CORRELATION = 0.59495691
```

```
                          FISHER (1936) IRIS DATA                                5

                        STEPWISE SELECTION:  STEP  4

                    STATISTICS FOR REMOVAL,   DF = 2,145

                  PARTIAL
        VARIABLE    R**2          F      PROB > F  LABEL

        SEPALWID   0.4295      54.577    0.0001  SEPAL WIDTH   IN MM.
        PETALLEN   0.3482      38.724    0.0001  PETAL LENGTH  IN MM.
        PETALWID   0.3229      34.569    0.0001  PETAL WIDTH   IN MM.

                       NO VARIABLES CAN BE REMOVED

                       ---------------------------

                    STATISTICS FOR ENTRY, DF = 2,144

               PARTIAL
    VARIABLE     R**2          F       PROB > F  TOLERANCE  LABEL

    SEPALLEN    0.0615       4.721     0.0103     0.0320   SEPAL LENGTH IN MM.

                    VARIABLE SEPALLEN WILL BE ENTERED

                    ALL VARIABLES HAVE BEEN ENTERED

                         MULTIVARIATE STATISTICS

WILKS' LAMBDA  = 0.02343863      F(8,288) =          199.145      PROB > F = 0.0000
PILLAI'S TRACE =    1.191899     F(8,290) =           53.466      PROB > F = 0.0000

              AVERAGE SQUARED CANONICAL CORRELATION = 0.59594941
-------------------------------------------------------------------------------------------

                        STEPWISE SELECTION:  STEP  5

                    STATISTICS FOR REMOVAL,   DF = 2,144

                  PARTIAL
        VARIABLE    R**2          F      PROB > F  LABEL

        SEPALLEN   0.0615       4.721    0.0103  SEPAL LENGTH  IN MM.
        SEPALWID   0.2335      21.936    0.0001  SEPAL WIDTH   IN MM.
        PETALLEN   0.3308      35.590    0.0001  PETAL LENGTH  IN MM.
        PETALWID   0.2570      24.904    0.0001  PETAL WIDTH   IN MM.

                       NO VARIABLES CAN BE REMOVED

                       NO FURTHER STEPS ARE POSSIBLE
```

```
                          FISHER (1936) IRIS DATA                                6

                        STEPWISE SELECTION:  SUMMARY

                                                                             AVERAGE
                                                                             SQUARED
          VARIABLE        NUMBER   PARTIAL       F       PROB >    WILKS'    PROB >  CANONICAL     PROB >
STEP  ENTERED   REMOVED     IN       R**2    STATISTIC     F       LAMBDA    LAMBDA CORRELATION    ASCC

  1   PETALLEN              1      0.9414    1180.161  0.0001  0.05862828  0.0000  0.47068586  0.0000
  2   SEPALWID              2      0.3709      43.035  0.0001  0.03688411  0.0000  0.55995394  0.0000
  3   PETALWID              3      0.3229      34.569  0.0001  0.02497554  0.0000  0.59495691  0.0000
  4   SEPALLEN              4      0.0615       4.721  0.0103  0.02343863  0.0000  0.59594941  0.0000
```

REFERENCES

Costanza, M.C. and Afifi, A.A. (1979), "Comparison of Stopping Rules in Forward Stepwise Discriminant Analysis," *Journal of the American Statistical Association*, 74, 777–785.

Jennrich, R.I. (1977), "Stepwise Discriminant Analysis," *Statistical Methods for Digital Computers*, eds. K. Enslein, A. Ralston, and H. Wilf, New York: John Wiley & Sons.

Klecka, W.R. (1980), *Discriminant Analysis*, Sage University Paper series on Quantitative Applications in the Social Sciences, series no. 07–019. Beverly Hills and London: Sage Publications.

CLUSTERING

Introduction to SAS Clustering Procedures

The procedures in this chapter can be used to cluster the observations or the variables in a SAS data set. Both hierarchical and disjoint clusters can be obtained. Only numeric variables are permitted.

The purpose of cluster analysis is to place objects into groups or clusters suggested by the data, not defined a priori, such that objects in a given cluster tend to be similar to each other in some sense, and objects in different clusters tend to be dissimilar. Cluster analysis can also be used for summarizing data rather than for finding "natural" or "real" clusters; this use of clustering is sometimes called *dissection* (Everitt, 1980).

Any generalization about cluster analysis must be vague because a vast number of clustering methods have been developed in several different fields, with different definitions of clusters and similarity among objects. The variety of clustering techniques is reflected by the variety of terms used for cluster analysis: botryology, classification, clumping, morphometrics, nosography, nosology, numerical taxonomy, partitioning, Q-analysis, systematics, taximetrics, taxonorics, typology, and unsupervised pattern recognition. Good (1977) has also suggested aciniformics and agminatics.

Several types of clusters are possible:

- Disjoint clusters place each object in one and only one cluster.
- Hierarchical clusters are organized so that one cluster may be entirely contained within another cluster, but no other kind of overlap between clusters is allowed.
- Overlapping clusters can be constrained to limit the number of objects that belong simultaneously to two clusters, or they can be unconstrained, allowing any degree of overlap in cluster membership.
- Fuzzy clusters are defined by a probability or grade of membership of each object in each cluster. Fuzzy clusters can be disjoint, hierarchical, or overlapping.

The data representations of objects to be clustered also take many forms. The most common are:

- A square distance or similarity matrix, in which both rows and columns correspond to the objects to be clustered. A correlation matrix is an example of a similarity matrix.
- A multivariate data matrix in which the rows are observations and the columns are variables, as in the usual SAS data set. The observations, or the variables, or both may be clustered.

The SAS procedures for clustering are oriented toward disjoint or hierarchical clusters from a multivariate data matrix. The procedures are:

CLUSTER does hierarchical clustering of observations using the centroid method, Ward's method, or average linkage on squared Euclidean distances.

FASTCLUS finds disjoint clusters of observations using a k-means method and is especially suitable for large data sets (as many as 100,000 observations).

VARCLUS is for both hierarchical and disjoint clustering of variables by oblique multiple-group component analysis.

TREE draws tree diagrams, also called *dendrograms* or *phenograms*, using output from CLUSTER or VARCLUS.

There are also clustering procedures described in the SAS Supplemental Library User's Guide:

IPFPHC hierarchically clusters the units of a transaction flow table (an asymmetric similarity matrix) and can be used for single linkage clustering.

OVERCLUS finds overlapping clusters from similarity data.

HIER draws hierarchical diagrams and can be used instead of TREE with the output data sets from CLUSTER and VARCLUS.

Everitt (1980) is the best introduction to cluster analysis. Anderberg (1973) and Hartigan (1975) are also useful references. Hartigan (1975) and Spath (1980) give numerous FORTRAN programs for clustering. Milligan (1980) is an excellent empirical study of hierarchical and k-means clustering algorithms.

Clustering Observations

CLUSTER and FASTCLUS both find compact clusters that can be separated by hyperplanes. DISCRIM can classify these clusters perfectly with POOL=NO. If you want to be able to find clusters with long, narrow, concave, or irregular shapes, you should use IPFPHC.

CLUSTER is easier to use than FASTCLUS because one run produces results from one cluster up to as many as you like. FASTCLUS must be run once for each number of clusters.

The time required by FASTCLUS is roughly proportional to the number of observations, whereas the time required by CLUSTER varies with the square of the number of observations. FASTCLUS can therefore be used with much larger data sets than CLUSTER. If you want to hierarchically cluster a data set that is too large to use with CLUSTER directly, you can have FASTCLUS produce, say, fifty clusters, and let CLUSTER analyze these fifty clusters instead of the entire data set.

The MEAN= data set produced by FASTCLUS contains two special variables:

- _FREQ_ gives the number of observations in the cluster.
- _RMSSTD_ gives the root-mean-square across variables of the cluster standard deviations.

These variables can be used with CLUSTER to give the correct results when clustering clusters:

```
PROC FASTCLUS MAXCLUSTERS=50 MEAN=TEMP;
PROC CLUSTER;
  FREQ _FREQ_;
  RMSSTD _RMSSTD_;
```

Clustering Variables

Factor rotation is often used to cluster variables, but the resulting clusters are fuzzy. VARCLUS is preferable if you want hard (non-fuzzy), disjoint clusters. Factor rotation is better if you want to be able to find overlapping clusters. It is often a good idea to try both VARCLUS and FACTOR with an oblique rotation, compare the variance explained by each, and see how fuzzy the factor loadings are and whether there seem to be overlapping clusters.

You can use VARCLUS to harden a fuzzy factor rotation by creating an output data set with FACTOR containing scoring coefficients and using this data set to initialize VARCLUS:

```
PROC FACTOR ROTATE=PROMAX SCORE OUTSTAT=FACT;
PROC VARCLUS INITIAL=INPUT PROPORTION=0;
```

Any rotation method could be used instead of PROMAX. Only the SCORE and OUTSTAT= options are necessary on the PROC FACTOR statement. VARCLUS reads the correlation matrix from the data set created by FACTOR. The INITIAL=INPUT option tells VARCLUS to read initial scoring coefficients from the data set. PROPORTION=0 keeps VARCLUS from splitting any of the clusters.

The Number of Clusters

There are no satisfactory methods for determining the number of clusters for any type of cluster analysis (Everitt, 1979, 1980). The number-of-clusters problem is, if anything, more difficult than the number-of-factors problem.

If your purpose in clustering is dissection, that is, to summarize the data without trying to uncover "real" clusters, it may suffice to look at R^2 for each variable and pooled over all variables. Plots of R^2 against the number of clusters are useful. This method is appropriate for clusters of observations or variables.

It is always a good idea to look at your data graphically. If you have only two or three variables, use PLOT to make scatterplots identifying the clusters. With more variables, use CANDISC to compute canonical variables for plotting.

Ordinary significance tests, such as analysis of variance *F* tests, are not valid for testing differences between clusters. Since clustering methods attempt to maximize the separation between clusters, the assumptions of the usual significance tests, parametric or nonparametric, are drastically violated. For example, if you take a sample of 100 observations from a single univariate normal distribution, have FASTCLUS divide it into two clusters, and run a *t* test between the clusters, you will usually obtain a probability level of less than .0001.

A variety of heuristic methods for determining the number of clusters have been suggested (for example, Mojena and Wishart, 1980), but methods without theoretical justification should be used cautiously.

Binder (1978, 1981) has taken a Bayesian approach to the number of clusters, but his method is not practical for large samples.

Most valid tests for clusters either have intractable sampling distributions or involve null hypotheses for which rejection is uninformative. For clustering methods based on distance matrices, a popular null hypothesis is that all permutations of the values in the distance matrix are equally likely (Ling, 1973; Hubert, 1974). Using this null hypothesis you can do a permutation test or a rank test. The trouble with the permutation hypothesis is that with any real data, the null hypothesis is totally implausible even if the data do not contain clusters. Rejecting the null hypothesis does not provide any useful information (Hubert and Baker, 1977).

Another common null hypothesis is that the data are a random sample from a multivariate normal distribution (Wolfe, 1970, 1978; Lee, 1979). The multivariate normal null hypothesis is better than the permutation null hypothesis, but it is not satisfactory because there is typically a high probability of rejection if the data are sampled from a distribution with lower kurtosis than a normal distribution, such as a uniform distribution. The tables in Englemann and Hartigan (1969), for example, generally lead to rejection of the null hypothesis when the data are sampled from a uniform distribution. Hartigan (1978) and Arnold (1979) discuss both normal and uniform null hypotheses, and the uniform null hypothesis seems preferable for most practical purposes. It is not clear, however, that the uniform distribution provides an adequate null hypothesis. The effects of long-tailed unimodal distributions, such as the Cauchy, on clustering methods have apparently not been studied, although Milligan (1980) examines the effect of outliers on several popular techniques. The basic difficulty in the number-of-clusters problem may be that cluster analysis is inherently non-robust, like outlier detection.

If you accept the uniform distribution as an acceptable null hypothesis, there are still serious difficulties in obtaining sampling distributions. Hartigan (1978) has obtained asymptotic distributions for the within-cluster sum of squares (WSS), the criterion that CLUSTER (with Ward's method) and FASTCLUS attempt to optimize, but only in one dimension. Minimizing WSS is equivalent to maximizing the R^2 for predicting the variable from the clusters. Hartigan's results can be used to evaluate clusters from FASTCLUS or CLUSTER-with-Ward's- method, providing the sample size is large enough. Empirical studies have indicated that Hartigan's expression for the expected value of WSS should be multiplied by roughly $(1-c/n)^2$ if there are fewer than 100 or so observations per cluster, where c is the number of clusters and n is the number of observations.

Arnold (1979) used Monte Carlo methods to derive tables of the distribution of a criterion based on the determinant of the within-cluster sum of squares matrix $|\mathbf{W}|$. Having obtained clusters with either FASTCLUS or CLUSTER, you can compute Arnold's criterion with ANOVA or CANDISC. Arnold's tables provide a conservative test because FASTCLUS and CLUSTER attempt to minimize the trace of **W** rather than the determinant. Marriott (1971, 1975) also gives useful information on $|\mathbf{W}|$ as a criterion for the number of clusters.

If you run FASTCLUS or CLUSTER with more than one variable, Hartigan's results are not applicable and Arnold's may be very conservative. We have derived an approximation to the expected value of WSS, or equivalently R^2, that appears to give useful results in small samples. The approximation is based on the assumption that a uniform distribution on a hyperrectangle will be divided into clusters shaped roughly like hypercubes. In large samples that can be divided into the appropriate number of hypercubes, this assumption gives very accurate results. In other cases the approximation is slightly conservative.

A criterion for the number of clusters based on this approximation is called the *cubic clustering criterion* (CCC). The CCC is printed by FASTCLUS and CLUSTER. The best way to use the CCC is to plot its value against the number of clusters, ranging from two clusters up to about one-tenth the number of observations. The CCC may become liberal if the average number of observations per cluster is less than ten. Peaks on the plot with the CCC greater than 2 or 3 indicate good clusterings. There may be several such peaks if the data have a hierarchical structure. Very distinct non-hierarchical clusters show a sharp rise before the peak followed by a gradual decline. Peaks with the CCC between 0 and 2 indicate possible clusters that should be interpreted cautiously. If all values of the CCC are negative and decreasing for 2 or more clusters, the distribution is probably unimodal. If the CCC increases continuously as the number of clusters increases, the distribution may be grainy or the data may have been excessively rounded or recorded with just a few digits. The power of the CCC seems to be at least as good as that of the human eye in two dimensions with 100 observations, although outliers may seriously reduce the power and cause highly negative values of the CCC.

Note: since research on the CCC is still in progress, the formula used to compute it may be slightly modified in future releases of SAS. The CCC will be fully described in a forthcoming technical report.

REFERENCES

Anderberg, M.R. (1973), *Cluster Analysis for Applications*, New York: Academic Press.

Arnold, S.J. (1979), "A Test for Clusters," *Journal of Marketing Research*, 16, 545-551.

Binder, D.A. (1978), "Bayesian Cluster Analysis," *Biometrika*, 65, 31-38.

Binder, D.A. (1981), "Approximations to Bayesian Clustering Rules," *Biometrika*, 68, 275-285.

Day, N.E. (1969), "Estimating the Components of a Mixture of Normal Distributions," *Biometrika*, 56, 463-474.

Duran, B.S. and Odell, P.L. (1974), *Cluster Analysis*, New York: Springer-Verlag.

Englemann, L. and Hartigan, J.A. (1969), "Percentage Points of a Test for Clusters," *Journal of the American Statistical Association*, 64, 1647-1648.

Everitt, B.S. (1980), *Cluster Analysis, 2nd ed.*, London: Heineman Educational Books Ltd.

Everitt, B.S. (1979), "Unresolved Problems in Cluster Analysis," *Biometrics*, 35, 169-181.

Fisher, R.A. (1936), "The Use of Multiple Measurements in Taxonomic Problems," *Annals of Eugenics*, 7, 179-188.

Good, I.J. (1977), "The Botryology of Botryology," in *Classification and Clustering*, ed. J. Van Ryzin, New York: Academic Press.

Harman, H.H. (1976), *Modern Factor Analysis, 3rd ed.*, Chicago: University of Chicago Press.

Hartigan, J.A. (1975), *Clustering Algorithms*, New York: John Wiley & Sons Inc.

Hartigan, J.A. (1977), "Distribution Problems in Clustering," in *Classification and Clustering*, ed. J. Van Ryzin, New York: Academic Press.

Hartigan, J.A. (1978), "Asymptotic Distributions for Clustering Criteria," *Annals of Statistics*, 6, 117-131.

Hubert, L. (1974), "Approximate Evaluation Techniques for the Single-link and Complete-link Hierarchical Clustering Procedures," *Journal of the American Statistical Association*, 69, 698-704.

Hubert, L.J. and Baker, F.B. (1977), "An Empirical Comparison of Baseline Models Goodness-of-Fit in r-Diameter Hierarchical Clustering," in *Classification and Clustering*, ed. J. Van Ryzin, New York: Academic Press.

Lee, K.L. (1979), "Multivariate Tests for Clusters," *Journal of the American Statistical Association*, 74, 708-714.

Ling, R.F (1973), "A Probability Theory of Cluster Analysis," *Journal of the American Statistical Association*, 68, 159-169.

MacQueen, J.B. (1967), "Some Methods for Classification and Analysis of Multivariate Observations," *Proceedings of the Fifth Berkeley Symposium on Mathematical Statistics and Probability*, 1, 281-297.

Marriott, F.H.C. (1971), "Practical Problems in a Method of Cluster Analysis," *Biometrics*, 27, 501-514.

Marriott, F.H.C. (1975), "Separating Mixtures of Normal Distributions," *Biometrics*, 31, 767-769.

Mezzich, J.E and Solomon, H. (1980), *Taxonomy and Behavioral Science*, New York: Academic Press.

Milligan, G.W. (1980), ''An Examination of the Effect of Six Types of Error Perturbation on Fifteen Clustering Algorithms,'' *Psychometrika*, 45, 325-342.

Mojena, R. and Wishart, D. (1980), ''Stopping Rules for Ward's Clustering Method,'' *COMPSTAT: 1980, Proceedings in Computational Statistics, 4th Symposium*. Wien: Physica-Verlag.

Scott, A.J. and Symons, M.J. (1971), ''Clustering Methods Based on Likelihood Ratio Criteria,'' *Biometrics*, 27, 387-397.

Spath, H. (1980), *Cluster Analysis Algorithms*, Chichester, England: Ellis Horwood.

Symons, M.J. (1981), ''Clustering Criteria and Multivariate Normal Mixtures,'' *Biometrics*, 37, 35-43.

Tou, J.T. and Gonzalez, R.C. (1974), *Pattern Recognition Principles*, Reading, MA: Addison-Wesley Publishing Co.

Wolfe, J.H. (1970), ''Pattern Clustering by Multivariate Mixture Analysis,'' *Multivariate Behavioral Research*, 5, 329-350.

Wolfe, J.H. (1978), ''Comparative Cluster Analysis of Patterns of Vocational Interest,'' *Multivariate Behavioral Research*, 13, 33-44.

The CLUSTER Procedure

ABSTRACT

The CLUSTER procedure hierarchically clusters the observations in a SAS data set using the centroid method, Ward's method, or average linkage on squared Euclidean distances. The variables must be numeric. CLUSTER creates an output data set from which the TREE procedure can draw a tree diagram.

INTRODUCTION

The CLUSTER procedure finds hierarchical clusters of the observations in a SAS data set. Three well-known algorithms can be used: the centroid method, Ward's method, or average linkage on squared Euclidean distances. CLUSTER prints a history of the clustering process, giving statistics useful for estimating the number of clusters. CLUSTER also creates an output data set that may be used by the TREE procedure to draw a tree diagram of the cluster hierarchy.

Background

CLUSTER uses three standard agglomerative hierarchical clustering algorithms. Each observation begins in a cluster by itself. The two closest clusters are merged to form a new cluster replacing the two old clusters. Merging of the two closest clusters is repeated until only one cluster is left. The algorithms used by CLUSTER differ in how the distance between two clusters is computed.

In the centroid method the distance between two clusters is defined as the distance (Euclidean distance for CLUSTER) between their centroids or means. The centroid method is more robust to outliers than most other hierarchical methods but in other respects does not perform as well as Ward's method or average linkage (Milligan, 1980).

In Ward's method the distance between two clusters is the sum of squares between the two clusters added up over all the variables. At each generation, the within-cluster sum of squares is minimized over all partitions (an exhaustive set of disjoint clusters) obtainable by merging two clusters from the previous generation. Ward's method tends to join clusters with a small number of observations, and is biased toward producing clusters with roughly the same number of observations.

In average linkage the distance between two clusters is defined as the average distance (squared Euclidean for CLUSTER) between pairs of observations, one in each cluster. Average linkage tends to join clusters with small variances, and is biased toward producing clusters with roughly the same variance.

Agglomerative hierarchical clustering algorithms are described in all standard references on clustering, for example, Anderberg (1973), Hartigan (1975), Everitt

(1980), and Spath (1980). Many studies of cluster analysis have indicated that Ward's method and average linkage are among the best available hierarchical clustering algorithms (for example, Milligan, 1980).

SPECIFICATIONS

The CLUSTER procedure is invoked by the following statements:

PROC CLUSTER *options;*
VAR *variables;*
ID *variable;*
COPY *variables:*
FREQ *variable;*
RMSSTD *variable;*
BY *variables;*

Usually only the VAR statement is needed in addition to the PROC CLUSTER statement.

PROC CLUSTER Statement

PROC CLUSTER *options;*

The following options may appear on the PROC statement:

DATA=*SASdataset* names the input data set containing observations to be clustered. If DATA= is omitted, the most recently created SAS data set is used.

OUTTREE=*SASdataset* names an output data set that can be used by the TREE procedure to draw a tree diagram. The data set must be given a two-level name if it is to be saved (see Chapter 12, "SAS Data Sets" in *SAS User's Guide: Basics, 1982 Edition*, for a discussion of permanent data sets.) If OUTTREE= is omitted, the data set is named using the DATA*n* convention and is not permanently saved. If you do not want to create an output data set, use OUTTREE=_NULL_.

SIMPLE
S prints means, standard deviations, skewness, kurtosis, and a coefficient of bimodality

$$b=(m_3^2 + 1)/(m_4 + 3)$$

where m_3 is skewness and m_4 is kurtosis. Values of b greater than 0.555 generally indicate bimodal or multimodal distributions that may yield distinct clusters.

STANDARD
STD standardizes the variables to mean 0 and standard deviation 1 before clustering.

PRINT=*n*
P=*n* specifies the number of generations of the cluster history to print. The default value is MAX(10,MIN(40,N/10)), where N is the number of observations in the data set. PRINT=0 suppresses the cluster history.

NOEIGEN suppresses computation of eigenvalues for the cubic clustering criterion. NOEIGEN saves time if the number of variables is large but should only be used if the variables are nearly uncorrelated or if you are not interested in the cubic clustering criterion. If you specify NOEIGEN and the variables are highly correlated, the cubic clustering criterion may be very liberal.

METHOD=*name*
M=*name* specifies what clustering method to use. METHOD=WARD (or W) requests Ward's method; specifying METHOD=AVERAGE (or A) requests average linkage on squared Euclidean distances; and specifying METHOD=CENTROID (or C) requests the centroid method.

The default method is Ward's method.

VAR Statement

VAR *variables*;

The VAR statement lists numeric variables to be used in the cluster analysis. If the VAR statement is omitted, all numeric variables not listed in other statements are used.

ID Statement

ID *variable*;

The values of the ID variable identify observations in the OUTTREE= data set and can be used in the tree diagram printed by the TREE procedure.

COPY Statement

COPY *variables*;

The variables in the COPY statement are copied from the input data set to the OUTTREE= data set. Observations in the OUTTREE= data set that correspond to clusters rather than observations in the input data set have missing values for the COPY variables.

FREQ Statement

FREQ *variable*;

If one variable in the input data set represents the frequency of occurrence for other values in the observation, specify the variable's name in a FREQ statement. CLUSTER then treats the data set as if each observation appeared *n* times, where *n* is the value of the FREQ variable for the observation.

If each observation in the DATA= data set represents a cluster (for example, clusters formed by PROC FASTCLUS) the variable specified in the FREQ statement should give the number of original observations in each cluster. When observations represent clusters, you must use the FREQ and RMSSTD (below) statements to obtain accurate statistics in the cluster histories.

RMSSTD Statement

RMSSTD *variable*;

If the observations in the DATA= data set represent clusters, (for example, formed by the FASTCLUS procedure), you can obtain accurate statistics in the cluster histories if the data set contains:

- a variable giving the number of original observations in each cluster (see the FREQ statement, above)
- a variable giving the root-mean-square standard deviation of each cluster.

Specify the name of the variable containing root-mean-square standard deviation in the RMSSTD statement.

BY Statement

BY *variables;*

A BY statement may be used with PROC CLUSTER to obtain separate analyses on observations in groups defined by the BY variables. When a BY statement appears, the procedure expects the input data set to be sorted in order of the BY variables. If your input data set is not sorted in ascending order, use the SORT procedure with a similar BY statement to sort the data, or, if appropriate, use the BY statement options NOTSORTED or DESCENDING. For more information, see the discussion of the BY statement in Chapter 8, ''Statements Used in the PROC Step,'' in *SAS User's Guide: Basics, 1982 Edition*.

DETAILS

Missing Values

Observations with missing values are excluded from the analysis.

Output Data Set

The OUTTREE= data set contains one observation for each observation in the input data set, plus one observation for each cluster of two or more observations, that is, one observation for each node of the cluster tree. The total number of output observations is $2n - 1$, where n is the number of input observations.

The variables in the OUTTREE= data set are:

- the BY variables, if any
- _NAME_, a character variable giving the name of the node. If the node is a cluster, the name will be CL*n* where *n* is the number of the cluster. If the node is an observation, the name will be OB*n* where *n* is the observation number, unless the ID statement is used, in which case the name is the formatted value of the first ID variable.
- _PARENT_, a character variable giving the value of _NAME_ of the parent of the node
- _NCL_, the number of clusters
- _FREQ_, the number of observations in the current cluster
- _RMSSTD_, the root-mean-square standard deviation of the current cluster
- _DIST_, the Euclidean distance between the last clusters joined divided by the root-mean-square distance among all observations
- _AVLINK_, the average linkage between the last clusters joined divided by the root-mean-square distance among all observations
- _SPRSQ_, the semi-partial squared multiple correlation

- __RSQ__, the squared multiple correlation
- __ERSQ__, the approximate expected value of the squared multiple correlation under the uniform null hypothesis
- __RATIO__, equal to (1 − __ERSQ__)/(1 − __RSQ__)
- __CCC__, the cubic clustering criterion
- the variables used in the cluster analysis
- the ID variables, if any.

__ERSQ__, __RATIO__, and __CCC__ have missing values for the original observations.

Computational Resources

Let

n = number of observations
v = number of variables

The overall time required by CLUSTER is roughly proportional to vn^2 for Ward's method or average linkage, or vn^3 for the centroid method.

The variables in the VAR statement should be listed in order of decreasing variance for greatest efficiency.

Printed Output

Optional output includes:

1. simple statistics, including the MEAN, ST DEV (standard deviation), SKEWNESS, KURTOSIS, and a coefficient of BIMODALITY for each variable
2. EIGENVALUEs of the correlation or covariance matrix, as well as the DIFFERENCE between successive eigenvalues, the PROPORTION of variance explained by each eigenvalue, and the CUMULATIVE proportion of variance explained
3. the ROOT-MEAN-SQUARE TOTAL-SAMPLE STANDARD DEVIATION of the variables
4. the ROOT-MEAN-SQUARE DISTANCE BETWEEN OBSERVATIONS

For the generations in the clustering process specified by the PRINT= option, CLUSTER prints:

5. the NUMBER OF CLUSTERS or NCL
6. FREQUENCY OF NEW CLUSTER or FREQ, the number of observations in the new cluster
7. RMS STD OF NEW CLUSTER or RMSSTD, the root-mean-square standard deviation of the new cluster
8. NORMALIZED CENTROID DISTANCE or DIST, the distance between the means of the clusters just joined divided by the root-mean-square distance among all observations
9. NORMALIZED AVERAGE LINKAGE or AVLINK, the root-mean-square distance between pairs of observations, each pair containing one observation from each of the clusters just joined, divided by the root-mean-square distance among all observations
10. SEMIPARTIAL R-SQUARED or SPRSQ, the squared semi-partial correlation, which is the sum of squares between the clusters just joined, divided by the corrected total sum of squares, or the decrease in R^2 caused by joining the clusters

11. R-SQUARED or RSQ, the squared multiple correlation R^2, which is the sum of squares between all clusters divided by the corrected total sum of squares
12. APPROXIMATE EXPECTED R-SQUARED or ERSQ, the approximate expected value of R^2 under the uniform null hypothesis
13. the CUBIC CLUSTERING CRITERION or CCC.

EXAMPLE

The iris data published by Fisher (1936) have been widely used for examples in discriminant analysis and cluster analysis. The sepal length, sepal width, petal length, and petal width were measured in millimeters on 50 iris specimens from each of three species, *Iris setosa, I. versicolor,* and *I. virginica*. Mezzich and Solomon (1980) discuss a variety of cluster analyses of the iris data.

The CLUSTER procedure uses Ward's method to cluster the irises.The PLOT procedure plots the cubic clustering criterion for the last fifteen clusters. There is a local peak of the CCC at three clusters, the actual number of species, but there is a higher peak at five or six clusters, suggesting the possibility of subspecies.

```
DATA IRIS;
  INPUT SEPALLEN SEPALWID PETALLEN PETALWID SPEC_NO @@;
  IF SPEC_NO=1 THEN SPECIES='SETOSA';
  IF SPEC_NO=2 THEN SPECIES='VERSICOLOR';
  IF SPEC_NO=3 THEN SPECIES='VIRGINICA ';
CARDS;
50 33 14 02 1 64 28 56 22 3 65 28 46 15 2
67 31 56 24 3 63 28 51 15 3 46 34 14 03 1
69 31 51 23 3 62 22 45 15 2 59 32 48 18 2
46 36 10 02 1 61 30 46 14 2 60 27 51 16 2
65 30 52 20 3 56 25 39 11 2 65 30 55 18 3
58 27 51 19 3 68 32 59 23 3 51 33 17 05 1
57 28 45 13 2 62 34 54 23 3 77 38 67 22 3
63 33 47 16 2 67 33 57 25 3 76 30 66 21 3
49 25 45 17 3 55 35 13 02 1 67 30 52 23 3
70 32 47 14 2 64 32 45 15 2 61 28 40 13 2
48 31 16 02 1 59 30 51 18 3 55 24 38 11 2
63 25 50 19 3 64 32 53 23 3 52 34 14 02 1
49 36 14 01 1 54 30 45 15 2 79 38 64 20 3
44 32 13 02 1 67 33 57 21 3 50 35 16 06 1
58 26 40 12 2 44 30 13 02 1 77 28 67 20 3
63 27 49 18 3 47 32 16 02 1 55 26 44 12 2
50 23 33 10 2 72 32 60 18 3 48 30 14 03 1
51 38 16 02 1 61 30 49 18 3 48 34 19 02 1
50 30 16 02 1 50 32 12 02 1 61 26 56 14 3
64 28 56 21 3 43 30 11 01 1 58 40 12 02 1
51 38 19 04 1 67 31 44 14 2 62 28 48 18 3
49 30 14 02 1 51 35 14 02 1 56 30 45 15 2
58 27 41 10 2 50 34 16 04 1 46 32 14 02 1
60 29 45 15 2 57 26 35 10 2 57 44 15 04 1
50 36 14 02 1 77 30 61 23 3 63 34 56 24 3
58 27 51 19 3 57 29 42 13 2 72 30 58 16 3
54 34 15 04 1 52 41 15 01 1 71 30 59 21 3
64 31 55 18 3 60 30 48 18 3 63 29 56 18 3
```

```
49 24 33 10 2 56 27 42 13 2 57 30 42 12 2
55 42 14 02 1 49 31 15 02 1 77 26 69 23 3
60 22 50 15 3 54 39 17 04 1 66 29 46 13 2
52 27 39 14 2 60 34 45 16 2 50 34 15 02 1
44 29 14 02 1 50 20 35 10 2 55 24 37 10 2
58 27 39 12 2 47 32 13 02 1 46 31 15 02 1
69 32 57 23 3 62 29 43 13 2 74 28 61 19 3
59 30 42 15 2 51 34 15 02 1 50 35 13 03 1
56 28 49 20 3 60 22 40 10 2 73 29 63 18 3
67 25 58 18 3 49 31 15 01 1 67 31 47 15 2
63 23 44 13 2 54 37 15 02 1 56 30 41 13 2
63 25 49 15 2 61 28 47 12 2 64 29 43 13 2
51 25 30 11 2 57 28 41 13 2 65 30 58 22 3
69 31 54 21 3 54 39 13 04 1 51 35 14 03 1
72 36 61 25 3 65 32 51 20 3 61 29 47 14 2
56 29 36 13 2 69 31 49 15 2 64 27 53 19 3
68 30 55 21 3 55 25 40 13 2 48 34 16 02 1
48 30 14 01 1 45 23 13 03 1 57 25 50 20 3
57 38 17 03 1 51 38 15 03 1 55 23 40 13 2
66 30 44 14 2 68 28 48 14 2 54 34 17 02 1
51 37 15 04 1 52 35 15 02 1 58 28 51 24 3
67 30 50 17 2 63 33 60 25 3 53 37 15 02 1
;
PROC CLUSTER DATA=IRIS SIMPLE TREE=TREE;
  VAR SEPALLEN SEPALWID PETALLEN PETALWID;
PROC PRINT DATA=TREE(OBS=10);
DATA TREE;
  SET TREE;
  IF _NCL_<=15;
PROC PLOT DATA=TREE;
  PLOT _CCC_*_NCL_=_NCL_;
```

STATISTICAL ANALYSIS SYSTEM 1

WARD'S HIERARCHICAL CLUSTER ANALYSIS

(1) SIMPLE STATISTICS

	MEAN	STD DEV	SKEWNESS	KURTOSIS	BIMODALITY
SEPALLEN	58.43333	8.28066	0.31491	-0.55206	0.43804
SEPALWID	30.57333	4.35866	0.31897	0.22825	0.33491
PETALLEN	37.58000	17.65298	-0.27488	-1.40210	0.64822
PETALWID	11.99333	7.62238	-0.10297	-1.34060	0.58730

(2) EIGENVALUES OF THE COVARIANCE MATRIX

EIGENVALUE	DIFFERENCE	PROPORTION	CUMULATIVE
420.0053	395.9000	0.9246	0.9246
24.1053	16.3365	0.0531	0.9777
7.7688	5.4012	0.0171	0.9948
2.3676	.	0.0052	1.0000

(3) ROOT-MEAN-SQUARE TOTAL-SAMPLE STANDARD DEVIATION = 10.6922

(4) ROOT-MEAN-SQUARE DISTANCE BETWEEN OBSERVATIONS = 21.3845

(5) NUMBER OF CLUSTERS	(6) FREQUENCY OF NEW CLUSTER	(7) RMS STD OF NEW CLUSTER	(8) NORMALIZED CENTROID DISTANCE	(9) NORMALIZED AVERAGE LINKAGE	(10) SEMIPARTIAL R-SQUARED	(11) R-SQUARED	(12) APPROXIMATE EXPECTED R-SQUARED	(13) CUBIC CLUSTERING CRITERION
15	15	2.15362	0.2559	0.3300	0.001641	0.970932	0.957871	5.8544
14	7	3.24771	0.4035	0.4845	0.001873	0.969059	0.955418	5.7798
13	15	2.6054	0.3186	0.3966	0.002271	0.966788	0.952670	5.6247
12	24	2.35205	0.2454	0.3501	0.002274	0.964514	0.949541	4.5819
11	12	3.25262	0.3574	0.4745	0.002500	0.962014	0.945886	4.6274
10	22	2.36245	0.2713	0.3612	0.002694	0.959320	0.941547	4.7662
9	29	2.07762	0.2392	0.3251	0.002702	0.956618	0.936296	5.0858
8	23	2.45754	0.2973	0.3780	0.003095	0.953523	0.929791	5.5064
7	26	2.99037	0.5058	0.5644	0.005811	0.947713	0.921496	5.4832
6	38	2.99017	0.3149	0.4529	0.006042	0.941671	0.910514	5.8580
5	50	2.78031	0.3627	0.4462	0.010753	0.930917	0.895232	5.8170
4	36	3.91834	0.5667	0.6726	0.017245	0.913673	0.872331	3.9867
3	64	4.11402	0.5386	0.6644	0.030051	0.883621	0.826664	4.3292
2	100	5.94155	0.8474	0.9971	0.111026	0.772595	0.696871	3.8329
1	150	10.6922	1.8584	1.9559	0.772595	0.000000	0.000000	0.0000

STATISTICAL ANALYSIS SYSTEM 2

OBS	_NAME_	_PARENT_	_NCL_	_FREQ_	_RMSSTD_	_DIST_	_AVLINK_	_SPRSQ_	_RSQ_	_ERSQ_	_RATIO_	_LOGR_	_CCC_	SEPALLEN	SEPALWID	PETALLEN	PETALWID
1	OB16	CL149	150	1	0	0	0	0	1	.	.	.	.	58	27	51	19
2	OB76	CL149	150	1	0	0	0	0	1	.	.	.	.	58	27	51	19
3	OB116	CL148	150	1	0	0	0	0	1	.	.	.	.	54	37	15	2
4	OB150	CL148	150	1	0	0	0	0	1	.	.	.	.	53	37	15	2
5	OB65	CL147	150	1	0	0	0	0	1	.	.	.	.	51	35	14	2
6	OB126	CL147	150	1	0	0	0	0	1	.	.	.	.	51	35	14	3
7	OB89	CL146	150	1	0	0	0	0	1	.	.	.	.	49	31	15	2
8	OB113	CL146	150	1	0	0	0	0	1	.	.	.	.	49	31	15	1
9	OB96	CL145	150	1	0	0	0	0	1	.	.	.	.	50	34	15	2
10	OB107	CL145	150	1	0	0	0	0	1	.	.	.	.	51	34	15	2

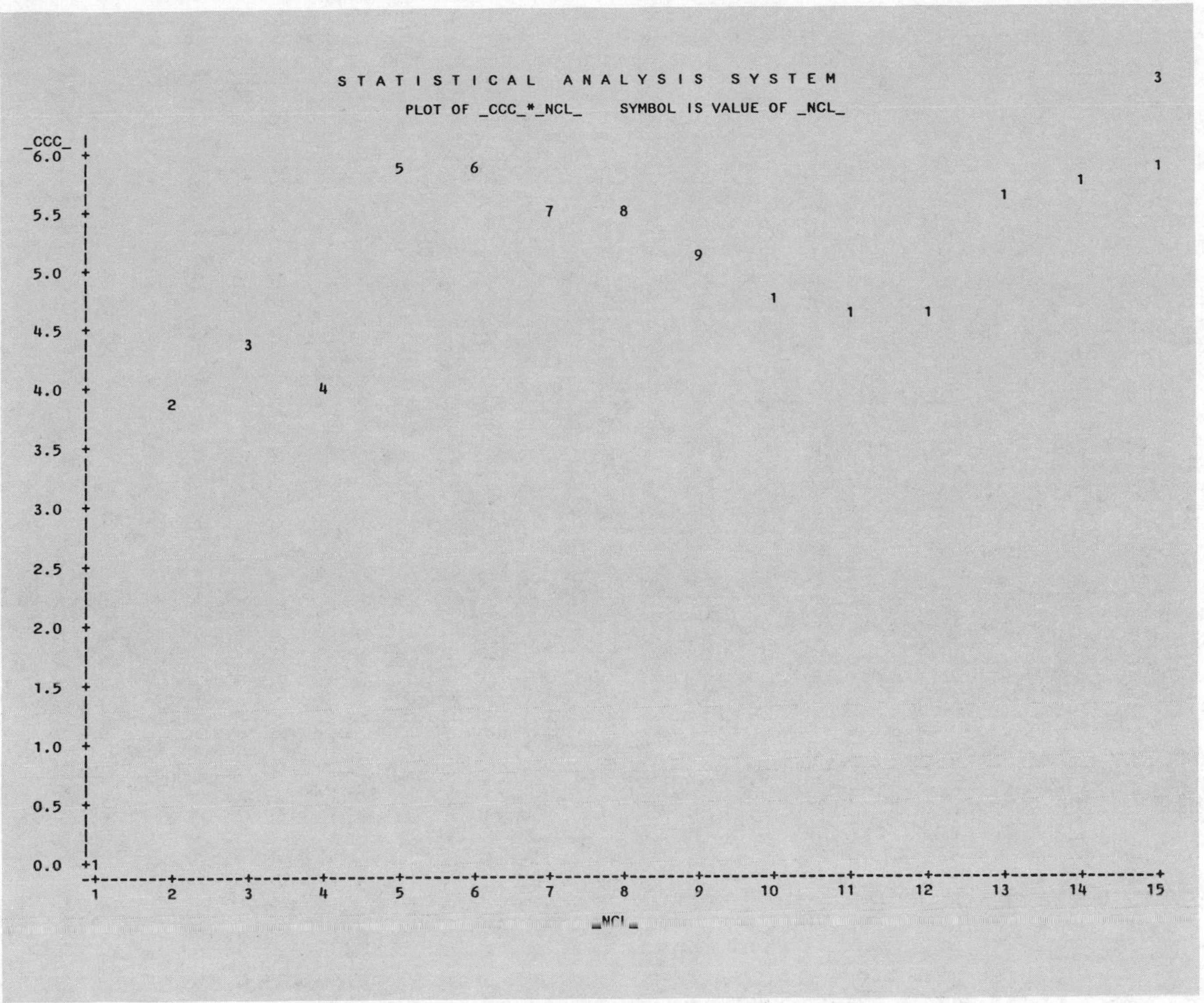

REFERENCES

Anderberg, M.R. (1973), *Cluster Analysis for Applications*, New York: Academic Press.

Everitt, B.S. (1980), *Cluster Analysis, 2nd ed.*, London: Heineman Educational Books Ltd.

Fisher, R.A. (1936), "The Use of Multiple Measurements in Taxonomic Problems," *Annals of Eugenics*, 7, 179–188.

Hartigan, J.A. (1975), *Clustering Algorithms*, New York: John Wiley & Sons Inc.

Mezzich, J.E and Solomon, H. (1980), *Taxonomy and Behavioral Science*, New York: Academic Press.

Milligan, G.W. (1980), "An Examination of the Effect of Six Types of Error Perturbation on Fifteen Clustering Algorithms," *Psychometrika*, 45, 325–342.

Spath, H. (1980), *Cluster Analysis Algorithms*, Chichester, England: Ellis Horwood.

REFERENCES

The FASTCLUS Procedure

ABSTRACT

FASTCLUS is designed for disjoint clustering of very large data sets and can find good clusters with only two or three passes over the data. You specify the maximum number of clusters and, optionally, the minimum radius of the clusters. An output data set containing a cluster membership variable may be produced, as well as an output data set containing cluster means.

INTRODUCTION

FASTCLUS performs a disjoint cluster analysis on the basis of Euclidean distances computed from one or more quantitative variables. The observations are divided into clusters such that every observation belongs to one and only one cluster (the clusters do not form a tree structure as they do in the CLUSTER procedure). If you want separate analyses for different numbers of clusters, you must run FASTCLUS once for each analysis.

FASTCLUS is intended for use with large data sets, from approximately 100 to 100,000 observations. With small data sets, the results may be sensitive to the order of the observations in the data set.

FASTCLUS prints brief summaries of the clusters it finds. For more extensive examination of the clusters, you can request an output data set containing a cluster membership variable.

Background

FASTCLUS combines an effective method for finding initial clusters with a standard iterative algorithm for minimizing the sum of squared distances from the cluster means. The result is an efficient procedure for disjoint clustering of large data sets. FASTCLUS was directly inspired by Hartigan's *leader* algorithm (1975) and MacQueen's *k-means* algorithm (1967).

FASTCLUS uses a method that Anderberg (1973) calls *nearest centroid sorting*. A set of points called *cluster seeds* is selected as a first guess of the means of the clusters. Each observation is assigned to the nearest seed to form temporary clusters. The seeds are then replaced by the means of the temporary clusters and the process is repeated until no further changes occur in the clusters. Similar techniques are described in most references on clustering (Anderberg, 1973; Hartigan, 1975; Everitt, 1980; Spath, 1980).

FASTCLUS differs from other nearest centroid sorting methods in how the initial cluster seeds are selected. The initialization method of FASTCLUS guarantees that if there are clusters such that all distances between observations in the same cluster are less than all distances between observations in different clusters, and if you tell FASTCLUS the correct number of clusters to find, then it always finds such a clustering without iterating. Even with clusters that are not as well separated, FASTCLUS usually finds sufficiently good initial seeds that few iterations are required. The importance of initial seed selection is demonstrated by Milligan (1980).

The initialization method used by FASTCLUS makes it sensitive to outliers. FASTCLUS can be an effective procedure for detecting outliers, since outliers often appear as clusters with only one member.

The clustering is done on the basis of Euclidean distances computed from one or more numeric variables. If there are missing values, FASTCLUS computes an adjusted distance using the nonmissing values. Observations that are very close to each other are usually assigned to the same cluster, while observations that are far apart are in different clusters.

FASTCLUS operates in four steps:

- observations called cluster seeds are selected.
- optionally, temporary clusters are formed by assigning each observation to the cluster with the nearest seed. Each time an observation is assigned, the cluster seed is updated as the current mean of the cluster.
- optionally, clusters are formed by assigning each observation to the nearest seed. After all observations are assigned, the cluster seeds are replaced by the cluster means. This step can be repeated until the changes in the cluster seeds become small or zero.
- final clusters are formed by assigning each observation to the nearest seed.

The initial cluster seeds must be observations with no missing values. You can specify the maximum number of seeds (and hence clusters) using the MAXCLUSTERS= option. You can also specify a minimum distance by which the seeds must be separated using the RADIUS= option.

FASTCLUS always selects the first complete (no missing values) observation as the first seed. The next complete observation that is separated from the first seed by at least the RADIUS becomes the second seed. Later observations are selected as new seeds if they are separated from all previous seeds by at least the RADIUS, as long as the maximum number of seeds is not exceeded.

If an observation is complete but fails to qualify as a new seed, FASTCLUS considers using it to replace one of the old seeds. Two tests are made to see if the observation can qualify as a new seed.

First, an old seed is replaced if the distance between the two closest seeds is less than the distance from the observation to the nearest seed. The seed that is replaced is selected from the two seeds that are closest to each other, and is the one of these two that is also closest to the observation.

If the observation fails the first test for seed replacement, a second test is made. The observation replaces the nearest seed if the smallest distance from the observation to all seeds other than the nearest one is greater than the shortest distance from the nearest seed to all other seeds. If this test is failed, FASTCLUS goes on to the next observation.

The REPLACE= option can be used to limit seed replacement. The second test for seed replacement can be omitted, causing FASTCLUS to run faster, but the seeds selected may not be as widely separated as those obtained by the default method. Seed replacement can also be suppressed entirely. In this case FASTCLUS runs much faster, but you must choose a good value for RADIUS in order to get good clusters. This method is similar to Hartigan's leader algorithm (1975, pp.74-78) and the *simple cluster-seeking* algorithm described by Tou and Gonzalez (1974, pp.90-92).

SPECIFICATIONS

The FASTCLUS procedure uses the following statements:

PROC FASTCLUS *options;*
VAR *variables;*
ID *variable;*
FREQ *variable;*
WEIGHT *variable;*
BY *variables;*

Usually only the VAR statement is used in addition to the PROC FASTCLUS statement.

PROC FASTCLUS Statement

PROC FASTCLUS *options;*

The following data set options may appear on the PROC statement:

DATA=*SASdataset* names the input data set containing observations to be clustered. If DATA= is omitted, the most recently created SAS data set is used.

SEED=*SASdataset* names an input data set containing initial cluster seeds. If SEED= is specified, seeds from the DATA= data set are not used. All options given below for initial seed selection can be used.

OUT=*SASdataset* names an output data set to contain all the original data, plus the new variables CLUSTER and DISTANCE. If you want to create a permanent SAS data set, you must specify a two-level name (see Chapter 12, "SAS Data Sets," in *SAS User's Guide: Basics, 1982 Edition*, for more information on permanent data sets).

MEAN=*SASdataset* names an output data set to contain the cluster means and the number of observations in each cluster. (See Chapter 12, "SAS Data Sets," in *SAS User's Guide: Basics, 1982 Edition*, for more information on permanent data sets).

The following options for the FASTCLUS statement control initial cluster seed selection:

MAXCLUSTERS=*n*
MAXC=*n* specifies the maximum number of clusters allowed. The MAXCLUSTERS= option must be specified. The value may not exceed 250.

RADIUS=*n* establishes minimum distance criterion for selecting new seeds. No observation is considered as a new seed (not a replacement seed) unless its minimum distance to previous seeds exceeds the value given by RADIUS=. The default value is 0.

REPLACE= specifies how seed replacement is performed. REPLACE=FULL requests default seed replacement as described above. REPLACE=PART requests seed replacement only when the minimum distance between the observation and any current seed is greater than the minimum distance between current seeds. REPLACE= NONE suppresses seed replacement.

The following options for the FASTCLUS statement control computation of final cluster seeds:

DRIFT causes cluster seeds to be updated by computing the mean of current cluster members. After initial seed selection, each observation is assigned to the cluster with the nearest seed. After an observation is processed, that cluster's seed is recalculated as the mean of observations currently assigned to the cluster.

MAXITER=*n* specifies a maximum number of iterations for recomputing cluster seeds. Cluster seeds are iteratively recomputed, with the specified number giving the maximum iterations allowed. In each iteration, each observation is assigned to the nearest seed, and the seeds are recomputed as the means of the clusters. The default value is 1.

CONVERGE=*n*
CONV=*n* iterations terminate when the maximum distance by which any seed has changed is less than or equal to the minimum distance between initial seeds times the CONVERGE value. The default is 0.02. CONVERGE= is only useful if you have specified a MAXITER value greater than 1.

Miscellaneous options for the FASTCLUS statement:

SHORT suppresses printing of the cluster means and standard deviations and distances between cluster means.

LIST lists all observations, giving the value of the ID variable (if any), the number of the cluster to which the observation is assigned, and the distance between the observation and the final cluster seed.

NOMISS causes observations with missing values to be excluded from the analysis.

VAR Statement

VAR *variables;*

The VAR statement lists the numeric variables to be used in the cluster analysis. If the VAR statement is omitted, all numeric variables not listed in other statements are used.

ID Statement

ID *variable;*

The ID variable is used to identify observations on the printout only if the LIST option is specified.

FREQ Statement

FREQ *variable*;

The FREQ variable gives the frequency of occurrence of each observation. The values of the FREQ variable should be non-negative integers. Each observation is treated as if it actually occurred as many times as indicated by the FREQ variable.

WEIGHT Statement

WEIGHT *variable*;

The values of the WEIGHT variable are used to compute weighted cluster means. The effect of WEIGHT is similar to that of FREQ except in determining the number of observations. The WEIGHT variable may take nonintegral values.

If the WEIGHT statement is specified, the divisor used to compute variances is the sum of the weights, rather than the number of observations minus one.

BY Statement

BY *variables*;

A BY statement may be used with PROC FASTCLUS to obtain separate analyses on observations in groups defined by the BY variables. When a BY statement appears, the procedure expects the input data set to be sorted in order of the BY variables. If your input data set is not sorted in ascending order, use the SORT procedure with a similar BY statement to sort the data, or, if appropriate, use the BY statement options NOTSORTED or DESCENDING. For more information, see the discussion of the BY statement in Chapter 8, "Statements Used in the PROC Step," in *SAS User's Guide: Basics, 1982 Edition*.

DETAILS

Missing Values

Observations with all missing values are excluded from the analysis. If you specify NOMISS, observations with any missing values are excluded. Observations with missing values cannot be cluster seeds.

The distance between an observation with missing values and a cluster seed is obtained by computing the squared distance based on the nonmissing values, multiplying by the ratio of the number of variables to the number of nonmissing values, and taking the square root:

$$(n/m\,\Sigma(x_i - s_i)^2)^{1/2}$$

where:

n = number of variables
m = number of nonmissing values
x_i = value of the ith variable for the observation
s_i = value of the ith variable for the cluster seed

and the summation is taken over variables with nonmissing values.

Output Data Sets

OUT= Data Set The OUT= data set contains:

- the original variables
- a new variable CLUSTER, taking values from 1 to MAXCLUSTERS, indicating the cluster to which each observation has been assigned
- a new variable DISTANCE giving the distance from the observation to its cluster seed.

MEAN= Data Set The MEAN= data set contains one observation for each cluster. The variables are:

- the BY variables, if any
- either the FREQ variable or a new variable called _FREQ_ giving the number of observations in the cluster
- a new variable, _RMSSTD_, giving the root-mean-square standard deviation for the cluster
- the VAR variables giving the cluster means.

Computational Resources

Let:

n = number of observations
v = number of variables
c = number of clusters
r = number of first level seed replacements
p = number of passes over the data set.

The overall time required by FASTCLUS is usually roughly proportional to *nvcp*.

Initial seed selection requires one pass over the data set with time roughly proportional to:

$$nvc + (r + c)vc^2/2$$

unless REPLACE=NONE is specified. In that case, a complete pass may not be necessary and the time is roughly proportional to *mvc*, where $c <= m <= n$.

The DRIFT option, each iteration, and the final assignment of cluster seeds each require one pass, with time for each pass roughly proportional to *nvc*.

The variables in the VAR statement should be listed in order of decreasing variance for greatest efficiency.

Usage Notes

Before using FASTCLUS, decide whether your variables should be standardized in some way. If all variables are measured in the same units, standardization may not be necessary. It is generally recommended that you use either STANDARD to produce standard scores or FACTOR to produce factor scores for submission to FASTCLUS.

The easiest way to use FASTCLUS is to specify MAXCLUSTERS= and the LIST option. It is usually desirable to try several values of MAXCLUSTERS=. The RADIUS= option is not used often.

FASTCLUS produces relatively little printed output. In most cases you should create an output data set and use other procedures such as PRINT, PLOT, CHART, MEANS, DISCRIM, or CANDISC to study the clusters. Macros are useful for running

FASTCLUS repeatedly with other procedures.

A simple application of FASTCLUS with two variables might proceed as follows:

```
PROC STANDARD MEAN=0 STD=1 OUT=STAN;
  VAR V1 V2;

PROC FASTCLUS DATA=STAN OUT=CLUST MAXCLUSTERS=2;
  VAR V1 V2;
PROC PLOT;
  PLOT V2*V1 = CLUSTER;

PROC FASTCLUS DATA=STAN OUT=CLUST MAXCLUSTERS=3;
  VAR V1 V2;
PROC PLOT;
  PLOT V2*V1 = CLUSTER;
```

If you have more than two variables, you can use CANDISC to compute canonical variables for plotting the clusters. If the data set is not too large, it may also be helpful to use

```
PROC SORT;
  BY CLUSTER DISTANCE;
PROC PRINT;
  BY CLUSTER;
```

to list the clusters. By examining the values of DISTANCE you can determine if any observations are unusually far from their clusters.

FASTCLUS is very sensitive to outliers and can be a useful procedure for outlier detection. If you have not screened your data for outliers before using FASTCLUS, try running FASTCLUS with twice as many clusters as you expect to find, or perhaps five to 20 clusters if you have no idea how many clusters to expect. Outliers often appear as clusters with only one member. You can then remove the outliers and run FASTCLUS again, specifying a smaller number of clusters.

Printed Output

If requested, FASTCLUS prints:

1. INITIAL SEEDS, initial cluster seeds
2. CHANGE IN CLUSTER SEEDS for each iteration, if MAXITER is specified.

FASTCLUS prints a CLUSTER SUMMARY, giving for each cluster:

3. MEMBERS, the number of members
4. RMS ST DEV, the root-mean-square across variables of the cluster standard deviations, which is equal to the root-mean-square distance between observations in the cluster
5. MAX DISTANCE FROM SEED, the maximum distance from the cluster seed to an observation in the cluster.

There is also a table of statistics for each variable giving:

6. TOTAL STD, the total standard deviation
7. WITHIN STD, the pooled within-cluster standard deviation
8. R-SQUARED, the R^2 for predicting the variable from the cluster
9. VAR RATIO, the ratio of between-cluster variance to within-cluster variance ($R^2/(1-R^2)$)

10. OVER-ALL, all of the above quantities, pooled across variables
11. the APPROXIMATE EXPECTED OVER-ALL R-SQUARED, the approximate expected value of the overall R^2 under the uniform null hypothesis assuming that the variables are uncorrelated. If you are interested in the expected R^2 but your variables are highly correlated, you should cluster principal component scores from PRINCOMP.
12. the CUBIC CLUSTERING CRITERION computed under the assumption that the variables are uncorrelated. If you are interested in the cubic clustering criterion but your variables are highly correlated, you should cluster principal component scores from PRINCOMP.
13. optionally, CLUSTER MEANS for each variable
14. optionally, CLUSTER STANDARD DEVIATIONS for each variable
15. optionally, DISTANCES BETWEEN CLUSTER MEANS.

EXAMPLE

The iris data published by Fisher (1936) have been widely used for examples in discriminant analysis and cluster analysis. The sepal length, sepal width, petal length, and petal width were measured in millimeters on 50 iris specimens from each of three species, *Iris setosa, I. versicolor,* and *I. virginica*. Mezzich and Solomon (1980) discuss a variety of cluster analyses of the iris data.

In this example FASTCLUS is used to find two and three clusters. An output data set is created and FREQ is invoked to compare the clusters with the species classification. For three clusters, CANDISC is used to compute canonical variables for plotting the clusters.

```
DATA IRIS;
  TITLE FISHER (1936) IRIS DATA;
  INPUT SEPALLEN SEPALWID PETALLEN PETALWID SPEC_NO @@;
  IF SPEC_NO=1 THEN SPECIES='SETOSA     ';
  ELSE IF SPEC_NO=2 THEN SPECIES='VERSICOLOR';
  ELSE SPECIES='VIRGINICA ';
  LABEL SEPALLEN=SEPAL LENGTH IN MM.
        SEPALWID=SEPAL WIDTH IN MM.
        PETALLEN=PETAL LENGTH IN MM.
        PETALWID=PETAL WIDTH IN MM.;
  CARDS;
50 33 14 02 1 64 28 56 22 3 65 28 46 15 2
67 31 56 24 3 63 28 51 15 3 46 34 14 03 1
69 31 51 23 3 62 22 45 15 2 59 32 48 18 2
46 36 10 02 1 61 30 46 14 2 60 27 51 16 2
65 30 52 20 3 56 25 39 11 2 65 30 55 18 3
58 27 51 19 3 68 32 59 23 3 51 33 17 05 1
57 28 45 13 2 62 34 54 23 3 77 38 67 22 3
63 33 47 16 2 67 33 57 25 3 76 30 66 21 3
49 25 45 17 3 55 35 13 02 1 67 30 52 23 3
70 32 47 14 2 64 32 45 15 2 61 28 40 13 2
48 31 16 02 1 59 30 51 18 3 55 24 38 11 2
63 25 50 19 3 64 32 53 23 3 52 34 14 02 1
49 36 14 01 1 54 30 45 15 2 79 38 64 20 3
44 32 13 02 1 67 33 57 21 3 50 35 16 06 1
58 26 40 12 2 44 30 13 02 1 77 28 67 20 3
63 27 49 18 3 47 32 16 02 1 55 26 44 12 2
```

```
50 23 33 10 2 72 32 60 18 3 48 30 14 03 1
51 38 16 02 1 61 30 49 18 3 48 34 19 02 1
50 30 16 02 1 50 32 12 02 1 61 26 56 14 3
64 28 56 21 3 43 30 11 01 1 58 40 12 02 1
51 38 19 04 1 67 31 44 14 2 62 28 48 18 3
49 30 14 02 1 51 35 14 02 1 56 30 45 15 2
58 27 41 10 2 50 34 16 04 1 46 32 14 02 1
60 29 45 15 2 57 26 35 10 2 57 44 15 04 1
50 36 14 02 1 77 30 61 23 3 63 34 56 24 3
58 27 51 19 3 57 29 42 13 2 72 30 58 16 3
54 34 15 04 1 52 41 15 01 1 71 30 59 21 3
64 31 55 18 3 60 30 48 18 3 63 29 56 18 3
49 24 33 10 2 56 27 42 13 2 57 30 42 12 2
55 42 14 02 1 49 31 15 02 1 77 26 69 23 3
60 22 50 15 3 54 39 17 04 1 66 29 46 13 2
52 27 39 14 2 60 34 45 16 2 50 34 15 02 1
44 29 14 02 1 50 20 35 10 2 55 24 37 10 2
58 27 39 12 2 47 32 13 02 1 46 31 15 02 1
69 32 57 23 3 62 29 43 13 2 74 28 61 19 3
59 30 42 15 2 51 34 15 02 1 50 35 13 03 1
56 28 49 20 3 60 22 40 10 2 73 29 63 18 3
67 25 58 18 3 49 31 15 01 1 67 31 47 15 2
63 23 44 13 2 54 37 15 02 1 56 30 41 13 2
63 25 49 15 2 61 28 47 12 2 64 29 43 13 2
51 25 30 11 2 57 28 41 13 2 65 30 58 22 3
69 31 54 21 3 54 39 13 04 1 51 35 14 03 1
72 36 61 25 3 65 32 51 20 3 61 29 47 14 2
56 29 36 13 2 69 31 49 15 2 64 27 53 19 3
68 30 55 21 3 55 25 40 13 2 48 34 16 02 1
48 30 14 01 1 45 23 13 03 1 57 25 50 20 3
57 38 17 03 1 51 38 15 03 1 55 23 40 13 2
66 30 44 14 2 68 28 48 14 2 54 34 17 02 1
51 37 15 04 1 52 35 15 02 1 58 28 51 24 3
67 30 50 17 2 63 33 60 25 3 53 37 15 02 1
;
PROC FASTCLUS DATA=IRIS MAXC=2 MAXITER=10 OUT=CLUS;
  VAR SEPALLEN SEPALWID PETALLEN PETALWID;
PROC FREQ;
  TABLES CLUSTER*SPECIES;
PROC FASTCLUS DATA=IRIS MAXC=3 MAXITER=10 OUT=CLUS;
  VAR SEPALLEN SEPALWID PETALLEN PETALWID;
PROC FREQ;
  TABLES CLUSTER*SPECIES;
PROC CANDISC UNI OUT=CAN;
  CLASS CLUSTER;
  VAR SEPALLEN SEPALWID PETALLEN PETALWID;
  TITLE2 CANONICAL DISCRIMINANT ANALYSIS OF IRIS CLUSTERS;
PROC PLOT;
  PLOT CAN2*CAN1=CLUSTER;
  TITLE2 PLOT OF CANONICAL VARIABLES IDENTIFIED BY CLUSTER;
```

```
                          FISHER (1936) IRIS DATA                                  1
                            FASTCLUS PROCEDURE

                  REPLACE = FULL  MAXCLUSTERS = 2  MAXITER = 10

             SEED REPLACEMENTS:  FIRST LEVEL = 0  SECOND LEVEL = 7

                           (1) INITIAL SEEDS

      CLUSTER       SEPALLEN       SEPALWID       PETALLEN       PETALWID

         1         43.00000       30.00000       11.00000        1.00000
         2         77.00000       26.00000       69.00000       23.00000

                 (2) ITERATION    CHANGE IN CLUSTER SEEDS
                          1       13.493       22.4136
                          2       4.22265      1.86994
                          3       1.23521      0.542766
                          4       0.45894      0.235033

                 (3) (4) CLUSTER SUMMARY   (5)

             CLUSTER   MEMBERS  RMS ST DEV  MAX DISTANCE FROM SEED

                1         53     3.704991     21.61969
                2         97     5.677897     24.84476

                 (6)        STATISTICS FOR VARIABLES     (8)          (9)

         VARIABLE       TOTAL STD  (7) WITHIN STD       R-SQUARED        VAR RATIO

         SEPALLEN       8.2806613       5.4931283       0.5628959        1.2877845
         SEPALWID       4.3586628       3.7039310       0.2827102        0.3941366
         PETALLEN      17.6529823       6.8033102       0.8524702        5.7782905
         PETALWID       7.6223767       3.5720039       0.7818685        3.5843896
    (10) OVER-ALL      10.6922367       5.0729135       0.7764096        3.4724631

(11) APPROXIMATE EXPECTED OVER-ALL R-SQUARED =  0.1341 (12) CUBIC CLUSTERING CRITERION = 129.555

                           (13) CLUSTER MEANS

      CLUSTER       SEPALLEN       SEPALWID       PETALLEN       PETALWID

         1         50.05660       33.69811       15.60377        2.90566
         2         63.01031       28.86598       49.58763       16.95876

                     (14) CLUSTER STANDARD DEVIATIONS

      CLUSTER       SEPALLEN       SEPALWID       PETALLEN       PETALWID

         1         3.427351       4.396611       4.404279       2.105525
         2         6.336887       3.267991       7.800578       4.155612
```

```
                   (15) FISHER (1936) IRIS DATA                                    2
                          FASTCLUS PROCEDURE

                    DISTANCES BETWEEN CLUSTER MEANS

                    CLUSTER            1              2

                       1               .          39.28791
                       2           39.28791          .
```

FISHER (1936) IRIS DATA 3

TABLE OF CLUSTER BY SPECIES

CLUSTER / FREQUENCY / PERCENT / ROW PCT / COL PCT	SPECIES: SETOSA	VERSICOLOR	VIRGINICA	TOTAL
1	50	3	0	53
	33.33	2.00	0.00	35.33
	94.34	5.66	0.00	
	100.00	6.00	0.00	
2	0	47	50	97
	0.00	31.33	33.33	64.67
	0.00	48.45	51.55	
	0.00	94.00	100.00	
TOTAL	50	50	50	150
	33.33	33.33	33.33	100.00

FISHER (1936) IRIS DATA 4

FASTCLUS PROCEDURE

REPLACE = FULL MAXCLUSTERS = 3 MAXITER = 10

SEED REPLACEMENTS: FIRST LEVEL = 1 SECOND LEVEL = 8

INITIAL SEEDS

CLUSTER	SEPALLEN	SEPALWID	PETALLEN	PETALWID
1	58.00000	40.00000	12.00000	2.00000
2	77.00000	38.00000	67.00000	22.00000
3	49.00000	25.00000	45.00000	17.00000

ITERATION	CHANGE IN CLUSTER SEEDS		
1	10.1409	12.2566	11.4148
2	0	1.75348	1.21228
3	0	0.69766	0.473303
4	0	0.497031	0.328075

CLUSTER SUMMARY

CLUSTER	MEMBERS	RMS ST DEV	MAX DISTANCE FROM SEED
1	50	2.780306	12.4803
2	38	4.016812	15.2971
3	62	4.03981	16.6064

STATISTICS FOR VARIABLES

VARIABLE	TOTAL STD	WITHIN STD	R-SQUARED	VAR RATIO
SEPALLEN	8.2806613	4.3948825	0.7220955	2.5983585
SEPALWID	4.3586628	3.2481625	0.4521016	0.8251560
PETALLEN	17.6529823	4.2143140	0.9437725	16.7848955
PETALWID	7.6223767	2.4524358	0.8978718	8.7916177
OVER-ALL	10.6922367	3.6619816	0.8842753	7.6411940

APPROXIMATE EXPECTED OVER-ALL R-SQUARED = 0.1754 CUBIC CLUSTERING CRITERION = 192.895

CLUSTER MEANS

CLUSTER	SEPALLEN	SEPALWID	PETALLEN	PETALWID
1	50.06000	34.28000	14.62000	2.46000
2	68.50000	30.73684	57.42105	20.71053
3	59.01613	27.48387	43.93548	14.33871

FISHER (1936) IRIS DATA 5

FASTCLUS PROCEDURE

CLUSTER STANDARD DEVIATIONS

CLUSTER	SEPALLEN	SEPALWID	PETALLEN	PETALWID
1	3.524897	3.790644	1.736640	1.053856
2	4.941550	2.900924	4.885896	2.798725
3	4.664101	2.962841	5.088950	2.974997

DISTANCES BETWEEN CLUSTER MEANS

CLUSTER	1	2	3
1	.	50.17569	33.56935
2	50.17569	.	17.97182
3	33.56935	17.97182	.

FISHER (1936) IRIS DATA 6

TABLE OF CLUSTER BY SPECIES

CLUSTER SPECIES

FREQUENCY PERCENT ROW PCT COL PCT	SETOSA	VERSICOLOR	VIRGINICA	TOTAL
1	50 33.33 100.00 100.00	0 0.00 0.00 0.00	0 0.00 0.00 0.00	50 33.33
2	0 0.00 0.00 0.00	2 1.33 5.26 4.00	36 24.00 94.74 72.00	38 25.33
3	0 0.00 0.00 0.00	48 32.00 77.42 96.00	14 9.33 22.58 28.00	62 41.33
TOTAL	50 33.33	50 33.33	5C 33.33	150 100.00

FISHER (1936) IRIS DATA 7
CANONICAL DISCRIMINANT ANALYSIS OF IRIS CLUSTERS

CANONICAL DISCRIMINANT ANALYSIS

150 OBSERVATIONS 149 DF TOTAL
4 VARIABLES 147 DF WITHIN CLASSES
3 CLASSES 2 DF BETWEEN CLASSES

CLUSTER	FREQUENCY	WEIGHT	PROPORTION
1	50	50	0.333333
2	38	38	0.253333
3	62	62	0.413333

(continued on next page)

(continued from previous page)

UNIVARIATE STATISTICS

	MEAN	TOTAL STD	WITHIN STD	BETWEEN STD	R SQUARED	VAR RATIO	F	PROB GT F
SEPALLEN	58.433333333	8.280661280	4.3948825149	7.036591544	.72209550862	2.598358541	190.9793527	1.33759E-41
SEPALWID	30.573333333	4.358662849	3.2481625282	2.930699622	.45210161986	0.825155971	60.6489639	6.22947E-20
PETALLEN	37.580000000	17.652982333	4.2143140381	17.149511294	.94377251183	16.784895478	1233.6898176	0
PETALWID	11.993333333	7.622376690	2.4524358102	7.222666803	.89787183007	8.791617735	646.1839036	0

AVERAGE R-SQUARED: UNWEIGHTED = 0.7539604 WEIGHTED BY VARIANCE = 0.8842753

CLASS MEANS

CLUSTER	SEPALLEN	SEPALWID	PETALLEN	PETALWID
1	50.060000000	34.280000000	14.620000000	2.460000000
2	68.500000000	30.736842105	57.421052632	20.710526316
3	59.016129032	27.483870968	43.935483871	14.338709677

CANONICAL CORRELATIONS AND TESTS OF HO: THE CANONICAL CORRELATION IN THE CURRENT ROW AND ALL THAT FOLLOW ARE ZERO

	CANONICAL CORRELATION	ADJUSTED CAN CORR	APPROX STD ERROR	VARIANCE RATIO	CANONICAL R-SQUARED	LIKELIHOOD RATIO	F STATISTIC	NUM DF	DEN DF	PROB>F
1	0.976613341	0.975487306	0.003787013	20.6327	0.953773617	0.032223371	164.5474	8	288	0.0000
2	0.550383964	0.526353703	0.057106813	0.4346	0.302922508	0.697077492	21.0038	3	145	0.0000

MULTIVARIATE TEST STATISTICS AND F APPROXIMATIONS

STATISTIC	VALUE	F	NUM DF	DEN DF	PROB>F
WILKS' LAMBDA	0.03222337	164.5474	8	288	0
PILLAI'S TRACE	1.256696	61.2875	8	290	6.21510E-58
HOTELLING-LAWLEY TRACE	21.06723	376.5767	8	286	0
ROY'S GREATEST ROOT	20.63267	747.9342	4	145	0

NOTE: F STATISTIC FOR WILKS' LAMBDA IS EXACT
NOTE: F STATISTIC FOR ROY'S GREATEST ROOT IS AN UPPER BOUND

FISHER (1936) IRIS DATA 8
CANONICAL DISCRIMINANT ANALYSIS OF IRIS CLUSTERS

CANONICAL DISCRIMINANT ANALYSIS

TOTAL CANONICAL STRUCTURE

	CAN1	CAN2	CAN3	CAN4
SEPALLEN	0.8320	0.4521	-0.1349	-0.2919
SEPALWID	-0.5151	0.8106	-0.1258	0.2485
PETALLEN	0.9935	0.0875	-0.0712	0.0137
PETALWID	0.9663	0.1547	0.1724	0.1121

WITHIN CANONICAL STRUCTURE

	CAN1	CAN2	CAN3	CAN4
SEPALLEN	0.3393	0.7161	-0.2560	-0.5537
SEPALWID	-0.1496	0.9144	-0.1699	0.3357
PETALLEN	0.9008	0.3081	-0.3003	0.0578
PETALWID	0.6501	0.4043	0.5393	0.3507

STANDARDIZED CANONICAL COEFFICIENTS

	CAN1	CAN2	CAN3	CAN4
SEPALLEN	0.0477	1.0215	0.7314	-2.3907
SEPALWID	-0.5776	0.8645	-0.6415	1.0606
PETALLEN	3.3413	-1.2830	-4.7447	2.5818
PETALWID	0.9965	0.9005	3.9067	-0.0309

RAW CANONICAL COEFFICIENTS

	CAN1	CAN2	CAN3	CAN4
SEPALLEN	0.0057661265	0.1233581748	0.0883204155	-.2887089130
SEPALWID	-.1325106494	0.1983303556	-.1471676272	0.2433317158
PETALLEN	0.1892773419	-.0726814163	-.2687774850	0.1462547079
PETALWID	0.1307270927	0.1181359305	0.5125272838	-.0040492204

CLASS MEANS ON CANONICAL VARIABLES

CLUSTER	CAN1	CAN2	CAN3	CAN4
1	-6.1315	0.2448	0.0000	-0.0000
2	4.9314	0.8620	0.0000	-0.0000
3	1.9223	-0.7257	0.0000	-0.0000

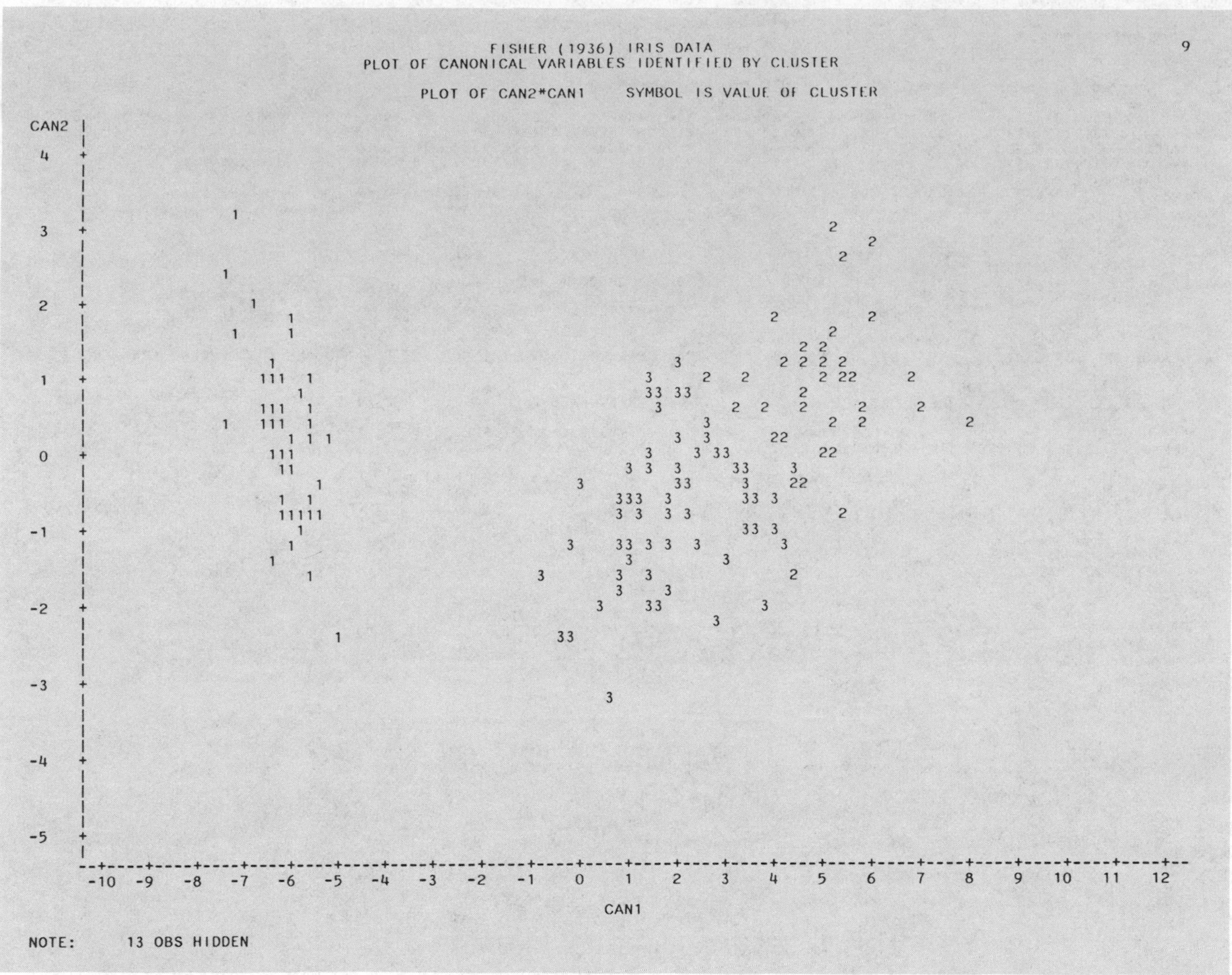

REFERENCES

Anderberg, M.R. (1973), *Cluster Analysis for Applications*, New York: Academic Press.

Everitt, B.S. (1980), *Cluster Analysis, 2nd ed.*, London: Heineman Educational Books Ltd.

Fisher, R.A. (1936), "The Use of Multiple Measurements in Taxonomic Problems," *Annals of Eugenics*, 7, 179–188.

Hartigan, J.A. (1975), *Clustering Algorithms*, New York: John Wiley & Sons.

MacQueen, J.B. (1967), "Some Methods for Classification and Analysis of Multivariate Observations," *Proceedings of the Fifth Berkeley Symposium on Mathematical Statistics and Probability*, 1, 281–297.

Mezzich, J.E and Solomon, H. (1980), *Taxonomy and Behavioral Science*, New York: Academic Press.

Milligan, G.W. (1980), "An Examination of the Effect of Six Types of Error Perturbation on Fifteen Clustering Algorithms," *Psychometrika*, 45, 325–342.

Spath, H. (1980), *Cluster Analysis Algorithms*, Chichester, England: Ellis Horwood.

Tou, J.T. and Gonzalez, R.C. (1974), *Pattern Recognition Principles*, Reading, MA: Addison-Wesley Publishing Co..

The TREE Procedure

ABSTRACT

The TREE procedure prints a tree diagram, also known as a dendrogram or phenogram, using a data set created by the CLUSTER or VARCLUS procedures. TREE can also create an output data set identifying disjoint clusters at a specified level in the tree.

INTRODUCTION

The CLUSTER and VARCLUS procedures create output data sets giving the results of hierarchical clustering as a tree structure. The TREE procedure uses the output data set to print a diagram of the tree structure in the style of Johnson (1967), with the root at the top. Any numeric variable in the output data set can be used to specify the heights of the clusters. TREE can also create an output data set containing a variable to indicate the disjoint clusters at a specified level in the tree and a variable specifying the cluster to which each object belongs.

Trees are discussed in the context of cluster analysis by Duran and Odell (1974), Hartigan (1975), and Everitt (1980). Knuth (1973) provides a general treatment of trees in computer programming.

The literature on trees contains a mixture of botanical and genealogical terminology. The objects that are clustered are *leaves*. The cluster containing all objects is the *root*. A cluster containing at least two objects but not all of them is a *branch*. The general term for leaves, branches, and roots is *node*. If a cluster *A* is the union of clusters *B* and *C*, then *A* is the *parent* of *B* and *C*, and *B* and *C* are *children* of *A*. A leaf is thus a node with no children, and a root is a node with no parent. If every cluster has at most two children, the tree is a *binary* tree. The CLUSTER procedure always produces binary trees. The VARCLUS procedure may produce trees with clusters that have many children.

SPECIFICATIONS

The TREE procedure is invoked by the following statements:

PROC TREE *options;*
NAME *variable;*
PARENT *variable;*
HEIGHT *variable;*
ID *variable;*
COPY *variables;*
BY *variables;*

If the input data set has been created by CLUSTER or VARCLUS, the only statement required is the PROC TREE statement.

PROC TREE Statement

PROC TREE *options*;

The following options may appear on the PROC TREE statement:

DATA=*SASdataset* names the input data set defining the tree. If DATA= is omitted, the most recently created SAS data set is used.

OUT=*SASdataset* names an output data set that contains one observation for each object and a variable called CLUSTER showing cluster membership at the level specified by the LEVEL= option. If you want to create a permanent SAS data set you must specify a two-level name (see Chapter 12, "SAS Data Sets," in *SAS User's Guide: Basics, 1982 Edition*).

LEVEL=*n* specifies the level of the tree defining disjoint clusters in the OUT= data set. The value *n* is a value of the HEIGHT variable.

If the HEIGHT variable is _NCL_ (number of clusters) and LEVEL=5 is specified, then the OUT= data set contains 5 disjoint clusters. If the HEIGHT variable is _RSQ_ (R^2) and LEVEL=.9 is specified, then the OUT= data set contains the smallest number of disjoint clusters that yields an R^2 of at least .9.

LIST lists all the nodes in the tree, printing the height, parent, and children of each node.

NOPRINT suppresses printing the tree if you only want to create an OUT= data set.

ROOT='*name*' specifies the value of the NAME variable for the root of a subtree to be printed if you do not wish to print the entire tree. Up to 16 characters may be specified, enclosed in single quotes.

PRUNE=*n* causes only that part of the tree between the root and the height specified by *n* to be printed.

SORT sorts the children of each node by the HEIGHT variable, in the order of cluster formation.

SPACES=*s*
S=*s* specifies the number of spaces between objects on the printout. The default depends on the number of objects and the line size used.

PAGES=*n* specifies the number of pages over which the tree (from root to leaves) is to extend. The default is chosen to make the tree diagram approximately square.

POS=*n* specifies the number of print positions on the height axis. The default depends on PAGES=.

TICPOS=*n* specifies the number of print positions per tic interval on the height axis. If the HEIGHT variable is _NCL_, then the default is 1. Otherwise, the default is usually

	5, although a different value may be used to be consistent with other options.
NTIC=*n*	specifies the number of tic intervals on the height axis. The default depends on the values of other options.
INC=*n*	specifies the increment between tic values on the height axis. If the HEIGHT variable is _NCL_, then the default is 1; otherwise the default is a power of 10 times 1, 2, or 5.
LEAFCHAR='*c*' LC='*c*'	specifies a character to represent clusters containing only one object. The character should be enclosed in single quotes. The default is a period.
TREECHAR='*c*' TC='*c*'	specifies a character to represent clusters containing more than one object. The character should be enclosed in single quotes. The default is X.

NAME Statement

NAME *variable*;

The NAME statement specifies a character variable identifying the node represented by each observation. The NAME variable and PARENT variable jointly define the tree structure. If the NAME statement is omitted, TREE looks for a variable called _NAME_. If the _NAME_ variable is not found in the data set, TREE issues an error message and stops.

PARENT Statement

PARENT *variable*;

The PARENT statement specifies a character variable identifying the node in the tree that is the parent of each observation. The PARENT variable must be the same length as the NAME variable. If the PARENT statement is omitted, TREE looks for a variable called _PARENT_. If the _PARENT_ variable is not found in the data set, TREE issues an error message and stops.

HEIGHT Statement

HEIGHT *variable*;

The HEIGHT statement specifies the name of a numeric variable to define the height of each node (cluster) in the tree. If the HEIGHT statement is omitted, the first numeric variable in the data set that is not mentioned in another statement is used. If CLUSTER or VARCLUS created the DATA= data set, the default HEIGHT variable is _NCL_, the number of clusters.

ID Statement

ID *variable*;

The ID variable is used to identify the objects (leaves) in the tree on the printout. The ID variable may be a character or numeric variable of any length. If the ID statement is omitted, the variable in the NAME statement is used instead. If both ID and NAME are omitted, TREE looks for a variable called _NAME_. If the _NAME_ variable is not found in the data set, TREE issues an error message and stops.

COPY Statement

COPY *variables;*

The COPY statement lists one or more character or numeric variables to be transferred to the OUT= data set.

BY Statement

BY *variables;*

A BY statement may be used with PROC TREE to obtain separate analyses on observations in groups defined by the BY variables. When a BY statement appears, the procedure expects the input data set to be sorted in order of the BY variables. If your input data set is not sorted in ascending order, use the SORT procedure with a similar BY statement to sort the data, or, if appropriate, use the BY statement options NOTSORTED or DESCENDING. For more information, see the discussion of the BY statement in Chapter 8, "Statements Used in the PROC Step," in *SAS User's Guide: Basics, 1982 Edition*.

DETAILS

Missing Values

An observation with a missing value for the NAME variable is omitted from processing. If the PARENT variable has a missing value but the NAME variable is present, the observation is treated as the root of a tree. A data set may contain several roots, hence several trees. Missing values of the HEIGHT variable are treated as zeros.

Output Data Set

The OUT= data set contains one observation for each object in the tree. The variables are:

- The BY variables, if any.
- The ID variable, or the NAME variable if the ID statement is not used.
- The COPY variables.
- A numeric variable CLUSTER taking values from 1 to *c*, where *c* is the number of disjoint clusters. The cluster to which the first observation belongs is given the number 1, the cluster to which the next observation belongs that does not belong to cluster 1 is given the number 2, and so on.
- A character variable CLUSNAME giving the value of the NAME variable of the cluster to which each observation belongs.

Printed Output

The printed output of TREE includes:

1. the names of the objects in the tree printed along the top of the tree diagram
2. the height axis printed along the left edge of the tree diagram
3. the tree diagram. The root (the cluster containing all the objects) is at the top, indicated by a solid line of the character specified by TREECHAR= (by default, Xs). At each horizontal level in the tree, clusters are shown by unbroken lines of the TREECHAR= symbol with blanks separating the clusters. The character specified by the LEAFCHAR= option represents single-member clusters (the default character is a period.)

EXAMPLES

Mammals' Teeth: Example 1

The data below give the numbers of different kinds of teeth for a variety of mammals. The mammals are clustered by average linkage using CLUSTER. The first PROC TREE uses the number of clusters as the height axis by default. The second PROC TREE sorts the clusters at each branch in order of formation and uses the average linkage between clusters joined for the height axis. The third PROC TREE produces no printed output but creates an output data set indicating the cluster to which each observation belongs at the 3-cluster level in the tree; this data set is printed by PRINT.

```
DATA TEETH;
  TITLE MAMMALS'' TEETH;
  INPUT MAMMAL $ 1-16 @21 (V1-V8) (1.);
  LABEL V1=TOP INCISORS
        V2=BOTTOM INCISORS
        V3=TOP CANINES
        V4=BOTTOM CANINES
        V5=TOP PREMOLARS
        V6=BOTTOM PREMOLARS
        V7=TOP MOLARS
        V8=BOTTOM MOLARS;
  CARDS;
BROWN BAT           23113333
MOLE                32103333
SILVER HAIR BAT     23112333
PIGMY BAT           23112233
HOUSE BAT           23111233
RED BAT             13112233
PIKA                21002233
RABBIT              21003233
BEAVER              11002133
GROUNDHOG           11002133
GRAY SQUIRREL       11001133
HOUSE MOUSE         11000033
PORCUPINE           11001133
WOLF                33114423
BEAR                33114423
RACCOON             33114432
MARTEN              33114412
WEASEL              33113312
WOLVERINE           33114412
BADGER              33113312
RIVER OTTER         33114312
SEA OTTER           32113312
JAGUAR              33113211
COUGAR              33113211
FUR SEAL            32114411
SEA LION            32114411
GREY SEAL           32113322
ELEPHANT SEAL       21114411
REINDEER            04103333
ELK                 04103333
```

```
DEER                    04003333
MOOSE                   04003333
;
PROC CLUSTER METHOD=AVERAGE STD OUTTREE=TREE;
  ID MAMMAL;
  VAR V1-V8;
PROC TREE;
PROC TREE SORT;
  HEIGHT _AVLINK_;
  ID MAMMAL;
PROC TREE NOPRINT OUT=PART LEVEL=3;
  HEIGHT _NCL_;
  ID MAMMAL;
  COPY V1-V8;
PROC PRINT;
  ID MAMMAL;
  VAR CLUSTER V1-V8;
```

To see how the first tree diagram is interpreted, consider the five-cluster level shown on the height axis. The five BATs are in a cluster indicated by an unbroken line of Xs. The next cluster is represented by a period (.) because it contains only one mammal, MOLE. REINDEER, ELK, DEER, and MOOSE form the next cluster. The mammals PIKA through HOUSE MOUSE are in the fourth cluster, and WOLF through ELEPHANT SEAL form the last cluster.

MAMMALS' TEETH 1

AVERAGE LINKAGE HIERARCHICAL CLUSTER ANALYSIS

EIGENVALUES OF THE CORRELATION MATRIX

EIGENVALUE	DIFFERENCE	PROPORTION	CUMULATIVE
4.593366	3.172257	0.592692	0.592692
1.421109	0.686109	0.183369	0.776061
0.735000	0.243604	0.094839	0.870900
0.491396	0.293190	0.063406	0.934306
0.198206	0.057406	0.025575	0.959881
0.140800	0.033423	0.018168	0.978049
0.107377	0.044630	0.013855	0.991904
0.062747	.	0.008096	1.000000

ROOT-MEAN-SQUARE TOTAL-SAMPLE STANDARD DEVIATION = 1
ROOT-MEAN-SQUARE DISTANCE BETWEEN OBSERVATIONS = 2.82843

NUMBER OF CLUSTERS	FREQUENCY OF NEW CLUSTER	RMS STD OF NEW CLUSTER	NORMALIZED CENTROID DISTANCE	NORMALIZED AVERAGE LINKAGE	SEMIPARTIAL R-SQUARED	R-SQUARED	APPROXIMATE EXPECTED R-SQUARED	CUBIC CLUSTERING CRITERION
10	6	0.344873	0.5666	0.6137	0.013807	0.933845	0.864293	6.8379
9	9	0.402139	0.6509	0.7172	0.021260	0.912586	0.845490	5.5656
8	4	0.446853	0.7740	0.7740	0.019324	0.893262	0.823796	5.0529
7	12	0.472311	0.6698	0.8016	0.032563	0.860699	0.798364	3.8708
6	7	0.425226	0.7563	0.8192	0.015813	0.844886	0.767852	3.8295
5	15	0.542155	0.7699	0.9364	0.045891	0.798995	0.729652	2.9924
4	6	0.492675	0.9684	1.0120	0.025209	0.773786	0.679090	3.8478
3	10	0.721392	1.0937	1.2443	0.092612	0.681174	0.605432	2.1974
2	17	0.864804	1.2268	1.4589	0.199923	0.481251	0.469716	0.2173
1	32	1	1.3683	1.6883	0.481251	0.000000	0.000000	0.0000

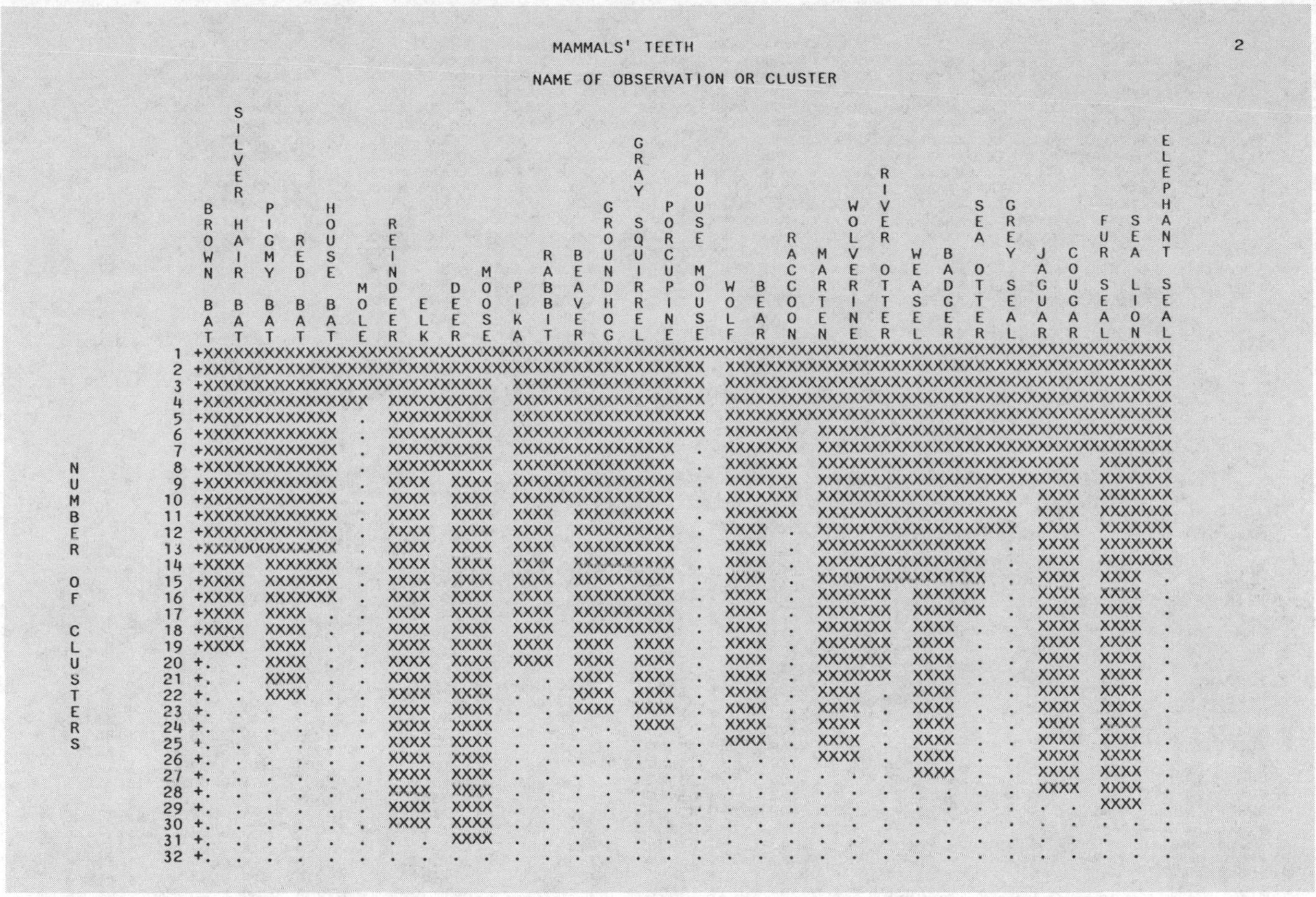

```
                                    MAMMALS' TEETH                                    2

                             NAME OF OBSERVATION OR CLUSTER

  BROWN BAT, SILVER HAIR BAT, PIGMY BAT, RED BAT, HOUSE BAT, MOLE, REINDEER, ELK, DEER,
  MOOSE, PIKA, RABBIT, BEAVER, GROUNDHOG, GRAY SQUIRREL, PORCUPINE, HOUSE MOUSE, WOLF,
  BEAR, RACCOON, MARTEN, WOLVERINE, RIVER OTTER, WEASEL, BADGER, SEA OTTER, GREY SEAL,
  JAGUAR, COUGAR, FUR SEAL, SEA LION, ELEPHANT SEAL

NUMBER OF CLUSTERS
  1 +XXXXXXXXXXXXXXXXXXXXXXXXXXXXXXXXXXXXXXXXXXXXXXXXXXXXXXXXXXXXXXXXXXXXXXXXXXXXXXXXXXXXXXXXXXXXXX
  2 +XXXXXXXXXXXXXXXXXXXXXXXXXXXXXXXXXXXXXXXXXXXXXXXXX  XXXXXXXXXXXXXXXXXXXXXXXXXXXXXXXXXXXXXXXXXXX
  3 +XXXXXXXXXXXXXXXXXXXXXXXXXXXX  XXXXXXXXXXXXXXXXXXX  XXXXXXXXXXXXXXXXXXXXXXXXXXXXXXXXXXXXXXXXXXX
  4 +XXXXXXXXXXXXXXXX  XXXXXXXXXX  XXXXXXXXXXXXXXXXXXX  XXXXXXXXXXXXXXXXXXXXXXXXXXXXXXXXXXXXXXXXXXX
  5 +XXXXXXXXXXXXX  .  XXXXXXXXXX  XXXXXXXXXXXXXXXXXXX  XXXXXXXXXXXXXXXXXXXXXXXXXXXXXXXXXXXXXXXXXXX
  6 +XXXXXXXXXXXXX  .  XXXXXXXXXX  XXXXXXXXXXXXXXXXXXX  XXXXXXX  XXXXXXXXXXXXXXXXXXXXXXXXXXXXXXXXXXXXX
  7 +XXXXXXXXXXXXX  .  XXXXXXXXXX  XXXXXXXXXXXXXXXX  .  XXXXXXX  XXXXXXXXXXXXXXXXXXXXXXXXXXXXXXXXXXXXX
  8 +XXXXXXXXXXXXX  .  XXXXXXXXXX  XXXXXXXXXXXXXXXX  .  XXXXXXX  XXXXXXXXXXXXXXXXXXXXXXXXXXXX  XXXXXXX
  9 +XXXXXXXXXXXXX  .  XXXX  XXXX  XXXXXXXXXXXXXXXX  .  XXXXXXX  XXXXXXXXXXXXXXXXXXXXXXXXXXXX  XXXXXXX
 10 +XXXXXXXXXXXXX  .  XXXX  XXXX  XXXXXXXXXXXXXXXX  .  XXXXXXX  XXXXXXXXXXXXXXXXXXXXXX  XXXX  XXXXXXX
 11 +XXXXXXXXXXXXX  .  XXXX  XXXX  XXXX  XXXXXXXXXX  .  XXXXXXX  XXXXXXXXXXXXXXXXXXXXXX  XXXX  XXXXXXX
 12 +XXXXXXXXXXXXX  .  XXXX  XXXX  XXXX  XXXXXXXXXX  .  XXXX  .  XXXXXXXXXXXXXXXXXXX  XXXX  XXXXXXX
 13 +XXXXXXXXXXXXX  .  XXXX  XXXX  XXXX  XXXXXXXXXX  .  XXXX  .  XXXXXXXXXXXXXXXX  .  XXXX  XXXXXXX
 14 +XXXX  XXXXXXX  .  XXXX  XXXX  XXXX  XXXXXXXXXX  .  XXXX  .  XXXXXXXXXXXXXXXX  .  XXXX  XXXXXXX
 15 +XXXX  XXXXXXX  .  XXXX  XXXX  XXXX  XXXXXXXXXX  .  XXXX  .  XXXXXXXXXXXXXXXX  .  XXXX  XXXX  .
 16 +XXXX  XXXXXXX  .  XXXX  XXXX  XXXX  XXXXXXXXXX  .  XXXX  .  XXXXXXX  XXXXXXX  .  XXXX  XXXX  .
 17 +XXXX  XXXX  .  .  XXXX  XXXX  XXXX  XXXXXXXXXX  .  XXXX  .  XXXXXXX  XXXXXXX  .  XXXX  XXXX  .
 18 +XXXX  XXXX  .  .  XXXX  XXXX  XXXX  XXXXXXXXXX  .  XXXX  .  XXXXXXX  XXXX  .  .  XXXX  XXXX  .
 19 +XXXX  XXXX  .  .  XXXX  XXXX  XXXX  XXXX  XXXX  .  XXXX  .  XXXXXXX  XXXX  .  .  XXXX  XXXX  .
 20 +.  .  XXXX  .  .  XXXX  XXXX  XXXX  XXXX  XXXX  .  XXXX  .  XXXXXXX  XXXX  .  .  XXXX  XXXX  .
 21 +.  .  XXXX  .  .  XXXX  XXXX  .  .  XXXX  XXXX  .  XXXX  .  XXXXXXX  XXXX  .  .  XXXX  XXXX  .
 22 +.  .  XXXX  .  .  XXXX  XXXX  .  .  XXXX  XXXX  .  XXXX  .  XXXX  .  XXXX  .  .  XXXX  XXXX  .
 23 +.  .  .  .  .  .  XXXX  XXXX  .  .  XXXX  XXXX  .  XXXX  .  XXXX  .  XXXX  .  .  XXXX  XXXX  .
 24 +.  .  .  .  .  .  XXXX  XXXX  .  .  .  .  XXXX  .  XXXX  .  XXXX  .  XXXX  .  .  XXXX  XXXX  .
 25 +.  .  .  .  .  .  XXXX  XXXX  .  .  .  .  .  .  .  XXXX  .  XXXX  .  XXXX  .  .  XXXX  XXXX  .
 26 +.  .  .  .  .  .  XXXX  XXXX  .  .  .  .  .  .  .  .  .  .  XXXX  .  XXXX  .  .  XXXX  XXXX  .
 27 +.  .  .  .  .  .  XXXX  XXXX  .  .  .  .  .  .  .  .  .  .  .  .  .  XXXX  .  .  XXXX  XXXX  .
 28 +.  .  .  .  .  .  XXXX  XXXX  .  .  .  .  .  .  .  .  .  .  .  .  .  .  .  .  .  XXXX  XXXX  .
 29 +.  .  .  .  .  .  XXXX  XXXX  .  .  .  .  .  .  .  .  .  .  .  .  .  .  .  .  .  .  .  XXXX  .
 30 +.  .  .  .  .  .  XXXX  XXXX  .  .  .  .  .  .  .  .  .  .  .  .  .  .  .  .  .  .  .  .  .  .
 31 +.  .  .  .  .  .  .  .  XXXX  .  .  .  .  .  .  .  .  .  .  .  .  .  .  .  .  .  .  .  .  .  .
 32 +.  .  .  .  .  .  .  .  .  .  .  .  .  .  .  .  .  .  .  .  .  .  .  .  .  .  .  .  .  .  .  .
```

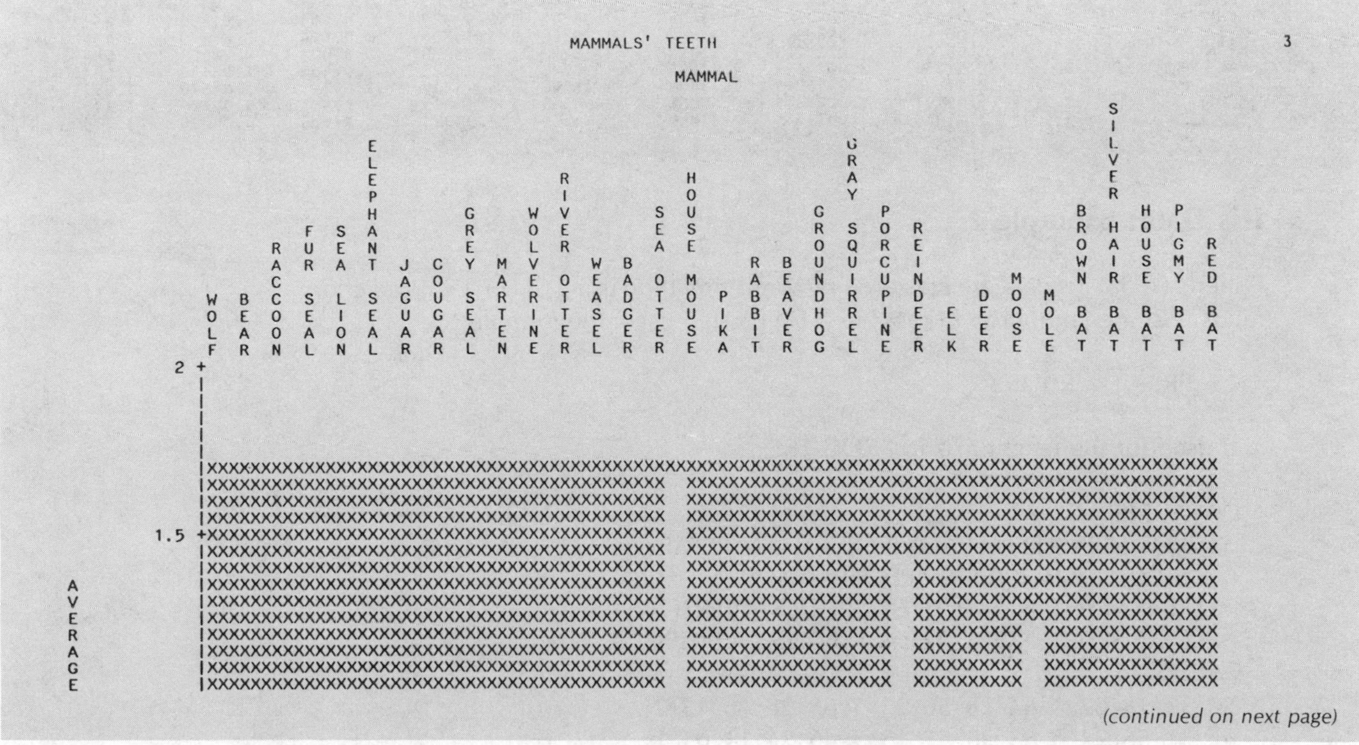

(continued on next page)

(continued from previous page)

```
           1 +XXXXXXXXXXXXXXXXXXXXXXXXXXXXXXXXXXXXXXXXXXX  XXXXXXXXXXXXXXXXXXX  XXXXXXXXXX  XXXXXXXXXXXXXXXX
 L           |XXXXXXXXXXXXXXXXXXXXXXXXXXXXXXXXXXXXXXXXXXX  XXXXXXXXXXXXXXXXXXX  XXXXXXXXXX  .  XXXXXXXXXXXXX
 I           |XXXXXXX  XXXXXXXXXXXXXXXXXXXXXXXXXXXXXXXXXX  XXXXXXXXXXXXXXXXXXX  XXXXXXXXXX  .  XXXXXXXXXXXXX
 N           |XXXXXXX  XXXXXXXXXXXXXXXXXXXXXXXXXXXXXXXXXX  XXXXXXXXXXXXXXXXXXX  XXXXXXXXXX  .  XXXXXXXXXXXXX
 K           |XXXXXXX  XXXXXXXXXXXXXXXXXXXXXXXXXXXXXXXXXX  XXXXXXXXXXXXXXXXXXX  XXXXXXXXXX  .  XXXXXXXXXXXXX
 A           |XXXXXXX  XXXXXXX  XXXXXXXXXXXXXXXXXXXXXXXXX  .  XXXXXXXXXXXXXXXX  XXXXXXXXXX  .  XXXXXXXXXXXXX
 G           |XXXXXXX  XXXXXXX  XXXXXXXXXXXXXXXXXXXXXXXXX  .  XXXXXXXXXXXXXXXX  XXXX  XXXX  .  XXXXXXXXXXXXX
 E           |XXXXXXX  XXXXXXX  XXXX  XXXXXXXXXXXXXXXXXXX  .  XXXXXXXXXXXXXXXX  XXXX  XXXX  .  XXXXXXXXXXXXX
             |XXXXXXX  XXXXXXX  XXXX  XXXXXXXXXXXXXXXXXXX  .  XXXXXXXXXXXXXXXX  XXXX  XXXX  .  XXXXXXXXXXXXX
             |XXXX  .  XXXXXXX  XXXX  .  XXXXXXXXXXXXXXXX  .  XXXX  XXXXXXXXXX  XXXX  XXXX  .  XXXXXXXXXXXXX
         0.5 +XXXX  .  XXXXXXX  XXXX  .  XXXXXXXXXXXXXXXX  .  XXXX  XXXXXXXXXX  XXXX  XXXX  .  XXXXXXXXXXXXX
             |XXXX  .  XXXX  .  XXXX  .  XXXXXXXXXXXXXXXX  .  XXXX  XXXXXXXXXX  XXXX  XXXX  .  XXXX  XXXXXXX
             |XXXX  .  XXXX  .  XXXX  .  XXXXXXX  XXXXXXX  .  XXXX  XXXXXXXXXX  XXXX  XXXX  .  XXXX  XXXXXXX
             |XXXX  .  XXXX  .  XXXX  .  XXXXXXX  XXXXXXX  .  XXXX  XXXXXXXXXX  XXXX  XXXX  .  XXXX  .  XXXX
             |XXXX  .  XXXX  .  XXXX  .  XXXXXXX  XXXX  .  .  .  .  XXXX  XXXX  XXXX  XXXX  .  .  .  .  XXXX
             |XXXX  .  XXXX  .  XXXX  .  XXXX  .  XXXX  .  .  .  .  XXXX  XXXX  XXXX  XXXX  .  .  .  .  .  .
             |XXXX  .  XXXX  .  XXXX  .  XXXX  .  XXXX  .  .  .  .  XXXX  XXXX  XXXX  XXXX  .  .  .  .  .  .
             |XXXX  .  XXXX  .  XXXX  .  XXXX  .  XXXX  .  .  .  .  XXXX  XXXX  XXXX  XXXX  .  .  .  .  .  .
             |XXXX  .  XXXX  .  XXXX  .  XXXX  .  XXXX  .  .  .  .  XXXX  XXXX  XXXX  XXXX  .  .  .  .  .  .
             |XXXX  .  XXXX  .  XXXX  .  XXXX  .  XXXX  .  .  .  .  XXXX  XXXX  XXXX  XXXX  .  .  .  .  .  .
           0 +XXXX  .  XXXX  .  XXXX  .  XXXX  .  XXXX  .  .  .  .  XXXX  XXXX  XXXX  XXXX  .  .  .  .  .  .
```

MAMMALS' TEETH 4

MAMMAL	CLUSTER	V1	V2	V3	V4	V5	V6	V7	V8
DEER	1	0.17119	3.97662	-0.85468	-0.64566	2.98928	2.97403	3.04870	3.17826
MOOSE	1	0.17119	3.97662	-0.85468	-0.64566	2.98928	2.97403	3.04870	3.17826
REINDEER	1	0.17119	3.97662	1.33444	-0.64566	2.98928	2.97403	3.04870	3.17826
ELK	1	0.17119	3.97662	1.33444	-0.64566	2.98928	2.97403	3.04870	3.17826
FUR SEAL	2	2.91836	2.00716	1.33444	1.38740	3.93209	3.89091	0.92403	0.54443
SEA LION	2	2.91836	2.00716	1.33444	1.38740	3.93209	3.89091	0.92403	0.54443
JAGUAR	2	2.91836	2.99189	1.33444	1.38740	2.98928	2.05714	0.92403	0.54443
COUGAR	2	2.91836	2.99189	1.33444	1.38740	2.98928	2.05714	0.92403	0.54443
WEASEL	2	2.91836	2.99189	1.33444	1.38740	2.98928	2.97403	0.92403	1.86135
BADGER	2	2.91836	2.99189	1.33444	1.38740	2.98928	2.97403	0.92403	1.86135
MARTEN	2	2.91836	2.99189	1.33444	1.38740	3.93209	3.89091	0.92403	1.86135
WOLVERINE	2	2.91836	2.99189	1.33444	1.38740	3.93209	3.89091	0.92403	1.86135
WOLF	2	2.91836	2.99189	1.33444	1.38740	3.93209	3.89091	1.98636	3.17826
BEAR	2	2.91836	2.99189	1.33444	1.38740	3.93209	3.89091	1.98636	3.17826
GRAY SQUIRREL	3	1.08691	1.02242	-0.85468	-0.64566	1.10366	1.14026	3.04870	3.17826
PORCUPINE	3	1.08691	1.02242	-0.85468	-0.64566	1.10366	1.14026	3.04870	3.17826
BEAVER	3	1.08691	1.02242	-0.85468	-0.64566	2.04647	1.14026	3.04870	3.17826
GROUNDHOG	3	1.08691	1.02242	-0.85468	-0.64566	2.04647	1.14026	3.04870	3.17826
PIGMY BAT	1	2.00263	2.99189	1.33444	1.38740	2.04647	2.05714	3.04870	3.17826
RED BAT	1	1.08691	2.99189	1.33444	1.38740	2.04647	2.05714	3.04870	3.17826
RIVER OTTER	2	2.91836	2.99189	1.33444	1.38740	3.93209	2.97403	0.92403	1.86135
PIKA	3	2.00263	1.02242	-0.85468	-0.64566	2.04647	2.05714	3.04870	3.17826
RABBIT	3	2.00263	1.02242	-0.85468	-0.64566	2.98928	2.05714	3.04870	3.17826
BROWN BAT	1	2.00263	2.99189	1.33444	1.38740	2.98928	2.97403	3.04870	3.17826
SILVER HAIR BAT	1	2.00263	2.99189	1.33444	1.38740	2.04647	2.97403	3.04870	3.17826
SEA OTTER	2	2.91836	2.00716	1.33444	1.38740	2.98928	2.97403	0.92403	1.86135
HOUSE BAT	1	2.00263	2.99189	1.33444	1.38740	1.10366	2.05714	3.04870	3.17826
ELEPHANT SEAL	2	2.00263	1.02242	1.33444	1.38740	3.93209	3.89091	0.92403	0.54443
GREY SEAL	2	2.91836	2.00716	1.33444	1.38740	2.98928	2.97403	1.98636	1.86135
RACCOON	2	2.91836	2.99189	1.33444	1.38740	3.93209	3.89091	3.04870	1.86135
HOUSE MOUSE	3	1.08691	1.02242	-0.85468	-0.64566	0.16085	0.22337	3.04870	3.17826
MOLE	1	2.91836	2.00716	1.33444	-0.64566	2.98928	2.97403	3.04870	3.17826

Iris Data: Example 2

Fisher's (1936) iris data are clustered by Ward's method using CLUSTER. Prediction ratios (PR) are computed from R^2 in a DATA step using the formula:

$$PR = (1 - R^2)^{1/2}$$

and used for the height axis in PROC TREE.

```
DATA IRIS;
   INPUT SEPALLEN SEPALWID PETALLEN PETALWID SPEC_NO@@;
   IF SPEC_NO=1 THEN SPECIES='SETOSA     ';
   ELSE IF SPEC_NO=2 THEN SPECIES='VERSICOLOR';
   ELSE IF SPEC_NO=3 THEN SPECIES='VIRGINICA  ';
CARDS;
50 33 14 02 1 64 28 56 22 3 65 28 46 15 2
67 31 56 24 3 63 28 51 15 3 46 34 14 03 1
```

```
69 31 51 23 3 62 22 45 15 2 59 32 48 18 2
46 36 10 02 1 61 30 46 14 2 60 27 51 16 2
65 30 52 20 3 56 25 39 11 2 65 30 55 18 3
58 27 51 19 3 68 32 59 23 3 51 33 17 05 1
57 28 45 13 2 62 34 54 23 3 77 38 67 22 3
63 33 47 16 2 67 33 57 25 3 76 30 66 21 3
49 25 45 17 3 55 35 13 02 1 67 30 52 23 3
70 32 47 14 2 64 32 45 15 2 61 28 40 13 2
48 31 16 02 1 59 30 51 18 3 55 24 38 11 2
63 25 50 19 3 64 32 53 23 3 52 34 14 02 1
49 36 14 01 1 54 30 45 15 2 79 38 64 20 3
44 32 13 02 1 67 33 57 21 3 50 35 16 06 1
58 26 40 12 2 44 30 13 02 1 77 28 67 20 3
63 27 49 18 3 47 32 16 02 1 55 26 44 12 2
50 23 33 10 2 72 32 60 18 3 48 30 14 03 1
51 38 16 02 1 61 30 49 18 3 48 34 19 02 1
50 30 16 02 1 50 32 12 02 1 61 26 56 14 3
64 28 56 21 3 43 30 11 01 1 58 40 12 02 1
51 38 19 04 1 67 31 44 14 2 62 28 48 18 3
49 30 14 02 1 51 35 14 02 1 56 30 45 15 2
58 27 41 10 2 50 34 16 04 1 46 32 14 02 1
60 29 45 15 2 57 26 35 10 2 57 44 15 04 1
50 36 14 02 1 77 30 61 23 3 63 34 56 24 3
58 27 51 19 3 57 29 42 13 2 72 30 58 16 3
54 34 15 04 1 52 41 15 01 1 71 30 59 21 3
64 31 55 18 3 60 30 48 18 3 63 29 56 18 3
49 24 33 10 2 56 27 42 13 2 57 30 42 12 2
55 42 14 02 1 49 31 15 02 1 77 26 69 23 3
60 22 50 15 3 54 39 17 04 1 66 29 46 13 2
52 27 39 14 2 60 34 45 16 2 50 34 15 02 1
44 29 14 02 1 50 20 35 10 2 55 24 37 10 2
58 27 39 12 2 47 32 13 02 1 46 31 15 02 1
69 32 57 23 3 62 29 43 13 2 74 28 61 19 3
59 30 42 15 2 51 34 15 02 1 50 35 13 03 1
56 28 49 20 3 60 22 40 10 2 73 29 63 18 3
67 25 58 18 3 49 31 15 01 1 67 31 47 15 2
63 23 44 13 2 54 37 15 02 1 56 30 41 13 2
63 25 49 15 2 61 28 47 12 2 64 29 43 13 2
51 25 30 11 2 57 28 41 13 2 65 30 58 22 3
69 31 54 21 3 54 39 13 04 1 51 35 14 03 1
72 36 61 25 3 65 32 51 20 3 61 29 47 14 2
56 29 36 13 2 69 31 49 15 2 64 27 53 19 3
68 30 55 21 3 55 25 40 13 2 48 34 16 02 1
48 30 14 01 1 45 23 13 03 1 57 25 50 20 3
57 38 17 03 1 51 38 15 03 1 55 23 40 13 2
66 30 44 14 2 68 28 48 14 2 54 34 17 02 1
51 37 15 04 1 52 35 15 02 1 58 28 51 24 3
67 30 50 17 2 63 33 60 25 3 53 37 15 02 1
;
PROC CLUSTER DATA=IRIS;
  VAR SEPALLEN SEPALWID PETALLEN PETALWID;
  COPY SPECIES;
DATA;
  SET;
  PRERATIO=SQRT(1-_RSQ_);
```

```
PROC TREE SORT PAGES=1;
   ID SPECIES;
   HEIGHT PRERATIO;
   LABEL PRERATIO=PREDICTION RATIO;
```

WARD'S HIERARCHICAL CLUSTER ANALYSIS 1

EIGENVALUES OF THE COVARIANCE MATRIX

EIGENVALUE	DIFFERENCE	PROPORTION	CUMULATIVE
420.0053	395.9000	0.9246	0.9246
24.1053	16.3365	0.0531	0.9777
7.7688	5.4012	0.0171	0.9948
2.3676	.	0.0052	1.0000

ROOT-MEAN-SQUARE TOTAL-SAMPLE STANDARD DEVIATION = 10.6922
ROOT-MEAN-SQUARE DISTANCE BETWEEN OBSERVATIONS = 21.3845

NUMBER OF CLUSTERS	FREQUENCY OF NEW CLUSTER	RMS STD OF NEW CLUSTER	NORMALIZED CENTROID DISTANCE	NORMALIZED AVERAGE LINKAGE	SEMIPARTIAL R-SQUARED	R-SQUARED	APPROXIMATE EXPECTED R-SQUARED	CUBIC CLUSTERING CRITERION
15	15	2.15362	0.2559	0.3300	0.001641	0.970932	0.957871	5.8544
14	7	3.24771	0.4035	0.4845	0.001873	0.969059	0.955418	5.7798
13	15	2.6054	0.3186	0.3966	0.002271	0.966788	0.952670	5.6247
12	24	2.35205	0.2454	0.3501	0.002274	0.964514	0.949541	4.5819
11	12	3.25262	0.3574	0.4745	0.002500	0.962014	0.945886	4.6274
10	22	2.36245	0.2713	0.3612	0.002694	0.959320	0.941547	4.7662
9	29	2.07762	0.2392	0.3251	0.002702	0.956618	0.936296	5.0858
8	23	2.45754	0.2973	0.3780	0.003095	0.953523	0.929791	5.5064
7	26	2.99037	0.5058	0.5644	0.005811	0.947713	0.921496	5.4832
6	38	2.99017	0.3149	0.4529	0.006042	0.941671	0.910514	5.8580
5	50	2.78031	0.3627	0.4462	0.010753	0.930917	0.895232	5.8170
4	36	3.91834	0.5667	0.6726	0.017245	0.913673	0.872331	3.9867
3	64	4.11402	0.5386	0.6644	0.030051	0.883621	0.826664	4.3292
2	100	5.94155	0.8474	0.9971	0.111026	0.772595	0.696871	4.3292
1	150	10.6922	1.8584	1.9559	0.772595	0.000000	0.000000	0.0000

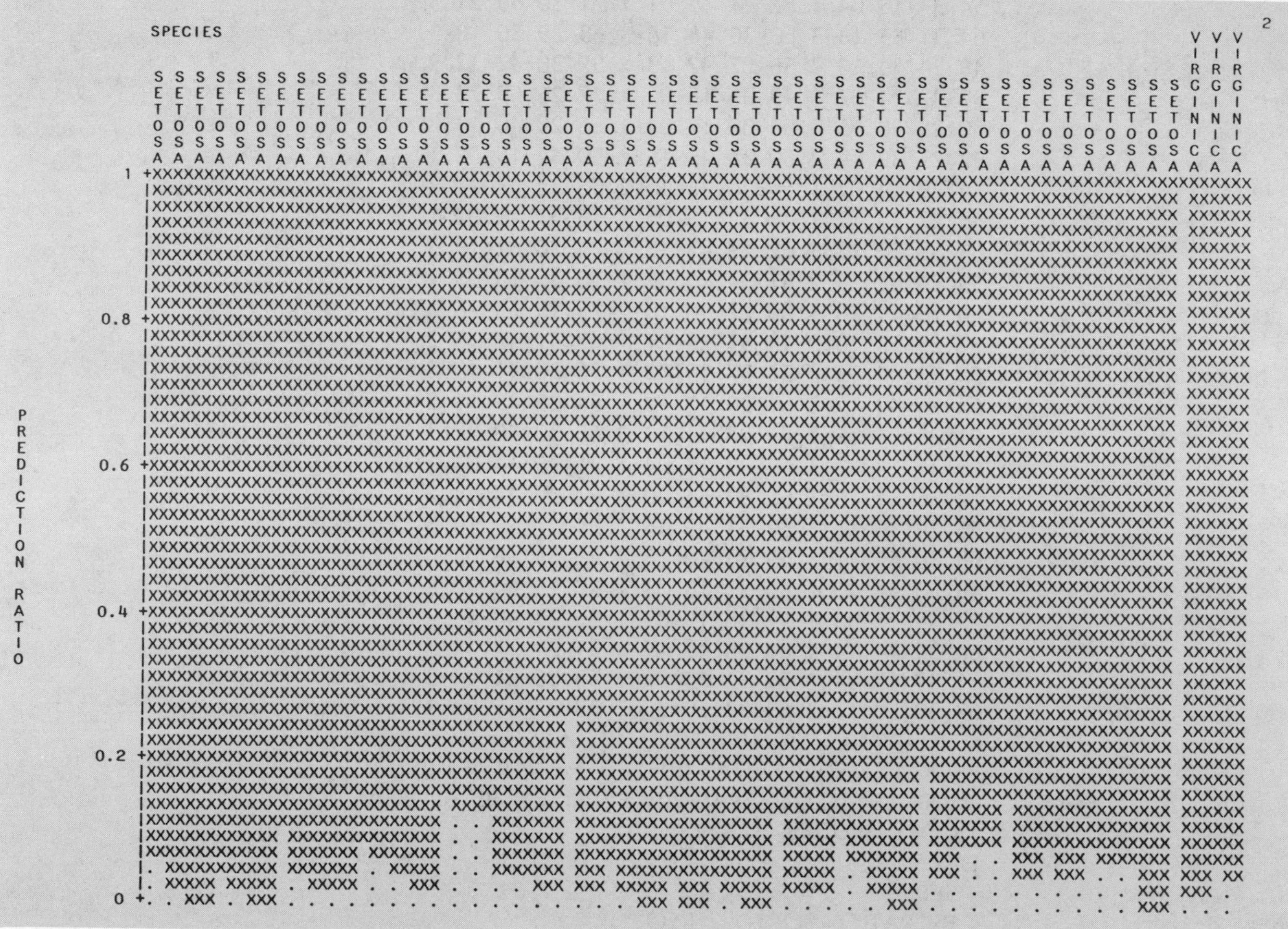

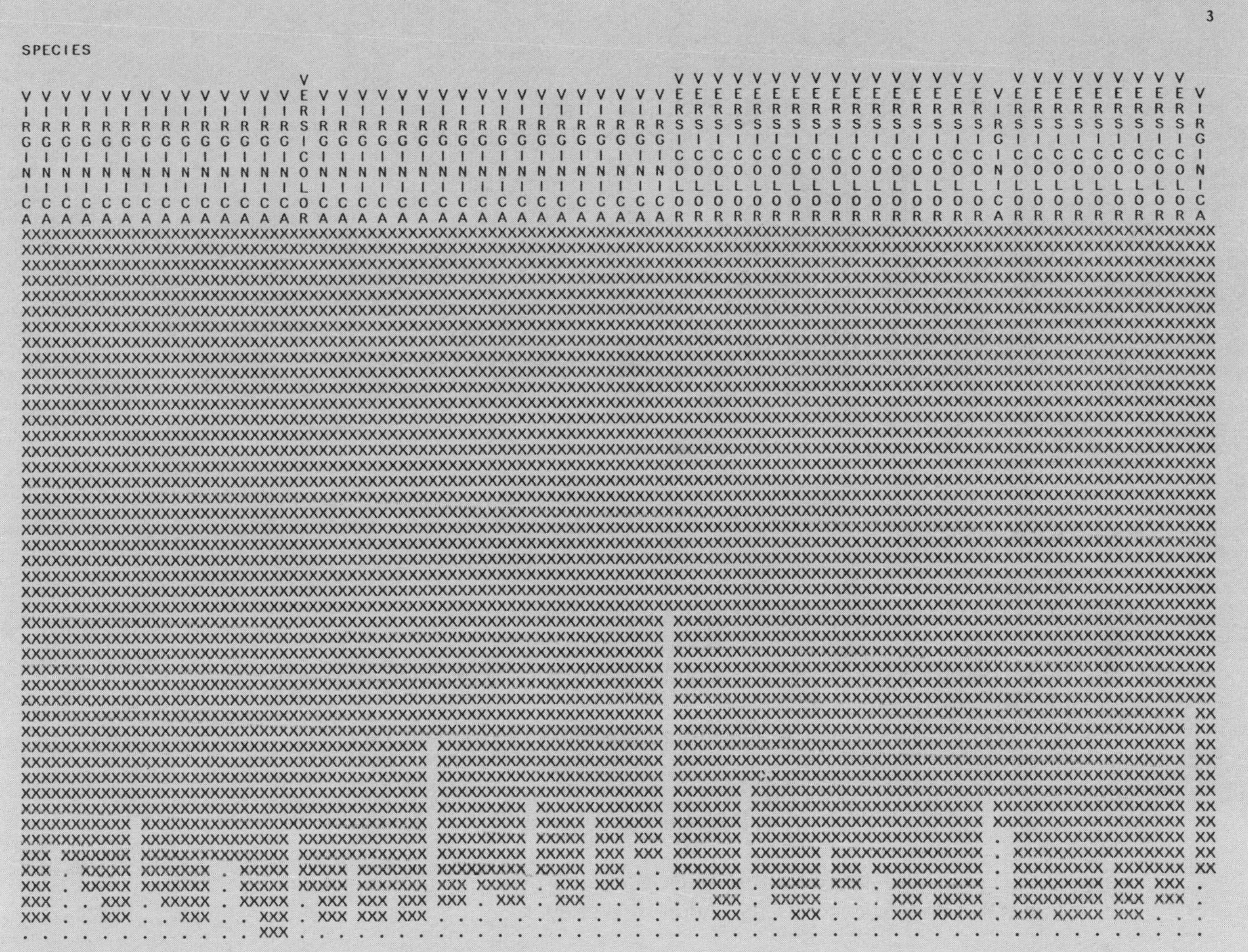

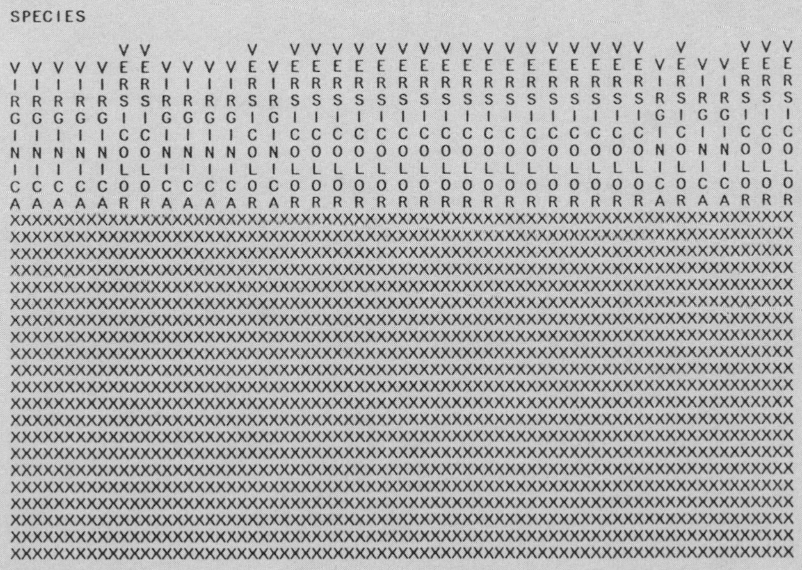

SPECIES

(continued on next page)

(continued from previous page)

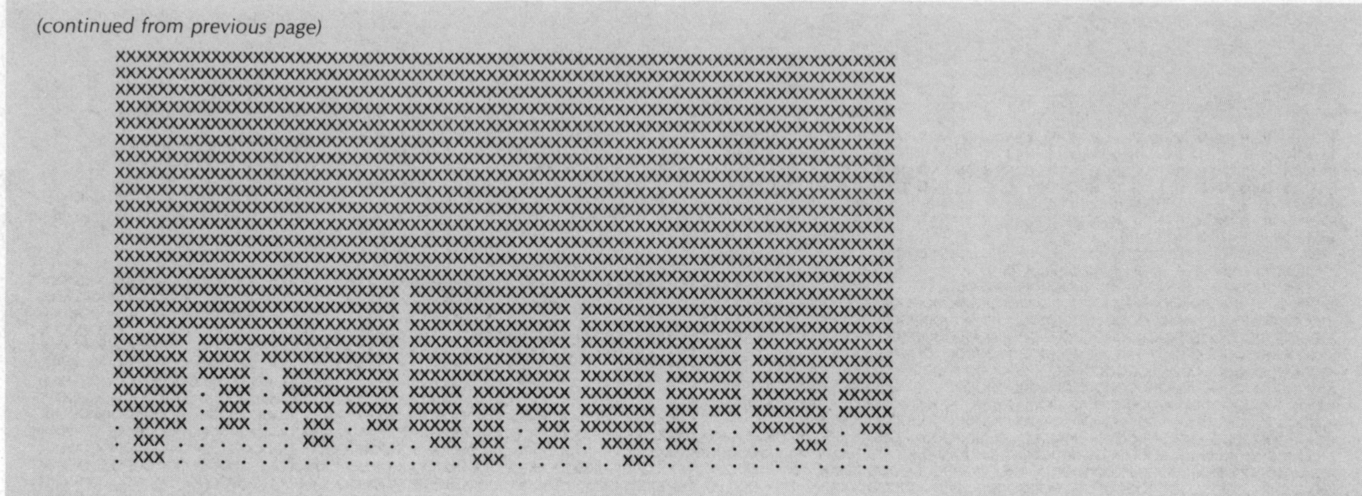

REFERENCES

Duran, B.S. and Odell, P.L. (1974), *Cluster Analysis*, New York: Springer-Verlag.

Everitt, B.S. (1980), *Cluster Analysis,* Second Edition, London: Heineman Educational Books Ltd.

Fisher, R.A. (1936), "The Use of Multiple Measurements in Taxonomic Problems," *Annals of Eugenics*, 7, 179-188.

Hartigan, J.A. (1975), *Clustering Algorithms*, New York: John Wiley & Sons.

Johnson, S.C. (1967), "Hierarchical Clustering Schemes," *Psychometrika*, 32, 241-254.

Knuth, D.E. (1973), *The Art of Computer Programming, Volume 1, Fundamental Algorithms*, Reading, Massachusetts: Addison-Wesley.

The VARCLUS Procedure

ABSTRACT

The VARCLUS procedure performs either disjoint or hierarchical clustering of variables based on a correlation or covariance matrix. The clusters are chosen to maximize the variation accounted for by either the first principal component or the centroid component of each cluster. An output data set containing the results of the analysis can be created and used with the SCORE procedure to compute cluster component scores.

INTRODUCTION

The VARCLUS procedure divides a set of numeric variables into either disjoint or hierarchical clusters. Associated with each cluster is a linear combination of the variables in the cluster, which may be either the first principal component or the centroid component. VARCLUS tries to maximize the sum across clusters of the variance of the original variables that is explained by the cluster components.

Either the correlation or the covariance matrix can be analyzed. If correlations are used, all variables are treated as equally important. If covariances are used, variables with larger variances have more importance in the analysis.

VARCLUS creates an output data set that may be used with the SCORE procedure to compute component scores for each cluster.

Background

The VARCLUS procedure attempts to divide a set of variables into non-overlapping clusters in such a way that each cluster can be interpreted as essentially unidimensional. VARCLUS computes a component for each cluster that can be either the first principal component or the centroid component, and tries to maximize the sum across clusters of the variation accounted for by the cluster components. VARCLUS is a type of oblique component analysis related to multiple group factor analysis (Harman, 1976).

VARCLUS can be used as a variable-reduction method. A large set of variables can often be replaced by the set of cluster components with little loss of information. A given number of cluster components does not generally explain as much variance as the same number of principal components, but the cluster components are usually easier to interpret than the principal components, even if the latter are rotated.

For example, an educational test might contain 50 items. VARCLUS could be used to divide the items into, say, five clusters. Each cluster could be treated as a subtest, and the subtest scores would be given by the cluster components. If the cluster components were centroid components of the covariance matrix, then each subtest score would simply be the sum of the item scores for that cluster.

By default, VARCLUS begins with all variables in a single cluster. It then repeats the following steps:

1. A cluster is chosen for splitting. The selected cluster has either the smallest percentage of variation explained by its cluster component or the largest second eigenvalue.
2. The chosen cluster is split into two clusters by finding the first two principal components, performing an orthoblique rotation (raw quartimax rotation on the eigenvectors), and assigning each variable to the rotated component with which it has the higher squared correlation.
3. Variables are iteratively reassigned to clusters to maximize the variance accounted for by the cluster components. The reassignment may be required to maintain a hierarchical structure.

The procedure stops when each cluster satisfies a user-specified criterion involving either the percentage of variation accounted for or the second eigenvalue of each cluster. By default, VARCLUS stops when each cluster has only a single eigenvalue greater than one, satisfying the most popular criterion for determining the sufficiency of a single underlying factor dimension.

The iterative reassignment of variables to clusters proceeds in two phases. First is a nearest component sorting (NCS) phase, similar in principle to the nearest centroid sorting algorithms described by Anderberg (1973). In each iteration the cluster components are computed and each variable is assigned to the component with which it has the highest squared correlation. The second phase involves a search algorithm in which each variable in turn is tested to see if assigning it to a different cluster increases the variance explained. If a variable is reassigned during the search phase, the components of the two clusters involved are recomputed before the next variable is tested. The alternating least-squares phase is much faster than the search phase but is more likely to be trapped by a local optimum.

If principal components are used, the NCS phase is an alternating least squares method and converges rapidly. The search phase is very time-consuming for a large number of variables and is omitted by default. If the default initialization method is used, the search phase is rarely able to improve the result of the NCS phase. If random initialization is used, the NCS phase may be trapped by a local optimum from which the search phase can escape.

If centroid components are used, the NCS phase may not increase the variance explained. It is therefore limited to one iteration by default.

SPECIFICATIONS

The VARCLUS procedure is invoked by the following statements:

PROC VARCLUS *options*;
VAR *variables*;
PARTIAL *variables*;
SEED *variables*;
FREQ *variable*;
WEIGHT *variable*;
BY *variables*;

Usually only the VAR statement is used in addition to the PROC VARCLUS statement.

PROC VARCLUS Statement

PROC VARCLUS *options*;

The following data set options may appear on the PROC statement:

DATA=*SASdataset* names the input data set to be analyzed. The data set may be an ordinary SAS data set or TYPE=CORR, COV, or FACTOR. If DATA= is omitted, the most recently created SAS data set is used.

OUTSTAT= *SASdataset* names an output data set to contain statistics including means, standard deviations, correlations, cluster scoring coefficients, and the cluster structure. If you want to create a permanent SAS data set, you must specify a two-level name. See Chapter 12, "SAS Data Sets," in *SAS User's Guide: Basics, 1982 Edition*, for more information on permanent SAS data sets.

OUTTREE= *SASdataset* names an output data set to contain information on the tree structure that can be used by the TREE procedure to print a tree diagram. The OUTTREE= option implies the HIERARCHY option. If you want to create a permanent SAS data set, you must specify a two-level name. See Chapter 12, "SAS Data Sets," in *SAS User's Guide: Basics, 1982 Edition*, for more information on permanent SAS data sets.

These options allow printing of descriptive statistics:

SIMPLE
S prints means and standard deviations.

CORR
C prints the correlation matrix.

The following options control the number of clusters:

MINCLUSTERS=*n*
MINC=*n* specifies the smallest number of clusters desired. The default value is 2 if INITIAL=RANDOM or INITIAL=SEED; otherwise the procedure begins with one cluster and tries to split it in accordance with the PROPORTION= or MAXEIGEN= options.

MAXCLUSTERS=*n*
MAXC=*n* specifies the largest number of clusters desired. The default value is the number of variables.

PROPORTION=*n*
PERCENT=*n* gives the proportion or percentage of variation that must be explained by the cluster component. PROPORTION=0.75 and PERCENT=75 are equivalent. If CENTROID is specified, the default value is 0.75; otherwise the default value is 0.

MAXEIGEN=*n* specifies the largest permissible value of the second eigenvalue in each cluster. If neither PROPORTION nor MAXCLUSTERS is specified, then the default value is 1 if the correlation matrix is analyzed, or the average variance of the variables if the covariance matrix is analyzed. Otherwise the default is 0. MAXEIGEN= may not be used with the CENTROID option.

The following options control the method of cluster formation:

COVARIANCE
COV analyzes the covariance matrix rather than the correlation matrix.

INITIAL= specifies the method for initializing the clusters. Values for INITIAL= can be RANDOM, SEED, INPUT, or GROUP. If INITIAL= is omitted and MINCLUSTERS= is greater than 1, the initial cluster components are obtained by extracting the required number of principal components and performing an orthoblique rotation.

INITIAL=RANDOM assigns variables randomly to clusters. If INITIAL=RANDOM is used without the CENTROID option, it is recommended that MAXSEARCH=5 be specified, although the CPU time required is substantially increased.

INITIAL=SEED is used in conjunction with the SEED statement. Each variable listed in the SEED statement becomes the sole member of a cluster, and the other variables remain unassigned. If the SEED statement is omitted, the first MINCLUSTERS= variables in the VAR statement are used as seeds.

INITIAL=INPUT may be used if the input data set is a TYPE=CORR, COV, or FACTOR data set, in which case scoring coefficients are read from the data set. Scoring coefficients from the FACTOR procedure or a previous run of VARCLUS can be used, or you can enter other coefficients in a DATA step.

INITIAL=GROUP may be used if the input data set is a TYPE=CORR, COV, or FACTOR data set. The cluster membership of each variable is obtained from an observation with _TYPE_='GROUP', which contains an integer for each variable ranging from one to the number of clusters. The data set may have been created by a previous run of VARCLUS or in a DATA step.

CENTROID uses centroid components rather than principal components. Centroid components should be used if you want the cluster components to be (unweighted) averages of the standardized or unstandardized (use the COV option) variables. It is possible to obtain locally optimal clusterings in which a variable is not assigned to the cluster component with which it has the highest squared correlation.

MAXITER=*n* specifies the maximum number of iterations during the alternating least squares phase. The default value is 1 if CENTROID is specified, 10 otherwise.

MAXSEARCH=*n* specifies the maximum number of iterations during the search phase. The default is 10 if CENTROID is specified, 0 otherwise.

HIERARCHY
HI requires the clusters at different levels to maintain a hierarchical structure.

MULTIPLEGROUP
MG performs a multiple group component analysis. The input data set must be TYPE=CORR, COV, or FACTOR and must contain an observation with _TYPE_='GROUP' defining the variable groups. Specifying MULTIPLEGROUP is equivalent to specifying all of the following options:

MING=1 MAXITER=0 MAXSEARCH=0
MAXEIGEN=0 PROPORTION=0 INITIAL=GROUP

These options control the amount of printed output:

TRACE lists the cluster to which each variable is assigned during the iterations.

SHORT suppresses printing of the cluster structure, scoring coefficient, and inter-cluster correlation matrices.

VAR Statement

VAR variables;

The VAR statement specifies the variables to be clustered. If it is omitted, all numeric variables not listed in another statement are processed. The VAR statement must be present if the SEED statement is used.

SEED Statement

SEEDS *variables;*
SEED *variables;*

The SEED statement specifies variables to be used as seeds to initialize the clusters. It is not necessary to use INITIAL=SEED if the SEED statement is present, but if any other INITIAL= option is specified, the SEED statement is ignored. The VAR statement must be used if the SEED statement is used.

PARTIAL Statement

PARTIAL *variables;*

If you wish to base the clustering on partial correlations, list the variables to be partialled out in the PARTIAL statement.

WEIGHT Statement

WEIGHT *variable;*

If you want to use relative weights for each observation in the input data set, place the weights in a variable in the data set and specify the name in a WEIGHT statement. This is often done when the variance associated with each observation is different and the values of the weight variable are proportional to the reciprocals of the variances.

If the WEIGHT statement is specified, the divisor used to compute variances is the sum of the weights, rather than the number of observations minus one.

FREQ Statement

FREQ *variable;*

If a variable in your data set represents the frequency of occurrence for the other values in the observation, include the variable's name in a FREQ statement. The procedure then treats the data set as if each observation appears *n* times, where *n* is the value of the FREQ variable for the observation. The total number of observations is considered equal to the sum of the FREQ variable when the procedure determines degrees of freedom for significance probabilities.

The WEIGHT and FREQ statements have a similar effect, except in the calculation of degrees of freedom.

BY Statement

BY *variables;*

A BY statement may be used with PROC VARCLUS to obtain separate analyses on observations in groups defined by the BY variables. When a BY statement appears, the procedure expects the input data set to be sorted in order of the BY variables. If your input data set is not sorted in ascending order, use the SORT procedure with a similar BY statement to sort the data, or, if appropriate, use the BY statement options NOTSORTED or DESCENDING. For more information, see the discussion of the BY statement in Chapter 8, "Statements Used in the PROC Step," in *SAS User's Guide: Basics, 1982 Edition*.

DETAILS

Missing Values

Observations containing missing values are omitted from the analysis.

Output Data Sets

OUTSTAT= Data Set The OUTSTAT= data set is of TYPE=CORR and can be used as input to the SCORE procedure or a subsequent run of VARCLUS. The variables it contains are:

- BY variables
- __NCL__, a numeric variable giving the number of clusters
- __TYPE__, a character variable indicating the type of statistic the observation contains
- __NAME__, a character variable containing a variable name or a cluster name
- the variables that were clustered.

The values of __TYPE__ are:

__TYPE__	Contents
MEAN	means
STD	standard deviations
N	number of observations
CORR	correlations
MEMBERS	number of members in each cluster
VAREXP	variance explained by each cluster
PROPOR	proportion of variance explained by each cluster
GROUP	number of the cluster to which each variable belongs
RSQUARED	squared multiple correlation of each variable with its cluster component
SCORE	standardized scoring coefficients
STRUCTUR	cluster structure
CCORR	correlations between cluster components

The observations with __TYPE__='MEAN', 'STD', 'N', and 'CORR' have missing values for __NCL__. All other values of __TYPE__ are repeated for each cluster solution, with different solutions distinguished by the value of __NCL__. If you wish to use the OUTSTAT= data set with the SCORE procedure, you must use a DATA step to select observations with __NCL__ missing or equal to the desired number of clusters.

OUTTREE= Data Set The OUTTREE= data set contains one observation for each variable clustered, plus one observation for each cluster of two or more variables; that is, one observation for each node of the cluster tree. The total number of output observations is between *n* and 2*n*-1, where *n* is the number of variables clustered.

The variables in the OUTTREE= data set are:

- the BY variables if any
- _NAME_, a character variable giving the name of the node. If the node is a cluster, the name is CL*n* where *n* is the number of the cluster. If the node is a single variable, the variable name is used.
- _PARENT_, a character variable giving the value of _NAME_ of the parent of the node
- _NCL_, the number of clusters
- _VAREXP_, the total variance explained by the clusters at the current level of the tree
- _PROPOR_, the total proportion of variance explained by the clusters at the current level of the tree.

Usage Notes

Default options for VARCLUS often provide satisfactory results. If you wish to change the final number of clusters, use the MAXCLUSTERS=, MAXEIGEN=, or PROPORTION= options. The MAXEIGEN= and PROPORTION= options usually produce similar results, but occasionally cause different clusters to be selected for splitting. MAXEIGEN= tends to choose clusters with a large number of variables, while PROPORTION= is more likely to select a cluster with a small number of variables.

VARCLUS usually requires more computer time than principal factor analysis, but can be faster than some of the iterative factoring methods. If you have more than 30 variables you may want to reduce execution time by one or more of the following methods:

- use the MINCLUSTERS= and MAXCLUSTERS= options if you know how many clusters you want.
- use the HIERARCHY option.
- use the SEED statement if you have some prior knowledge of what clusters to expect.

If you have sufficient computer time, you may want to try one of the following methods to obtain a better solution:

- with principal components, use the MAXSEARCH= option with a value of 5 or 10.
- try several factoring and rotation methods with FACTOR to use as input to VARCLUS.
- run VARCLUS several times with INITIAL=RANDOM.

Computational Resources

The time required for VARCLUS to analyze a given data set varies greatly depending on the number of clusters requested, the number of iterations in both the alternating least squares and search phases, and whether centroid or principal components are used.

Let:

n = number of observations
v = number of variables
c = number of clusters.

It is assumed that at each stage of clustering, the clusters all contain the same number of variables.

The time required to compute the correlation matrix is roughly proportional to nv^2.

Default cluster initialization requires time roughly proportional to v^3. Any other method of initialization requires time roughly proportional to cv^2.

In the alternating least squares phase, each iteration requires time roughly proportional to cv^2 if centroid components are used, or:

$$(c + 5v/c^2)v^2$$

if principal components are used.

In the search phase, each iteration requires time roughly proportional to v^3/c if centroid components are used, or v^4/c^2 if principal components are used. The HIERARCHY option speeds up each iteration after the first split by as much as $c/2$.

Interpreting VARCLUS Output

Since VARCLUS is a type of oblique component analysis, its output is similar to the output from PROC FACTOR for oblique rotations. The scoring coefficients have the same meaning in both VARCLUS and FACTOR; they are coefficients applied to the standardized variables to compute component scores. The cluster structure is analogous to the factor structure, containing the correlations between each variable and each cluster component. A cluster pattern is not printed because it would be the same as the cluster structure, except that zeros would appear in the same places that zeros appear in the scoring coefficients. The inter-cluster correlations are analogous to inter-factor correlations; they are the correlations among cluster components.

VARCLUS also has a cluster summary and a cluster listing. The cluster summary gives the number of variables in each cluster and the variation explained by the cluster component. The latter is similar to the variation explained by a factor, but includes contributions from only the variables in that cluster rather than from all variables, as in FACTOR. The PROPORTION is obtained by dividing the variance explained by the total variance of variables in the cluster. If the cluster contains two or more variables and the CENTROID option is not used, the second largest eigenvalue of the cluster is also printed.

The cluster listing gives the variables in each cluster. Two squared correlations are printed for each cluster. The column labeled OWN CLUSTER gives the squared correlation of the variable with its own cluster component. This value should be higher than the squared correlation with any other cluster unless an iteration limit has been exceeded or the CENTROID option has been used. The larger the squared correlation is, the better. The column labeled NEXT HIGHEST contains the next highest squared correlation of the variable with a cluster component. This value is low if the clusters are well separated. The column headed R**2 RATIO gives the ratio of NEXT HIGHEST to OWN CLUSTER. A small R**2 RATIO indicates a good clustering.

Printed Output

The output described below is repeated for each cluster solution. VARCLUS prints CLUSTER SUMMARY information, including:

1. the CLUSTER number
2. MEMBERS, the number of members in the cluster
3. CLUSTER VARIATION of the variables in the cluster

4. VARIATION EXPLAINED by the cluster component. This statistic is based only on the variables in the cluster rather than all variables.
5. PROPORTION EXPLAINED, the result of dividing the variation explained by the total variation.
6. SECOND EIGENVALUE, the second largest eigenvalue of the cluster. This is printed if the cluster contains more than one variable and the CENTROID option is not specified.

VARCLUS also prints:

7. TOTAL VARIATION EXPLAINED, the sum across clusters of the variation explained by each cluster.
8. PROPORTION, the total variation explained divided by the total variation of all the variables.

Cluster listings include:

9. VARIABLE, the variables in each cluster
10. R-SQUARED WITH OWN CLUSTER, the squared correlation of the variable with its own cluster component; and R-SQUARED WITH NEXT HIGHEST, the next highest squared correlation of the variable with a cluster component. OWN CLUSTER values should be higher than the R^2 with any other cluster unless an iteration limit has been exceeded or the CENTROID option was used. NEXT HIGHEST should be a low value if the clusters are well separated.
11. R**2 RATIO, the ratio of NEXT HIGHEST to OWN CLUSTER. Low ratios indicate well-separated clusters.

If SHORT is not specified, VARCLUS also prints:

12. STANDARDIZED SCORING COEFFICIENTS, standardized regression coefficients for predicting clusters from variables
13. CLUSTER STRUCTURE, the correlations between each variable and each cluster component
14. INTER-CLUSTER CORRELATIONS, the correlations between the cluster components.

EXAMPLE

The data are correlations among eight physical variables as given by Harman (1976). VARCLUS is run twice: the first time with principal cluster components; the second time with centroid cluster components.

```
DATA PHYS8(TYPE=CORR);
  TITLE EIGHT PHYSICAL VARIABLES MEASURED ON 305 SCHOOL GIRLS;
  TITLE2 SEE PAGE 22 OF HARMAN: MODERN FACTOR ANALYSIS, 3RD ED;
  INPUT _NAME_ $ HEIGHT ARM_SPAN FOREARM LOW_LEG
     WEIGHT BIT_DIAM GIRTH WIDTH;
  _TYPE_ = 'CORR';
  LABEL HEIGHT=HEIGHT
    ARM_SPAN=ARM SPAN
    FOREARM=LENGTH OF FOREARM
    LOW_LEG=LENGTH OF LOWER LEG
    WEIGHT=WEIGHT
    BIT_DIAM=BITROCHANTERIC DIAMETER
    GIRTH=CHEST GIRTH
    WIDTH=CHEST WIDTH;
  CARDS;
```

```
HEIGHT     1.0  .846 .805 .859 .473 .398 .301 .382
ARM_SPAN   .846 1.0  .881 .826 .376 .326 .277 .415
FOREARM    .805 .881 1.0  .801 .380 .319 .237 .345
LOW_LEG    .859 .826 .801 1.0  .436 .329 .327 .365
WEIGHT     .473 .376 .380 .436 1.0  .762 .730 .629
BIT_DIAM   .398 .326 .319 .329 .762 1.0  .583 .577
GIRTH      .301 .277 .237 .327 .730 .583 1.0  .539
WIDTH      .382 .415 .345 .365 .629 .577 .539 1.0
;
PROC VARCLUS;
PROC VARCLUS CENTROID;
```

```
                    EIGHT PHYSICAL VARIABLES MEASURED ON 305 SCHOOL GIRLS                    1
                    SEE PAGE 22 OF HARMAN: MODERN FACTOR ANALYSIS, 3RD ED

                         OBLIQUE PRINCIPAL COMPONENT CLUSTER ANALYSIS

                       10000 OBSERVATIONS          PROPORTION  = 0.000000
                           8 VARIABLES             MAXEIGEN    =   1.0000

                              CLUS(3)SUMMARY (4) 1 CLUSTER(5)          (6)
           (1)      (2)
                                CLUSTER      VARIATION PROPORTION     SECOND
         CLUSTER MEMBERS      VARIATION      EXPLAINED  EXPLAINED  EIGENVALUE
            1        8         8.000000       4.672880  (8) 841     1.770983

        (7)TOTAL VARIATION EXPLAINED = 4.67288      PROPORTION = 0.58411

CLUSTER    1 WILL BE SPLIT
```

```
                    EIGHT PHYSICAL VARIABLES MEASURED ON 305 SCHOOL GIRLS                    2
                    SEE PAGE 22 OF HARMAN: MODERN FACTOR ANALYSIS, 3RD ED

                         OBLIQUE PRINCIPAL COMPONENT CLUSTER ANALYSIS

                              CLUSTER SUMMARY FOR 2 CLUSTERS

                               CLUSTER     VARIATION PROPORTION        SECOND
         CLUSTER MEMBERS      VARIATION    EXPLAINED  EXPLAINED     EIGENVALUE
            1        4         4.000000     3.509218    0.8773        0.236135
            2        4         4.000000     2.917284    0.7293        0.476418

         TOTAL VARIATION EXP(10)ED = 6.426502      PROPORTION = 0.803313

                   (9)         R-SQUARED WITH        (11)
                                 OWN       NEXT       R**2
                    VARIABLE   CLUSTER   HIGHEST      RATIO
CLUSTER      1-----------------------------------------
                    HEIGHT      0.8777    0.2088     0.2378     HEIGHT
                    ARM_SPAN    0.9002    0.1658     0.1842     ARM SPAN
                    FOREARM     0.8661    0.1413     0.1631     LENGTH OF FOREARM
                    LOW_LEG     0.8652    0.1829     0.2115     LENGTH OF LOWER LEG
CLUSTER      2-----------------------------------------
                    WEIGHT      0.8477    0.1974     0.2329     WEIGHT
                    BIT_DIAM    0.7386    0.1341     0.1816     BITROCHANTERIC DIAMETER
                    GIRTH       0.6981    0.0929     0.1331     CHEST GIRTH
                    WIDTH       0.6329    0.1619     0.2559     CHEST WIDTH

                          (12) STANDARDIZED SCORING COEFFICIENTS

         CLUSTER                     1                2

         HEIGHT               .2669773         .0000000        HEIGHT
         ARM_SPAN             .2703773         .0000000        ARM SPAN
         FOREARM              .2651940         .0000000        LENGTH OF FOREARM
         LOW_LEG              .2650569         .0000000        LENGTH OF LOWER LEG
         WEIGHT               .0000000         .3155971        WEIGHT
         BIT_DIAM             .0000000         .2945905        BITROCHANTERIC DIAMETER
         GIRTH                .0000000         .2864066        CHEST GIRTH
         WIDTH                .0000000         .2727100        CHEST WIDTH

                               (13) CLUSTER STRUCTURE

         CLUSTER                     1                2

         HEIGHT               .9368815         .4569081        HEIGHT
         ARM_SPAN             .9488130         .4072103        ARM SPAN
         FOREARM              .9306237         .3758646        LENGTH OF FOREARM
         LOW_LEG              .9301424         .4277147        LENGTH OF LOWER LEG
         WEIGHT               .4442806         .9206865        WEIGHT
         BIT_DIAM             .3662006         .8594042        BITROCHANTERIC DIAMETER
         GIRTH                .3047792         .8355294        CHEST GIRTH
         WIDTH                .4024296         .7955724        CHEST WIDTH
```

```
                    EIGHT PHYSICAL VARIABLES MEASURED ON 305 SCHOOL GIRLS                    3
                    SEE PAGE 22 OF HARMAN: MODERN FACTOR ANALYSIS, 3RD ED

                         OBLIQUE PRINCIPAL COMPONENT CLUSTER ANALYSIS

                              (14) INTER-CLUSTER CORRELATIONS

                           CLUSTER                 1              2

                           1                1.000000       0.445130
                           2                0.445130       1.000000


NO CLUSTER MEETS THE CRITERION FOR SPLITTING
```

```
                    EIGHT PHYSICAL VARIABLES MEASURED ON 305 SCHOOL GIRLS                    4
                    SEE PAGE 22 OF HARMAN: MODERN FACTOR ANALYSIS, 3RD ED

                         OBLIQUE CENTROID COMPONENT CLUSTER ANALYSIS

                     10000 OBSERVATIONS        PROPORTION  = 0.750000
                         8 VARIABLES           MAXEIGEN    =   0.0000


                              CLUSTER SUMMARY FOR 1 CLUSTER

                               CLUSTER     VARIATION PROPORTION       SECOND
              CLUSTER MEMBERS  VARIATION   EXPLAINED EXPLAINED    EIGENVALUE
                 1        8    8.000000    4.631000    0.5789

                TOTAL VARIATION EXPLAINED = 4.631     PROPORTION = 0.578875

CLUSTER    1 WILL BE SPLIT
```

```
                    EIGHT PHYSICAL VARIABLES MEASURED ON 305 SCHOOL GIRLS                    5
                    SEE PAGE 22 OF HARMAN: MODERN FACTOR ANALYSIS, 3RD ED

                         OBLIQUE CENTROID COMPONENT CLUSTER ANALYSIS

                              CLUSTER SUMMARY FOR 2 CLUSTERS

                               CLUSTER     VARIATION PROPORTION       SECOND
              CLUSTER MEMBERS  VARIATION   EXPLAINED EXPLAINED    EIGENVALUE
                 1        4    4.000000    3.509000    0.8772
                 2        4    4.000000    2.910000    0.7275

                TOTAL VARIATION EXPLAINED = 6.419     PROPORTION = 0.802375

                               R-SQUARED WITH
                                OWN       NEXT       R**2
                   VARIABLE  CLUSTER  HIGHEST      RATIO
         CLUSTER    1-----------------------------------
                   HEIGHT      0.8778    0.2075    0.2364     HEIGHT
                   ARM_SPAN    0.8994    0.1669    0.1856     ARM SPAN
                   FOREARM     0.8663    0.1410    0.1627     LENGTH OF FOREARM
                   LOW_LEG     0.8658    0.1824    0.2106     LENGTH OF LOWER LEG
         CLUSTER    2-----------------------------------
                   WEIGHT      0.8368    0.1975    0.2360     WEIGHT
                   BIT_DIAM    0.7335    0.1341    0.1828     BITROCHANTERIC DIAMETER
                   GIRTH       0.6988    0.0929    0.1330     CHEST GIRTH
                   WIDTH       0.6473    0.1618    0.2499     CHEST WIDTH

                              STANDARDIZED SCORING COEFFICIENTS

               CLUSTER                 1               2

               HEIGHT           .2669183        .0000000        HEIGHT
               ARM_SPAN         .2669183        .0000000        ARM SPAN
               FOREARM          .2669183        .0000000        LENGTH OF FOREARM
               LOW_LEG          .2669183        .0000000        LENGTH OF LOWER LEG
               WEIGHT           .0000000        .2931052        WEIGHT
               BIT_DIAM         .0000000        .2931052        BITROCHANTERIC DIAMETER
               GIRTH            .0000000        .2931052        CHEST GIRTH
               WIDTH            .0000000        .2931052        CHEST WIDTH


                                     CLUSTER STRUCTURE

               CLUSTER                 1               2

               HEIGHT           .9368832        .4554855        HEIGHT
               ARM_SPAN         .9483607        .4085886        ARM SPAN
               FOREARM          .9307440        .3754677        LENGTH OF FOREARM
               LOW_LEG          .9304771        .4270543        LENGTH OF LOWER LEG
               WEIGHT           .4444189        .9147813        WEIGHT
               BIT_DIAM         .3662119        .8564534        BITROCHANTERIC DIAMETER
               GIRTH            .3048207        .8359360        CHEST GIRTH
               WIDTH            .4022459        .8045737        CHEST WIDTH
```

```
                 EIGHT PHYSICAL VARIABLES MEASURED ON 305 SCHOOL GIRLS                  6
                 SEE PAGE 22 OF HARMAN: MODERN FACTOR ANALYSIS, 3RD ED

                      OBLIQUE CENTROID COMPONENT CLUSTER ANALYSIS

                            INTER-CLUSTER CORRELATIONS

                         CLUSTER              1             2

                         1               1.000000      0.444845
                         2               0.444845      1.000000

CLUSTER   2 WILL BE SPLIT
```

```
                 EIGHT PHYSICAL VARIABLES MEASURED ON 305 SCHOOL GIRLS                  7
                 SEE PAGE 22 OF HARMAN: MODERN FACTOR ANALYSIS, 3RD ED

                      OBLIQUE CENTROID COMPONENT CLUSTER ANALYSIS

                          CLUSTER SUMMARY FOR 3 CLUSTERS

                            CLUSTER     VARIATION PROPORTION      SECOND
            CLUSTER MEMBERS  VARIATION   EXPLAINED  EXPLAINED   EIGENVALUE
               1       4     4.000000    3.509000    0.8772
               2       3     3.000000    2.383333    0.7944
               3       1     1.000000    1.000000    1.0000

             TOTAL VARIATION EXPLAINED = 6.892333     PROPORTION = 0.861542

                            R-SQUARED WITH
                             OWN       NEXT       R**2
                  VARIABLE  CLUSTER  HIGHEST     RATIO
       CLUSTER   1------------------------------------
                  HEIGHT     0.8778    0.1921    0.2189     HEIGHT
                  ARM_SPAN   0.8994    0.1722    0.1915     ARM SPAN
                  FOREARM    0.8663    0.1225    0.1414     LENGTH OF FOREARM
                  LOW_LEG    0.8658    0.1668    0.1926     LENGTH OF LOWER LEG
       CLUSTER   2------------------------------------
                  WEIGHT     0.8685    0.3956    0.4555     WEIGHT
                  BIT_DIAM   0.7691    0.3329    0.4329     BITROCHANTERIC DIAMETER
                  GIRTH      0.7482    0.2905    0.3883     CHEST GIRTH
       CLUSTER   3------------------------------------
                  WIDTH      1.0000    0.4259    0.4259     CHEST WIDTH

                         STANDARDIZED SCORING COEFFICIENTS

     CLUSTER             1              2              3

     HEIGHT          0.266918       0.000000       0.000000        HEIGHT
     ARM_SPAN        0.266918       0.000000       0.000000        ARM SPAN
     FOREARM         0.266918       0.000000       0.000000        LENGTH OF FOREARM
     LOW_LEG         0.266918       0.000000       0.000000        LENGTH OF LOWER LEG
     WEIGHT          0.000000       0.373979       0.000000        WEIGHT
     BIT_DIAM        0.000000       0.373979       0.000000        BITROCHANTERIC DIAMETER
     GIRTH           0.000000       0.373979       0.000000        CHEST GIRTH
     WIDTH           0.000000       0.000000       1.000000        CHEST WIDTH

                               CLUSTER STRUCTURE

     CLUSTER             1              2              3

     HEIGHT          0.936883       0.438303       0.382000        HEIGHT
     ARM_SPAN        0.948361       0.366125       0.415000        ARM SPAN
     FOREARM         0.930744       0.350044       0.345000        LENGTH OF FOREARM
     LOW_LEG         0.930477       0.408385       0.365000        LENGTH OF LOWER LEG
     WEIGHT          0.444419       0.931955       0.629000        WEIGHT
     BIT_DIAM        0.366212       0.876980       0.577000        BITROCHANTERIC DIAMETER
     GIRTH           0.304821       0.865013       0.539000        CHEST GIRTH
     WIDTH           0.402246       0.652593       1.000000        CHEST WIDTH
```

```
                 EIGHT PHYSICAL VARIABLES MEASURED ON 305 SCHOOL GIRLS                  8
                 SEE PAGE 22 OF HARMAN: MODERN FACTOR ANALYSIS, 3RD ED

                      OBLIQUE CENTROID COMPONENT CLUSTER ANALYSIS

                            INTER-CLUSTER CORRELATIONS

                CLUSTER              1              2              3

                1               1.000000       0.417155       0.402246
                2               0.417155       1.000000       0.652593
                3               0.402246       0.652593       1.000000


NO CLUSTER MEETS THE CRITERION FOR SPLITTING
```

REFERENCES

Anderberg, M.R. (1973), *Cluster Analysis for Applications*, New York: Academic Press.

Harman, H.H. (1976), *Modern Factor Analysis*, (3rd ed.), Chicago: University of Chicago Press.

SCORING

Introduction to Scoring Procedures

Scoring procedures are utilities that produce an output data set with new variables that are transformations of data in the old data set. PROC STANDARD transforms each variable individually. PROC SCORE constructs functions across the variables. PROC RANK produces rank scores across observations. All three procedures produce an output data set, but no printed output.

STANDARD	standardizes variables to a given mean and standard deviation.
RANK	ranks the observations of each numeric variable from low to high and outputs ranks or rank scores.
SCORE	constructs new variables that are a linear combination of old variables according to a scoring data set. This procedure is used with PROC FACTOR and other procedures that output scoring coefficients.

The RANK Procedure

ABSTRACT

The RANK procedure computes ranks for one or more numeric variables across the observations of a SAS data set. The ranks are output to a new SAS data set. Alternatively, PROC RANK produces normal scores or other rank scores.

INTRODUCTION

The RANK procedure ranks values from smallest to largest, assigning the rank 1 to the smallest number, 2 to the next largest, and so on up to rank *n*, the number of nonmissing observations. Tied values are given averaged ranks. Several options are available to request other ranking and tie-handling rules.

Many nonparametric statistical methods use ranks rather than the original values of a variable. For example, a set of data might be passed through PROC RANK to obtain the ranks for a response variable, which could then be fit to an analysis-of-variance model using the ANOVA or GLM procedures.

Ranks are also useful for investigating the distribution of values for a variable. The ranks divided by n or $n+1$ form values in the range 0 to 1, and they estimate the cumulative distribution function. Inverse cumulative distribution functions can be applied to these fractional ranks to obtain probability quantile scores, which can be compared to the original values to judge the fit to the distribution. For example, if a set of data has a normal distribution, the normal scores should be a linear function of the original values, and a plot of scores vs. original values should be a straight line.

PROC RANK is also useful for grouping continuous data into ranges. The GROUPS= feature can break a population into approximately equal-sized groups.

SPECIFICATIONS

The following statements can be specified to invoke PROC RANK:

PROC RANK *options;*
VAR *variables;*
RANKS *names;*
BY *variables;*

PROC RANK Statement

PROC RANK *options;*

The options below may appear in the PROC RANK statement:

DATA=*SASdataset* names the SAS data set to be used by PROC RANK. If DATA= is omitted, the most recently created SAS data set is used.

TIES=MEAN
TIES=HIGH
TIES=LOW
specifies which rank to report for tied values. TIES=MEAN requests that tied values receive the mean of the corresponding ranks (midranks). TIES=HIGH requests that the largest of the corresponding ranks be used. TIES=LOW requests that the smallest of the corresponding ranks be used. The default method is TIES=MEAN. To illustrate the three options available for handling tied values, consider the values of the variable WEIGHT and the ranks that are assigned for each TIES value:

WEIGHT	TIES=MEAN	TIES=HIGH	TIES=LOW
107	1	1	1
110	2.5	3	2
110	2.5	3	2
121	4	4	4
125	6	7	5
125	6	7	5
125	6	7	5
132	8	8	8

DESCENDING
reverses the ranking from largest to smallest. The largest value is given a rank of 1, the next smallest a rank of 2, and so on. When DESCENDING is omitted, values are ranked from smallest to largest.

GROUPS=*n*
requests grouping scores, where *n* is the number of groups. The scores are the integers 0 to (*n*−1). The groups have equal or nearly equal numbers of observations. The lowest values are in the first group; the highest values are in the last group. Common GROUP values: 100 produces percentile ranks, 10 produces deciles, and 4 produces quartiles.

For example, if you want quartile ranks, you specify GROUPS=4. RANK then separates the values of the ranking variable into four groups according to size. The values in the group containing the smallest values receive a quartile value of 0, the values in the next group receive a value of 1, the values in the next group a value of 2, and the largest values receive the value 3.

The formula used to calculate the quantile rank of a value is

FLOOR(*rank***k*/(*n*+1))

where *rank* is the value's rank, *k* is the number of groups specified with the GROUPS= option, and *n* is the number of observations having non-missing values of the ranking variable.

FRACTION
F
requests fractional ranks. RANK divides each rank by the number of observations having non-missing values of the ranking variable and expresses the ranks as fractions. If the TIES= option is omitted or if TIES=HIGH is specified, these fractional ranks can be considered values of a right-continuous empirical cumulative distribution function.

PERCENT
P
asks RANK to divide each rank by the number of observations having non-missing values of the ranking variable and then to multiply the result by 100 to get percentages. Like FRACTION, the PERCENT option implies TIES=HIGH unless another TIES value is specified.

Note: the PERCENT option does not give what are usually called *percentile ranks*, which are produced by GROUPS=100.

NORMAL=BLOM
NORMAL=TUKEY
NORMAL=VW
requests normal scores to be computed from the ranks. The resulting variables appear normally distributed. If NORMAL= is specified, either BLOM, TUKEY, or VW must be given. The formulas are:

BLOM	$y_i = \Psi(r_i - 3/8)/(n + 1/4)$
TUKEY	$y_i = \Psi(r_i - 1/3)/(n + 1/3)$
VW	$y_i = \Psi(r_i)/(n + 1)$

where Ψ is the inverse cumulative normal (PROBIT) function, r_i is the rank, and n is the number of non-missing observations for the ranking variable. VW stands for van der Waerden, whose scores are used for a nonparametric location test.

These normal scores are approximations to the exact expected order statistics for the normal distribution, which are also called *normal scores*.

The BLOM version appears to fit slightly better than the others (Blom, 1958, 145; Tukey, 1962, 22).

SAVAGE
requests Savage (or exponential) scores to be computed from the ranks. The scores are computed by this formula (Lehman, 1975):

$$y_i = \left[\sum_{j=n-r_i+1}^{n} (1/j) \right] - 1$$

OUT=*SASdataset*
names the output data set created by PROC RANK to contain the resulting ranks. If you do not specify OUT=, the default name is OUT=_DATA_, which produces a name like DATA*n*. If you want to create a permanent SAS data set, you must specify a two-level name (see Chapter 12, "SAS Data Sets" for more information on permanent SAS data sets). For details on the data set created by RANK, see **Output Data Set** below.

VAR Statement

VAR *variables*;

RANK computes ranks for the variables given in the VAR statement. These variables must be numeric. If the VAR statement is omitted, ranks are computed for all numeric variables in the data set. The VAR statement must be included if a RANKS statement (below) is used.

RANKS Statement

RANKS *names;*

If you want the original variables included in the output data set in addition to the ranks, use the RANKS statement to assign variable names to the ranks. First, name the rank corresponding to the first variable in the VAR statement, next name the rank that corresponds to the second variable in the VAR statement, and so on.

BY Statement

BY *variables;*

A BY statement may be used with PROC RANK to obtain separate analyses on observations in groups defined by the BY variables. When a BY statement appears, the procedure expects the input data set to be sorted in order of the BY variables. If your input data set is not sorted in ascending order, use the SORT procedure with a similar BY statement to sort the data, or, if appropriate, use the BY statement options NOTSORTED or DESCENDING. For more information, see the discussion of the BY statement in Chapter 8, "Statements Used in the PROC Step," in *SAS User's Guide: Basics, 1982 Edition.*

DETAILS

Missing Values

Missing values are not ranked and are left missing when ranks or rank scores replace the other values of the ranking variable.

Limitations

The RANK procedure should be used only for data sets containing 32,767 or fewer observations.

PROC RANK uses the single-precision representation of a number for ranking (although it is very rare for SAS to use single precision). Thus RANK is unable to distinguish differences in numbers past approximately seven significant digits.

Output Data Set

RANK creates a new SAS data set containing the ranks or rank scores, but no printed output.

The new output data set contains all the variables from the input data set plus the variable named on the RANKS statement if one is specified. If a RANKS statement is used, a VAR statement must also be included. If no RANKS statement is given, then the procedure stores the ranks in the output data set into the original variables that were ranked. If no VAR statement is included, the procedure ranks all numeric variables.

Nonparametric Statistics

Many nonparametric methods are based on taking the ranks of a variable and analyzing these ranks instead of the original values.

- A two-sample t test applied to the ranks is equivalent to a Wilcoxon rank sum test using the t approximation for the significance level. If the *t* test is applied to the normal scores rather than to the ranks, the test becomes equivalent to the van der Waerden test. If the *t* test is applied to median scores (GROUPS= 2), the test becomes the median test.

- A one-way analysis of variance applied to ranks is equivalent to the Kruskal-Wallis *k*-sample test; the *F* test generated by the parametric procedure applied to the ranks is often better than the χ^2 approximation used by Kruskal-Wallis. This test can be extended to other rank scores. (Quade, 1966)
- Friedman's two-way analysis for block designs can be obtained by ranking within blocks (using the BY feature of PROC RANK) and then performing a main-effects analysis of variance on these ranks. (Conover, 1980)
- Regression relationships can be investigated using rank transformations with a method described by Iman and Conover (1979).

EXAMPLE

This example uses PROC RANK to get the ranks of the variable GAIN. The RANKS statement assigns the name RANKGAIN to the variable containing the ranks in the output data set. Since OUT= is not specified, RANK creates a new data set using the DATA*n* naming rules. The PRINT procedure prints the contents of the output data set. The second execution of PROC RANK uses a BY variable.

```
DATA A;
  INPUT LOCATION GAIN;
  CARDS;
1 7.2
1 7.9
1 7.6
1 6.3
1 8.4
1 8.1
2 8.1
2 7.3
2 7.7
2 7.7
;
PROC RANK;
  VAR GAIN;
  RANKS RANKGAIN;
PROC PRINT;
  TITLE RANK THE GAIN VALUES;
PROC RANK DATA=A OUT=B;
  BY LOCATION;
  RANKS RGAIN;
  VAR GAIN;
PROC PRINT;
  BY LOCATION;
  TITLE RANKINGS WITHIN LOCATIONS;
```

```
                         RANK THE GAIN VALUES                                  1

              OBS     LOCATION      GAIN     RANKGAIN

                1         1          7.2        2.0
                2         1          7.9        7.0
                3         1          7.6        4.0
                4         1          6.3        1.0
                5         1          8.4       10.0
                6         1          8.1        8.5
                7         2          8.1        8.5
                8         2          7.3        3.0
                9         2          7.7        5.5
               10         2          7.7        5.5
```

```
                         RANKINGS WITHIN LOCATIONS                             2
---------------------------------- LOCATION=1 ----------------------------------

                        OBS     GAIN     RGAIN

                         1       7.2       2
                         2       7.9       4
                         3       7.6       3
                         4       6.3       1
                         5       8.4       6
                         6       8.1       5

---------------------------------- LOCATION=2 ----------------------------------

                        OBS     GAIN     RGAIN

                         7       8.1      4.0
                         8       7.3      1.0
                         9       7.7      2.5
                        10       7.7      2.5
```

REFERENCES

Blom, G. (1958), *Statistical Estimates and Transformed Beta Variables*, New York: John Wiley and Sons, Inc.

Conover, W.J. (1980), *Practical Nonparametric Statistics*, Second Edition, New York: John Wiley and Sons, Inc.

Conover, W.J. and Iman, R.L. (1976), ''On Some Alternative Procedures Using Ranks for the Analysis of Experimental Designs,'' *Communications in Statistics*, A5, 14, 1348-1368.

Conover, W.J. and Iman, R.L. (1981), ''Rank Transformations as a Bridge Between Parametric and Nonparametric Statistics,'' *The American Statistician*, 35, 124-129.

Iman, R.L. and Conover, W.J. (1979), ''The Use of the Rank Transform in Regression,'' *Technometrics*, 21, 499-509.

Lehman, E.L. (1975), *Nonparametrics: Statistical Methods Based on Ranks*, San Francisco: Holden-Day.

Quade, D. (1966), ''On Analysis of Variance for the *k*-Sample Problem,'' *Annals of Mathematical Statistics*, 37, 1747-1758.

Tukey, John W. (1962), ''The Future of Data Analysis,'' *Annals of Mathematical Statistics*, 33, 22.

The SCORE Procedure

ABSTRACT

The SCORE procedure multiplies values from two SAS data sets, one containing coefficients (for example, factor scoring coefficients or regression coefficients) and the other containing the original data used to calculate the coefficients. The result of this multiplication is an output SAS data set containing the linear combinations of the coefficients and the original data values.

INTRODUCTION

Many statistical procedures output coefficients that PROC SCORE can apply to raw data to produce scores. The new score variable is formed as a linear combination of old variables, where the linear combination uses the scoring coefficients. In other words, SCORE crossmultiplies part of one data set with another.

The data set containing scoring coefficients must also have two special variables: the _TYPE_ variable identifies the observations that contain scoring coefficients; the _NAME_ or _MODEL_ variable contains a SAS name that can be used to name the new variables output from PROC SCORE.

For example, PROC FACTOR produces an output data set that contains factor scoring coefficients. These scoring coefficients are identified on the data set output from FACTOR by _TYPE_='SCORE', which is the TYPE that PROC SCORE searches for unless the TYPE= option on the PROC SCORE statement specifies otherwise. The _NAME_ variable on the data set from FACTOR gives each score a name like FACTOR1, FACTOR2, and so forth. The data set output from PROC FACTOR also has _TYPE_= MEAN and _TYPE_=STD records, which PROC SCORE uses to standardize the data before scores are calculated. Thus, with the FACTOR procedure you do not produce factor scores directly, but rather you invoke PROC SCORE using the FACTOR output data set to obtain scores.

Several other multivariate procedures in SAS involve methods for forming linear combinations of variables and produce coefficients in a form convenient for PROC SCORE to use. Consult the descriptions of PRINCOMP, CANCORR, CANDISC, and VARCLUS.

Some regression procedures, like REG, can output the parameter estimates to an output data set specified by OUTEST=. The _TYPE_ variable is given a value like OLS, which corresponds to the estimation method. The _MODEL_ variable contains the label of the MODEL statement. When PROC SCORE is invoked using the TYPE=OLS option, the procedure first reads in the parameter estimates, then multiplies them into the original data to produce scores. Since the dependent variable is coded with coefficient -1, the scores are the negative residuals. If SCORE is told to ignore the -1 coefficient (PREDICT option), the scores are predicted values.

When the data set containing the coefficients also includes observations containing means and standard deviations (_TYPE_='MEAN' and _TYPE_='STD'), the data are standardized before scoring.

SPECIFICATIONS

PROC SCORE is controlled by the statements:

PROC SCORE *options;*
VAR *variables;*

Since SCORE has not been programmed to synchronize BY groups across both input data sets, the BY statement may **not** be used with PROC SCORE.

PROC SCORE Statement

PROC SCORE *options;*

The options below may appear in the PROC SCORE statement.

DATA=*SASdataset* names the input SAS data set containing the raw data to score. This specification is **required**.

SCORE=*SASdataset* names the data set containing scoring coefficients. If SCORE= is omitted, the most recently created SAS data set is used. This data set must have two special variables: __TYPE__ and either __NAME__ or __MODEL__.

OUT=*SASdataset* names the output SAS data set to be created by SCORE. This data set has all the variables from the DATA= data set, plus score variables named from the __NAME__ or __MODEL__ values in the SCORE= data set. If you want to create a permanent SAS data set, you must specify a two-level name (see Chapter 12, "SAS Data Sets," in *SAS User's Guide: Basics, 1982 Edition* for more information on permanent SAS data sets).

TYPE=*value* specifies the observations in the SCORE= data set that have scoring coefficients. The TYPE procedure option is unrelated to the data set parameter that has the same name. The procedure examines the values of the special variable __TYPE__ in the SCORE= data set. When the value matches the TYPE= value, the observation is used to score a new variable. The default is TYPE=SCORE, which selects scoring coefficients for which the variable __TYPE__ has the value of SCORE. Since this is what is desired from data sets produced by FACTOR, you need not specify TYPE= for factor scoring. When regression coefficients are used, TYPE= is specified as a regression method, usually TYPE=OLS.

PREDICT signals PROC SCORE that it should treat coefficients of -1 in the SCORE= data set as 0. In regression applications, the dependent variable is coded with a coefficient of -1. Applied directly to regression results, PROC SCORE produces (negative) residuals; the PREDICT option changes this so that predicted values are produced instead.

NOSTD suppresses centering and scaling the raw data. Ordinarily, if SCORE finds MEAN and STD observations in the SCORE= data set, the procedure uses these to standardize the data before scoring.

VAR Statement

VAR *variables;*

The VAR statement specifies the variables to be used in computing a score. These variables must be in both the DATA= and SCORE= input data sets. If no VAR statement is given, the procedure uses all numeric variables in the SCORE= data set. You should almost always use a VAR statement with PROC SCORE, since you rarely want to score all the numeric variables.

DETAILS

Missing Values

If one of the original variables has a missing value for an observation, all the scores have missing values for that observation. Exception: if the PREDICT option is specified, the variable with a coefficient of -1 can tolerate a missing value and still produce a prediction score.

Output Data Set

PROC SCORE produces an output data set but no printed output.

EXAMPLE

The code below shows examples of three kinds of scoring.

First, PROC FACTOR produces an output data set containing scoring coefficients in observations identified by __TYPE__='SCORE'. These data, together with the original data set FITNESS, are supplied to PROC SCORE, resulting in a data set containing scores FACTOR1 and FACTOR2. (See the chapters for CANCORR, CANDISC, PRINCOMP and VARCLUS for more examples.)

```
* THIS DATA SET CONTAINS ONLY THE FIRST 12 OBSERVATIONS FROM
* THE FULL DATA SET USED IN THE REGRESSION CHAPTERS.;
DATA FITNESS;
  INPUT AGE WEIGHT OXY RUNTIME RSTPULSE RUNPULSE @@;
  CARDS;

44 89.47 44.609 11.37 62 178    40 75.07 45.313 10.07 62 185
44 85.84 54.297  8.65 45 156    42 68.15 59.571  8.17 40 166
38 89.02 49.874  9.22 55 178    47 77.45 44.811 11.63 58 176
40 75.98 45.681 11.95 70 176    43 81.19 49.091 10.85 64 162
44 81.42 39.442 13.08 63 174    38 81.87 60.055  8.63 48 170
44 73.03 50.541 10.13 45 168    45 87.66 37.388 14.03 56 186
;
PROC FACTOR DATA=FITNESS OUT=FACTOUT
  METHOD=PRIN ROTATE=VARIMAX SCORE;
  VAR AGE WEIGHT RUNTIME RUNPULSE RSTPULSE;
  TITLE FACTOR SCORING EXAMPLE;
```

```
PROC PRINT DATA=FACTOUT;
  TITLE2 DATASET FROM PROC  FACTOR;
PROC SCORE DATA=FITNESS SCORE=FACTOUT OUT=FSCORE;
  VAR AGE WEIGHT RUNTIME RUNPULSE RSTPULSE;
PROC PRINT DATA=FSCORE;
  TITLE2 DATASET FROM PROC SCORE;
```

```
                              FACTOR SCORING EXAMPLE                                         1
INITIAL FACTOR METHOD: PRINCIPAL COMPONENTS

                                   PRIOR COMMUNALITY ESTIMATES:  ONE

                EIGENVALUES OF THE CORRELATION MATRIX:  TOTAL =  5.000000   AVERAGE =  1.000000

                                     1          2          3          4          5
                  EIGENVALUE    2.309306   1.192200   0.882227   0.502567   0.113700
                  DIFFERENCE    1.117107   0.309972   0.379660   0.388867
                  PROPORTION      0.4619     0.2384     0.1764     0.1005     0.0227
                  CUMULATIVE      0.4619     0.7003     0.8767     0.9773     1.0000

                      2 FACTORS WILL BE RETAINED BY THE MINEIGEN CRITERION

                                      FACTOR PATTERN

                                      FACTOR1   FACTOR2

                          AGE         0.29795   0.93675
                          WEIGHT      0.43282  -0.17750
                          RUNTIME     0.91983   0.28782
                          RUNPULSE    0.72671  -0.38191
                          RSTPULSE    0.81179  -0.23344

                              VARIANCE EXPLAINED BY EACH FACTOR

                                    FACTOR1    FACTOR2
                                   2.309306   1.192200

                         FINAL COMMUNALITY ESTIMATES: TOTAL =    3.501506

                         AGE     WEIGHT    RUNTIME   RUNPULSE   RSTPULSE
                    0.966284   0.218834   0.928933   0.673962   0.713493
```

```
                              FACTOR SCORING EXAMPLE                                         2
ROTATION METHOD: VARIMAX

                               ORTHOGONAL TRANSFORMATION MATRIX

                                                1          2

                             1            0.92536    0.37908
                             2           -0.37908    0.92536

                                  ROTATED FACTOR PATTERN

                                      FACTOR1   FACTOR2

                          AGE        -0.07939   0.97979
                          WEIGHT      0.46780  -0.00018
                          RUNTIME     0.74207   0.61503
                          RUNPULSE    0.81725  -0.07792
                          RSTPULSE    0.83969   0.09172

                              VARIANCE EXPLAINED BY EACH FACTOR

                                    FACTOR1    FACTOR2
                                   2.148775   1.352731

                         FINAL COMMUNALITY ESTIMATES: TOTAL =    3.501506

                         AGE     WEIGHT    RUNTIME   RUNPULSE   RSTPULSE
                    0.966284   0.218834   0.928933   0.673962   0.713493

                SQUARED MULTIPLE CORRELATIONS OF THE VARIABLES WITH EACH FACTOR

                                    FACTOR1    FACTOR2
                                   1.000000   1.000000

                              STANDARDIZED SCORING COEFFICIENTS

                                      FACTOR1    FACTOR2

                          AGE        -0.17846    0.77600
                          WEIGHT      0.22987   -0.06672
                          RUNTIME     0.27707    0.37440
                          RUNPULSE    0.41263   -0.17714
                          RSTPULSE    0.39952   -0.04793
```

```
                         FACTOR SCORING EXAMPLE                                           3
                         DATASET FROM PROC FACTOR

OBS     TYPE       NAME         AGE       WEIGHT     RUNTIME    RUNPULSE    RSTPULSE

  1     MEAN                   42.4167    80.5125    10.6483    172.917     55.6667
  2     STD                     2.8431     6.7660     1.8444      8.918      9.2769
  3     N                      12.0000    12.0000    12.0000     12.000     12.0000
  4     CORR       AGE          1.0000     0.0128     0.5005     -0.095     -0.0080
  5     CORR       WEIGHT       0.0128     1.0000     0.2637      0.173      0.2396
  6     CORR       RUNTIME      0.5005     0.2637     1.0000      0.556      0.6620
  7     CORR       RUNPULSE    -0.0953     0.1731     0.5555      1.000      0.4853
  8     CORR       RSTPULSE    -0.0080     0.2396     0.6620      0.485      1.0000
  9     COMMUNAL                0.9663     0.2188     0.9289      0.674      0.7135
 10     PRIORS                  1.0000     1.0000     1.0000      1.000      1.0000
 11     EIGENVAL                2.3093     1.1922     0.8822      0.503      0.1137
 12     UNROTATE   FACTOR1      0.2980     0.4328     0.9198      0.727      0.8118
 13     UNROTATE   FACTOR2      0.9368    -0.1775     0.2878     -0.382     -0.2334
 14     TRANSFOR   FACTOR1      0.9254    -0.3791      .           .          .
 15     TRANSFOR   FACTOR2      0.3791     0.9254      .           .          .
 16     PATTERN    FACTOR1     -0.0794     0.4678     0.7421      0.817      0.8397
 17     PATTERN    FACTOR2      0.9798    -0.0002     0.6150     -0.078      0.0917
 18     SCORE      FACTOR1     -0.1785     0.2299     0.2771      0.413      0.3995
 19     SCORE      FACTOR2      0.7760    -0.0667     0.3744     -0.177     -0.0479
```

```
                         FACTOR SCORING EXAMPLE                                              4
                         DATASET FROM PROC SCORE

OBS   AGE   WEIGHT    OXY      RUNTIME   RSTPULSE   RUNPULSE    FACTOR1    FACTOR2

  1    44   89.47    44.609    11.37        62         178       0.8213     0.3566
  2    40   75.07    45.313    10.07        62         185       0.7117    -0.9961
  3    44   85.84    54.297     8.65        45         156      -1.4606     0.3651
  4    42   68.15    59.571     8.17        40         166      -1.7609    -0.2766
  5    38   89.02    49.874     9.22        55         178       0.5582    -1.6768
  6    47   77.45    44.811    11.63        58         176      -0.0011     1.4071
  7    40   75.98    45.681    11.95        70         176       0.9532    -0.4860
  8    43   81.19    49.091    10.85        64         162      -0.1295     0.3672
  9    44   81.42    39.442    13.08        63         174       0.6627     0.8574
 10    38   81.87    60.055     8.63        48         170      -0.4450    -1.5310
 11    44   73.03    50.541    10.13        45         168      -1.1183     0.5535
 12    45   87.66    37.388    14.03        56         186       1.2084     1.0595
```

In this continued example, regression coefficients are output by the REG procedure. The __TYPE__ is marked OLS and the names of the variables are found in the variable __MODEL__ which gets its values from the label of a model. If the scoring variables include only the independent variables, the resulting score variable contains predicted values. If the dependent variable is included (with a score coefficient of -1), the resulting scores are negative residuals. You can also use the PREDICT option to ignore a scoring coefficient of -1.

```
PROC REG DATA=FITNESS OUTEST=REGOUT;
  OXYHAT: MODEL OXY=AGE WEIGHT RUNTIME RUNPULSE RSTPULSE;
  TITLE REGRESSION SCORING EXAMPLE;
PROC PRINT DATA=REGOUT;
  TITLE2 OUTEST DATASET FROM PROC REG;
PROC SCORE DATA=FITNESS SCORE=REGOUT OUT=RSCOREP TYPE=OLS;
  VAR AGE WEIGHT RUNTIME RUNPULSE RSTPULSE;
PROC PRINT DATA=RSCOREP;
  TITLE2 PREDICTED SCORES FOR REGRESSION;
PROC SCORE DATA=FITNESS SCORE=REGOUT
  OUT=RSCORER TYPE=OLS;
  VAR OXY AGE WEIGHT RUNTIME RUNPULSE RSTPULSE;
PROC PRINT DATA=RSCORER;
  TITLE2 RESIDUAL SCORES FOR  REGRESSION;
```

```
                              REGRESSION SCORING EXAMPLE                                   5

MODEL: OXYHAT
DEP VARIABLE: OXY

                       SUM OF           MEAN
SOURCE       DF       SQUARES         SQUARE        F VALUE        PROB>F

MODEL         5       509.622        101.924         15.802        0.0021
ERROR         6     38.700603       6.450101
C TOTAL      11       548.323

     ROOT MSE        2.539705       R-SQUARE         0.9294
     DEP MEAN       48.389417       ADJ R-SQ         0.8706
     C.V.            5.248472

                    PARAMETER       STANDARD     T FOR HO:
VARIABLE     DF      ESTIMATE          ERROR   PARAMETER=0    PROB > |T|

INTERCEP      1       151.916      31.047376         4.893        0.0027
AGE           1     -0.630450       0.425027        -1.483        0.1885
WEIGHT        1     -0.105862       0.118688        -0.892        0.4068
RUNTIME       1     -1.756978       0.938441        -1.872        0.1103
RUNPULSE      1     -0.228910       0.121686        -1.881        0.1090
RSTPULSE      1     -0.179102       0.130050        -1.377        0.2176
```

```
                              REGRESSION SCORING EXAMPLE                                   6
                             OUTEST DATASET FROM PROC REG

OBS    TYPE    MODEL     SIGMA    OXY      AGE       WEIGHT     RUNTIME   RUNPULSE   RSTPULSE   INTERCEP

 1     OLS     OXYHAT    2.5397   -1    -0.63045   -0.10586    -1.757   -0.22891    -0.1791    151.916
```

```
                              REGRESSION SCORING EXAMPLE                                   7
                          PREDICTED SCORES FOR REGRESSION

       OBS    AGE    WEIGHT      OXY      RUNTIME    RSTPULSE    RUNPULSE     OXYHAT

         1    44     89.47     44.609      11.37        62          178       42.8771
         2    40     75.07     45.313      10.07        62          185       47.6050
         3    44     85.84     54.297       8.65        45          156       56.1211
         4    42     68.15     59.571       8.17        40          166       58.7044
         5    38     89.02     49.874       9.22        55          178       51.7386
         6    47     77.45     44.811      11.63        58          176       42.9756
         7    40     75.98     45.681      11.95        70          176       44.8329
         8    43     81.19     49.091      10.85        64          162       48.6020
         9    44     81.42     39.442      13.08        63          174       41.4613
        10    38     81.87     60.055       8.63        48          170       56.6171
        11    44     73.03     50.541      10.13        45          168       52.1299
        12    45     87.66     37.388      14.03        56          186       37.0080
```

```
                              REGRESSION SCORING EXAMPLE                                   8
                           RESIDUAL SCORES FOR REGRESSION

       OBS    AGE    WEIGHT      OXY      RUNTIME    RSTPULSE    RUNPULSE     OXYHAT

         1    44     89.47     44.609      11.37        62          178       -1.7319
         2    40     75.07     45.313      10.07        62          185        2.2920
         3    44     85.84     54.297       8.65        45          156        1.8241
         4    42     68.15     59.571       8.17        40          166       -0.8666
         5    38     89.02     49.874       9.22        55          178        1.8646
         6    47     77.45     44.811      11.63        58          176       -1.8354
         7    40     75.98     45.681      11.95        70          176       -0.8481
         8    43     81.19     49.091      10.85        64          162       -0.4890
         9    44     81.42     39.442      13.08        63          174        2.0193
        10    38     81.87     60.055       8.63        48          170       -3.4379
        11    44     73.03     50.541      10.13        45          168        1.5889
        12    45     87.66     37.388      14.03        56          186       -0.3800
```

Finally, we create a custom scoring data set. The first scoring coefficient creates a variable that is AGE-WEIGHT; the second evaluates RUNPULSE-RSTPULSE, and the third totals all six variables.

```
DATA A;
  INPUT _TYPE_ $ _NAME_ $
    AGE WEIGHT RUNTIME RUNPULSE RSTPULSE;
  CARDS;
  SCORE AGE_WGT  1 -1 0 0 0
  SCORE RUN_RST  0 0 0 1 -1
  SCORE TOTAL    1 1 1 1 1
PROC PRINT DATA=A;
  TITLE CONSTRUCTED SCORING EXAMPLE;
  TITLE2 SCORING COEFFICIENTS;
PROC SCORE DATA=FITNESS SCORE=A OUT=B ;
  VAR AGE WEIGHT RUNTIME RUNPULSE RSTPULSE;
PROC PRINT DATA=B;
  TITLE2 SCORED DATA;
```

CONSTRUCTED SCORING EXAMPLE 9
SCORING COEFFICIENTS

OBS	TYPE	NAME	AGE	WEIGHT	RUNTIME	RUNPULSE	RSTPULSE
1	SCORE	AGE_WGT	1	-1	0	0	0
2	SCORE	RUN_RST	0	0	0	1	-1
3	SCORE	TOTAL	1	1	1	1	1

CONSTRUCTED SCORING EXAMPLE 10
SCORED DATA

OBS	AGE	WEIGHT	OXY	RUNTIME	RSTPULSE	RUNPULSE	AGE_WGT	RUN_RST	TOTAL
1	44	89.47	44.609	11.37	62	178	-45.47	116	384.84
2	40	75.07	45.313	10.07	62	185	-35.07	123	372.14
3	44	85.84	54.297	8.65	45	156	-41.84	111	339.49
4	42	68.15	59.571	8.17	40	166	-26.15	126	324.32
5	38	89.02	49.874	9.22	55	178	-51.02	123	369.24
6	47	77.45	44.811	11.63	58	176	-30.45	118	370.08
7	40	75.98	45.681	11.95	70	176	-35.98	106	373.93
8	43	81.19	49.091	10.85	64	162	-38.19	98	361.04
9	44	81.42	39.442	13.08	63	174	-37.42	111	375.50
10	38	81.87	60.055	8.63	48	170	-43.87	122	346.50
11	44	73.03	50.541	10.13	45	168	-29.03	123	340.16
12	45	87.66	37.388	14.03	56	186	-42.66	130	388.69

The STANDARD Procedure

ABSTRACT

The STANDARD procedure standardizes some or all of the variables in a SAS data set to a given mean and standard deviation and produces a new SAS data set to contain the standardized values.

INTRODUCTION

Standardizing is a technique for removing location and scale attributes from a set of data. Sometimes you need to center the values on a variable to a mean of 0 and a standard deviation of 1. Some statistical techniques begin the analysis by standardizing the data in this way. If your data are normally distributed, standardizing is also "studentizing," since the result has Student's *t* distribution.

SPECIFICATIONS

The statements that control PROC STANDARD are:

PROC STANDARD *options;*
VAR *variables;*
BY *variables;*

PROC STANDARD Statement

PROC STANDARD *options;*

The options below may appear in the PROC STANDARD statement. Note that unless at least one of the last three options is specified, your output SAS data set is an exact copy of the input SAS data set.

DATA=*SASdataset* gives the name of the data set to be used by STANDARD. If it is omitted, STANDARD uses the most recently created SAS data set.

OUT=*SASdataset* gives the name of the new SAS data set to contain the standardized variables. If OUT= is omitted, SAS names the new data set using the DATA*n* convention. The OUT= data set contains all the variables from the input data set, including those not standardized. If you want to create a permanent SAS data set, you must supply a two-level name (see Chapter 12, "SAS Data Sets," in *SAS User's Guide: Basics, 1982 Edition,* for more information on permanent SAS data sets).

MEAN=*m*
M=*m*
: requests that all the variables in the VAR statement (or all the numeric variables if the VAR statement is omitted) be standardized to a mean of *m*. Without the MEAN option, the mean of the output values is the same as the mean of the input values.

STD=*s*
S=*s*
: requests that all the variables in the VAR statement (or all the numeric variables if the VAR statement is omitted) be standardized to a standard deviation of *s*. Without the STD option, the standard deviation of the output values is the same as the standard deviation of the input values. The REPLACE option, below, does not affect STD's action.

REPLACE
: requests that all missing values be replaced with the variable mean. If MEAN=*m* is also specified, missing values are set instead to *m*.

VAR Statement

VAR *variables;*

List the variables that you want standardized in the VAR statement. If the VAR statement is omitted, all the numeric variables in the data set are standardized.

BY Statement

BY *variables;*

A BY statement may be used with PROC STANDARD to obtain separate analyses on observations in groups defined by the BY variables. When a BY statement appears, the procedure expects the input data set to be sorted in order of the BY variables. If your input data set is not sorted in ascending order, use the SORT procedure with a similar BY statement to sort the data, or, if appropriate, use the BY statement options NOTSORTED or DESCENDING. For more information, see the discussion of the BY statement in Chapter 8, "Statements Used in the PROC Step," in *SAS User's Guide: Basics, 1982 Edition.*

DETAILS

Missing Values

Missing values are excluded from the standardization process. Unless the REPLACE option is specified, missing values are left as missing in the output data set. When REPLACE is specified, missing values are replaced with the variable mean or with the mean specified by the MEAN option in the PROC STANDARD statement.

Output Data Set

STANDARD produces an output SAS data set to contain the standardized variables but no printed output.

EXAMPLE

The data in the example below consist of three test scores for students in two sections of a course. For each section, we want to standardize all three test scores to a mean of 80 and a standard deviation of 5. To keep the original test scores with the standardized scores in the same SAS data set, create the three variables STEST1-STEST3 to contain the test scores.

The PROC STANDARD statement includes the MEAN and STD options and gives the name NEW to the new SAS data set containing the standardized values. Since only STEST1-STEST3 appear in the VAR statement, they are the only variables standardized—the original variables TEST1-TEST3 are not standardized. The BY statement asks that the standardization be done for each section of the course separately. The PROC PRINT statement prints the new data set, and the PROC MEANS output shows that STEST1-STEST3 in both sections have (1) means of 80 and (2) standard deviations of 5.

```
DATA A;
  INPUT STUDENT SECTION TEST1-TEST3;
  STEST1=TEST1;
  STEST2=TEST2;
  STEST3=TEST3;
  CARDS;
238900545 1 94 91 87
254701167 1 95 96 97
238806445 2 91 86 94
999002527 2 80 76 78
263924860 1 92 40 85
459700886 2 75 76 80
416724915 2 66 69 72
999001230 1 82 84 80
242760674 1 75 76 70
990001252 2 51 66 91
;
PROC SORT;
  BY SECTION;
PROC STANDARD MEAN=80 STD=5 OUT=NEW;
  BY SECTION;
  VAR STEST1-STEST3;
PROC PRINT DATA=NEW;
  BY SECTION;
  TITLE STANDARDIZED TEST SCORES;
PROC MEANS DATA=NEW MAXDEC=2 N MEAN STD;
  BY SECTION;
  DROP STUDENT;
```

```
                         STANDARDIZED TEST SCORES                                    1

-------------------------------------------- SECTION=1 ---------------------------------------------

     OBS    STUDENT     TEST1    TEST2    TEST3    STEST1     STEST2     STEST3

      1    238900545     94       91       87     83.6634    83.0601    81.6187
      2    254701167     95       96       97     84.2358    84.1851    86.6772
      3    263924860     92       40       85     82.5186    71.5848    80.6070
      4    999001230     82       84       80     76.7945    81.4850    78.0778
      5    242760674     75       76       70     72.7876    79.6850    73.0193
```

(continued on next page)

(continued from previous page)

```
------------------------------------------------------- SECTION=2 -------------------------------------------------------

          OBS     STUDENT      TEST1     TEST2     TEST3     STEST1      STEST2      STEST3

            6    238806445      91        86        94      86.1022     87.3710     85.9656
            7    999002527      80        76        78      82.4542     80.9052     77.2884
            8    459700886      75        76        80      80.7959     80.9052     78.3730
            9    416724915      66        69        72      77.8112     76.3792     74.0344
           10    990001252      51        66        91      72.8365     74.4394     84.3386
```

```
                  STANDARDIZED TEST SCORES                               2

          VARIABLE          N         MEAN          STANDARD
                                                    DEVIATION
          ------------------ SECTION=1 -------------------

          TEST1             5         87.60           8.73
          TEST2             5         77.40          22.22
          TEST3             5         83.80           9.88
          STEST1            5         80.00           5.00
          STEST2            5         80.00           5.00
          STEST3            5         80.00           5.00

          ------------------ SECTION=2 -------------------

          TEST1             5         72.60          15.08
          TEST2             5         74.60           7.73
          TEST3             5         83.00           9.22
          STEST1            5         80.00           5.00
          STEST2            5         80.00           5.00
          STEST3            5         80.00           5.00
```

MATRIX

Introduction

MATRIX

Introduction to the MATRIX Language

This chapter introduces MATRIX, which is both a SAS procedure and a programming language. The purpose of this section is to explain how MATRIX is used and to compare MATRIX to other programming languages.

MATRIX is a programming language. While most SAS procedures are prepackaged routines where the specifications merely control the details of an analysis, MATRIX is a complete programming language, and MATRIX statements are executable programming statements.

MATRIX makes SAS open-ended—suitable for custom work as well as packaged analyses. Analysts need two kinds of tools. They need packaged analysis programs for routine work and a general-purpose language for custom programming of methods too specialized or too new to be packaged. MATRIX provides a high-level programming language to serve the latter need. If a statistical method has not been implemented directly in a SAS procedure, you can program it using the MATRIX procedure.

MATRIX data elements are matrices. Most programming languages deal with single data elements; the fundamental data element in MATRIX is a matrix, a two-dimensional (row by column) array of numeric (double-precision real floating-point) values.

MATRIX expressions use operators that are applied to entire matrices. For example, the expression A+B in MATRIX adds the elements of the two matrices A and B. The expression A*B performs a matrix multiplication, while A#B does elementwise multiplication of the values in A and B.

MATRIX incorporates a powerful vocabulary of operators. Matrix operations that require calls to math-library subroutines in other languages are built into the language of MATRIX. Most programming languages have six to ten operators and a small function library. MATRIX offers 26 operators, 29 built-in functions, and all the functions of the DATA step; and MATRIX allows you to define your own functions.

MATRIX avoids housekeeping work typical in other programming languages. In many languages you have to explicitly declare, dimension, or allocate storage for a data item, and you cannot change its attributes once it is declared. In contrast, MATRIX does all the housekeeping automatically. The attributes of a matrix are determined when the matrix is given a value (this is called *late binding*). The procedure automatically finds space as needed for a matrix of any size. A matrix can change its size and attributes at any time.

MATRIX operations allow more direct programming. MATRIX eliminates most iterative specifications, calls to math subroutines, declarations, and allocations. For example, finding the sum of all positive elements of a matrix **X** requires one line in MATRIX:

```
S = SUM(X # (X>0)) .
```

The expression involves the sum of an elementwise product of two items. The first item is the matrix **X**; the second item is an indicator matrix showing the positive values of **X**. Once you have learned the matrix notation, this operation is clearer and more direct than an iterative specification that would be used in another programming language such as PL/I:

```
SUM=0;
DO I=1 TO N;
  DO J=1 TO M;
    IF X(I,J)>0 THEN SUM=SUM+X(I,J);
    END;
  END;
```

MATRIX notation is compact. Since MATRIX operations deal with arrays of numbers and the most commonly used mathematical and matrix operations are built directly into the language, programs that take hundreds of lines of code in other languages may take only a few lines in MATRIX.

MATRIX allows you to think directly in matrix algebra terms. If you are a statistician, you already know the formulas for many statistical methods. But in most languages it is no easy job to transcribe these formulas into a program. In MATRIX, you can transcribe almost directly from matrix algebra notation into MATRIX program statements. For example, the formula for the least-squares estimates of parameters in a linear model in matrix notation is written:

$$\mathbf{b} = (\mathbf{X}'\mathbf{X})^{-1}\mathbf{X}'\mathbf{y} .$$

It can be written in MATRIX as:

```
B=INV(X'*X)*X'*Y;
```

MATRIX is an alternative to APL. APL is another language that deals with matrices of values. Unlike MATRIX, however, APL works with a specialized character set that takes time to learn and is available only on special terminals. MATRIX can be learned more quickly than APL, since it builds on traditional and more familiar notation.

MATRIX has a rich set of control statements. MATRIX allows you to program with almost all of the control statements that exist in the DATA step, including LINK, GOTO, IF-THEN/ELSE, DO/END, iterative DO, and STOP.

MATRIX is preferred to DATA step programs for applications that must work across observations. The SAS DATA step is fine for programs that work with one observation at a time. The LAG function even allows you to work with a short set of observations. But to process across observations in ways that are impossible or difficult in the DATA step, use PROC MATRIX.

MATRIX is not always the method to choose:

- MATRIX is not an interactive programming language. You must execute the statements as a unit.
- MATRIX does not support character data except in a primitive form for names.
- MATRIX does not recognize missing values except for certain comparison operations.
- MATRIX is inefficient for certain highly iterative applications. These problems, especially those with subscripting, require overhead for interpretive execution and housekeeping activities. Fortunately, however, highly iterative applications can usually be recoded into matrix operations.
- MATRIX needs the memory to hold all the data, while the DATA step needs space for only one observation at a time.

MATRIX allows you to stay close to the original problem. Every time you make housekeeping specifications in another programming language, you are distracted from the original problem. Since MATRIX code is easier and clearer than most programming languages, you stay closer to the heart of the problem and have opportunities for greater insights. Shorter MATRIX programs allow you a better overall perspective of the work instead of an endless stream of detail.

The MATRIX Procedure

ABSTRACT

The MATRIX procedure implements an interpretive programming language in which data elements are matrices of values and operations are performed on entire matrices of values.

INTRODUCTION

MATRIX is a complete programming language that performs operations on matrices of values. The language is patterned directly after matrix algebra notation. For example, the familiar least-squares formula

$$\mathbf{b} = (\mathbf{X'X})^{-1}\mathbf{X'Y}$$

can be easily translated into the MATRIX assignment statement

```
B=INV(X'*X)*X'*Y;
```

If a statistical method has not been implemented directly in a SAS procedure, you may be able to program it using the MATRIX procedure. Because the operations in MATRIX deal with arrays of numbers rather than with one number at a time and the most commonly used mathematical and matrix operations are built directly into the language, programs that take hundreds of lines of code in FORTRAN or other languages may take only a few lines in MATRIX. Since MATRIX is built around traditional matrix algebra notation, you can often transcribe statistical methods from matrix algebraic expressions in textbooks into executable MATRIX statements.

Take the problem of solving these two simultaneous equations:

$$2X_1 + 3X_2 = 5$$
$$X_1 - 4X_2 = -3 \quad .$$

To solve the problem, these equations may be written in matrix form:

$$\begin{bmatrix} 2 & 3 \\ 1 & -4 \end{bmatrix} \begin{bmatrix} X_1 \\ X_2 \end{bmatrix} = \begin{bmatrix} 5 \\ -3 \end{bmatrix} ,$$

which can be written symbolically as

$$\mathbf{Ax} = \mathbf{c} \quad .$$

A is nonsingular, and the solution is

$$\mathbf{x} = \mathbf{A}^{-1}\mathbf{c} \quad .$$

MATRIX can be used to solve the problem with only four statements. First is the PROC MATRIX statement, which includes the PRINT option for printing all results:

```
PROC MATRIX PRINT;
```

Then you define the matrices A and C. The slash (/) indicates the beginning of a new row in a matrix:

```
A= 2   3 /
   1  -4;
C= 5 / -3;
```

Then you write the solution equation $\mathbf{x}=\mathbf{A}^{-1}\mathbf{c}$ as a matrix statement. As in SAS program statements, two asterisks denote a power, and one asterisk denotes multiplication:

```
X=INV(A)*C;
```

After MATRIX executes this statement, the first row of the matrix **X** consists of the X_1 value for which you are solving; the second row contains the X_2 value. Your entire program looks like this:

```
PROC MATRIX PRINT;
  A=2 3 / 1 -4;
  C=5 / -3;
  X=INV(A)*C;
```

and produces this output:

```
A              COL1         COL2                        1

ROW1              2            3
ROW2              1           -4


               C            COL1

               ROW1            5
               ROW2           -3


               X            COL1

               ROW1            1
               ROW2            1
```

Note: in this chapter most matrix names are shown as they occur in PROC MATRIX statements (Roman type) rather than in mathematical notation (boldface).

Data in PROC MATRIX

MATRIX is a programming language in which the fundamental data entity is not a single numeric or character value but a two-dimensional (row by column) numeric (double-precision real floating-point) matrix. This matrix is usually given a name and is stored by rows.

MATRIX deals with expressions in terms of these matrices. For example, the expression A+B in MATRIX adds the elements of the two matrices A and B.

The dimension of a matrix is described by the number of rows and columns. Thus, an *m* by *n* ($m \times n$) matrix has $m*n$ numbers arranged in *m* rows and *n* columns. In this chapter, matrices of dimensions

$1 \times n$ are called *row vectors*,
$n \times 1$ are called *column vectors*, and
1×1 are called *scalars*.

Matrix Literals and Matrix Names

A matrix literal, which is a matrix referred to by its values, is written as a series of numbers separated by one or more blanks. If more than one row of a matrix is described, each row is separated from the next by a slash (/).

For example, the statement

```
X=1 2 / 3 4 / 5 6;
```

assigns a 3-row by 2-column matrix literal to the matrix X. The numbers, which represent elements of the matrix, can be written with minus signs, with decimal points, and with exponents in scientific notation.

Matrices are usually referred to by name. The rules for naming matrices are the same as for other SAS names: 1 to 8 characters long beginning with a letter or an underscore. Matrix names refer to matrices that have been set up by an assignment statement or command. A matrix assumes the dimensions of the result of an expression and can be redefined later with different dimensions.

A matrix literal can be used in expressions anywhere a matrix name is used. The MATRIX statements

```
D=J(2,3,0);
C=(1 2)*(3/4);
```

are equivalent to the statements

```
A=2;  B=3;  C=0;  D=J(A,B,C);
A=1;  B=3/4;  C=A*B;
```

Caution It is easy to confuse negative literal values with the minus sign used for subtraction or to change a sign. **The sign on a literal value always takes priority**. For example,

```
X=1-2*A;        is   X=(1-2)*A;
                not  X=(1)-(2*A);
X=-1 2 3;       is   X=-1 2 3;
                not  X=-(1 2 3);
X=Y**2-1;       is   X=Y**(2-1);  (which is illegal)
                not  X=(Y**2)-1;
```

Use parentheses to insure that the expression means what you want it to mean.

Missing Values

Missing values can be brought in from SAS data sets or specified in literals with a single period. However, most matrix operations do not distinguish between missing values and zeros. Missing values can be moved around and printed by various matrix operations, but they are not propagated by arithmetic operations, nor do they compare low as in DATA step programming statements. The only two operators that distinguish between zero and missing are the equal (=) and the not equal (¬=) comparison operators.

SPECIFICATIONS

PROC MATRIX *options;*
statements;

Statements used with PROC MATRIX are much like programming statements used in a DATA step. Programming statements in a DATA step, however, work with one value at a time, while MATRIX statements work with matrices made up of many values. Three kinds of statements are used in MATRIX:

- **Assignment statements** evaluate expressions and assign the results to a name.
- **Control statements** direct the flow of execution.
- **Commands** perform special processing such as printing or special mathematical operations.

Each type of MATRIX statement is described below.

Other SAS statements like COMMENT, TITLE, and OPTIONS can also be used within the MATRIX procedure, although the BY statement is not permitted.

PROC MATRIX Statement

PROC MATRIX *options;*

Typically, using MATRIX requires no options. However, the following options may be specified in the PROC MATRIX statement:

PRINT requests MATRIX to print the result of each matrix statement. Using it may be more convenient than using many PRINT matrix commands.

FLOW causes MATRIX to list each operation as it is performed, including the names of the operands and results. The FLOW option is useful for finding errors in your MATRIX program.

FW=*width* defines the number of print positions used to print elements of matrices. The value specified for FW must lie between 4 and 20. If FW is not specified, MATRIX uses a field width of 10 for batch jobs, 7 for interactive sessions.

ERRMAX=*number* gives the maximum number of errors MATRIX can encounter before it stops executing. If ERRMAX= is not specified, MATRIX stops after encountering six errors.

Several MATRIX operations (SQRT, LOG, EXP, RECIP) tolerate improper arguments and produce only warnings. These warnings do not add to the ERRMAX count.

PRINTALL causes MATRIX to print the intermediate results of each matrix statement, as well as the final results printed when the PRINT option is specified.

LIST requests MATRIX to print, for each operation in each MATRIX statement, the arguments, the results, and the line and column of the source in which the operation appears. LIST is useful in determining how MATRIX interprets matrix statements.

FUZZ requests that all numbers with absolute values less than 1E-8 be printed as zeros. Such values are usually the result of accumulation of numerical error and are effectively zero.

DUMP requests that the values of all defined matrices be printed if an error is encountered.

Assignment Statements

Assignment statements can be used in PROC MATRIX to set up new matrices or to assign new values to existing matrices. An assignment statement consists of a result variable, an equal sign (=), an expression to be evaluated, and a semicolon (;):

result=*expression*;

The expression is evaluated and assigned to the result name. You do not need any special declaration or dimensioning of the result matrix, since the result automatically acquires the dimensions and values of the expression.

Details on writing expressions are described under **Matrix Expressions**.

The statement

```
Y=4 5 6;
```

sets up a 1×3 matrix named Y, whose elements are 4, 5, and 6.

The statements

```
A=8 5 3 / 5 8 5 / 3 5 8;
B=INV(A);
```

set up a 3×3 matrix A. The slashes separate one row from the next. The second statement computes the inverse of matrix A and assigns that result to the matrix B.

Control Statements

Control statements direct the flow of execution of the other statements in MATRIX. The control statements in MATRIX work like the corresponding control statements in the SAS DATA step. The control statements are:

GOTO
LINK and RETURN
STOP
IF-THEN and ELSE
DO and END

GOTO Statement

GOTO *label*;

The GOTO (or GO TO) statement directs MATRIX to jump immediately to the statement with the given label and begin executing statements from that point. Any MATRIX statement may have a label, which is a name followed by a colon preceding any executable statement.

GOTO statements are usually clauses of conditional IF statements. For example,

```
      IF X>Y THEN GOTO SKIP;
      Y=LOG(Y-X);
      YY=Y-20;
SKIP: more statements
```

The function of GOTO statements is usually better performed by DO groups, described below. For example, the statements above could be better written:

```
IF X<=Y THEN DO;
  Y=LOG(Y-X);
  YY=Y-20;
  END;
  more statements
```

You should avoid GOTO statements when they refer to a label above the GOTO statement; in this case an infinite loop is possible.

LINK and RETURN Statements

```
  LINK label;
    statements
label: statements
  RETURN;
```

The LINK statement, like the GOTO statement, directs MATRIX to jump to the statement with the specified label. Unlike the GOTO statement, MATRIX remembers where the LINK was issued and returns to that point when a RETURN statement is executed.

LINK provides a way of calling subroutines. The LINK statement calls the routine. The routine begins with the label and ends with a RETURN statement. LINKs can be nested within other LINKs up to 30 levels deep. A RETURN statement without a LINK is executed just like the STOP statement described below.

STOP Statement

```
STOP;
```

The STOP statement stops the MATRIX program, and no further matrix statements are executed. The program exits from MATRIX and SAS begins executing the next SAS DATA or PROC step.

IF-THEN and ELSE Statements

```
IF expression THEN statement;
ELSE statement;
```

IF statements contain an expression to be evaluated, the keyword THEN, and an action to be taken when the result of the evaluation has a true value.

The ELSE statement follows the IF statement. ELSE gives an action to be taken when the IF expression is false. The expression to be evaluated is often a comparison. For example

```
IF MAX(A)<20 THEN P=0;
ELSE P=1;
```

The IF statement results in the evaluation of the condition (MAX(A)<1). If the largest value found in matrix A is less than 20, P is set to 0. Otherwise, P is set to 1. (More about MAX and other functions appears below in **Functions**.)

When the condition to be evaluated is a matrix expression, the result of the evaluation is another matrix. If all values of the result are nonzero, the condition is true; if any element is zero, it is false. Such evaluation is like using the ALL function.

For example, writing

```
IF X<Y THEN DO;
```

produces the same result as writing

```
IF ALL(X<Y) THEN DO;
```

IF statements can be nested within the clauses of other IF or ELSE statements. Any number of nesting levels is allowed. For example:

```
IF X=Y THEN IF ABS(Y)=Z THEN DO;
```

Caution These expressions

```
IF A¬=B THEN ...;
IF ¬(A=B) THEN ...;
```

are valid ones, but the THEN clause in each case is only executed when all corresponding elements of A and B are unequal.

Evaluation of the expression

```
IF ANY(A¬=B) THEN ...
```

requires only one element of A and B to be unequal for the expression to be true.

The expression

```
IF 1<A<2 THEN ... ;
```

is always true (or 1), since 1<A is evaluated first; the result (0 or 1) is always less than 2.

DO and END Statements

```
DO;
  statements
  END;
```

The DO statement specifies that the statements following the DO are to be executed as a group until a matching END statement appears. DOs usually appear in IF-THEN and ELSE statements, where they designate groups of statements to be performed when the IF condition is true or false.

For example, the statements

```
IF X=Y THEN DO;
  I=I+L;
  PRINT X;
  END;
PRINT Y;
```

specify that the statements between the DO and the END (the DO group) are to be performed only if X=Y. If X is not equal to Y, the statements in the DO group are skipped and the next statement executed is

```
PRINT Y;
```

DO groups can be nested. Any number of nested DO groups is allowed. Here is an example of nested DO statements:

```
IF Y>Z THEN DO;
  IF Z=0 THEN DO;
    Z=B+C;
    END;
  X=Z#R;
  END;
```

It is good practice to indent the statements in a DO group as shown so that their positions indicate their levels of nesting.

Iterative Execution of DO Groups

DO *variable*=*start* TO *stop*;
DO *variable*=*start* TO *stop* BY *increment*;

When the DO group has one of these forms, the statements between the DO and the END are executed repetitively. The number of times the statements are executed depends on the evaluation of the expressions given in the DO statement.

The *start*, *stop*, and *increment* values should be scalars or expressions whose evaluation yields scalars. The *variable* is given a new value for each repetition of the group. The index variable starts with the start value, then is incremented by the increment value each time. The iterations continue as long as the index variable is less than or equal to the stop value. If a negative increment is used, then the rules reverse so that the index variable decrements to a lower bound.

To study how the iterations are controlled, look at these equivalent sections of code. The first section is done with an iterative DO group; the second performs the same task using other statements such as IF and GOTO.

```
DO I=start TO stop BY increment;

  matrix statements

  END;

Z: more matrix statements
```

is equivalent to

```
  I=start;                        * STARTING VALUE;
  DOLIM=stop;                     * REMEMBER BOUND;
  DOINC=increment;                * REMEMBER INCREMENT;
L:IF I>DOLIM THEN GO TO Z;        * CHECK INDEX>BOUND?;

matrix statements

  I=I+DOINC;                      * INCREMENTING;
  GO TO L;                        * ITERATE BACK;
Z: more matrix statements         * CONTINUE AFTER THE LOOP;
```

Note that the start, stop, and increment expressions are evaluated only once before the looping is started.

DO Statement with WHILE Clause

DO WHILE(*expression*);
DO *variable*=*start* TO *stop* WHILE(*expression*);
DO *variable*=start TO *stop* BY *increment* WHILE(*expression*);

Using a WHILE expression makes it possible to conditionally execute a set of statements iteratively. The WHILE expression is evaluated at the top of the loop, and the statements inside the loop are executed repeatedly as long as the expression yields a nonzero value.

The statements

```
L: DO I=start TO stop BY increment WHILE(expression);
   matrix statements
   END;
Z: more matrix statements
```

are equivalent to the statements

```
     L: I=start;                        * STARTING VALUE;
        DOLIM=stop;                     * REMEMBER BOUND;
        DOINC=increment;                * REMEMBER INCREMENT;
        GO TO IN;                       * SKIP INCREMENTING FIRST TIME;
REPEAT: I=I+DOINC;                      * INCREMENTING;
    IN: IF I>DOLIM THEN GO TO Z;        * CHECK INDEX> BOUND;
        IF ¬expression THEN GO TO Z;    * CHECK WHILE EXPRESSION;
        matrix statements
        GO TO REPEAT;
     Z: more matrix statements          * continue after the loop;
```

Note that the incrementing is done before the WHILE expression is tested.

Commands

Matrix statements can also take the form of commands. Most matrix commands input and output matrices or show the results of matrix operations. These commands are:

FETCH	to input a SAS data set
FREE	to free matrix storage space
LIST	to list dimensions of active matrices
NOTE	to annotate printouts
OUTPUT	to output a matrix to a SAS data set
PRINT	to print a matrix.

Other commands in MATRIX are mathematical commands that generate result matrices. These mathematical commands are:

EIGEN	to create eigenvalues and eigenvectors of symmetric matrices
GS	to compute the Gram-Schmidt orthonormal factorization of a matrix
SVD	for singular value decomposition of a matrix.

All of the matrix commands are described in detail in the **Encyclopedia of Operators** section below. They appear in alphabetical order along with the matrix functions.

MATRIX EXPRESSIONS

Operators

MATRIX provides prefix, postfix, and infix operators to be used in expressions involving matrices.

For example, the expression

```
-A
```

uses the prefix operator minus (–) in front of the operand A to reverse the sign of each element of matrix A.

The expression

```
A+B
```

uses the infix operator plus (+) between its operands A and B to add corresponding elements of matrices A and B.

The expression

```
A'
```

uses the postfix operator prime (') after its operand A to get the transpose of matrix A.

All of these matrix operators are shown in Table I and are described in detail in the section **Encyclopedia of Operators** below.

Compound Expressions

Compound expressions involving several matrix operators and operands can be written in MATRIX. For example, the statements

```
A=X+Y+Z;
A=X+Y*Z;
A=X#/Y#/Z;
A=X#/(Y#/Z);
```

are valid matrix assignment statements.

The rules for evaluating such expressions are:

- Evaluation follows the order of operator precedence, according to the precedence table shown below. For example, the MATRIX expression

  ```
  A=X+Y*Z;
  ```

  first multiplies matrices Y and Z, since the * operator has higher precedence than +; it then adds the result of this multiplication to the matrix X, and assigns the new matrix to A.
- If neighboring operators in an expression have equal precedence, the expression is evaluated from left to right, except for the highest priority operators. For example, the expression

  ```
  A=X #/ Y #/ Z;
  ```

  first divides each element of matrix X by the corresponding element of matrix Y. Then, using the result of this division, each element of the resulting matrix is divided by the corresponding element of matrix Z. The operators in the highest group of the precedence table below are evaluated from right to left. For example

```
-X**2
```

is evaluated as

```
-(X**2) .
```

When multiple prefix or postfix operators are juxtaposed, their order is from the inside to the outside. For example

```
¬-A
```

is evaluated as ¬(-A), and

```
A'(I,J)
```

is evaluated as (A')(I,J).

- All expressions enclosed in parentheses are evaluated first, using the two rules above. Thus, the MATRIX statement

```
A=X #/ (Y#/Z);
```

is evaluated by first dividing elements of matrix Y by the elements of Z, then dividing the elements of X by the corresponding elements of the matrix resulting from the first division.

Operator precedence table The operators in PROC MATRIX have the following priority from highest to lowest:

```
¬ ' subscripts -(prefix) ## **
# <> >< #/ @| @
*
+ -
|| // :
< <= > >= = ¬=
&
|
```

Functions

Functions are expressed by the name of the function followed by a list of arguments enclosed in parentheses, with arguments separated by commas:

result=*function* (*argument1, argument2*);

The arguments can be matrix names, literals, or expressions.

All the functions that can be used in DATA step expressions can also be used in MATRIX. The function is called for each element of the matrix argument. If several arguments are used, they must all have the same dimensions (except that scalar values are always permitted). These functions have the same form and function in MATRIX as in SAS, with the exception of the MAX, MIN, and SUM functions, which have different meanings in MATRIX.

For example, to create a 10×1 matrix of random numbers, use

```
X=UNIFORM (J(10,1,0));
```

The J function produces a 10×1 matrix of zeros. Since the UNIFORM function is applied to each element of J(10,1,0), ten numbers result.

Other functions including EIGENVAL, INV, and others are built into MATRIX and are described in detail in **Encyclopedia of Operators** below. They appear in alphabetical order along with the commands.

It is also possible for you to add your own functions to the library. Simple scalar functions for both MATRIX and the DATA step are written in FORTRAN. Matrix library functions can be written in either FORTRAN or PL/I.

Elementwise Binary Extensions

The following operators are called *elementwise binary operators* because they work in an elementwise fashion, composing a result matrix from element-by-element operations of two argument matrices:

+	addition	<	less than
−	subtraction	<=	less than or equal to
#	elementary multiplication	>	greater than
#/	elementary division	>=	greater than or equal to
<>	elementary maximum	¬=	not equal to
><	elementary minimum	=	equal to
|	logical or	MOD(m,n)	modulus remainder
&	logical and		

All these operators can also work in a one-to-many or many-to-one manner as well as in an element-to-element manner. For example, they allow you to perform operations like adding a scalar to a matrix or dividing a matrix by a scalar.

For example, the MATRIX statement

```
X=X # (X>0);
```

replaces each negative element of the matrix X with zero. (X>0) is a matrix of logical values, 1 if the corresponding element of X is positive, zero otherwise. The expression (X>0) is a many-to-one operation that compares each element of X with 0. When the expression is true (the element is positive), the element is multiplied by one. When the expression is false (the element is negative or zero), the element is multiplied by zero.

Subscripts

Subscripts are special parenthetical operators placed after a matrix operand. You can use subscripts to:

- refer to a single element of a matrix
- refer to an entire row or column of a matrix
- refer to any submatrix contained within a matrix
- perform a reduction across rows or columns of a matrix.

Subscript operations are of the form:

operand (*rows*, *columns*)

where *rows* is an expression to select one or more rows from the operand, and *columns* is an expression to select one or more columns from the operand. The operand is usually a matrix name, but it can also be an expression or literal. The row and column arguments are scalars or vectors. In expressions, subscripts have the same high precedence that the transpose postfix operator (′) has.

Here is an example:

```
PROC MATRIX;
  X=1 2 3/
     4 5 6/
     7 8 9;
  A=X(2,3);
```

The second statement above sets up the 3×3 matrix X. The next statement assigns to matrix A the element in the second row, third column of X. A is thus a 1×1 matrix whose only element has the value 6.

A matrix with subscripts can also appear on the left side of the equal sign in assignment statements.

Here is an example:

```
PROC MATRIX;
  X=1 2 3/
     4 5 6/
     7 8 9;
  X(1,3)=0;
```

The last statement assigns a value of 0 to the element in the first row, third column of X. The values of X are:

$$\begin{bmatrix} 1 & 2 & 0 \\ 4 & 5 & 6 \\ 7 & 8 & 9 \end{bmatrix}$$

Note that when subscripts are used on the left side of an equal sign, the matrix that is subscripted must already be defined and have values.

When subscripts are used, the first selects the row and the second the column. While other languages with which you may be familiar allow single subscripts, MATRIX always looks for both subscripts, although one may be empty.

Entire row or column To refer to an entire row or column of a matrix, write the subscript with the row or column number, omitting the other subscript. For example, the statement

```
A=X(3,);
```

assigns the third row of X to A. The statement

```
B=X(,2);
```

assigns the second column of X to B. The statement

```
X(,1)=5 / 6 / 7;
```

changes the elements in the first column of X to 5, 6, and 7.

Submatrices Submatrices within matrices can also be defined with subscripts. For example, to refer to the matrix formed by the four elements in the upper right-hand corner of matrix X above, use the statement

```
R=X(1 2, 2 3);
```

The first subscript, 1 2, selects the rows of X to be included in the new matrix R. The second subscript, 2 3, selects the columns to be included. The new matrix R is

$$\begin{bmatrix} 2 & 3 \\ 5 & 6 \end{bmatrix} .$$

Index vectors Index vectors generated using the colon (:) operator are often used in subscripts. For example, the statement

```
Y=A(1:7,4);
```

is equivalent to

```
Y=A(1 2 3 4 5 6 7,4);
```

and defines Y to be the 7×1 matrix formed from the first 7 elements of A's fourth column.

Note that the number of elements in the first subscript defines the number of rows in the new matrix; the number of elements in the second subscript defines the number of columns. For example, in the statement

```
P=C(2 4 5, 1 3);
```

the first subscript has 3 elements: 2, 4, and 5. The second subscript has 2 elements: 1 and 3. Thus, P has 3 rows and 2 columns.

Expressions as subscripts Subscripts can also contain other expressions whose results are either row or column vectors. For example, consider these statements:

```
PROC MATRIX;
 A=3 1;
 X=1 2 3/
    4 5 6/
    7 8 9;
 M=X(2,A);
```

A is a row vector with 2 elements. It appears as a subscript in the last statement, which can also be written

```
M=X(2, 3 1);
```

M is a 1×2 matrix with the values

$$\begin{bmatrix} 6 & 4 \end{bmatrix} .$$

If a noninteger value is used as a subscript, it is truncated to the next lower integer. Using a subscript value less than one or greater than the dimension of the matrix results in an error.

Subscript operators are general postfix operators that can be applied to any valid expression. For example:

(A+B)(1,2) and A(2,)(1,2)

are valid expressions. The first expression selects the element in the first row, second column of the matrix formed by the sum of matrices A and B; the second expression first selects the second row of A, and then selects the second element of this row vector.

Subscript reduction operators Reduction operators can be used in place of numbers in subscripts to get reductions across all rows or across all columns or across both rows and columns.

Eight operators are available in MATRIX subscript reduction:

+	addition
#	multiplication
<>	maximum
><	minimum
<:>	index of maximum
>:<	index of minimum
:	mean
##	sum of squares

For example, to get column sums of the matrix X (sum of the row elements, which reduces the row dimension to 1), specify X(+,). The elements in each column are added, and the new matrix consists of one row containing the column sums.

Since the second subscript is omitted, the column dimension is not changed. The first subscript (+) means that summation reduction takes place across the rows.

These operators can be used to reduce either rows or columns or both. When both rows and columns are reduced, the row reduction is done first.

For example, the expression A(+,<>) results in the maximum of the column sums. To get the sum of the row maxima, specify A(,<>) (+,).

Subscript reduction can be combined with regular subscripts. For example, A(2 3,+) first selects the second and third rows of A, and then finds the row sums of that matrix.

Examples:

If $A = \begin{bmatrix} 0 & 1 & 2 \\ 5 & 4 & 3 \\ 7 & 6 & 8 \end{bmatrix}$

Then A(2 3,+)=	12 21	sum for rows 2 and 3
A(+,<>)=	13	maximum column sum
A(<>,+)=	21	sum of column maxima
A(,><)(+,)=	9	sum of row minima
A(,<:>)=	3 1 3	for each row, the column of the maximum
A(>:<,)=	1 1 1	for each column, the row of the column minimum

Functions and subscripts: rule of first use When a subscripted matrix name is the same as the name of a function, references to the name are ambiguous. For example, J(2,2) can refer to the (2,2) element of a matrix named J or to the J function.

If the first use of the name in the statements after the PROC MATRIX statement is followed by information in parentheses, this and later uses of the name followed by a parenthetical field refer to the function.

If the first use of the name within the MATRIX procedure is not followed by a parenthetical field, later uses of the name followed by information in parentheses refer to the subscripted matrix.

For example, consider these statements:

```
PROC MATRIX;
  J=3 4 5;
  K=J(3,1);
```

Since J first appears alone, MATRIX assumes that the last statement refers to a submatrix of J. An error results, since J has only one row.

To explicitly request the J function in such a case, follow the function name with a period. For example, the statement

```
  K=J.(3,1);
```

requests that the J function be used to set up a 3×1 matrix of 1s.

To refer explicitly to a subscripted matrix, parenthesize the matrix name before writing the subscripts. For example,

```
  (J)(1,3)
```

unambiguously refers to the (1,3) element of the matrix J.

Note: if you use names that are not also names of functions, you can avoid these problems.

LEARNING EXAMPLES

Example 1

Here is an example that uses MATRIX statements to find means and standard deviations.

```
DATA D1;
  INPUT X1-X6;
  CARDS;
1 21 11 10 34 41
2 12 11 30 31 42
1 23 12 10 32 43
4 14 12 30 33 42
2 23 11 10 31 43
5 12 13 33 32 43
2 26 14 10 33 44
3 27 12 33 31 45
2 23  9 10 32 46
5 18 16 33 33 47
```

```
2 29 11 10 31 47
7 23 11 33 32 49
;
PROC MATRIX PRINT FW=7;              * INVOKE PROCEDURE WITH
                                       OPTIONS;
  FETCH X DATA=D1;                   * INPUT D1 INTO MATRIX X;
  NOBS=NROW(X);                      * FIND NUMBER OF OBSERVATIONS;
  NVAR=NCOL(X);                      * FIND NUMBER OF VARIABLES;
  SUM=X(+, );                        * GET COLUMN SUMS;
  MEAN=SUM#/NOBS;                    * COMPUTE COLUMN MEANS;
  SS=(X#X)(+, );                     * COLUMN SUM OF  SQUARES;
  CSS=SS-SUM#SUM#/NOBS;              * SUM OF SQUARES CORRECTED
                                       FOR MEAN;
  IF NOBS>1 THEN
    STD=SQRT(CSS#/(NOBS-1));         * ESTIMATE OF STANDARD
                                       DEVIATIONS;
```

```
                                                                    1
X          COL1     COL2     COL3     COL4     COL5     COL6

ROW1          1       21       11       10       34       41
ROW2          2       12       11       30       31       42
ROW3          1       23       12       10       32       43
ROW4          4       14       12       30       33       42
ROW5          2       23       11       10       31       43
ROW6          5       12       13       33       32       43
ROW7          2       26       14       10       33       44
ROW8          3       27       12       33       31       45
ROW9          2       23        9       10       32       46
ROW10         5       18       16       33       33       47
ROW11         2       29       11       10       31       47
ROW12         7       23       11       33       32       49


                NOBS      COL1

                ROW1        12


                NVAR      COL1

                ROW1         6


SUM        COL1     COL2     COL3     COL4     COL5     COL6

ROW1         36      251      143      252      385      532


MEAN       COL1     COL2     COL3     COL4     COL5     COL6

ROW1          3  20.9167  11.9167       21  32.0833  44.3333


SS         COL1     COL2     COL3     COL4     COL5     COL6

ROW1        146     5611     1739     6756    12363    23652


CSS        COL1     COL2     COL3     COL4     COL5     COL6

ROW1         38  360.917  34.9167     1464  10.9167  66.6667
```

```
                                                                    2
STD        COL1     COL2     COL3     COL4     COL5     COL6

ROW1    1.85864  5.72805  1.78164  11.5365  .996205  2.46183
```

Example 2

This example shows the code in MATRIX for an *ANOVA*-regression problem.

```
PROC MATRIX PRINT;                    * INVOKE THE PROCEDURE;
  X=1 1 1/
    1 1 0/
    1 0 1/
    1 0 0;                            * DEFINE THE DESIGN MATRIX;
  Y = 2.8/1.5/2.1/1.9;                * DEFINE THE RESPONSE VECTOR;
  XPXI=INV(X'*X);                     * INVERSE CROSSPRODUCTS
                                        MATRIX;
  BETA=XPXI*X'*Y;                     * PARAMETER ESTIMATES;
  YHAT=X*BETA;                        * PREDICTED VALUES;
  RESID=Y-YHAT;                       * RESIDUALS;
  SSE=RESID'*RESID;                   * SUM OF SQUARES ERROR;
  DFE=NROW(X)-NCOL(X);                * DEGREES OF FREEDOM ERROR;
  MSE=SSE#/DFE;                       * MEAN SQUARED ERROR;
  COV=MSE*XPXI;                       * COVARIANCE OF ESTIMATES;
  STDB=SQRT(VECDIAG(COV));            * STANDARD ERRORS OF
                                        ESTIMATES;
  TTEST=BETA#/STDB;                   * TTESTS FOR EACH PARAMETER;
  PROB=2#(1-PROBT(ABS(TTEST),
    DFE));                            * 2-TAIL SIGNIFICANCE PROB;
```

```
X          COL1       COL2       COL3                1

ROW1          1          1          1
ROW2          1          1          0
ROW3          1          0          1
ROW4          1          0          0


           Y          COL1

           ROW1        2.8
           ROW2        1.5
           ROW3        2.1
           ROW4        1.9


XPXI       COL1       COL2       COL3

ROW1       0.75       -0.5       -0.5
ROW2       -0.5          1          0
ROW3       -0.5          0          1


           BETA       COL1

           ROW1      1.625
           ROW2       0.15
           ROW3       0.75


           YHAT       COL1

           ROW1      2.525
           ROW2      1.775
           ROW3      2.375
           ROW4      1.625


           RESID      COL1

           ROW1      0.275
           ROW2     -0.275
           ROW3     -0.275
           ROW4      0.275
```

```
                          SSE            COL1                          2

                          ROW1         0.3025


                          DFE            COL1

                          ROW1              1


                          MSE            COL1

                          ROW1         0.3025


              COV          COL1          COL2          COL3

              ROW1     0.226875      -0.15125      -0.15125
              ROW2     -0.15125        0.3025             0
              ROW3     -0.15125             0        0.3025


                          STDB           COL1

                          ROW1       0.476314
                          ROW2           0.55
                          ROW3           0.55


                          TTEST          COL1

                          ROW1        3.41162
                          ROW2       0.272727
                          ROW3        1.36364


                          PROB           COL1

                          ROW1       0.181519
                          ROW2       0.830499
                          ROW3        0.40282
```

ENCYCLOPEDIA OF OPERATORS, FUNCTIONS, AND COMMANDS

Table I Operators

Operation	Symbol	Type
sign reverse	–	prefix
addition	+	infix
subtraction	–	infix
index creation	:	infix
matrix multiplication	*	infix
element multiplication	#	infix
direct product	@	infix
matrix power	**	infix
element power	##	infix
division	#/	infix
horizontal direct product	@|	infix
horizontal concatenation	||	infix
vertical concatenation	//	infix
element maximum	<>	infix
element minimum	><	infix
and	&	infix
or	|	infix

not	¬	prefix
less than	<	infix
greater than	>	infix
equal to	=	infix
less or equal	<=	infix
greater or equal	>=	infix
not equal	¬=	infix
transpose	′	postfix
subscript	(,)	postfix

Table II Operator Precedence Table (highest to lowest)

¬ ′ subscripts − (prefix) ## **

<> >< #/ @| @

*

\+ −

|| // :

< <= > >= = ¬=

&

|

Table III Subscript Reduction Operators

+	addition
#	multiplication
<>	maximum
><	minimum
<:>	index of maximum
>:<	index of minimum
.	mean
##	sum of squares

Table IV Functions

Function	Example
are all elements nonzero?	B= ALL(A);
are any elements nonzero?	I= ANY(A);
combine diagonally	C= BLOCK(A,B);
bivariate ranks	F= BRANKS(X);
create design matrix	X= DESIGN(A);
determinant	D= DET(A);
diagonal	D= DIAG(A);
row-echelon form	M= ECHELON(A);
eigenvalues	M= EIGVAL(A);
eigenvectors	E= EIGVEC(A);
generalized inverse	G= GINV(A);
Cholesky root	U= HALF(A);
identity matrix	B= I(A);
inverse	I= INV(A);
inverse update	B= INVUPDT(A,X,W);
iterative proportional fitting	CALL IPF(FIT);
location of nonzeros	B= LOC(A);

matrix of identical values	B=J(NROW,NCOL,VALUE);
evaluates marginal totals	CALL MARG(LOCMAR);
maximum value	B=MAX(A);
minimum value	B=MIN(A);
number of columns	N=NCOL(A);
number or rows	K=NROW(A);
orthogonal polynomials	A=ORPOL(X,MD,W);
ranking values	R=RANK(A);
ranking of values (ties averaged)	R=RANKTIE(X);
reshape	B=SHAPE(A,NCOL);
solve linear system	X=SOLVE(A,C);
sum of squares	S=SSQ(A);
sum	S=SUM(A);
sweep	B=SWEEP(A,1:5);
trace	T=TRACE(X);
diagonal to vector	V=VECDIAG(A);

Table V Frequently Used SAS Functions See Chapter 5, "SAS Functions," *SAS User's Guide: Basics, 1982 Edition*, for a complete description of functions in the SAS function library.

Function	Usage
absolute value	A=ABS(X);
exponential	B=EXP(X);
fuzz to integer	F=FUZZ(X);
integer value (truncation)	I=INT(X);
natural logarithm	Y=LOG(X);
modulo (remainder)	Y=MOD(X,D);
normal random number	N=NORMAL(SEED);
square root	S=SQRT(X);
uniform random number	U=UNIFORM(SEED);

Table VI Commands

Command	Example
eigenvalues and vectors	EIGEN M E A;
input from SAS data set	FETCH X DATA=A;
free storage	FREE A B;
Gram-Schmidt	GS P T LINDEP A;
list dimensions	LIST;
print message	NOTE 'CONVERGED OK';
output to SAS data set	OUTPUT X OUT=B;
print a matrix	PRINT A B;
singular value decomposition	SVD U Q V A;

Table VII Control Statements

IF-THEN/ELSE
DO-END
iterative DO-END

GOTO
LINK
RETURN
STOP

Table VIII Options

Option	Effect
PRINT	automatically prints all results
FLOW	traces flow of execution with messages
FW=	specifies field width for printing matrices
ERRMAX=	specifies number of execution errors to tolerate
PRINTALL	automatically prints intermediate results
LIST	lists the parsed code
FUZZ	rounds values near zero to zero for printing
DUMP	dumps all values when there is an error

Operators

Sign reverse: –

Form: –*matrix*
Example: X= – Y;

The minus (–) prefix operator produces a new matrix whose elements are formed by reversing the sign of each element in *matrix*. For example, let

```
A=-1  7  6/
   2  0 -8;
B=-A;
```

Then the result in B is

$$\begin{bmatrix} 1 & -7 & -6 \\ -2 & 0 & 8 \end{bmatrix}$$

Addition: +

Form: *matrix1* + *matrix2*
matrix1 + *scalar*
Example: A=X+Y;

The plus (+) infix operator produces a new matrix whose elements are the sums of the corresponding elements of *matrix1* and *matrix2*. For example, the element in the first row, first column of the first matrix is added to the element in the first row, first column of the second matrix. The sum becomes the element in the first row, first column of the new matrix.

When the *matrix* + *scalar* (or *scalar* + *matrix*) form is used, the scalar value is added to each element of the matrix, producing a new matrix.

For example, let

```
A=1 2/
  3 4;
B=1 1/
  1 1;
C=A+B;
```

Then the result in C is

$$\begin{bmatrix} 2 & 3 \\ 4 & 5 \end{bmatrix}$$

The same value results from C=A+1.

Subtraction: –

Form:	*matrix1* – *matrix2*
	matrix1 – *scalar*
Example:	A=B–C;

The minus (–) infix operator produces a new matrix whose elements are formed by subtracting the corresponding elements of the second matrix from those of the first matrix.

When either argument is a scalar, the operation is performed by using the scalar against each element of the matrix argument.

When you are writing MATRIX statements that include subtraction of literal numbers, enclose the numbers in parentheses. Otherwise, MATRIX will interpret a minus sign before a number as the number's sign. For example, MATRIX interprets

```
1-4*3
```

to be the 1×2 vector (1 -4) multiplied by 3:

```
(1 -4)*3
```

Index creation: :

Form:	*value1* : *value2*
Example:	I=7:10;
	A=B(1:5,);

The colon (:) operator creates a row vector whose first element is *value1*. The second element is *value1*+1, and so on while the elements are less than or equal to *value2*. For example, the statement

```
I=7:10;
```

results in I= 7 8 9 10 .

If *value1*>*value2*, an empty matrix results, in other words, 0 rows and 0 columns. (Note: this is different from previous editions of MATRIX.)

You can use a multiplication operator (* or #) with the colon (:) operator to create vectors whose increments between elements are more than 1. For example,

```
J=2*(1:10);
```

creates a row vector J of the even numbers between 2 and 20.

Matrix multiplication: *

Form:	*matrix1* * *matrix2*
Example:	R=A*B

The asterisk (*) infix operator produces a new matrix by performing matrix multiplication. The first matrix must have the same number of columns as the second matrix has rows. The new matrix has the same number of rows as the first matrix and the same number of columns as the second matrix.

If either matrix is a scalar, the operator performs scalar multiplication.
For example, let

```
A=1 2/
  3 4;
B=1 2;

C=B*A;    results in  7 10
C=A*B';   results in  5
                     11
E=B*5;    results in  5 10
```

Element multiplication: #

Form: *matrix1 # matrix2*
matrix1 # scalar

Example: Z=X#Y;

The pound sign (#) operator produces a new matrix whose elements are the products of the corresponding elements of *matrix1* and *matrix2*. When either argument is a scalar, the scalar value is multiplied by each element in *matrix1* to form the new matrix.

For example,

```
A=1 2/
  3 4;
B=4 8/
  0 5;
C=A#B;    results in
```

$$\begin{bmatrix} 4 & 16 \\ 0 & 20 \end{bmatrix}$$

Element multiplication with the # operator should not be confused with matrix multiplication (*). The matrix produced by the # operator is also known as the Schur or Hadamard product.

Direct product: @

Form: *matrix1 @ matrix2*

Example: K=A@B;

The at sign (@) operator produces a new matrix that is the Kronecker product (also called the *direct product*) of *matrix1* and *matrix2*, usually denoted by A⊗B. The number of rows in the new matrix equals the product of the number of rows in *matrix1* and the number of rows in *matrix2*; the number of columns in the new matrix equals the product of the number of columns in *matrix1* and the number of columns in *matrix2*.

For example, let

```
A=1 2/
  3 4;
B=0 2;
```

C=A@B; results in

$$\begin{bmatrix} 0 & 2 & 0 & 4 \\ 0 & 6 & 0 & 8 \end{bmatrix}$$

C=B@A; results in

$$\begin{bmatrix} 0 & 0 & 2 & 4 \\ 0 & 0 & 6 & 8 \end{bmatrix}$$

Matrix power: **

Form: *matrix1 ** scalar*
Example: ACUBE=A**3

The ** operator creates a new matrix that is *matrix1* multiplied by itself *scalar* times. *matrix1* must be square; *scalar* must be an integer, greater than or equal to –1. Large *scalar* values cause numerical problems. If the *scalar* is not an integer, it is truncated to an integer.

For example, let

```
A=1 2/
  1 1;
C=A**2;
```

results in

```
3 4
2 3
```

If the matrix is symmetric, it is preferable to power its eigenvalues rather than using ** on the matrix directly (see EIGEN). Note that the expression

```
A**(-1)
```

is permitted and is equivalent to INV(A).

Element power: ##

Form: *matrix1 ## scalar*
Example: ZSQ=Z##2;

The ## operator creates a new matrix whose elements are the elements of *matrix1* raised to the *scalar* power. If any value in *matrix1* is negative, the *scalar* must be an integer.

For example,

```
A=1 2 3;
B=A##3;    results in  1 8 27
B=A##.5;   results in  1 1.414 1.732
```

Division: #/

Form: *matrix1 #/ matrix2*
matrix1 #/ scalar
Example: A=P#/Q;

The #/ operator divides each element of *matrix1* by the corresponding element of *matrix2* or by *scalar* and places the result in the corresponding element of the new matrix that is produced.

If an attempt to divide by zero is made, the result is set to 0 and no warning is printed.

For example, if division by zero is attempted,

```
A=10 8 3;
B=5 4 2;
C=A#/B;    results in  2 2 1.5
```

Horizontal direct product: @|

Form: *matrix1 @| matrix2*
Example: C=A@|B;

The @| operator performs a direct product on all rows of *matrix1* and *matrix2* and creates a new matrix by stacking these row vectors into a matrix. This operation is useful in constructing design matrices of interaction effects. *matrix1* and *matrix2* must have the same number of rows. The result has the same number of rows as *matrix1* and *matrix2*. The number of columns in the resulting matrix will be equal to the product of the number of columns in *matrix1* and *matrix2*.

For example:

```
A= 1 2/
   2 4/
   3 6;

B= 0  2/
   1  1/
   0 -1;

C=A@|B;
```

results in

$$\begin{bmatrix} 0 & 2 & 0 & 4 \\ 2 & 2 & 4 & 4 \\ 0 & -3 & 0 & -6 \end{bmatrix}$$

Horizontal concatenation: ||

Form: *matrix1* || *matrix2*
Example: Z=C||D;

The || operator produces a new matrix by horizontally joining *matrix1* and *matrix2*. *matrix1* and *matrix2* must have the same number of rows, which is also the number of rows in the new matrix. The number of columns in the new matrix will be the number of columns in *matrix1* plus the number of columns in *matrix2*.

For example,

```
A= 1 1 1/
   7 7 7;
B= 0 0 0/
   8 8 8;
C=A||B;
```

results in

$$\begin{bmatrix} 1 & 1 & 1 & 0 & 0 & 0 \\ 7 & 7 & 7 & 8 & 8 & 8 \end{bmatrix}$$

You can use the horizontal concatenation operator when one of the arguments has no value. For example, if A has not been defined, and B is a matrix, A||B results in a new matrix equal to B.

Vertical concatenation: //

Form: *matrix1* // *matrix2*
Example: X=R//M;

The // operator produces a new matrix by vertically joining *matrix1* and *matrix2*. *matrix1* and *matrix2* must have the same number of columns which is also the number of columns in the new matrix. For example, if A has three rows and two columns, and if B has four rows and two columns, then A//B produces a matrix with seven rows and two columns. Rows one, two, and three of the new matrix correspond to A; rows four through seven correspond to B. For example, if

```
A= 1 1 1/
   7 7 7;
```

and

```
B= 0 0 0/
   8 8 8;
```

A//B produces the matrix

$$\begin{bmatrix} 1 & 1 & 1 \\ 7 & 7 & 7 \\ 0 & 0 & 0 \\ 8 & 8 & 8 \end{bmatrix}$$

You can use the vertical concatenation operator when one of the arguments has not been assigned a value. For example, if A has not been defined and B is a matrix, A//B results in a new matrix equal to B.

Note: the // operator should not appear in columns 1 and 2 of a line, since the computer assumes the line is job control language for IBM OS batch jobs.

Element maximum: <>

Form: *matrix1 <> matrix2*
Example: C=X<>Y;

The element maximum operator (<>) compares each element of *matrix1* to the corresponding element of *matrix2*. The larger of the two values becomes the corresponding element of the new matrix that is produced.

When either argument is a scalar, the comparison above is between each of the elements of the matrix and the scalar.

Element minimum: ><

Form: *matrix1 >< matrix2*
Example: V=R><E;

The element miminum operator (><) compares each element of *matrix1* with the corresponding element of *matrix2*. The smaller of the values becomes the corresponding element of the new matrix that is produced.

When either argument is a scalar, the comparison above is between the scalar and each element of the matrix.

Logical operators: & | ¬

Form:
matrix1 & matrix2
matrix1 & scalar
matrix1 | matrix2
matrix1 | scalar
¬matrix

Example:
```
Z=X&R;
Z=X AND Y;
IF A | B THEN PRINT C;
IF A OR B THEN PRINT C;
IF ¬M THEN LINK X1;
```

The logical operator ampersand (&) compares two matrices, element-by-element, to produce a new matrix. An element of the new matrix is 1 if the corresponding elements of *matrix1* and *matrix2* are both nonzero; otherwise, it is a zero. The & operator can also be written AND, with blanks on both sides of the word AND.

An element of the new matrix produced by the vertical bar(|) operator is 1 if either of the corresponding elements of *matrix1* and *matrix2* is nonzero. If both are zero, the element is zero. The | operator can also be written OR, with blanks on both sides of the word OR.

The not (¬) prefix operator examines each element of a matrix and produces a new matrix whose elements are ones and zeros. If an element of *matrix* equals zero, the corresponding element in the new matrix is a 1. If an element of *matrix* is nonzero, the corresponding element in the new matrix is 0. If either argument is a scalar, the operation is performed with the scalar for all elements.

Comparison operators: < > = <= >= ¬=

Form: *matrix1* < *matrix2*
matrix1 > *matrix2*
matrix1 = *matrix2*
matrix1 <= *matrix2*
matrix1 >= *matrix2*
matrix1 ¬= *matrix2*

Example: IF X>=Y THEN GO TO LOOP1;
B=Z LT T;
IF A=5 THEN DO;

Scalar values may be used instead of matrices in any of the forms shown above.

The comparison operators compare two matrices, element-by-element, and produce a new matrix that contains only zeros and ones. If an element comparison is true, the corresponding element of the new matrix is 1. If the comparison is not true, the corresponding element is 0.

If either argument is a scalar, the comparison is between each element of the matrix and the scalar.

For example, let

A= 1 7 3/
6 2 4;
B= 0 8 2/
4 1 3;

Evaluation of the expression

A>B

results in the matrix

$$\begin{bmatrix} 1 & 0 & 1 \\ 1 & 1 & 1 \end{bmatrix}.$$

The following synonyms are available for the six comparison operators:

LT for <
LE for <=
GT for >
GE for >=
EQ for =
NE for ¬=

Transpose: '

Form: *matrix'*
Example: TR=A';

The transpose operator prime (′) exchanges the rows and columns of *matrix*, producing the transpose of *matrix*. For example, if an element in *matrix* is in the first row and second column, it is in the second row and first column of the transpose; an element in the first row and third column of *matrix* is in the third row and first column of the transpose. If *matrix* contains three rows and two columns, its transpose has two rows and three columns.

For example, if A is

$$\begin{bmatrix} 1 & 2 \\ 3 & 4 \\ 5 & 6 \end{bmatrix}$$

then B=A′ becomes

$$\begin{bmatrix} 1 & 3 & 5 \\ 2 & 4 & 6 \end{bmatrix}.$$

Functions and commands

ALL: check for all elements nonzero

Form: ALL(*matrix*)
Example: I=ALL(X);
IF ALL(A<=Y) THEN LIST;

ALL returns a value of 1 if all elements in *matrix* are nonzero. If any element of *matrix* is zero, ALL returns a value of 0.

You can use ALL to express the results of a comparison operator as a single one or zero. For example, the comparison operation A>B yields a matrix whose elements can be either ones or zeros. All the elements of the new matrix are ones only if each element of A is greater than the corresponding element of B.

In the matrix statement

```
IF ALL(A>B) THEN GOTO LOOP;
```

the GOTO is executed only if each element of A is greater than the corresponding element of B.

ANY: check for any nonzero element

Form: ANY(*matrix*)
Example: IF ANY(A=B) THEN PRINT A B;

ANY returns a value of 1 if any of the elements in *matrix* are nonzero. If all the elements of *matrix* are zeros, ANY returns a value of 0.

BLOCK: blocking

Form: BLOCK(*matrices*)
Example: A=BLOCK(B,C);
D=BLOCK(A,B,C);

The BLOCK function creates a new block-diagonal matrix from all the matrices specified in *matrices*. Up to 15 matrices can be listed. The matrices are combined diagonally to form a new matrix. For example:

```
BLOCK(A,B)= A 0
            0 B
```

and

```
BLOCK(A,B,C)= A 0 0
              0 B 0
              0 0 C
```

Let

```
A= 2 2/
   4 4;
B= 6 6/
   8 8;
C=BLOCK(A,B);
```

results in

$$\begin{bmatrix} 2 & 2 & 0 & 0 \\ 4 & 4 & 0 & 0 \\ 0 & 0 & 6 & 6 \\ 0 & 0 & 8 & 8 \end{bmatrix}$$

BRANKS: bivariate ranks

Form: F=BRANKS(X)
Example: X 2×*n* matrix (with n<32768)
F 3×*n* matrix

BRANKS calculates the tied ranks and the bivariate ranks for a 2×*n* matrix X. The tied ranks of the first row of X comprise the first row of F; the tied ranks of the second row of X comprise the second row of F; and the bivariate ranks of X comprise the third row of F. The tied rank of an element, x_i, of a vector is defined as:

$$\mathbf{R}_i = 1/2 + \Sigma_j u(x_i - x_j)$$

where u(t) is defined as 1 if t>0, 0 if t<0, and 1/2 if t=0. The bivariate rank of a pair of elements, (x_i, y_i), is defined as

$$\mathbf{Q}_i = 3/4 + \Sigma_j u(x_i - x_j) u(y_i - y_j) \quad .$$

For example:

```
X= 1 4 3 5 6/
   0 2 4 3 3;

4|         *
 |
3|               *  *
 |
2|            *
 |
1|
 |
0|__*___________________
  0  1  2  3  4  5  6
```

Then

```
F=1 3 2 4 5/
  1 2 5 3.5 3.5/
  1 2 2 3 3.5;
```

This function computes the bivariate ranks in order n*log(n) operations. The bivariate ranks allow the efficient computation of Kendall's τ and Hoeffding's test of independence, as well as their components. For example, to compute Kendall's τ_a:

```
F=BRANKS(X);
N=NCOL(X);
TAU=(4#SUM(F(3,))-N#N-3#N)#/(N#(N-1));
```

To compute Hoeffding's test of independence:

```
F=BRANKS(X);  N=NCOL(X);
INDEP=(SUM((F(1,)-1)#(F(1,)-2)#(F(2,)-1)#(F(2,)-2))
        -2#(N-2)#SUM((F(1,)-2)#(F(2,)-2)#(F(3,)-1))+
        (N-2)#(N-3)#SUM((F(3,)-1)#(F(3,)-2)))
        #/(N#(N-1)#(N-2)#(N-3)#(N-4));
```

DESIGN: design matrices

Form: DESIGN(*columnvector*)
Example: D=DESIGN(V);

DESIGN creates a full design matrix from *columnvector*. Each unique value of the vector generates a column of the design matrix: this column contains ones in elements whose corresponding elements in the vector are the current value; zeros elsewhere. The columns are arranged in the sort order of the original values.

For example:

$$A = \begin{bmatrix} 1 \\ 1 \\ 2 \\ 2 \\ 3 \\ 1 \end{bmatrix} \qquad \text{DESIGN(A)} = \begin{bmatrix} 1 & 0 & 0 \\ 1 & 0 & 0 \\ 0 & 1 & 0 \\ 0 & 1 & 0 \\ 0 & 0 & 1 \\ 1 & 0 & 0 \end{bmatrix}$$

DET: determinant

Form: DET(*matrix*)
Example: D=DET(A);

DET computes the determinant of *matrix*, which must be square. The determinant, the product of the eigenvalues, is a single numeric value.

DIAG: diagonal

Form: DIAG(*squarematrix*)
DIAG(vector)
Example: D=DIAG(A);

DIAG creates a matrix with diagonal elements equal to the corresponding diagonal elements in *squarematrix*. All off-diagonal elements in the new matrix are zeros.

If the argument is a vector, DIAG creates a matrix whose diagonal elements are the values in the vector. All off-diagonal elements are zero.

For example,

```
A=4 3/
  2 1;
B=1 2 3;
C=DIAG(A);     results in [4 0]
                          [0 1]
D=DIAG(B);     results in [1 0 0]
                          [0 2 0]
                          [0 0 3]
```

ECHELON: row-echelon form

Form: ECHELON(*matrix*)
Example: E=ECHELON(A);

ECHELON uses elementary row operations to reduce a matrix to row-echelon form. This normal form is upper triangular; see Noble (1969) for a precise definition.

If A is square, the row-echelon form can be converted to Hermite normal form by rearranging rows, as shown below:

```
N=NROW(A);
RE=ECHELON(A);
K=1;
DO L=1 TO N;
  IF RE(K,L)=1 THEN DO;
    H=H//RE(K, );
    K=K+1;
    END;
  ELSE H=H//J(1,N,0);
  END;
```

After these statements have been executed, H contains the Hermite normal form of A. The Hermite normal form is idempotent.

The computational method is described in Noble (1969, pp. 80–90).

EIGEN: eigenvalues and eigenvectors

Form: EIGEN *eigenvalues eigenvectors symmetricmatrix;*
Example: EIGEN M E A;

The EIGEN command creates *eigenvalues*, a column vector containing the eigenvalues of *symmetricmatrix* arranged in descending order. EIGEN also creates *eigenvectors*, a matrix containing the orthonormal column eigenvectors of *symmetricmatrix* arranged so that the matrices are correspondent.

Using EIGEN is more efficient than using EIGVAL and EIGVEC when both the eigenvalues and eigenvectors are needed. The results of EIGEN have the properties:

```
A*E=E*diag(M)
E'*E=I(N)
```

that is,

```
E'=INV(E) .
```

Thus also,

```
A=E*diag(M)*E' .
```

The QL method is used (Wilkinson and Reinsch, 1971).

MATRIX cannot directly compute the eigenvalues of a general nonsymmetric matrix because some of the eigenvalues may be imaginary. In statistical applications, nonsymmetric matrices for which eigenvalues are desired are usually of the form $\mathbf{E}^{-1}\mathbf{H}$, where **E** and **H** are symmetric. The eigenvalues **L** and eigenvectors **V** of $\mathbf{E}^{-1}\mathbf{H}$ can be obtained as follows:

```
F=HALF(EINV);
A=F*H*F';
EIGEN L W A;
V=F'*W;
```

The computation can be checked by forming the residuals:

```
R=EINV*H*V-V*DIAG(L);
```

The values in R should be of the order of round-off error.

EIGVAL: eigenvalues

Form: EIGVAL(*symmetricmatrix*)
Example: M=EIGVAL(A);

EIGVAL creates a column vector of the eigenvalues of *symmetricmatrix*. The eigenvalues are arranged in descending order. See EIGEN for discussion.

EIGVEC: eigenvectors

Form: EIGVEC(*symmetricmatrix*)
Example: D=EIGVEC(A);

EIGVEC creates a matrix containing the orthonormal eigenvectors of *symmetricmatrix*. The columns of the new matrix are the eigenvectors. See EIGEN for discussion.

FETCH: input a SAS data set

Form: FETCH *matrix* DATA=*SASdataset*(*options*);
Example: FETCH X;
FETCH Y DATA=APRICOTS;
FETCH P 25 DATA=PRUNES;

The FETCH command creates a new matrix, whose rows correspond to the observations in the SAS data set specified by DATA=*SASdataset*. The columns in the new matrix correspond to the numeric variables in the input SAS data set. If DATA= is omitted, the most recently created SAS data set is used.

The KEEP= data set name option can be used to FETCH only specified variables from the SAS dataset. See Chapter 12, "SAS Data Sets," in *SAS User's Guide: Basics, 1982 Edition* for information about the KEEP= option.

The FETCH command may be executed any number of times.

An optional feature of the FETCH command is to input a specified number of observations. In this case the command is written:

FETCH *matrix obs* DATA=*SASdataset*(*options*);

The *obs* value must be positive (either literal or symbol), and there must be at least this number of observations left in the data set. No rewinding of the data set is performed when this optional feature is specified, so you may read in the data a few observations at a time. When fetching new data repeatedly, you must execute the same FETCH statement using control statements.

Different FETCH statements work independently.

FETCH character features

The following options may appear in the FETCH command:

TYPE=CHAR requests MATRIX to fetch columns corresponding to the character variables in the input data set. If the character variables have lengths greater than 8, the lengths are truncated to 8. Values of variables with lengths less than 8 are expanded with blanks on the right to 8 characters.

COLNAME=*matrix* specifies the name of a new character matrix to be created. This COLNAME= matrix is created in addition to the matrix created with the FETCH command, and contains the variable names from the input data set corresponding to the columns of the fetched matrix. The COLNAME= matrix has dimensions 1×*nvar*, where *nvar* is the number of variables in the fetched matrix.

The COLNAME= matrix can be used with the COLNAME= option on the PRINT command to provide more descriptive column names than the default names COL1, COL2,...COL*n*.

The COLNAME= option can be used to keep variable names with data that are input to PROC MATRIX, manipulated, and written out as a data set again. (See the COLNAME= parameter in the OUTPUT command.)

ROWNAME=*variable* specifies the name of a character variable in the input data set. The values of this variable are put in a character matrix with the same name as the variable. This matrix has dimensions *nobs*×1, where *nobs* is the number of observations in the input data set.

The ROWNAME= matrix can be used with the ROWNAME= option on the PRINT command to provide more descriptive row names than the default names ROW1, ROW2,...ROW*n*.

For example, the command

```
FETCH X ROWNAME=R COLNAME=C DATA=A;
```

fetches all the numeric variables from data set A into the matrix X. The names of the variables from the data set are stored in the matrix C. The values of the character variable R from the data set are stored in a matrix called R. To print the data, use:

```
PRINT X ROWNAME=R COLNAME=C;
```

FREE: free matrix storage space

Form: FREE *matrices;*
Example: FREE X;
FREE M N A;

The FREE command creates additional work space for MATRIX operations by freeing the space used to store the specified matrices. After execution of FREE, the contents of the matrices are no longer available. The NROW and NCOL functions return 0 when the argument is a matrix that has been freed.

The FREE command should be unnecessary for small problems. Consult the section on **Storage Requirements** below for more information.

GINV: generalized inverse

Form: GINV(matrix)
Example: G=GINV(A);

GINV creates the Moore-Penrose generalized inverse of *matrix*. This inverse, known as the four-condition inverse, has these four properties:

AGA=A
GAG=G
(AG)′=AG
(GA)′=GA

The inverse is also known as the *pseudoinverse*, usually denoted by A^+. It is computed using the singular value decomposition (Wilkinson and Reinsch, 1971).

Least-squares regression for the model

$$Y=X\beta+\varepsilon$$

may be performed by using B=GINV(X)*Y; this solution has minimum B′B among all solutions minimizing $\varepsilon'\varepsilon$.

Projection matrices can be formed by specifying GINV(X)*X (row space) or X*GINV(X) (column space).

See Rao and Mitra (1971) for a discussion of properties of this function.

GS: Gram-Schmidt orthonormalization

Form: GS *matrix matrix scalar matrix*;
Example: GS P T LINDEP A;

The GS matrix command computes the Gram-Schmidt orthonormal factorization of the $m \times n$ matrix A; that is, GS computes the column-orthonormal $m \times n$ matrix P and the upper triangular $n \times n$ matrix T such that

A=P*T

If the columns of A are linearly independent (that is, rank (A)=n) then P is full-rank column-orthonormal: $P'*P=I_m$, T is nonsingular, and the value of LINDEP (a scalar) is set to 0. If the columns of A are linearly dependent (say rank (A)=$k<n$) then $n-k$ columns of P are set to 0, the corresponding rows of T are set to 0 (T is singular), and LINDEP is set to 1. The pattern of zero columns in P corresponds to the pattern of linear dependencies of the columns of A, when columns are considered in left-to-right order.

GS is not recommended for the construction of matrices of values of orthogonal polynomials; use the ORPOL function instead.

If LINDEP=1, you can rearrange the columns of P and rows of T so that the zero columns of P are right-most, that is, P=(P(,1), P(2), ..., P(,k), 0, ..., 0) where k=column rank of A and A=P*T is preserved. The following statements make this rearrangement:

```
D=RANK((NCOL(T)-(1:NCOL(T)))#(VECDIAG(T)=0));
TEMP=P;  P(D,)=TEMP;
TEMP=T;  T(,D)=TEMP;
```

Note: the GS function was contributed by Ronald Helms of the University of North Carolina at Chapel Hill based on a method described by Golub (1969).

HALF: Cholesky decomposition

Form: HALF(*matrix*)
Example: U=HALF(A);

The HALF function performs the Cholesky decomposition of A such that

U'U=A

where U is upper triangular. A must be symmetric and nonnegative definite.

I: identity matrix

Form: I(*dimension*)
Example: ID=I(M);

I creates an identity matrix with *dimension* rows and columns. The diagonal elements of an identity matrix are ones; all other elements are zeros. The value of *dimension* must be an integer greater than or equal to 1.

INV: matrix inverse

Form: INV(*matrix*)
Example: G=INV(A);

INV produces a matrix that is the inverse of *matrix*, which must be square and non-singular.

The inverse has the properties:

GA=AG=*identity*

To solve a system of linear equations AX=B for X, you can use

X=INV(A)*B;

However, the SOLVE function is more accurate and efficient for this use.

See Forsythe and Moler (1967) for the method used in INV.

INVUPDT: matrix inverse update

Form: INVUPDT(*matrix,vector,scalar*)
Example: R=INVUPDT(A,X);
R=INVUPDT(A,X,W);

where A is an $n \times n$ positive definite matrix; x is an $n \times 1$ (or $1 \times n$) vector; w is a scalar (if not specified, w is 1); and R is an $n \times n$ matrix.

The INVUPDT function updates a matrix inverse. INVUPDT computes the matrix expression

$$R = A - wAX(1 - wX'AX)^{-1}X'A$$

or, in MATRIX,

R=A−W*A*X*INV(1−W*X'*A*X)*X'*A;

INVUPDT is used primarily to update a matrix inverse, since the function has the property

$$\text{INVUPDT}(B^{-1},X,W) = (B + wXX')^{-1} \;.$$

If Z is a design matrix and X a new observation to be used in estimating the parameters of a linear model, then the inverse crossproducts matrix that includes the new observation can be updated from the old inverse by

C2=INVUPDT(C,X);

where C=INV (Z'*Z).

Note that

C2=INV((Z//X)'*(Z//X)) .

If W is 1, the function adds an observation to the inverse; if W is −1, the function removes an observation from the inverse. If weighting is used, W is the weight.

To perform the computation, INVUPDT uses about $2n^2$ multiplications and additions, where n is the row dimension of the positive definite matrix argument.

IPF: iterative proportional fitting

Form: CALL IPF(FIT,STATUS,DIM,TABLE,CONFIG,INITAB,MOD)

IPF performs an iterative proportional fit of the marginal totals of a contingency table. The arguments used with the IPF function are:

DIM is a vector specifying the number of variables and the number of their possible levels in a contingency table. If DIM is $1\times l$ there are l variables, and the value of the ith element is the number of levels of the ith variable.

TABLE specifies an array of the number of observations at each level of each variable. Variables are nested across columns and then across rows.

CONFIG gives an array specifying which marginal totals to fit. Each column specifies a distinct marginal in the model under consideration. Since the model is hierarchical, all subsets of specified marginals are included in fitting.

INITAB is an optional array of starting values for iterative procedures. If values are not specified, 1s are used. For incomplete tables, INITAB is set to 1 if the cell is included in the design, and 0 if it is not.

MOD is an optional two-element vector specifying stopping criteria. If MOD=(MAXDEV,MAXIT) then the procedure iterates until either MAXIT iterations are completed or the maximum difference between estimates of the last two iterations is <=MAXDEV. Default values are MAXDEV=.25; MAXIT=15.

FIT returns an array of the estimates of the expected number in each cell under the model specified in CONFIG. This matrix conforms to TABLE.

STATUS returns a row vector of length 3. If

STATUS=(ERROR,OBS_MAXDEV,NO_ITERATE)

then

ERROR=0 if normal convergence to desired accuracy

ERROR=3 if no convergence within specified limits

OBS_MAXDEV=maximum of differences between estimates on last two iterations

NO_ITERATE =number of iterations performed.

The matrix TABLE must conform in size to the contingency table as specified in DIM. In particular, if TABLE is $n\times m$, the product of the entries in DIM must equal *nm*. Furthermore, there must be some integer *k* such that the product of the first *k* entries in DIM equals *m*. If INITAB is specified, then it must be the same size as TABLE. For the algorithm, see Haberman (1972) and Deming and Stephan (1940).

For example, consider the no-three-factor-effect model for interpreting Bartlett's data as described in Bishop et al. (1975).

```
DIM=2 2 2;
TABLE=156  84  84  156 /
      107 133  31  209;
CONFIG=1 1 2 /
       2 3 3;
CALL IPF(FIT,STATUS,DIM,TABLE,CONFIG);
```

Returned are:

```
FIT = 161.097  78.9062 78.9026 161.094
      101.903 138.094  36.0974 203.906
STATUS = 0 .166966 4
```

Equivalent results are obtained by:

```
TABLE=156  84 /
       84 156 /
      107 133 /
       31 209;
```

or

```
TABLE=156 84 84 156 107 133 31 209;
```

In the first specification, TABLE is interpreted as:

	variable 2	1		2	
variable 3	variable 1	1	2	1	2
1		156	84	84	156
2		107	133	31	209

In the second specification, TABLE is interpreted as:

variable 3	variable 2	variable 1	1	2
1	1		156	84
	2		84	156
2	1		107	133
	2		31	209

and in the third specification as:

variable 3	1				2			
variable 2	1		2		1		2	
variable 1	1	2	1	2	1	2	1	2
	156	84	84	156	107	133	31	209

J: matrix of identical values

Form: J(*rows, columns, value*);

Example:
```
B=J(3,2);
Y=J(4,5,0);
X=J(2);
```

J creates a matrix whose elements are all equal to *value*, which must be a scalar or an expression that evaluates to a scalar. If the value is omitted, the matrix contains all ones. The number of rows is defined by the *rows* value, and the number of columns is defined by the *columns* value. If the column value is omitted, the matrix is square.

For example,

```
A=J(2,3,4);
```

results in

$$\begin{bmatrix} 4 & 4 & 4 \\ 4 & 4 & 4 \end{bmatrix}$$

LIST: list dimensions of active matrices

Form: LIST;
Example: LIST;

The LIST command prints the dimensions of all defined matrices, the size of the current work space, and the amount of work space in use. Matrix names that have not been assigned values are reported as having zero rows and columns.

LOC: find nonzero elements of a vector

Form: LOC(*vector*)
Example: B=LOC(A);

LOC creates a $1 \times n$ row vector, where n is the number of nonzero elements in the argument. The argument must be a row or column vector. The values in the resulting row vector are the locations of the nonzero elements in the argument. For example, suppose

```
A=(1 0 2 3 0);
```

then,

```
B=LOC(A);
```

results in

```
B=(1 3 4) ,
```

since the first, third, and fourth elements of A are nonzero. If every element of the argument vector is zero, the result is empty, that is, B will have 0 rows and 0 columns.

The LOC function is useful for subscripting parts of a matrix that satisfy some condition. For example, suppose you want to create a matrix Y that contains the rows of X having a positive element in the diagonal of X.

Let

$$X = \begin{bmatrix} 1 & 1 & 0 \\ 0 & -2 & 2 \\ 0 & 0 & 3 \end{bmatrix}.$$

Then

```
Y=X(LOC(VECDIAG(X)>0),);
```

results in

$$Y = X(1\ 3,) = \begin{bmatrix} 1 & 1 & 0 \\ 0 & 0 & 3 \end{bmatrix}$$

since the first and third rows of X have positive elements on the diagonal of X.

To select all positive elements of a column vector

$$A = \begin{bmatrix} 0 \\ -1 \\ 2 \\ 0 \end{bmatrix}$$

the statement

```
Y=A(LOC(A>0),);
```

results in

Y=A(3,)=2 .

MARG: evaluates marginal totals in a multiway contingency table

Form: CALL MARG (LOCMAR,MARGINAL,DIM,TABLE,CONFIG)

The arguments used with MARG are:

DIM specifies a vector containing the number of variables and the number of their possible levels in a contingency table. If DIM is 1×I, there are I variables and the value of the ith element is the number of levels of the ith variable.

TABLE specifies an array containing the number of observations at each level of each variable. Variables are nested across columns and then across rows.

CONFIG specifies an array containing the marginal totals to be evaluated. Each column specifies a distinct marginal.

LOCMAR returns a vector of indexes to each new set of marginal totals specified by CONFIG. A marginal total is exhibited for each level of the specified marginal. These indexes help locate particular totals.

MARGINAL returns a vector of marginal totals.

The matrix TABLE must conform in size to the contingency table specified in DIM. In particular, if TABLE is $n \times m$, the product of the entries in the DIM vector must equal nm. In addition, there must be some integer k such that the product of the first k entries in DIM equals m. See the IPF function for more information on specifying TABLE.

For example, consider the no-three-factor-effect model for Bartlett's data as described in Bishop et al. (1975).

```
DIM= 2 2 2;

TABLE= 156  84 84 156 /
       107 133 31 209;

CONFIG= 1 1 2 /
        2 3 3;

MARG(LOCMAR,MARGINAL,DIM,TABLE,CONFIG);
```

Returned are:

```
LOCMAR= 1 5 9

MARGINAL= 263 217 115 365 240 240
          138 342 240 240 240 240 .
```

MAX: maximum value of matrix

Form: MAX(*matrix*)
Example: I=MAX(D);

MAX produces a single numeric value that is the largest element in *matrix*. To find the elementwise maximum of two matrices, use the <> operator.

MIN: minimum value of matrix

Form: MIN(*matrix*)
Example: IF MIN(Z)<0 THEN GO TO LOOP;

MIN produces a single numeric value that is the smallest element in *matrix*. To find the elementwise minimum of two matrices, use the >< operator.

NCOL: number of columns

Form: NCOL(*matrix*)
Example: N=NCOL(A);

NCOL returns a single numeric value that is the number of columns in *matrix*. If the matrix has not been defined, NCOL returns a value of 0.

NOTE: annotating printout

```
                PAGE
Form:     NOTE  SKIP  message;
                SKIP=n
Example:  NOTE CONVERGED SUCCESSFULLY;
```

The NOTE command allows you to write comments on the printed output from the MATRIX procedure. The keyword PAGE causes the printout to skip to the next page; SKIP causes one line to be skipped; and SKIP=*n*, where *n* is a number, causes *n* lines to be skipped.

The message may be enclosed in single quotes; it **must** be enclosed in single quotes if it contains a semicolon. If a single quote appears in *message*, it must appear as two single quotes. If the *message* immediately follows the keyword NOTE, no lines are skipped before the message is printed.

NROW: number of rows

Form: NROW(*matrix*)
Example: N=NROW(A);

NROW returns a single numeric value that is the number of rows in matrix. If the matrix has not been defined, NROW returns a value of 0.

ORPOL: orthogonal polynomials

Form: ORPOL(*vector,maxdegree,weights*)
ORPOL(*vector,maxdegree*)
ORPOL(*vector*)
Example: P=ORPOL(X,MD,W);
P=ORPOL(X,MD);
P=ORPOL(X);

The ORPOL matrix function generates orthogonal polynomials.

If *maxdegree* or *weights* is omitted, the default values below are used:

maxdegree (maximum degree polynomial to be computed) = *min(n,20)*

If the third argument (*weights*) is specified, the second argument (*maxdegree*) **must** be specified also.

The result is a column-orthonormal matrix P with the same number of rows as the vector and with *maxdegree*+1 columns:

P'*DIAG(*weights*)*P=I ,

which is computed such that P(I,J) is the value of a polynomial of degree J−1 evaluated at the ith element of the vector.

Vector is an $n \times 1$ (or $1 \times n$) vector of values over which the polynomials are to be defined. This argument must always be given.

Maxdegree specifies the maximum degree polynomial to be computed. Note that the number of columns in the computed result is 1+*maxdegree*, whether *maxdegree* is specified or the default value is used.

Weights specifies an $n \times 1$ (or $1 \times n$) vector of nonnegative weights to be used in defining orthogonality: P'*DIAG(*weights*)*P=I. If *weights* is specified, *maxdegree* must be specified also. If *maxdegree* is not specified, or is specified erroneously, the default weights (all weights=1) are used.

The maximum number of nonzero orthogonal polynomials, which can be computed from the vector and the weights, is:

r = the number of distinct values in the vector,
ignoring any value associated with a zero weight.

The polynomial of maximum degree has degree=$r-1$. If the value of *maxdegree* exceeds $r-1$, then columns $r+1$, $r+2$, ..., *maxdegree*+1 of the result are set to zero. In this case,

$$P'*DIAG(weights)*P = \begin{matrix} I_{(r)} & 0 \\ 0 & 0*J(maxdegree+1-r) \end{matrix}$$

Note: the ORPOL function was contributed by Ronald Helms of the University of North Carolina at Chapel Hill based on a method described by Emerson (1968).

OUTPUT: output a matrix to a SAS data set

Form: OUTPUT *matrix* OUT=*SASdataset options*;
Example: OUTPUT X OUT=Y;
OUTPUT MAT;

The OUTPUT command creates a SAS data set from *matrix*. Each row of *matrix* corresponds to an observation, and each column corresponds to a variable. If you want to create a permanent SAS data set, you must specify a two-level name (see Chapter 12, "SAS Data Sets," in *SAS User's Guide: Basics, 1982 Edition*, for more information on permanent SAS data sets).

The first variable is a character variable of length 8 named ROW, and its values are ROW1, ROW2, ..., ROW*n*. The remaining variables are named COL1, COL2, ..., COL*n*.

The RENAME= data set name option can be used for the data set named by OUT= to give more descriptive variable names than COL1, COL2, ..., COL*n*, as shown below:

```
OUTPUT A OUT=ADATA(RENAME=(COL1=X COL2=Y));
```

The data set created by OUTPUT is available for input using the FETCH command in later executions of the MATRIX procedure.

The OUTPUT statement may be executed many times; each time more observations are written to the output data set. The number of columns in the matrix being written must be the same for each execution.

Looping through the OUTPUT statement one observation at a time is not as efficient as writing the entire matrix with a single OUTPUT statement.

The options below may appear in the OUTPUT command:

TYPE=CHAR requests that all the variables created be character, rather than one character variable and the rest numeric.

The TYPE=CHAR option is only effective when a character-valued matrix is used for output; using it with a numeric matrix does not produce printable representations of the numbers.

COLNAME=*matrix* gives names to the variables in the output data set, rather than the default names COL1, COL2, ..., COL*n*.

The COLNAME option specifies the name of a character matrix. The first *ncol* values from this matrix provide the variable names in the output data set where *ncol* is the number of columns in the matrix to be output. The procedure uses the first *ncol* elements of *matrix* in row-major order.

ROWNAME=*matrix* provides more descriptive row names than the default names ROW1, ROW2, ..., ROW*n*.

The ROWNAME option specifies the name of a character matrix. The first *nrow* values of this matrix become values of the variable named ROW in the output data set where *nrow* is the number of rows in the matrix to be output, and the scan to find the first *nrow* elements goes across row 1, then across row 2, and so on.

For example, to create a data set called B with two numeric variables X and Y and a character variable __ROW__ with values A B C, then use:

```
R='A' 'B' 'C';
C='X' 'Y';
X=1 2/3 4/5 6;
OUTPUT X ROWNAME=R COLNAME=C OUT=B;
```

PRINT: print a matrix

Form: PRINT *matricesoptions*;
Example: PRINT X;
PRINT A B C;

The PRINT command prints the specified matrices, which may include any defined matrix.

If a matrix row is too long for the print line, it is extended to the next line. The FW (field width) option of the PROC MATRIX statement can be used to control the number of print positions for printing each matrix element.

Using the PRINT option of the PROC MATRIX statement may be preferable to using many PRINT statements.

The options below may appear in the PRINT command:

FORMAT=*format* specifies a format to be used in printing the values of the matrix. For character-valued matrices, the format should be $8.

Examples:

```
PRINT X FORMAT=5.3;
PRINT CHARMAT FORMAT=$8.;
PRINT Y FORMAT=FRACT.;
```

ROWNAME=*matrix* specifies the name of a character matrix whose first *nrow* elements are to be used for the row labels of the matrix to be printed where *nrow* is the number of rows in the matrix to be printed, and where the scan to find the first *nrow* elements goes across ROW 1, then across ROW 2, ..., ROW*n*.

COLNAME=*matrix* specifies the name of a character matrix whose first *ncol* elements are to be used for the column labels of the matrix to be printed where *ncol* is the number of columns in the matrix to be printed and where the scan to find the first *ncol* elements goes across column 1, then across column 2, ..., column*n*.

For example, if you wanted to print a matrix called X in format 12.2 in columns headed 'AMOUNT' and 'NET PAY', and rows labeled 'DIV A' and 'DIV B', then code:

```
R='DIV A' 'DIV B';
C='AMOUNT' 'NET PAY';
PRINT X ROWNAME=R COLNAME=C FORMAT=12.2;
```

RANK: ranking

Form: RANK(*matrix*)
Example: R=RANK(A);

RANK creates a new matrix whose elements are the ranks of the corresponding elements of *matrix*. The ranks of tied values are not averaged. If

X= 2 2 1 0 5;

then RANK(X) produces the vector

(3 4 2 1 5)

This function can be used to sort a column vector:

```
B=A; A(RANK(A),)=B;
```

To find anti-ranks of A use:

```
R=RANK(A); I=R; I(R,)=1:NROW(A);
```

RANKTIE: ranking

Form: RANKTIE(*matrix*)
Example: R=RANKTIE(A);

RANKTIE creates a new matrix whose elements are the ranks of the corresponding elements of *matrix*. The ranks of tied values are averaged. If

X=(2 2 1 0 5)

then RANKTIE(X) produces the vector

(3.5 3.5 2 1 5)

RANKTIE differs from the RANK function in that RANKTIE averages the ranks of tied values, while RANK arbitrarily breaks ties.

SHAPE: shape a matrix

Form: SHAPE(*matrix, columns*)
Example: N=SHAPE(X,3);

The SHAPE function shapes a new matrix from a matrix with different dimensions or from a vector. *Columns* gives the number of columns in the new matrix. For example, the statements

```
PROC MATRIX;
  X= 1 1 /
     2 2 /
     3 3 ;
  A=SHAPE(X,3);
```

set up X as a 3×2 matrix. The last statement creates A from the elements of X. A has 3 columns, and thus 2 rows:

```
1 1 2
2 3 3
```

The number of elements in the original matrix must be a whole number multiple of the column value given.

Another example:

```
PROC MATRIX;
  Y= 3 3 3 /
     2 2 2 /
     1 1 1 ;
  V=SHAPE(Y,9);
```

Y is a 3×3 matrix. The last statement creates the row vector

```
3 3 3 2 2 2 1 1 1
```

and assigns it to V.

The statement

```
V=SHAPE(Y,1);
```

creates V as a column vector of nine elements.

SOLVE: solve system of linear equations

Form: SOLVE(*matrix1,matrix2*)
Example: X=SOLVE(A,B);

SOLVE solves the set of linear equations AX=B for X. A must be square and nonsingular.

X=SOLVE(A,B) is equivalent to using the INV function as X=INV(A)*B. However, SOLVE is recommended over INV because it is both more efficient and more accurate.

The method used is discussed in Forsythe and Moler (1967).

SSQ: sum of squares of all elements

Form: SSQ(*matrix*)
Example: S=SSQ(A);

SSQ returns as a single numeric value the uncorrected sum of squares for all the elements of *matrix*.

SUM: sum of all elements

Form: SUM(*matrix*)
Example: TOT=SUM(X);

SUM returns as a single numeric value the sum of all the elements in *matrix*.

SVD: singular value decomposition

Example: SVD U Q V A;

SVD decomposes a real $m \times n$ (say $m>=n$) matrix, denoted by A in the example above, into the form

A=U*diag(Q)*V′

where

$U'U=V'V=VV'=I_n$

and Q contains the singular values. U is $m \times n$, Q is $n \times 1$, and V is $n \times n$.

When $m>=n$, U consists of the orthonormal eigenvectors of AA′ and V consists of the eigenvectors of A′A. Q contains the square roots of the eigenvalues of A′A and AA′, except for some zeros.

If $m<n$, a corresponding decomposition is done where U and V switch roles:

A=U*diagvec(Q)*V′

but

$U'U=UU'=V'V=I_m$.

For information about the method used in SVD, see Wilkinson and Reinsch (1971).

To sort the singular values, use

```
SVD U Q V A;
R=RANK(Q);
TEMP=U;  U(,R)=TEMP;
TEMP=Q;  Q(R,)=TEMP;
TEMP=V;  V(,R)=TEMP;
```

SWEEP: sweep operator

Form: SWEEP(*matrix,indexvector*)
Example: S=SWEEP(A,I);

SWEEP sweeps *matrix* on the pivots indicated in *index vector* to produce a new matrix. The values of the index vector must be less than or equal to the number of rows or the number of columns in *matrix*, whichever is smaller.

For example, suppose that A is partitioned into

$$A = \begin{bmatrix} R & S \\ T & U \end{bmatrix}$$

such that

R is q by q, U is $m-q$ by $n-q$.

Let

$$I = 1{:}q = 1\ 2 \ldots q,$$

then

S=SWEEP(A,I)

becomes

$$S = \begin{bmatrix} R^{-1} & R^{-1}S \\ -TR^{-1} & U-TR^{-1}S \end{bmatrix}$$

The index vector may be omitted. In this case, the function sweeps the matrix on all pivots on the main diagonal 1:MIN(*nrow*,*ncol*).

The SWEEP function has sequential and reversibility properties when the submatrix swept is positive definite. For example:

```
SWEEP(SWEEP(A,1),2)=SWEEP(A,1 2)
SWEEP(SWEEP(A,I),I)=A
```

See Beaton (1964) for more information about these properties.

To use SWEEP for regression, suppose

$$A = \begin{bmatrix} X'X & X'Y \\ Y'X & Y'Y \end{bmatrix}$$

where X'X is k by k.

Then B=SWEEP(A,1...k) forms

$$B = \begin{bmatrix} (X'X)^{-1} & (X'X)^{-1}X'Y \\ -Y'X(X'X)^{-1} & Y'(I-X(X'X)^{-1}X')Y \end{bmatrix}$$

whose partitions form the beta values, SSE, and a matrix proportional to the covariance of the beta values for the least-squares estimates of B in the linear model:

$$Y = X*B+\varepsilon$$

If any pivot becomes very close to zero ($<=1E^{-12}$), the row and column for that pivot are zeroed. (See Goodnight (1979) for more information.)

TRACE: sum of diagonal elements

Form: TRACE(*matrix*)
Example: DISUM=TRACE(X);

TRACE produces a single numeric value that is the sum of the diagonal elements of matrix.

VECDIAG: vector from diagonal

Form: VECDIAG(*squarematrix*)
Example: V=VECDIAG(A);

VECDIAG creates a column vector whose elements are the main diagonal elements of *squarematrix*.

Extended features

SOLVIT: solve system of linear equations

This function can only be used on machines that have the extended-precision floating point feature.

Form: SOLVIT(*matrix1,matrix2*)
Example: SOLVIT(A,X);

SOLVIT works like the SOLVE function, but is more accurate.

SOLVIT computes a solution to the linear system like SOLVE, then attempts to improve the solution iteratively. (See Forsythe and Moler (1967) for more information about this method.)

The number of iterations required and a measure of the accuracy attained in decimal digits are always printed. A maximum of 16 iterations is permitted.

SOLVIT computes an inverse if the second argument is an identity matrix.

The normal SOLVE routine is accurate enough for most problems and should be used unless great accuracy is required. If the problem is very ill-conditioned, iterative refinement will not always improve the solution. The epsilon used by SOLVIT is 1E−14.

XMULT: very accurate multiplication

Form: XMULT(*matrix1,matrix2*)
Example: M=XMULT(A,B);

XMULT computes the matrix product like the * operator, but uses extended precision to accumulate sums of products and thus is more accurate. XMULT should only be used when great accuracy is needed.

Character values

The MATRIX procedure has some limited character value features. Their primary use in MATRIX is in labeling rows and columns on a matrix that is printed or for specifying variable names on output data sets. Character values can be up to 8 characters long and are stored in matrices just as numbers are stored.

MATRIX does not keep track of which matrices contain character values and which contain numbers. A character format should always be specified when a character value is to be printed. For example,

```
PRINT NAME FORMAT=$8.;
```

can be used to print the character matrix NAME.

All basic MATRIX operations can be performed on character matrices, although arithmetic operations on character values are meaningless.

Character literals

Simple assignments of character literals to matrices are allowed. For example, this statement creates a character-valued matrix RNAMES:

```
RNAMES='SSE' 'DF' 'MSE';
```

The literal values, which have a maximum length of 8, must be enclosed in single quotes. Although character literals are not allowed in complex expressions (because the single quote is also used for the transpose), character-valued matrices can be manipulated with matrix operators in later statements.

Only moving operations such as transposing, subscripting, and concatenation should be used on a character matrix. Although mathematical operations may not produce error messages, the character representations are unreadable.

Other character value features are discussed in the descriptions of the FETCH, OUTPUT, and PRINT commands.

DETAILS

Storage Requirements

No single matrix should contain more than 32,767 elements, since many operators do not work properly above this limit.

The MATRIX procedure stores matrices in a work space. To find the size of the work space and the amount taken up by matrices, use the LIST command.

Each active matrix needs 2+*nrow***ncol* doublewords, where *nrow* is the number of rows in the matrix and *ncol* is the number of columns. (A doubleword is 8 bytes.)

From a 200K region, about 100K should be available for work space, enough for 125 10×10 matrices or one 110×110 matrix. Many matrix operations need additional storage to perform the computations.

If you receive an error message saying that not enough space is available, try one or more of these alternatives:

- increase region size
- free intermediate results when they are no longer needed by using the FREE command or by using the same matrix name to hold another matrix
- use the OUTPUT command to store matrices in SAS data sets.

Accuracy

In the MATRIX procedure, all numbers are stored and all arithmetic is done in double precision. The algorithms used are generally very accurate numerically. However, when many operations will be performed, or when the matrices are ill-conditioned, matrix operations should be used in a numerically responsible manner.

Error Diagnostics

When an error occurs, several lines of messages are printed. Under batch SAS, messages appear on both the log and the procedure output. The error description, the operation being performed, and the line and column of the source for that operation are printed.

When an error occurs with PROC MATRIX, the names of the operation's arguments are printed. The error message may include matrix names beginning with #; these are temporary names assigned by the MATRIX procedure. When an error occurs, the operation is not completed and nothing is assigned to the result.

The most common errors are:

- not enough work space (See **Storage Requirements** above)
- referencing a matrix that has not been set to a value

- indexing error: trying to refer to a row or column not present in the matrix
- matrix arguments not conformable: for example, multiplying two matrices together that do not conform, or using a function that requires a special scalar or vector argument
- matrix not square (for example, INV, DET, SOLVE)
- matrix not symmetric (for example, EIGEN)
- matrix singular (INV, SOLVE)
- matrix not positive definite or positive semidefinite (HALF, SWEEP).

These errors result from the actual dimensions or values of matrices and are caught only after the procedure has started to execute. Other errors, such as incorrect number of arguments, unbalanced parentheses, and so on, are syntactical errors and prevent the procedure from running at all.

Efficiency

The MATRIX procedure is an interpretive language executor, which can be characterized as:

- inexpensive to compile
- expensive for the number of operations executed
- inexpensive within each operation.

You should try to substitute matrix operations instead of iterative loops. There is high overhead involved in executing each instruction; however, within the instructions, MATRIX runs very efficiently.

Consider four methods of summing the elements of a matrix ranked in order from most expensive (a) to least expensive (d):

```
a) S=0;
   DO I=1 TO M;  DO J=1 TO N;
     S=S+X(I,J);  END;  END;
b) S=J(1,M)*X*J(N,1);
c) S=X(+,+);
d) S=SUM(X);
```

For some programs that can be written with a few matrix statements, MATRIX reduces programming time and executes even faster than FORTRAN because MATRIX:

- has less to compile and compiles faster
- uses very fast assembler code inside most operations.

The MATRIX Library Interface

Special functions and CALL routines can be written in FORTRAN and PL/I and invoked from PROC MATRIX as with built-in MATRIX functions. For complete documentation, see the *SAS Programmer's Guide*.

EXAMPLES

Correlation and Standardization: Example 1

In this example, MATRIX is invoked twice: first to compute a correlation matrix from an input data set and then to standardize the data (subtract by column means and divide by column standard deviations). The expression

```
J(N,1)*MEAN;
```

constructs a matrix of *n* rows where each row is MEAN. The results obtained in this example can be verified by using the CORR and STANDARD procedures in SAS.

```
DATA A; INPUT X Y Z @@; CARDS;
1 2 3 3 2 1 4 2 1 0 4 1 24 1 0 1 3 8
;
PROC PRINT;

PROC MATRIX;
  FETCH X DATA=A;           * GET THE DATA INTO MATRIX X;
  N=NROW(X);                * DIMENSION OF X;
  SUM=X(+,);                * COLUMN SUMS BY REDUCING ROWS;
  XPX=X'*X-SUM'*SUM#/N;     * CROSSPRODUCTS CORRECTED FOR MEAN;
  S=1#/SQRT(DIAG(XPX));     * SCALING MATRIX;
  CORR=S*XPX*S;             * CORRELATION MATRIX;
  NOTE CORRELATION MATRIX;
  PRINT CORR;

PROC MATRIX;
  FETCH X DATA=A;           * GET THE DATA INTO MATRIX X;
  N=NROW(X);                * DIMENSION OF X;
  MEAN=X(+,)#/N;            * MEANS FOR COLUMNS;
  X=X-J(N,1,1)*MEAN;        * CENTER X TO MEAN ZERO;
  SS=X(##,);                * SUM OF SQUARES FOR COLUMNS;
  STD=SQRT(SS'#/(N-1));     * STANDARD DEVIATION ESTIMATE;
  X=X*DIAG(1#/STD);         * SCALING TO STD DEV 1;
  NOTE STANDARDIZED DATA MATRIX;
  PRINT X;                  * PRINT STANDARDIZED DATA;
  OUTPUT X OUT=B;           * MAKE SAS DATA SET;

PROC PRINT DATA=B;
```

```
                     OBS    X    Y    Z                        1

                      1     1    2    3
                      2     3    2    1
                      3     4    2    1
                      4     0    4    1
                      5    24    1    0
                      6     1    3    8
```

```
                       CORRELATION MATRIX                       2

CORR            COL1          COL2          COL3

ROW1               1     -0.717102     -0.436558
ROW2       -0.717102             1      0.350823
ROW3       -0.436558      0.350823             1
```

```
                    STANDARDIZED DATA MATRIX                          3

          X              COL1          COL2          COL3

          ROW1       -0.490116     -0.322749      0.226455
          ROW2       -0.272287     -0.322749     -0.452911
          ROW3       -0.163372     -0.322749     -0.452911
          ROW4        -0.59903       1.61374     -0.452911
          ROW5         2.01492      -1.29099     -0.792594
          ROW6       -0.490116      0.645497       1.92487
```

```
OBS     ROW        COL1        COL2        COL3                         4

 1      ROW1     -0.49012     -0.3227     0.22646
 2      ROW2     -0.27229     -0.3227    -0.45291
 3      ROW3     -0.16337     -0.3227    -0.45291
 4      ROW4     -0.59903      1.6137    -0.45291
 5      ROW5      2.01492     -1.2910    -0.79259
 6      ROW6     -0.49012      0.6455     1.92487
```

Newton's Method for Solving Non-Linear Systems of Equations: Example 2

Let the non-linear system be represented by

$$F(\mathbf{x}) = 0$$

where **x** is a vector and F is vector-valued.

In order to find **x** such that F goes to zero, an initial estimate $\mathbf{x}_0$ is chosen, and Newton's iterative method for converging to the solution is used:

$$\mathbf{x}_{n+1} = \mathbf{x}_n - J^{-1}(\mathbf{x}_n)F(\mathbf{x}_n)$$

where J(**x**) is the Jacobian matrix of partial derivatives of F with respect to **x**.

For optimization problems, the same method is used, where F(**x**) is the gradient of the objective function, and J(**x**) becomes the Hessian (Newton-Raphson).

In this example, the system to be solved is:

$$x_1 + x_2 - x_1x_2 + 2 = 0$$
$$x_1e^{-x_2} - 1 = 0$$

```
PROC MATRIX;
*——— SOLVE A NONLINEAR SYSTEM BY NEWTON'S METHOD ————*;

  NOTE SOLVE: X1+X2-X1*X2+2=0, X1*EXP(-X2)-1=0 ;

  X= .1 / -2;                          * STARTING VALUES;
  X1=X(1,); X2=X(2,);                  * EXTRACT THE PARAMETERS;
  F=(X1+X2-X1*X2+2)//
     (X1*EXP(-X2)-1);                  * INITIAL VALUE OF THE FUNCTION;

  DO WHILE(MAX(ABS(F))>.000001);  * ITERATE UNTIL CONVERGENCE;
    J=((1-X2)||(1-X1))//
       (EXP(-X2)||(-X1*EXP(-X2)));     * EVALUATE JACOBIAN;
    DELTA=-SOLVE(J,F);                 * SOLVE FOR CORRECTION VECTOR;
```

```
   X =X+DELTA;                          * THE NEW APPROXIMATION;
   X1=X(1,*);  X2=X(2,*);               * EXTRACT THE TWO VALUES;

   F=(X1+X2-X1*X2+2)//
      (X1*EXP(-X2)-1);                  * EVALUATE THE FUNCTION;
   END;

 PRINT X F;                             * PRINT SOLUTION AND RESIDUAL;
```

```
                                                              1

        SOLVE:  X1+X2-X1*X2+2=0,  X1*EXP(-X2)-1=0

                X              COL1

                ROW1       0.0977731
                ROW2        -2.32511


                F              COL1

                ROW1       5.3523E-09
                ROW2       6.1501E-08
```

Alpha Factor Analysis: Example 3

This example shows how an algorithm for computing alpha factor patterns (Kaiser, 1965) is transcribed into MATRIX code.

For later reference, you could store the alpha subroutine in this code on a source library (possibly as a MACRO).

```
PROC MATRIX;

  *—CORRELATION MATRIX FROM HARMON, MODERN FACTOR ANALYSIS,
       2ND EDITION, PAGE 124, ''EIGHT PHYSICAL VARIABLES'' ;
  R= 1.000  .846  .805  .859  .473  .398  .301  .382 /
      .846 1.000  .881  .826  .376  .326  .277  .415 /
      .805  .881 1.000  .801  .380  .319  .237  .345 /
      .859  .826  .801 1.000  .436  .329  .327  .365 /
      .473  .376  .380  .436 1.000  .762  .730  .629 /
      .398  .326  .319  .329  .762 1.000  .583  .577 /
      .301  .277  .237  .327  .730  .583 1.000  .539 /
      .382  .415  .345  .365  .629  .577  .539 1.000;

  LINK ALPHA;
  NOTE EIGENVALUES;  PRINT M;
  NOTE COMMUNALITIES;  PRINT H;
  NOTE FACTOR PATTERN;  PRINT F;
  STOP;
```

```
* ----------------------------------ALPHA FACTOR ANALYSIS---------------------------------- *
|  REF: KAISER ET AL., 1965 PSYCHOMETRIKA, PP. 12-13                                        |
|                                                                                           |
|  R CORRELATION MATRIX (N.S.) ALREADY SET UP                                               |
|  P NUMBER OF VARIABLES                                                                    |
|  Q NUMBER OF FACTORS                                                                      |
|  H COMMUNALITIES                                                                          |
|  M EIGENVALUES                                                                            |
|  E EIGENVECTORS                                                                           |
|  F FACTOR PATTERN                                                                         |
|  (IQ,H2,HI,G,MM) TEMPORARY USE. FREED UP                                                  |
* ----------------------------------------------------------------------------------------- * ;
 ALPHA:
  P=NCOL(R);  Q=0;  H=0;                     * INITIALIZE;
  H2=I(P)-DIAG(1#/VECDIAG(INV(R)));          * SMCS;

  DO WHILE(MAX(ABS(H-H2))>.001);             * ITERATE UNTIL CONVERGES;
    H=H2;  HI=DIAG(SQRT(1#/VECDIAG(H)));
    G=HI*(R-I(P))*HI+I(P);
    EIGEN M E G;                             * GET EIGENVALUES AND VECS;
    IF Q=0 THEN DO;  Q=SUM(M>1);             * NUMBER OF FACTORS;
      IQ=1:Q;  END;                          * INDEX VECTOR;
    MM=DIAG(SQRT(M(IQ, )));                  * COLLAPSE EIGVALS;
    E=E( ,IQ);                               * COLLAPSE EIGVECS;
    H2=H*DIAG((E*MM)( ,##));                 * NEW COMMUNALITIES;
    END;

  HI=SQRT(H);  H=VECDIAG(H2);
  F=HI*E*MM;                                 * RESULTING PATTERN;
  FREE IQ H2 HI G MM;                        * FREE TEMPORARIES;
  RETURN;
```

```
                              EIGENVALUES                              1

                       M               COL1

                       ROW1         5.93785
                       ROW2          2.0622
                       ROW3        0.139018
                       ROW4       0.0821054
                       ROW5        0.018097
                       ROW6      -0.0474866
                       ROW7      -0.0914798
                       ROW8       -0.100304

                            COMMUNALITIES

                       H               COL1

                       ROW1         0.83812
                       ROW2        0.890572
                       ROW3         0.81893
                       ROW4        0.806729
                       ROW5        0.880215
                       ROW6        0.639198
                       ROW7        0.582158
                       ROW8        0.499813

                            FACTOR PATTERN

              F             COL1            COL2

              ROW1     -0.813386        0.420147
              ROW2     -0.802836         0.49601
              ROW3     -0.757909        0.494474
              ROW4     -0.787446        0.432039
              ROW5     -0.805144        -0.48162
              ROW6     -0.680413       -0.419805
              ROW7     -0.620623        -0.44383
              ROW8     -0.644942        -0.28959
```

Categorical Linear Models: Example 4

In the following example, replace the N and X matrix literals with any frequency count matrix and design matrix to use this MATRIX program for another set of data.

This example uses PROC MATRIX to fit a linear model to a function of the response probabilities

$$K \, log \, \pi = X\beta$$

where *K* compares each response category with the last. Data were obtained from Kastenbaum and Lamphiear (1959). First, the Grizzle-Starmer-Koch (1969) approach is used to obtain generalized least-squares estimates of β. These form the initial values for the Newton-Raphson solution for the maximum-likelihood estimates. PROC FUNCAT could also be used to analyze these binary data (see Cox, 1970).

```
 TITLE CATEGORICAL LINEAR MODELS BY LEAST SQUARES AND MAXIMUM
   LIKELIHOOD;

PROC MATRIX;
 * PREPARE FREQUENCY DATA AND DESIGN MATRIX ;
  N=58 11 05/
    75 19 07/
    49 14 10/
    58 17 08/
    33 18 15/
    45 22 10/
    15 13 15/
    39 22 18/
    04 12 17/
    05 15 08;                       * FREQUENCY COUNTS;

  X=1  1  1  0  0  0/
    1 -1  1  0  0  0/
    1  1  0  1  0  0/
    1 -1  0  1  0  0/
    1  1  0  0  1  0/
    1 -1  0  0  1  0/
    1  1  0  0  0  1/
    1 -1  0  0  0  1/
    1  1 -1 -1 -1 -1/
    1 -1 -1 -1 -1 -1;               * DESIGN MATRIX;

 * FIND DIMENSIONS -------------------------------------------------------*;
  S=NROW(N);                        * NUMBER OF POPULATIONS;
  R=NCOL(N);                        * NUMBER OF RESPONSES;
  Q=R-1;                            * NUMBER OF FUNCTION VALUES;
  D=NCOL(X);                        * NUMBER OF DESIGN PARAMETERS;
  QD=Q*D;                           * TOTAL NUMBER OF PARAMETERS;

 * GET PROBABILITY ESTIMATES---------------------------------------------*;
  ROWN=N(,+);                       * ROW TOTALS;
  PR=N#/(ROWN*J(1,R));              * PROBABILITY ESTIMATES;
  P=SHAPE(PR(,1:Q),1);              * CUT AND SHAPED TO VECTOR;
  NOTE INITIAL PROBABILITY ESTIMATES; PRINT PR;
```

```
  * ESTIMATE BY THE GSK METHOD --------------------------------------------------------*;
    F=LOG(P)-LOG(PR(,R))@J(Q,1);      * FUNCTION OF PROBABILITIES;
    SI=(DIAG(P)-P*P')#
      (DIAG(ROWN)@J(Q,Q));            * INV COVARIANCE OF F;
    Z=X@I(Q);                         * EXPANDED DESIGN MATRIX;
    H=Z'*SI*Z;                        * CROSSPRODUCTS MATRIX;
    G=Z'*SI*F;                        * CROSS WITH F;
    BETA=SOLVE(H,G);                  * LEAST SQUARES SOLUTION;
    STDERR=SQRT(VECDIAG(INV(H))); * STANDARD ERRORS;

    LINK PROB;  NOTE GSK ESTIMATES;  PRINT BETA STDERR PI;

  * ITERATIONS FOR ML SOLUTION --------------------------------------------------------*;
    CRIT=1;

    DO IT=1 TO 8
      WHILE(CRIT>.0005);              * ITERATE UNTIL CONVERGE;
      SI=(DIAG(PI)-PI*PI') #
          (DIAG(ROWN)@J(Q,Q));        * BLOCK DIAGONAL WEIGHTING;
      G=Z'*(ROWN@J(Q,1)#(P-PI));      * GRADIENT;
      H=Z'*SI*Z;                      * HESSIAN;
      DELTA=SOLVE(H,G);               * SOLVE FOR CORRECTION;
      BETA=BETA+DELTA;                * APPLY THE CORRECTION;
      LINK PROB;                      * COMPUTE PROB ESTIMATES;
      CRIT=MAX(ABS(DELTA));           * CONVERGENCE CRITERION;
      END;

    STDERR=SQRT(VECDIAG(INV(H))); * STANDARD ERRORS;
    NOTE ML ESTIMATES;  PRINT BETA STDERR PI;
    NOTE ITERATIONS AND CONVERGENCE CRITERION;  PRINT IT CRIT;
    RETURN;

  * SUBROUTINE TO COMPUTE NEW PROB ESTIMATES @ PARAMETERS---------*;
PROB: LA=EXP(X*SHAPE(BETA,Q));
    PI=LA#/((1+LA(,+))*J(1,Q));
    PI=SHAPE(PI,1);
    RETURN;
```

```
          CATEGORICAL LINEAR MODELS BY LEAST SQUARES AND MAXIMUM LIKELIHOOD          1

                          INITIAL PROBABILITY ESTIMATES

                 PR              COL1          COL2          COL3

                 ROW1        0.783784      0.148649     0.0675676
                 ROW2        0.742574      0.188119     0.0693069
                 ROW3        0.671233      0.191781      0.136986
                 ROW4        0.698795      0.204819     0.0963855
                 ROW5             0.5      0.272727      0.227273
                 ROW6        0.584416      0.285714       0.12987
                 ROW7        0.348837      0.302326      0.348837
                 ROW8        0.493671      0.278481      0.227848
                 ROW9        0.121212      0.363636      0.515152
                 ROW10       0.178571      0.535714      0.285714

                                GSK ESTIMATES

                          BETA          COL1

                          ROW1      0.945443
                          ROW2      0.400326
                          ROW3     -0.277777
```

(continued on next page)

(continued from previous page)

ROW4	-0.278472
ROW5	1.41469
ROW6	0.474136
ROW7	0.84647
ROW8	0.152609
ROW9	0.195239
ROW10	0.0723489
ROW11	-0.514488
ROW12	-0.400831

STDERR	COL1
ROW1	0.129092
ROW2	0.128487
ROW3	0.11647
ROW4	0.125592
ROW5	0.267351
ROW6	0.294943
ROW7	0.236264
ROW8	0.263305
ROW9	0.221444
ROW10	0.23666
ROW11	0.217199
ROW12	0.228578

CATEGORICAL LINEAR MODELS BY LEAST SQUARES AND MAXIMUM LIKELIHOOD 2

PI	COL1
ROW1	0.740287
ROW2	0.167447
ROW3	0.770406
ROW4	0.174502
ROW5	0.662481
ROW6	0.191774
ROW7	0.706162
ROW8	0.204703
ROW9	0.516981
ROW10	0.264887
ROW11	0.569745
ROW12	0.292328
ROW13	0.39887
ROW14	0.25891
ROW15	0.466792
ROW16	0.30342
ROW17	0.132036
ROW18	0.395802
ROW19	0.165191
ROW20	0.495878

ML ESTIMATES

BETA	COL1
ROW1	0.95336
ROW2	0.406934
ROW3	-0.279081
ROW4	-0.280699
ROW5	1.44232
ROW6	0.499312
ROW7	0.841159
ROW8	0.148587
ROW9	0.188338
ROW10	0.0667313
ROW11	-0.527163
ROW12	-0.414965

CATEGORICAL LINEAR MODELS BY LEAST SQUARES AND MAXIMUM LIKELIHOOD 3

STDERR	COL1
ROW1	0.128618
ROW2	0.128459
ROW3	0.115622
ROW4	0.125282
ROW5	0.266936

(continued on next page)

(continued from previous page)

```
                    ROW6       0.294344
                    ROW7       0.236309
                    ROW8       0.263516
                    ROW9       0.220275
                    ROW10      0.236031
                    ROW11      0.216581
                    ROW12      0.229962

                    PI             COL1

                    ROW1       0.743176
                    ROW2       0.167316
                    ROW3       0.772327
                    ROW4       0.174442
                    ROW5       0.662727
                    ROW6       0.191665
                    ROW7       0.706277
                    ROW8       0.204922
                    ROW9       0.517078
                    ROW10      0.264686
                    ROW11      0.569777
                    ROW12      0.292607
                    ROW13      0.398421
                    ROW14      0.257665
                    ROW15      0.466682
                    ROW16       0.30279
                    ROW17      0.132324
                    ROW18      0.396311
                    ROW19      0.165475
                    ROW20      0.497204

              ITERATIONS AND CONVERGENCE CRITERION

                    IT             COL1

                    ROW1              3


                    CRIT           COL1

                    ROW1    0.000409248
```

Large Program Example—Variable Selection in Regression: Example 5

The following example performs regression with variable selection like the STEPWISE and RSQUARE procedures in SAS. It is a large program that illustrates many programming techniques.

```
*---------------------DATA ON PHYSICAL FITNESS---------------------*
|   THESE DATA ARE EXPLAINED WITH THE REG AND STEPWISE PROCEDURES.  |
*-------------------------------------------------------------------* ;

  DATA FITNESS;
    INPUT AGE WEIGHT OXY RUNTIME RSTPULSE
        RUNPULSE MAXPULSE @@;
    CARDS;
  ...data records...(see REG procedure for data)
;

  PROC MATRIX FW=8;
  TITLE VARIABLE SELECTION ALGORITHMS;

*--------------------------SET UP X AND Y---------------------------*
*   ONLY THE FOLLOWING TWO STATEMENTS ARE SPECIAL TO                *
*   THE DATA AND REGRESSION MODEL. YOU MUST SET UP                  *
*   X, Y, AND VARNAMES. THE REST OF THE CODE WORKS                  *
*   FOR ANY DATA AND MODEL.                                         *
*-------------------------------------------------------------------* ;
```

```
FETCH Y DATA=FITNESS(KEEP=OXY);
FETCH X COLNAME=VARNAMES
  DATA=FITNESS(KEEP=AGE WEIGHT RUNTIME RSTPULSE
      RUNPULSE MAXPULSE);

*---------------------------------STEPWISE METHODS---------------------------------*
*   AFTER A LINK TO THE INITIAL ROUTINE, WHICH SETS UP                              *
*   THE DATA, FOUR DIFFERENT ROUTINES CAN BE CALLED                                 *
*   TO DO FOUR DIFFERENT MODEL-SELECTION METHODS.                                   *
*-----------------------------------------------------------------------------------*  ;

LINK INITIAL;                    * INITIALIZATION;
LINK ALL;                        * ALL POSSIBLE MODELS;
LINK FORWARD;                    * FOREWARD SELECTION METHOD;
LINK BACKWARD;                   * BACKWARD ELIMINATION
                                   METHOD;
LINK STEPWISE;                   * STEPWISE METHOD;
STOP;

*---------------------------------- INITIALIZATION ---------------------------------*
*     C,CSAVE  THE CROSSPRODUCTS MATRIX                                             *
*     N        NUMBER OF OBSERVATIONS                                               *
*     K        TOTAL NUMBER OF VARIABLES TO CONSIDER                                *
*     L        NUMBER OF VARIABLES CURRENTLY IN MODEL                               *
*     IN       A 0-1 VECTOR OF WHETHER VARIABLE IS IN                               *
*     B        PRINT COLLECTS RESULTS (L MSE RSQ BETAS)                             *
*-----------------------------------------------------------------------------------*  ;

INITIAL:
  N=NROW(X);  K=NCOL(X);  K1=K+1;  IK=1:K;
  BNAMES='NPARM' 'MSE' 'RSQUARE' ||VARNAMES;

  * CORRECT BY MEAN, ADJUST OUT INTERCEPT PARAMETER;
  Y=Y-Y(.,);                       * CORRECT Y BY MEAN;
  X=X-J(N,1,1)*X(.,);              * CORRECT X BY MEAN;
  XPY=X'*Y;                        * CROSSPRODUCTS;
  YPY=Y'*Y;
  XPX=X'*X;
  FREE X Y;                        * NO LONGER NEED THE DATA;
  CSAVE=(XPX || XPY) //
        (XPY' || YPY);             * SAVED COPY OF CROSSPRODUCTS;
RETURN;

* FORWARD METHOD -------------------------------------------------------------------*;

FORWARD:
 NOTE 'FORWARD SELECTION METHOD';
 FREE BPRINT;
 C=CSAVE;  IN=J(K,1,0);  L=0;      *NO VARIABLES ARE IN;
 DFE=N-1;  MSE=YPY#/DFE;
 SPROB=0;

 DO WHILE(SPROB<.15 & L<K);
   INDX=LOC(¬IN);                  * WHERE ARE THE VARIABLES NOT IN?;
   CD=VECDIAG(C)(INDX,);           * XPX DIAGONALS;
```

```
     CB=C(INDX,K1);                    * ADJUSTED XPY;
     TSQR=CB#CB#/(CD#MSE);             * SQUARES OF T TESTS;
     IMAX=TSQR(<:>,);                  * LOCATION OF MAXIMUM IN INDX;
     SPROB=(1-PROBT(SQRT(TSQR(IMAX,)),DFE))*2;
     IF SPROB<.15 THEN DO;             * IF T-TEST SIGNIFICANT;
       II=INDX(,IMAX);                 * PICK MOST SIGNIFICANT;
       LINK SWP;                       * ROUTINE TO SWEEP;
       LINK BPR;                       * ROUTINE TO COLLECT RESULTS;
       END;
     END;

   PRINT BPRINT COLNAME=BNAMES;
   RETURN;

* BACKWARD METHOD --------------------------------------------------------------*;
BACKWARD:
   NOTE 'BACKWARD ELIMINATION METHOD';
   FREE BPRINT;
   C=CSAVE;   IN=J(K,1,0);
   II=1:K;   LINK SWP;   LINK BPR;      * START WITH ALL VARIABLES IN;
   SPROB=1;

   DO WHILE(SPROB>.15 & L>0);
     INDX=LOC(IN);                     * WHERE ARE THE VARIABLES IN?;
     CD=VECDIAG(C)(INDX,);             * XPX DIAGONALS;
     CB=C(INDX,K1);                    * BVALUES;
     TSQR=CB#CB#/(CD#MSE);             * SQUARES OF T TESTS;
     IMIN=TSQR(>:<,);                  * LOCATION OF MINIMUM IN INDX;
     SPROB=(1-PROBT(SQRT(TSQR(IMIN,)),DFE))*2;
     IF SPROB>.15 THEN DO;             * IF T-TEST NONSIGNIFICANT;
       II=INDX(,IMIN);                 * PICK LEAST SIGNIFICANT;
       LINK SWP;                       * ROUTINE TO SWEEP IN VARIABLE;
       LINK BPR;                       * ROUTINE TO COLLECT RESULTS;
       END;
     END;

   PRINT BPRINT COLNAME=BNAMES;
   RETURN;

* STEPWISE METHOD --------------------------------------------------------------*
STEPWISE:
   NOTE 'STEPWISE METHOD';
   FREE BPRINT;
   C=CSAVE;   IN=J(K,1,0);   L=0;
   DFE=N-1;   MSE=YPY#/DFE;
   SPROB=0;

   DO WHILE(SPROB<.15 & L<K);
     INDX=LOC(¬IN);                    * WHERE ARE THE VARIABLES NOT IN?;
     NINDX=LOC(IN);                    * WHERE ARE THE VARIABLES IN?;
     CD=VECDIAG(C)(INDX,);             * XPX DIAGONALS;
     CB=C(INDX,K1);                    * ADJUSTED XPY;
     TSQR=CB#CB#/CD#/MSE;              * SQUARES OF T TESTS;
     IMAX=TSQR(<:>,);                  * LOCATION OF  MAXIMUM IN INDX;
     SPROB=(1-PROBT(SQRT(TSQR(IMAX,)),DFE))*2;
```

```
      IF SPROB<.15 THEN DO;              * IF T-TEST SIGNIFICANT;
        II=INDX(,IMAX);                  * FIND INDEX INTO C;
        LINK SWP;                        * ROUTINE TO SWEEP;
        LINK BACKSTEP;                   * CHECK IF REMOVE ANY TERMS;
        LINK BPR;                        * ROUTINE TO COLLECT RESULTS;
        END;
      END;

   PRINT  BPRINT  COLNAME=BNAMES;
   RETURN;

 * ROUTINE TO BACKWARDS-ELIMINATE FOR STEPWISE;
 BACKSTEP:
   IF  NROW(NINDX)=0  THEN  RETURN;
   BPROB=1;

   DO  WHILE(BPROB>.15  &  L<K);
     CD=VECDIAG(C)(NINDX,);             * XPX DIAGONALS;
     CB=C(NINDX,K1);                    * BVALUES;
     TSQR=CB#CB#/(CD#MSE);              * SQUARES OF T TESTS;
     IMIN=TSQR(>:<,);                   * LOCATION OF  MINIMUM IN NINDX;
     BPROB=(1-PROBT(SQRT(TSQR(IMIN,)),DFE))*2;
     IF BPROB>.15 THEN DO;
       II=NINDX(,IMIN);
       LINK SWP;
       LINK BPR;
       END;
     END;
   RETURN;

* SEARCH ALL POSSIBLE MODELS --------------------------------------------------------------*;
ALL: *—USE METHOD OF SCHATZOFF ET AL. FOR SEARCH TECHNIQUE—;
    BETAK=J(K,K,0);            * RECORD ESTIMATES FOR BEST L-PARAM
                                 MODEL;
    MSEK=J(K,1,1E50);          * RECORD BEST MSE PER # PARMS;
    RSQK=J(K,1,0);             * RECORD BEST RSQUARE;
    INK=J(K,K,0);              * RECORD BEST SET PER # PARMS;
    LIMIT=2**K-1;              * NUMBER OF MODELS TO EXAMINE;
    C=CSAVE;   IN=J(K,1,0);    * START OUT WITH NO  VARIABLES IN MODEL;

    DO  KK=1  TO  LIMIT;
      LINK ZTRAIL;                      * FIND WHICH ONE TO SWEEP;
      LINK SWP;                         * SWEEP IT IN;
      BB=BB//(L||MSE||RSQ||(C(IK,K1)#IN)');
      IF MSE<MSEK(L,) THEN DO;          * WAS THIS BEST FOR L PARMS?;
        MSEK(L,)=MSE;                   * RECORD MSE;
        RSQK(L,)=RSQ;                   * RECORD RSQUARE;
        INK(,L)=IN;                     * RECORD WHICH PARMS IN MODEL;
        BETAK(L,)=(C(IK,K1)#IN)';       * RECORD ESTIMATES;
        END;
    END;

    NOTE  'ALL  POSSIBLE  MODELS  IN  SEARCH  ORDER';
    PRINT  BB  COLNAME=BNAMES;   FREE BB;
```

```
    BPRINT=IK'||MSEK||RSQK||BETAK;
    NOTE PAGE 'THE BEST MODEL FOR EACH NUMBER OF PARAMETERS';
    PRINT BPRINT COLNAME=BNAMES;

    * MALLOWS CP PLOT ;
    CP=MSEK#(N-IK'-1)#/MIN(MSEK)-(N-2#IK');
    CP=IK'||CP;  CPNAME='NPARM' 'CP';
    OUTPUT CP OUT=CP COLNAME=CPNAME;

* ----------SUBROUTINE TO FIND NUMBER OF TRAILING ZEROS IN BINARY NUMBER---------- *
*   ON ENTRY: KK IS THE NUMBER TO EXAMINE;                                          *
*   ON EXIT: II HAS THE RESULT;                                                     *
* --------------------------------------------------------------------------------- * ;
ZTRAIL: II=1;  ZZ=KK;
    DO WHILE(MOD(ZZ,2)=0);  II=II+1;  ZZ=ZZ#/2;
      END;
    RETURN;

* --------------------------SUBROUTINE TO SWEEP IN A PIVOT-------------------------- *
*   ON ENTRY: II HAS THE POSITION(S) TO PIVOT                                       *
*   ON EXIT: IN, L, DFE, MSE, RSQ RECALCULATED                                      *
* --------------------------------------------------------------------------------- * ;
SWP: IF ABS(C(II,II))<1E-9 THEN DO;
    NOTE FAILURE;
    PRINT C;
    STOP;
    END;
    C=SWEEP(C,II);
    IN(II,)=□IN(II,);
    L=SUM(IN);
    DFE=N-1-L;
    SSE=C(K1,K1);
    MSE=SSE#/DFE;
    RSQ=1-SSE#/YPY;
    RETURN;

* --------------------SUBROUTINE TO COLLECT BPRINT RESULTS-------------------- *
*   ON ENTRY: L,MSE,RSQ, AND C SET UP TO COLLECT                                *
*   ON EXIT: BPRINT HAS ANOTHER ROW                                             *
* ----------------------------------------------------------------------------- * ;
BPR:
    BPRINT=BPRINT//(L||MSE||RSQ||(C(IK,K1)#IN)');
    RETURN;

PROC PLOT DATA=CP;
    TITLE2 'PLOT OF MALLOWS C(P) VS P';
    PLOT CP*NPARM='C' NPARM*NPARM='P'
        /OVERLAY VPOS=30 HPOS=60;
```

```
                         VARIABLE SELECTION ALGORITHMS                              1

                        ALL POSSIBLE MODELS IN SEARCH ORDER

BB          NPARM     MSE  RSQUARE      AGE    WEIGHT  RUNTIME RSTPULSE RUNPULSE MAXPULSE

ROW1            1 26.6343 .0927765 -0.31136         0        0        0        0        0
ROW2            2 25.8262 0.150635 -.370416 -.158232        0        0        0        0
```

(continued on next page)

(continued from previous page)

ROW3	1	28.5803	.0264885	0	-.104102	0	0	0	0
ROW4	2	7.75564	0.744935	0	-.025484	-3.2886	0	0	0
ROW5	3	7.22632	0.770831	-.173877	-.054437	-3.14039	0	0	0
ROW6	2	7.16842	0.764247	-.150366	0	-3.20395	0	0	0
ROW7	1	7.53384	0.74338	0	0	-3.31056	0	0	0
ROW8	2	7.79826	0.743533	0	0	-3.28661	-.009682	0	0
ROW9	3	7.33609	0.767349	-.167547	0	-3.07925	-.045492	0	0
ROW10	4	7.36665	0.775033	-.196026	-.059152	-2.9889	-.053255	0	0
ROW11	3	8.03731	0.745111	0	-.025685	-3.26268	-.010409	0	0
ROW12	2	24.9149	0.180607	0	-.093049	0	-.274742	0	0
ROW13	3	20.2803	0.356847	-.446982	-.156472	0	-.321863	0	0
ROW14	2	21.2763	0.30027	-.388821	0	0	-.322856	0	0
ROW15	1	24.6758	0.159485	0	0	0	-.279215	0	0
ROW16	2	23.26	0.235031	0	0	0	-.206838	-.152617	0
ROW17	3	16.818	0.466648	-.523382	0	0	-.225238	-.237688	0
ROW18	4	16.2615	0.503398	-.563168	-0.12697	0	-.229807	-.224602	0
ROW19	3	23.8182	0.244651	0	-.063813	0	-.208432	-.142789	0
ROW20	4	7.78515	0.762252	0	-.012313	-3.16759	.0166688	-.074897	0
ROW21	5	6.21317	0.817556	-.285277	-.051844	-2.70392	-.027109	-.126278	0
ROW22	4	6.16694	0.81167	-.262126	0	-2.77733	-.019814	-.128741	0
ROW23	3	7.50797	0.761898	0	0	-3.17665	.0176163	-.076576	0
ROW24	2	7.25426	0.761424	0	0	-3.14019	0	-.073509	0
ROW25	3	5.95669	0.811094	-.256398	0	-2.82538	0	-.130909	0
ROW26	4	6.00903	0.816493	-.276417	-.049323	-2.77237	0	-.129324	0
ROW27	3	7.51016	0.761829	0	-0.01315	-3.13261	0	-.071892	0
ROW28	2	25.333	0.166855	0	-.059868	0	0	-.197971	0
ROW29	3	18.6318	0.409126	-.544083	-.120495	0	0	-.282478	0
ROW30	2	18.9738	0.375995	-.506648	0	0	0	-.293816	0
ROW31	1	24.7082	0.158383	0	0	0	0	-.206799	0
ROW32	2	21.6063	0.289419	0	0	0	0	-0.6818	0.571538
ROW33	3	18.2172	0.422273	-.421398	0	0	0	-.579662	0.361557
ROW34	4	17.2988	0.47172	-.452427	-.149436	0	0	-0.61723	0.426862
ROW35	3	21.4176	0.320779	0	-0.11815	0	0	-.717449	0.635395
ROW36	4	6.03011	0.815849	0	-.051587	-2.9255	0	-.395293	0.38537
ROW37	5	5.17634	0.848002	-.219621	-.072302	-2.68252	0	-.373401	0.304908
ROW38	4	5.34346	0.836818	-.197735	0	-2.76758	0	-.348108	0.270513
ROW39	3	5.99157	0.809988	0	0	-2.97019	0	-.375114	0.354219
ROW40	4	6.20852	0.8104	0	0	-3.00426	0.016412	-.377784	0.353998
ROW41	5	5.54994	0.837031	-0.20154	0	-2.7386	-.012078	-.345624	0.269064
ROW42	6	5.36825	0.848672	-.226974	-.074177	-2.62865	-.021534	-.369628	0.303217
ROW43	5	6.26335	0.816083	0	-.050907	-2.95182	0.01239	-.397042	0.384793
ROW44	4	20.1124	0.385797	0	-0.1194	0	-.190917	-.645842	0.609632
ROW45	5	15.1864	0.554066	-.479225	-0.1527	0	-.215547	-.530449	0.385424
ROW46	4	16.2925	0.502451	-.447169	0	0	-.212659	-.493234	0.319267
ROW47	3	20.3773	0.353772	0	0	0	-.189933	-.610187	0.545236
ROW48	2	25.1146	0.174039	0	0	0	-0.25219	0	-.073642
ROW49	3	19.2347	0.390007	-.527358	0	0	-0.26492	0	-.200243
ROW50	4	18.8088	0.425607	-.558811	-.126038	0	-.270558	0	-0.17799
ROW51	3	25.5972	0.188232	0	-.078736	0	-.255238	0	-.055024
ROW52	4	8.3115	0.746179	0	-0.02053	-3.25232	-.003933	0	-.020639

VARIABLE SELECTION ALGORITHMS 2

BB	NPARM	MSE	RSQUARE	AGE	WEIGHT	RUNTIME	RSTPULSE	RUNPULSE	MAXPULSE
ROW53	5	7.19584	0.788701	-.257952	-0.04936	-2.86147	-.041206	0	-.081534
ROW54	4	7.09161	0.783432	-.239284	0	-2.92597	-.033905	0	-.087769
ROW55	3	8.03367	0.745227	0	0	-3.26805	-.001934	0	-.025261
ROW56	2	7.74693	0.745221	0	0	-3.27232	0	0	-.025605
ROW57	3	6.88263	0.78173	-.229232	0	-3.01222	0	0	-.090944
ROW58	4	7.00018	0.786224	-.244362	-.045249	-2.97011	0	0	-.085854
ROW59	3	8.00441	0.746155	0	-0.02027	-3.26114	0	0	-.021394
ROW60	2	28.3536	.0675159	0	-.070738	0	0	0	-.121588
ROW61	3	22.3815	0.290212	-.540763	-.116048	0	0	0	-0.24445
ROW62	2	22.5014	0.259982	-.512102	0	0	0	0	-.263695
ROW63	1	27.7126	.0560459	0	0	0	0	0	-.137621

VARIABLE SELECTION ALGORITHMS 3

THE BEST MODEL FOR EACH NUMBER OF PARAMETERS

BPRINT	NPARM	MSE	RSQUARE	AGE	WEIGHT	RUNTIME	RSTPULSE	RUNPULSE	MAXPULSE
ROW1	1	7.53384	0.74338	0	0	-3.31056	0	0	0
ROW2	2	7.16842	0.764247	-.150366	0	-3.20395	0	0	0
ROW3	3	5.95669	0.811094	-.256398	0	-2.82538	0	-.130909	0
ROW4	4	5.34346	0.836818	-.197735	0	-2.76758	0	-.348108	0.270513
ROW5	5	5.17634	0.848002	-.219621	-.072302	-2.68252	0	-.373401	0.304908
ROW6	6	5.36825	0.848672	-.226974	-.074177	-2.62865	-.021534	-.369628	0.303217

(continued on next page)

(continued from previous page)

FORWARD SELECTION METHOD

BPRINT	NPARM	MSE	RSQUARE	AGE	WEIGHT	RUNTIME	RSTPULSE	RUNPULSE	MAXPULSE
ROW1	1	7.53384	0.74338	0	0	-3.31056	0	0	0
ROW2	2	7.16842	0.764247	-.150366	0	-3.20395	0	0	0
ROW3	3	5.95669	0.811094	-.256398	0	-2.82538	0	-.130909	0
ROW4	4	5.34346	0.836818	-.197735	0	-2.76758	0	-.348108	0.270513

BACKWARD ELIMINATION METHOD

BPRINT	NPARM	MSE	RSQUARE	AGE	WEIGHT	RUNTIME	RSTPULSE	RUNPULSE	MAXPULSE
ROW1	6	5.36825	0.848672	-.226974	-.074177	-2.62865	-.021534	-.369628	0.303217
ROW2	5	5.17634	0.848002	-.219621	-.072302	-2.68252	0	-.373401	0.304908
ROW3	4	5.34346	0.836818	-.197735	0	-2.76758	0	-.348108	0.270513

STEPWISE METHOD

BPRINT	NPARM	MSE	RSQUARE	AGE	WEIGHT	RUNTIME	RSTPULSE	RUNPULSE	MAXPULSE
ROW1	1	7.53384	0.74338	0	0	-3.31056	0	0	0
ROW2	2	7.16842	0.764247	-.150366	0	-3.20395	0	0	0
ROW3	3	5.95669	0.811094	-.256398	0	-2.82538	0	-.130909	0
ROW4	4	5.34346	0.836818	-.197735	0	-2.76758	0	-.348108	0.270513

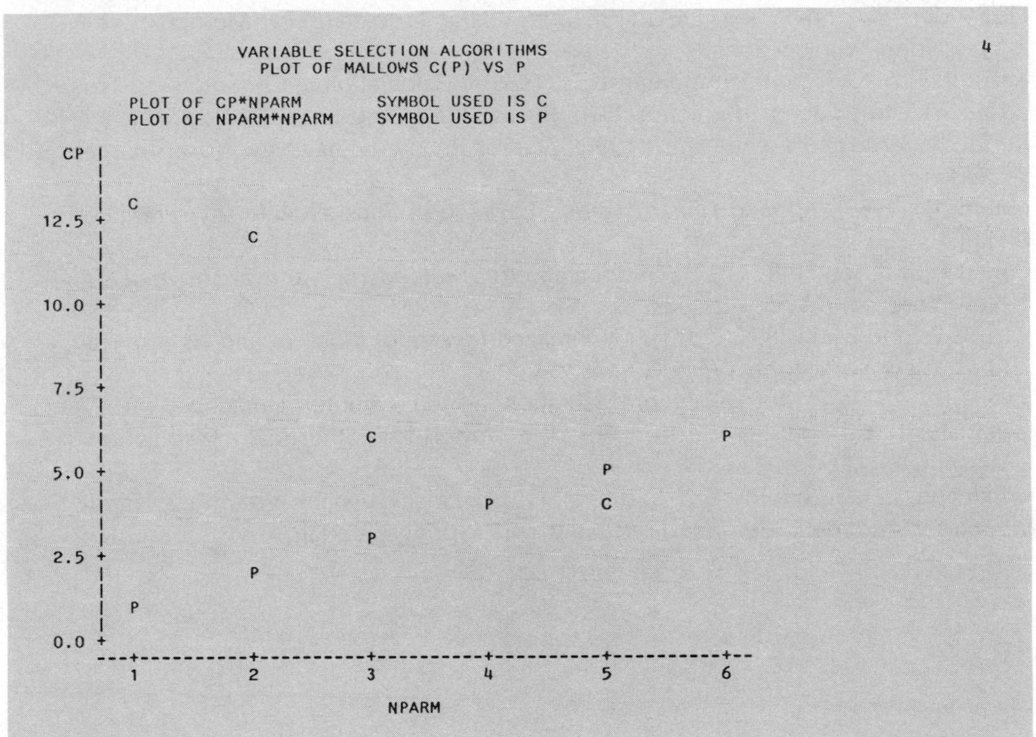

REFERENCES

Beaton, A. (1964), "The Use of Special Matrix Operators in Statistical Calculus," Research Bulletin, Princeton: Educational Testing Service.

Bishop, Y.M., Fienberg, S.E., and Holland, P.W. (1975), *Discrete Multivariate Analysis: Theory and Practice*, Cambridge: MIT Press.

Cox, D.R. (1970), *The Analysis of Binary Data*, New York: Halsted Press.

Deming, M.C. and Stephan, F.F. (1940), "On a Least Squares Adjustment of a Sampled Frequency Table when the Expected Marginal Totals Are Known," *Annals of Mathematical Statistics*, 11, 427–444.

Emerson, P.L. (1968), "Numerical Construction of Orthogonal Polynomials from a General Recurrence Formula," *Biometrics*, 24, 695.

Forsythe, G.E., Malcolm, M.A., and Moler, C.B. (1967), *Computer Solution of Linear Algebraic Systems*, Chapter 17, Englewood Cliffs, New Jersey: Prentice-Hall.

Golub, G.H. (1969), "Matrix Decompositions and Statistical Calculations," in *Statistical Computation*, R.C. Milton and J.A. Nelder, eds., New York: Academic Press.

Goodnight, J.H. (1979), "A Tutorial on the SWEEP Operator," *The American Statistician*, 33, 149–158. Also SAS Technical Report Series, R-106, Cary, NC: SAS Institute.

Grizzle, J.E., Starmer, C.F., and Koch, G.G. (1969), "Analysis of Categorical Data by Linear Models,"*Biometrics*, 25, 489–504.

Haberman, S.J. (1972), "Log-Linear Fit for Contingency Tables," *Applied Statistician*, 21, 218–225.

Hoeffding, W. (1948), "A Non-Parametric Test of Independence," *Annals of Mathematical Statistics*, 19, 546–557.

Hollander, M. and Wolfe, D.A. (1973), *Nonparametric Statistical Methods*, New York: John Wiley & Sons.

Kastenbaum, M.A., and Lamphiear, D.E. (1959), "Calculation of Chi-Square to Test the No Three-Factor Interaction Hypothesis," *Biometrics*, 15, 107–122.

Kennedy, W.J. and Gentle, J.E. (1980), *Statistical Computing*, New York: Marcel Dekker.

Nobel, B. (1969), *Applied Linear Algebra*, Englewood Cliffs, New Jersey: Prentice-Hall.

Puri, M. and Sen, P.K. (1971), *Nonparametric Methods in Multivariate Analysis*, New York: John Wiley & Sons.

Rao, C.R. and Mitra, S.K. (1971), *Generalized Inverse of Matrices and Its Applications*, New York: John Wiley & Sons.

Sall, J.P. (1977), "Matrix Algebra Notation as a Computer Language," 1977 Statistical Computing Section of the American Statistical Association, Washington, D.C., 342–344.

Wilkinson, J.H. and Reinsch, C. (Editors), (1971), *Linear Algebra*, Volume 2, *Handbook for Automatic Computation*, New York: Springer-Verlag.

Index

D

Q

R

T

U

V

Your Turn

If you have comments about SAS or the *SAS User's Guide: Statistics, 1982 Edition,* please let us know by writing your ideas in the space below. If you include your name and address, we will reply to you.

Please return this sheet to Publications Division, SAS Institute Inc., P.O. Box 8000, Cary, NC 27511.